Maximum Entropy and Bayesian Methods

Fundamental Theories of Physics

An International Book Series on The Fundamental Theories of Physics:
Their Clarification, Development and Application

Editor: ALWYN VAN DER MERWE
University of Denver, U.S.A.

Volume 50

Maximum Entropy and Bayesian Methods

Seattle, 1991

*Proceedings of the Eleventh International Workshop on
Maximum Entropy and Bayesian Methods
of Statistical Analysis*

edited by

C. Ray Smith
*Research, Development and Engineering Center,
U.S. Missile Command,
Redstone, Alabama,
U.S.A.*

Gary J. Erickson
and
Paul O. Neudorfer
*Department of Electrical Engineering,
Seattle University,
Seattle, Washington,
U.S.A.*

KLUWER ACADEMIC PUBLISHERS
DORDRECHT / BOSTON / LONDON

Library of Congress Cataloging-in-Publication Data

International Workshop on Maximum Entropy and Bayesian Methods of
  Statistical Analysis (11th : 1991 : Seattle, Wash.)
    Maximum entropy and Bayesian methods : proceedings of the Eleventh
  International Workshop on Maximum Entropy and Bayesian Methods of
  Statistical Analysis, Seattle, 1991 / edited by C. Ray Smith, Gary
  J. Erickson, and Paul O. Neudorfer.
      p.   cm. -- (Fundamental theories of physics ; v. 50)

    1. Maximum entropy method--Congresses. 2. Bayesian statistical
  decision theory--Congresses.   I. Smith, C. Ray, 1933-   .
  II. Erickson, Gary J.   III. Neudorfer, Paul O.   IV. Title.
  V. Series.
  Q370.I58   1991
  502.8--dc20   92-36593

ISBN 978-90-481-4220-0

Published by Kluwer Academic Publishers,
P.O. Box 17, 3300 AA Dordrecht, The Netherlands.

Kluwer Academic Publishers incorporates
the publishing programmes of
D. Reidel, Martinus Nijhoff, Dr W. Junk and MTP Press.

Sold and distributed in the U.S.A. and Canada
by Kluwer Academic Publishers,
101 Philip Drive, Norwell, MA 02061, U.S.A.

In all other countries, sold and distributed
by Kluwer Academic Publishers Group,
P.O. Box 322, 3300 AH Dordrecht, The Netherlands.

Printed on acid-free paper

TABLE OF CONTENTS

PREFACE

The Eleventh Workshop on maximum entropy and Bayesian methods—"MAXENT XI" for short—was held at Seattle University, Seattle, Washington, June 16 – 23, 1991. As in the past, the workshop was intended to bring together researchers from different fields to critically examine maximum entropy and Bayesian methods in science, engineering, medicine, economics, and other disciplines. Now applications are proliferating very rapidly, some in areas that hardly existed a decade ago, and in light of this it is gratifying to observe that due attention is still being paid to the foundations of the field.

We have made no distinction here between invited and contributed papers in that both have been gratefully received by the participants and organizers of MAXENT XI. We should also express our appreciation to those authors who presented work in progress and subsequently put forth the additional time and effort required to provide us with a finished product. A final note on the assembly of these proceedings: In spite of the advanced state of word processing, we encountered numerous difficulties with the electronic versions of a number of manuscripts and have had to re-typeset several of them. For their considerable assistance in this regard we acknowledge Dr. Larry Bretthorst and Mr. James Smith. The editors apologize for any typographical errors which have been created or overlooked in the assembly process.

An experimental one-day course was presented at MAXENT XI, entitled "Bayesian Probability Theory and its Applications," by Drs. G. Larry Bretthorst, Thomas J. Loredo, and C. Ray Smith. Unfortunately, the lecture notes for this course could not be included in this volume.

C.R.S. thanks Dr. Jay Loomis and Mr. Robert Eison for their support of his participation in this undertaking. Dr. David Larner's positive influence on these workshops is gratefully acknowledged.

ACKNOWLEDGMENTS

Special thanks to the following organizations who provided partial financial support for MAXENT XI:

Kluwer Academic Publishers, Dordrecht, The Netherlands

Puget Power and Light Company, Bellevue, Washington

THE GIBBS PARADOX

E. T. Jaynes
Department of Physics
Washington University
St. Louis, Missouri 63130 USA

ABSTRACT. We point out that an early work of J. Willard Gibbs (1875) contains a correct analysis of the "Gibbs Paradox" about entropy of mixing, free of any elements of mystery and directly connected to experimental facts. However, it appears that this has been lost for 100 years, due to some obscurities in Gibbs' style of writing and his failure to include this explanation in his later *Statistical Mechanics*. This "new" understanding is not only of historical and pedagogical interest; it gives both classical and quantum statistical mechanics a different status than that presented in our textbooks, with implications for current research.

1. Introduction

J. Willard Gibbs' *Statistical Mechanics* appeared in 1902. Recently, in the course of writing a textbook on the subject, the present writer undertook to read his monumental earlier work on *Heterogeneous Equilibrium* (1875–78). The original purpose was only to ensure the historical accuracy of certain background remarks; but this was superseded quickly by the much more important unearthing of deep insights into current technical problems.

Some important facts about thermodynamics have not been understood by others to this day, nearly as well as Gibbs understood them over 100 years ago. Other aspects of this "new" development have been reported elsewhere (Jaynes 1986, 1988, 1989). In the present note we consider the "Gibbs Paradox" about entropy of mixing and the logically inseparable topics of reversibility and the extensive property of entropy.

For 80 years it has seemed natural that, to find what Gibbs had to say about this, one should turn to his *Statistical Mechanics*. For 60 years, textbooks and teachers (including, regrettably, the present writer) have impressed upon students how remarkable it was that Gibbs, already in 1902, had been able to hit upon this paradox which foretold – and had its resolution only in – quantum theory with its lore about indistinguishable particles, Bose and Fermi statistics, etc.

It was therefore a shock to discover that in the first Section of his earlier work (which must have been written by mid–1874 at the latest), Gibbs displays a full understanding of this problem, and disposes of it without a trace of that confusion over the "meaning of entropy" or "operational distinguishability of particles" on which later writers have stumbled. He goes straight to the heart of the matter as a simple technical detail, easily understood as soon as one has grasped the full meanings of the words "state" and "reversible" as they are

1

C. R. Smith et al. (eds.), Maximum Entropy and Bayesian Methods, Seattle, 1991, 1–21.
© 1992 *Kluwer Academic Publishers.*

used in thermodynamics. In short, quantum theory did not resolve any paradox, because there was no paradox.

Why did Gibbs fail to give this explanation in his *Statistical Mechanics*? We are inclined to see in this further support for our contention (Jaynes, 1967) that this work was never finished. In reading Gibbs, it is important to distinguish between early and late Gibbs. His *Heterogeneous Equilibrium* of 1875–78 is the work of a man at the absolute peak of his intellectual powers; no logical subtlety escapes him and we can find no statement that appears technically incorrect today. In contrast, his *Statistical Mechanics* of 1902 is the work of an old man in rapidly failing health, with only one more year to live. Inevitably, some arguments are left imperfect and incomplete toward the end of the work.

In particular, Gibbs failed to point out that an "integration constant" was not an arbitrary constant, but an arbitrary function. But this has, as we shall see, nontrivial physical consequences. What is remarkable is not that Gibbs should have failed to stress a fine mathematical point in almost the last words he wrote; but that for 80 years thereafter all textbook writers (except possibly Pauli) failed to see it.

Today, the universally taught conventional wisdom holds that "Classical mechanics failed to yield an entropy function that was extensive, and so statistical mechanics based on classical theory gives qualitatively wrong predictions of vapor pressures and equilibrium constants, which was cleared up only by quantum theory in which the interchange of identical particles is not a real event". We argue that, on the contrary, phenomenological thermodynamics, classical statistics, and quantum statistics are all in just the same logical position with regard to extensivity of entropy; they are silent on the issue, neither requiring it nor forbidding it.

Indeed, in the phenomenological theory Clausius defined entropy by the integral of dQ/T over a reversible path; but in that path the size of a system was not varied, therefore the dependence of entropy on size was not defined. This was perceived by Pauli (1973), who proceeded to give the correct functional equation analysis of the necessary conditions for extensivity. But if this is required already in the phenomenological theory, the same analysis is required *a fortiori* in both classical and quantum statistical mechanics. As a matter of elementary logic, no theory can determine the dependence of entropy on the size N of a system unless it makes some statement about a process where N changes.

In Section 2 below we recall the familiar statement of the mixing paradox, and Sec. 3 presents the explanation from Gibbs' *Heterogeneous Equilibrium* in more modern language. Sec. 4 discusses some implications of this, while Sections 5 and 6 illustrate the points by a concrete scenario. Sec. 7 then recalls the Pauli analysis and Sec. 8 reexamines the relevant parts of Gibbs' *Statistical Mechanics* to show how the mixing paradox disappears, and the issue of extensivity of entropy is cleared up, when the aforementioned minor oversight is corrected by a Pauli type analysis. The final result is that entropy is just as much, and just as little, extensive in classical statistics as in quantum statistics. The concluding Sec. 9 points out the relevance of this for current research.

2. The Problem

We repeat the familiar story, already told hundreds of times; but with a new ending. There are n_1 moles of an ideal gas of type 1, n_2 of another noninteracting type 2, confined in two volumes V_1, V_2 and separated by a diaphragm. Choosing $V_1/V_2 = n_1/n_2$, we may have them initially at the same temperature $T_1 = T_2$ and pressure $P_1 = P_2 = n_1RT/V_1$. The diaphragm is then removed, allowing the gases to diffuse through each other. Eventually we reach a new equilibrium state with $n = n_1 + n_2$ moles of a gas mixture, occupying the total volume $V = V_1 + V_2$ with uniform composition, the temperature, pressure and total energy remaining unchanged.

If the gases are different, the entropy change due to the diffusion is, by standard thermodynamics,

$$\Delta S = S_{final} - S_{initial} = nR \log V - (n_1 R \log V_1 + n_2 R \log V_2)$$

or,

$$\Delta S = -nR[f \log f + (1-f)\log(1-f)] \tag{1}$$

where $f = n_1/n = V_1/V$ is the mole fraction of component 1. Gibbs [his Eq. (297)] considers particularly the case $f = 1/2$, whereupon

$$\Delta S = nR \log 2.$$

What strikes Gibbs at once is that this is independent of the nature of the gases, "... except that the gases which are mixed must be of different kinds. If we should bring into contact two masses of the same kind of gas, they would also mix, but there would be no increase of entropy." He then proceeds to explain this difference, in a very cogent way that has been lost for 100 years. But to understand it, we must first surmount a difficulty that Gibbs imposes on his readers.

Usually, Gibbs' prose style conveys his meaning in a sufficiently clear way, using no more than twice as many words as Poincaré or Einstein would have used to say the same thing. But occasionally he delivers a sentence with a ponderous unintelligibility that seems to challenge us to make sense out of it. Unfortunately, one of these appears at a crucial point of his argument; and this may explain why the argument has been lost for 100 years. Here is that sentence:

"*Again, when such gases have been mixed, there is no more impossibility of the separation of the two kinds of molecules in virtue of their ordinary motions in the gaseous mass without any especial external influence, than there is of the separation of a homogeneous gas into the same two parts into which it has once been divided, after these have once been mixed.*"

The decipherment of this into plain English required much effort, sustained only by faith in Gibbs; but eventually there was the reward of knowing how Champollion felt when he realized that he had mastered the Rosetta stone. Suddenly, *everything made sense*; when put back into the surrounding context there appeared an argument so clear and simple that it seemed astonishing that it has not been rediscovered independently dozens of times. Yet a library search has failed to locate any trace of this understanding in any modern work on thermodynamics or statistical mechanics.

We proceed to our rendering of that explanation, which is about half direct quotation from Gibbs, but with considerable editing, rearrangement, and exegesis. For this reason we call it "the explanation" rather than "Gibbs' explanation". We do not believe that we deviate in any way from Gibbs' intentions, which must be judged partly from other sections of his work; in any event, his original words are readily available for comparison. However, our purpose is not to clarify history, but to explain the technical situation as it appears today in the light of this clarification from history; therefore we carry the explanatory remarks slightly beyond what was actually stated by Gibbs, to take note of the additional contributions of Boltzmann, Planck, and Einstein.

3. The Explanation

When unlike gases mix under the conditions described above, there is an entropy increase $\Delta S = nR \log 2$ independent of the nature of the gases. When the gases are identical they still mix, but there is no entropy increase. But we must bear in mind the following.

When we say that two unlike gases mix and the entropy increases, we mean that the gases could be separated again and brought back to their original states by means that would leave changes in external bodies. These external changes might consist, for example, in the lowering of a weight, or a transfer of heat from a hot body to a cold one.

But by the "original state" we do not mean that every molecule has been returned to its original position, but only a state which is indistinguishable from the original one in the macroscopic properties that we are observing. For this we require only that a molecule originally in V_1 returns to V_1. In other words, we mean that we can recover the original *thermodynamic state*, defined for example by specifying only the chemical composition, total energy, volume, and number of moles of a gas; and nothing else. It is to the states of a system thus incompletely defined that the propositions of thermodynamics relate.

But when we say that two identical gases mix without change of entropy, we do not mean that the molecules originally in V_1 can be returned to V_1 without external change. The assertion of thermodynamics is that when the net entropy change is zero, then the original *thermodynamic state* can be recovered without external change. Indeed, we have only to reinsert the diaphragm; since all the observed macroscopic properties of the mixed and unmixed gases are identical, there has been no change in the thermodynamic state. It follows that there can be no change in the entropy or in any other thermodynamic function.

Trying to interpret the phenomenon as a discontinuous change in the physical nature of the gases (*i. e.,* in the behavior of their microstates) when they become exactly the same, misses the point. The principles of thermodynamics refer not to any properties of the hypothesized microstates, but to the observed properties of macrostates; there is no thought of restoring the original microstate. We might put it thus: when the gases become exactly the same, the discontinuity is in what you and I mean by the words "restore" and "reversible".

But if such considerations explain why mixtures of like and unlike gases are on a different footing, they do not reduce the significance of the fact that the entropy change with unlike gases is independent of the nature of the gases. We may, without doing violence to the general principles of thermodynamics, imagine other gases than those presently known, and there does not appear to be any limit to the resemblance which there might be between two of them; but ΔS would be independent of it.

We may even imagine two gases which are absolutely identical in all properties which come into play while they exist as gases in the diffusion cell, but which differ in their behavior in some other environment. In their mixing an increase of entropy would take place, although the process, dynamically considered, might be absolutely identical in its minutest details (even the precise path of each atom) with another process which might take place without any increase of entropy. In such respects, entropy stands strongly contrasted with energy.

A thermodynamic state is defined by specifying a small number of macroscopic quantities such as pressure, volume, temperature, magnetization, stress, etc. – denote them by $\{X_1, X_2, \ldots, X_n\}$ – which are observed and/or controlled by the experimenter; n is seldom greater than 4. We may contrast this with the physical state, or microstate, in which we imagine that the positions and velocities of all the individual atoms (perhaps 10^{24} of them) are specified.

All thermodynamic functions – in particular, the entropy – are by definition and construction properties of the thermodynamic state; $S = S(X_1, X_2, \ldots, X_n)$. A thermodynamic variable may or may not be also a property of the microstate. We consider the total mass and total energy to be "physically real" properties of the microstate; but the above considerations show that entropy cannot be.

To emphasize this, note that a "thermodynamic state" denoted by $X \equiv \{X_1 \ldots X_n\}$ defines a large class $C(X)$ of microstates compatible with X. Boltzmann, Planck, and Einstein showed that we may interpret the entropy of a macrostate as $S(X) = k \log W(C)$, where $W(C)$ is the phase volume occupied by all the microstates in the chosen reference class C. From this formula, a large mass of correct results may be explained and deduced in a neat, natural way (Jaynes, 1988). In particular, one has a simple explanation of the reason for the second law as an immediate consequence of Liouville's theorem, and a generalization of the second law to nonequilibrium conditions, which has useful applications in biology (Jaynes, 1989).

This has some interesting implications, not generally recognized. The thermodynamic entropy $S(X)$ is, by definition and construction, a property not of any one microstate, but of a certain reference class $C(X)$ of microstates; it is a measure of the size of that reference class. Then if two different microstates are in C, we would ascribe the same entropy to systems in those microstates. But it is also possible that two experimenters assign different entropies S, S' to what is in fact the same microstate (in the sense of the same position and velocity of every atom) without either being in error. That is, they are contemplating a different set of possible macroscopic observations on the same system, embedding that microstate in two different reference classes C, C'.

Two important conclusions follow from this. In the first place, it is necessary to decide at the outset of a problem which macroscopic variables or degrees of freedom we shall measure and/or control; and within the context of the thermodynamic system thus defined, entropy will be some function $S(X_1, \ldots, X_n)$ of whatever variables we have chosen. We can expect this to obey the second law $T dS \geq dQ$ only as long as all experimental manipulations are confined to that chosen set. If someone, unknown to us, were to vary a macrovariable X_{n+1} outside that set, he could produce what would appear to us as a violation of the second law, since our entropy function $S(X_1, \ldots, X_n)$ might decrease spontaneously, while his $S(X_1, \ldots, X_n, X_{n+1})$ increases. [We demonstrate this explicitly below].

Secondly, even within that chosen set, deviations from the second law are at least conceivable. Let us return to the mixing of identical gases. From the fact that they mix without change of entropy, we must not conclude that they can be separated again without external change. On the contrary, the "separation" of identical gases is entirely impossible *with or without external change*. If "identical" means anything, it means that there is no way that an "unmixing" apparatus could determine whether a molecule came originally from V_1 or V_2, short of having followed its entire trajectory.

It follows *a fortiori* that there is no way we could accomplish this separation reproducibly by manipulation of any macrovariables $\{X_i\}$. Nevertheless it might happen without any intervention on our part that in the course of their motion the molecules which came from V_1 all return to it at some later time. Such an event is not impossible; we consider it only improbable.

Now a separation that Nature can accomplish already in the case of identical molecules, she can surely accomplish at least as easily for unlike ones. The spontaneous separation of mixed unlike gases is just as possible as that of like ones. In other words, the impossibility of an uncompensated decrease of entropy seems to be reduced to improbability.

4. Discussion

The last sentence above is the famous one which Boltzmann quoted twenty years later in his reply to Zermelo's *Wiederkehreinwand* and took as the motto for the second (1898) volume of his *Vorlesungen über Gastheorie*. Note the superiority of Gibbs' reasoning. There is none of the irrelevancy about whether the interchange of identical particles is or is not a "real physical event", which has troubled so many later writers – including even Schrödinger. As we see, the strong contrast between the "physical" nature of energy and the "anthropomorphic" nature of entropy, was well understood by Gibbs before 1875.

Nevertheless, we still see attempts to "explain irreversibility" by searching for some entropy function that is supposed to be a property of the microstate, making the second law a theorem of dynamics, a consequence of the equations of motion. Such attempts, dating back to Boltzmann's paper of 1866, have never succeeded and never ceased. But they are quite unnecessary; for the second law that Clausius gave us was not a statement about any property of microstates. The difference in ΔS on mixing of like and unlike gases can seem paradoxical only to one who supposes, erroneously, that entropy is a property of the microstate.

The important constructive point that emerges from this is that thermodynamics has a greater flexibility in useful applications than is generally recognized. The experimenter is at liberty to choose his macrovariables as he wishes; whenever he chooses a set within which there are experimentally reproducible connections like an equation of state, the entropy appropriate to the chosen set will satisfy a second law that correctly accounts for the macroscopic observations that the experimenter can make by manipulating the macrovariables *within his set*.

We may draw two conclusions about the range of validity of the second law. In the first place, if entropy can depend on the particular way you or I decide to define our thermodynamic states, then obviously, the statement that entropy tends to increase but not decrease can remain valid only as long as, having made our choice of macrovariables, we stick to that choice.

Indeed, as soon as we have Boltzmann's $S = k \log W$, the reason for the second law is seen immediately as a proposition of macroscopic phenomenology, true "almost always" simply from considerations of phase volume. Two thermodynamic states of slightly different entropy $S_1 - S_2 = 10^{-6}/393$ cal/deg, corresponding to one microcalorie at room temperature, exhibit a phase volume ratio

$$W_1/W_2 = \exp[(S_1 - S_2)/k] = \exp[10^{16}]. \tag{2}$$

Macrostates of higher entropy are sometimes said to 'occupy' overwhelmingly greater phase volumes; put more accurately, macrostates of higher entropy may be *realized* by an over-

whelmingly greater number, or range, of microstates. Because of this, not only all reproducible changes between equilibrium thermodynamic states, but the overwhelming majority
of all the macrostate changes that could possibly occur – equilibrium or nonequilibrium –
are to ones of higher entropy, simply because there are overwhelmingly more microstates to
go to, by factors like (2). We do not see why any more than this is needed to understand
the second law as a fact of macroscopic phenomenology.

However, the argument just given showing how entropy depends, not on the microstate,
but on human choice of the reference class in which it is to be embedded, may appear so
abstract that it leaves the reader in doubt as to whether we are describing real, concrete
facts or only a particular philosophy of interpretation, without physical consequences.

This is so particularly when we recall that after the aforementioned paper of 1866,
Boltzmann spent the remaining 40 years of his life in inner turmoil and outward controversy
over the second law, repeatedly changing his position. The details are recounted by Klein
(1973). In the end this degenerated into nit–picking arguments over the exact conditions
under which his H–function did or did not decrease. But none of this was ever settled or
related to the real experimental facts about the second law – which make no reference to
any velocity distribution! It behooves us to be sure that we are not following a similar
course.

Fortunately, the concrete reality and direct experimental relevance of Gibbs' arguments
are easily shown. The actual calculation following is probably the most common one found
in elementary textbooks, but the full conditions of its validity have never, to the best of
our knowledge, been recognized. The scenario in which we set it is only slightly fanciful;
the discovery of isotopes was very nearly a realization of it. As examination of several
textbooks shows, that discovery prompted a great deal of confusion over whether entropy
of mixing of isotopes should or should not be included in thermodynamics.

5. The Gas Mixing Scenario Revisited

Presumably, nobody doubts today that the measurable macroscopic properties of argon (i.e.
equation of state, heat capacity, vapor pressure, heat of vaporization, etc.) are described
correctly by conventional thermodynamics which ascribes zero entropy change to the mixing
of two samples of argon at the same temperature and pressure. But suppose that, unknown
to us today, there are two different kinds of argon, $A1$ and $A2$, identical in all respects except
that $A2$ is soluble in Whifnium, while $A1$ is not (Whifnium is one of the rare superkalic
elements; in fact, it is so rare that it has not yet been discovered).

Until the discovery of Whifnium in the next Century, we shall have no way of preparing
argon with controllably different proportions of $A1$ and $A2$. And even if, by rare chance,
we should happen to get pure $A1$ in volume $V1$, and pure $A2$ in $V2$, we would have no way
of knowing this, or of detecting any difference in the resulting diffusion process. Thus all
the thermodynamic measurements we can actually make today are accounted for correctly
by ascribing zero entropy of mixing to argon.

Now the scene shifts to the next Century, when Whifnium is readily available to experimenters. What could happen before only by rare chance, we can now bring about by
design. We may, at will, prepare bottles of pure $A1$ and pure $A2$. Starting our mixing
experiment with $n_1 = fn$ moles of $A1$ in the volume $V_1 = fV$, and $n_2 = (1 - f)n$ moles of
$A2$ in $V_2 = (1 - f)V$, the resulting actual diffusion may be identical in every detail, down
to the precise path of each atom, with one that could have happened by rare chance before

the discovery of Whifnium; but because of our greater knowledge we shall now ascribe to that diffusion an entropy increase ΔS given by Eq (1), which we write as:

$$\Delta S = \Delta S_1 + \Delta S_2 \tag{3}$$

where

$$\Delta S_1 = -nRf \log f \tag{4a}$$

$$\Delta S_2 = -nR(1 - f) \log(1 - f) \tag{4b}$$

But if this entropy increase is more than just a figment of our imagination, it ought to have observable consequences, such as a change in the useful work that we can extract from the process.

There is a school of thought which militantly rejects all attempts to point out the close relation between entropy and information, claiming that such considerations have nothing to do with energy; or even that they would make entropy "subjective" and it could therefore could have nothing to do with experimental facts at all. We would observe, however, that the number of fish that you can catch is an "objective experimental fact"; yet it depends on how much "subjective" information you have about the behavior of fish.

If one is to condemn things that depend on human information, on the grounds that they are "subjective", it seems to us that one must condemn all science and all education; for in those fields, human information is all we have. We should rather condemn this misuse of the terms "subjective" and "objective", which are descriptive adjectives, not epithets. Science does indeed seek to describe what is "objectively real"; but our hypotheses about that will have no testable consequences unless it can also describe what human observers can see and know. It seems to us that this lesson should have been learned rather well from relativity theory.

The amount of useful work that we can extract from any system depends – obviously and necessarily – on how much "subjective" information we have about its microstate, because that tells us which interactions will extract energy and which will not; this is not a paradox, but a platitude. If the entropy we ascribe to a macrostate did not represent some kind of human information about the underlying microstates, it could not perform its thermodynamic function of determining the amount of work that can be extracted reproducibly from that macrostate.

But if this is not obvious, it is easily demonstrated in our present scenario. The diffusion will still take place without any change of temperature, pressure, or internal energy; but because of our greater information we shall now associate it with a free energy decrease $\Delta F = -T\Delta S$. Then, according to the principles of thermodynamics, if instead of allowing the uncontrolled irreversible mixing we could carry out the same change of state reversibly and isothermally, we should be able to obtain from it the work

$$W = -\Delta F = T\Delta S. \tag{5}$$

Let us place beside the diaphragm a movable piston of Whifnium. When the diaphragm is removed, the $A2$ will then diffuse through this piston until its partial pressure is the same on both sides, after which we move the piston slowly (to maintain equal $A2$ pressure and to allow heat to flow in to maintain constant temperature), in the direction of increasing V_1. From this expansion of $A1$ we shall obtain the work

$$W_1 = \int_{V_1}^{V} P_1 dV = n_1 RT \log(V/V_1)$$

or from (4a),

$$W_1 = T\Delta S_1 \tag{6}$$

The term ΔS_1 in the entropy of mixing therefore indicates the work obtainable from reversible isothermal expansion of component $A1$ into the full volume $V = V_1 + V_2$. But the initial diffusion of $A2$ still represents an irreversible entropy increase ΔS_2 from which we obtain no work.

Spurred by this partial success, the other superkalic element Whafnium is discovered, which has the opposite property that it is permeable to $A1$ but not to $A2$. Then we can make an apparatus with two superkalic pistons; the Whifnium moves to the right, giving the work $W_1 = T\Delta S_1$, while the Whafnium moves to the left, yielding $W_2 = T\Delta S_2$. We have succeeded in extracting just the work $W = T\Delta S$ predicted by thermodynamics. The entropy of mixing does indeed represent human information; *just the information needed to predict the work available from the mixing.*

In this scenario, our greater knowledge resulting from the discovery of the superkalic elements leads us to assign a different entropy change to what may be in fact the identical physical process, down to the exact path of each atom. But there is nothing "unphysical" about this, since that greater knowledge corresponds exactly to – because it is due to – our greater capabilities of control over the physical process. Possession of a superkalic piston gives us the ability to control a new thermodynamic degree of freedom X_{n+1}, the position of the piston. It would be astonishing if this new technical capability did not enable us to extract more useful work from the system.

This scenario has illustrated the aforementioned greater versatility of thermodynamics – the wider range of useful applications – that we get from recognizing the strong contrast between the natures of entropy and energy, that Gibbs pointed out so long ago.

To emphasize this, note that even after the discovery of superkalic elements, we still have the option not to use them and stick to the old macrovariables $\{X_1 \ldots X_n\}$ of the 20'th Century. Then we may still ascribe zero entropy of mixing to the interdiffusion of $A1$ and $A2$, and we shall predict correctly, just as was done in the 20'th Century, all the thermodynamic measurements that we can make on Argon without using the new technology. Both before and after discovery of the superkalic elements, the rules of thermodynamics are valid and correctly describe the measurements that it is possible to make by manipulating the macrovariables *within the set that we have chosen to use.*

This useful versatility – a direct result of, and illustration of, the "anthropomorphic" nature of entropy – would not be apparent to, and perhaps not believed by, someone who thought that entropy was, like energy, a physical property of the microstate.

6. Second Law Trickery

Our scenario has verified another statement made above; a person who knows about this new degree of freedom and manipulates it, can play tricks on one who does not know about it, and make him think that he is seeing a violation of the second law. Suppose there are two observers, one of whom does not know about $A1$, $A2$, and superkalic elements and one who does, and we present them with two experiments.

In experiment 1, mixing of a volume V_1 of $A1$ and V_2 of $A2$ takes place spontaneously, without superkalic pistons, from an initial thermodynamic state X_i to a final one X_f without any change of temperature, pressure, or internal energy and without doing any work; so it causes no heat flow between the diffusion cell and a surrounding heat bath of temperature T. To both observers, the initial and final states of the heat bath are the same, and to the ignorant observer this is also true of the argon; nothing happens at all.

In experiment 2 we insert the superkalic pistons and perform the same mixing reversibly, starting from the same initial state X_i. Again, the final state X_f of the argon has the same temperature, pressure, and internal energy as does X_i. But now work W is done, and so heat $Q = W$ flows into the diffusion cell from the heat bath. Its existence and magnitude could be verified by calorimetry. Therefore, for both observers, the initial and final states of the heat bath are now different. To the ignorant observer, an amount of heat Q has been extracted from the heat bath and converted entirely into work: $W = Q$, while the total entropy of the system and heat bath has decreased spontaneously by $\Delta S = -Q/T$, in flagrant violation of the second law!

To the informed observer, there has been no violation of the second law in either experiment. In experiment 1 there is an irreversible increase of entropy of the argon, with its concomitant loss of free energy; in experiment 2, the increase in entropy of the argon is exactly compensated by the decrease in entropy of the heat bath. For him, since there has been no change in total entropy, the entire process of experiment 2 is reversible. Indeed, he has only to move the pistons back slowly to their original positions. In this he must give back the work W, whereupon the argon is returned to its original unmixed condition and the heat Q is returned to the heat bath.

Both of these observers can in turn be tricked into thinking that they see a violation of the second law by a third one, still better informed, who knows that $A2$ is actually composed of two components $A2a$ and $A2b$ and there is a subkalic element Whoofnium – and so on *ad infinitum*. A physical system always has more macroscopic degrees of freedom beyond what we control or observe, and by manipulating them a trickster can always make us see an apparent violation of the second law.

Therefore the correct statement of the second law is not that an entropy decrease is impossible in principle, or even improbable; rather that it *cannot be achieved reproducibly by manipulating the macrovariables* $\{X_1, \ldots, X_n\}$ *that we have chosen to define our macro-state*. Any attempt to write a stronger law than this will put one at the mercy of a trickster, who can produce a violation of it.

But recognizing this should increase rather than decrease our confidence in the future of the second law, because it means that if an experimenter ever sees an apparent violation, then instead of issuing a sensational announcement, it will be more prudent to search for that unobserved degree of freedom. That is, the connection of entropy with information works both ways; seeing an apparent decrease of entropy signifies ignorance of what were the relevant macrovariables.

7. The Pauli Analysis

Consider now the phenomenological theory. The Clausius definition of entropy determines the difference of entropy of two thermodynamic states of a closed system (no particles enter or leave) that can be connected by a reversible path:

$$S_2 - S_1 = \int_1^2 \frac{dQ}{T} \tag{7}$$

Many are surprised when we claim that this is not necessarily extensive; the first reaction is: "Surely, two bricks have twice the heat capacity of one brick; so how could the Clausius entropy possibly not be extensive?" To see how, note that the entropy *difference* is indeed proportional to the number N of molecules whenever the heat capacity is proportional to N and the pressure depends only on V/N; but that is not necessarily true, and when it is true it is far from making the entropy itself extensive.

For example, let us evaluate this for the traditional ideal monoatomic gas of N molecules and consider the independent thermodynamic macrovariables to be (T, V, N). This has an equation of state $PV = NkT$ and heat capacity $C_v = (3/2)Nk$, where k is Boltzmann's constant. From this all elementary textbooks find, using the thermodynamic relations $(\partial S/\partial V)_T = (\partial P/\partial T)_V$ and $T(\partial S/\partial T)_V = C_v$:

$$S(T_2, V_2, N) - S(T_1, V_1, N) = \int_1^2 \left[\left(\frac{\partial S}{\partial V} \right)_T dV + \left(\frac{\partial S}{\partial T} \right)_V dT \right] \tag{8}$$

$$= Nk \log \frac{V_2}{V_1} + \frac{3}{2} Nk \log \frac{T_2}{T_1}$$

It is evident that this is satisfied by any entropy function of the form

$$S(T, V, N) = k \left[N \log V + \frac{3}{2} N \log T \right] + k f(N) \tag{9}$$

where $f(N)$ is not an arbitrary constant, but an arbitrary function. The point is that in the reversible path (7) we varied only T and V; consequently the definition (7) determines only the dependence of S on T and V. Indeed, if N varied on our reversible path, then (7) would not be correct (an extra 'entropy of convection' term $\int \mu dN/T$ would be needed).

Pauli (1973) noticed this incompleteness of (7) and saw that if we wish entropy to be extensive, then that is logically an additional condition, that we must impose separately. The extra condition is that entropy should satisfy the scaling law

$$S(T, qV, qN) = qS(T, V, N), \qquad 0 < q < \infty \tag{10}$$

Then, substituting (9) into (10), we find that $f(N)$ must satisfy the functional equation

$$f(qN) = qf(N) - qN \log q \tag{11}$$

Differentiating with respect to q and setting $q = 1$ yields a differential equation $Nf'(N) = f(N) - N$, whose general solution is

$$f(N) = Nf(1) - N \log N . \tag{12}$$

[alternatively, just set $N = 1$ in (11) and we see the general solution]. Thus the most general extensive entropy function for this gas has the form

$$S(T, V, N) = Nk \left[\log \frac{V}{N} + \frac{3}{2} \log T + f(1) \right]. \tag{13}$$

It contains one arbitrary constant, $f(1)$, which is essentially the chemical constant. This is not determined by either the Clausius definition (7) or the condition of extensivity (10); however, one more fact about it can be inferred from (13). Writing $f(1) = (3/2) \log(kC)$, we have

$$S(T, V, N) = Nk \log \left[\frac{CV(kT)^{3/2}}{N} \right]. \tag{14}$$

The quantity in the square brackets must be dimensionless, so C must have the physical dimensions of $(volume)^{-1}(energy)^{-3/2} = (mass)^{3/2}(action)^{-3}$. Thus on dimensional grounds we can rewrite (13) as

$$S(T, V, N) = Nk \left\{ \log \frac{V}{N} + \frac{3}{2} \log \left[\frac{mkT}{\zeta^2} \right] \right\}. \tag{15}$$

where m is the molecular mass and ζ is an undetermined quantity of the dimensions of action. It might appear that this is the limit of what can be said about the entropy function from phenomenological considerations; however, there may be further cogent arguments that have escaped our attention.

8. Would Gibbs Have Accepted It?

Note that the Pauli analysis *has not demonstrated from the principles of physics that entropy actually should be extensive*; it has only indicated the form our equations must take if it is. But this leaves open two possibilities:

(1) All this is a tempest in a teapot; the Clausius definition indicates that only entropy differences are physically meaningful, so we are free to define the arbitrary additive terms in any way we please. This is the view that was taught to the writer, as a student many years ago.

(2) The variation of entropy with N is not arbitrary; it is a substantive matter with experimental consequences. Therefore *the Clausius definition of entropy is logically incomplete*, and it needs to be supplemented either by experimental evidence or further theoretical considerations.

The original thermodynamics of Clausius considered only closed systems, and was consistent with conclusion (1). Textbooks for physicists have been astonishingly slow to move beyond it; that of Callen (1960) is almost the only one that recognizes the more complete and fundamental nature of Gibbs' work.

In the thermodynamics of Gibbs, the variation of entropy with N is just the issue that is involved in the prediction of vapor pressures, equilibrium constants, or any conditions of equilibrium with respect to exchange of particles. His invention of the chemical potential μ has, according to all the evidence of physical chemistry, solved the problem of equilibrium with respect to exchange of particles, leading us to conclusion (2); chemical thermodynamics could not exist without it.

Gibbs could predict the final equilibrium state of a heterogeneous system as the one with maximum total entropy subject to fixed total values of energy, mole numbers, and other conserved quantities. All of the results of his *Heterogeneous Equilibrium* follow from this variational principle. He showed that by a mathematical (Legendre) transformation this could be stated equally well as minimum Gibbs free energy subject to fixed temperature and chemical potentials, which form physical chemists have generally found more convenient (however, it is also less general; in extensions to nonequilibrium phenomena there is no temperature or Gibbs free energy, and one must return to the original maximum entropy principle, from which many more physical predictions may be derived).

This preference for one form of the variational principle over the other is largely a matter of our conditioning from previous training. For example, from our mechanical experience it seems intuitively obvious that a liquid sphere has a lower energy than any other shape with the same entropy; yet to most of us it seems far from obvious that the sphere has higher entropy than any other shape with the same energy.

It is interesting to note how the various classical textbooks deal with the question of extensivity; and then how Gibbs dealt with it. Planck (1926, p. 92), Epstein (1937; p. 136) and Zemansky (1943; p. 311) do not give the scaling law (10), but simply write entropy as extensive from the start, apparently considering this too obvious to be in need of any discussion. Callen (1960; p. 26) does write the scaling law, but does not derive it from anything or derive anything from it; thereafter he proceeds to write entropy as extensive without further discussion.

In other words, all of these works simply *assume* entropy to be extensive without investigating the question whether that extensivity follows from, or is even consistent with, the Clausius definition of entropy. More recent textbooks tend to be even more careless in the logic; the only exception known to us is that of Pauli. But let us note how much is lost thereby:

(a) The question of extensivity cannot have any universally valid answer; for there are systems, for example some spin systems and systems with electric charge or gravitational forces, for which the scaling law (10) does not hold because of long–range interactions; yet the Clausius definition of entropy is still valid as long as N is held constant. Such systems cannot be treated at all by a formalism that starts by assuming extensivity.

(b) This reminds us that for any system extensivity of entropy is, strictly speaking, only an approximation which can hold only to the extent that the range of molecular forces is small compared to the dimensions of the system. Indeed, for virtually all systems small deviations from exact extensivity are observable in the laboratory; the experimentalist calls them "surface effects" and we note that Gibbs' *Heterogeneous Equilibrium* gives beautiful treatments of surface tension and electrocapillarity, all following from his single variational principle.

Obviously, then, Gibbs did not assume extensivity as a general property, as did the others noted above. But of course, Gibbs was in agreement with them, that entropy is very nearly extensive for most single homogeneous systems of macroscopic size. But he is careful in saying this, adding the qualification "- - - , for many substances at least, - - -". Then he gives such relations as $G = \sum \mu_i n_i$ which hold in the cases where entropy can be considered extensive to sufficient accuracy.

But Gibbs never does give an explicit discussion of the circumstances in which entropy is or is not extensive! He appears at first glance to evade the issue by the device of talking only about the total entropy of a heterogeneous system, not the entropies of its separate parts. Movement of particles from one part to another is disposed of by the observation that, if it is reversible, then the total entropy is unchanged; and that is all he needs in order to discuss all his phenomena, including surface effects, fully.

But on further meditation we realize that Gibbs did not evade any issue; rather, his far deeper understanding enabled him to see that there was no issue. He had perceived that, when two systems interact, *only the entropy of the whole is meaningful.* Today we would say that the interaction induces correlations in their states which makes the entropy of the whole less than the sum of entropies of the parts; and it is the entropy of the whole that contains full thermodynamic information. This reminds us of Gibbs' famous remark, made in a supposedly (but perhaps not really) different context: "The whole is simpler than the sum of its parts." How could Gibbs have perceived this long before the days of quantum theory?

This is one more example where Gibbs had more to say than his contemporaries could absorb from his writings; over 100 years later, we still have a great deal to learn by studying how Gibbs manages to do things. For him, entropy is not a local property of the subsystems; it is a "global" property like the Lagrangian of a mechanical system; *i.e.* it presides over the whole and determines, by its variational properties, all the conditions of equilibrium – just as the Lagrangian presides over all of mechanics and determines, by its variational properties, all the equations of motion.

Up to this point we have been careful to consider only the phenomenology and experimental facts of thermodynamics, in order to make their independent logical status clear. Most discussions of these matters mix up the statistical and phenomenological aspects from the start, in a way that we think generates and perpetuates confusion. In particular, this has obscured the fact that the fundamental operational definitions of such terms as equilibrium, temperature, and entropy – and the statements of the first and second laws – involve only the macrovariables observed in the laboratory. They make no reference to microstates, much less to any velocity distributions, probability distributions, or correlations. As Helmholtz and Planck stressed, this much of the field has a validity and usefulness quite independent of whether atoms and microstates exist.

9. Gibbs' Statistical Mechanics

Now we turn to the final work of Gibbs, which appeared 27 years after his *Heterogeneous Equilibrium*, and examine the parts of it which are relevant to these issues. We expect that a statistical theory might supplement the phenomenology in two ways. Firstly, if our present microstate theory is correct, we would get a deeper interpretation of the basic reason for the phenomenology, and a better understanding of its range of validity; if it is not correct, we might find contradictions that would provide clues to a better theory. Secondly, the theoretical explanation would predict generalizations; the range of possible nonequilibrium conditions is many orders of magnitude greater than that of equilibrium conditions, so if new reproducible connections exist they would be almost impossible to find without the guidance of a theory that tells the experimenter where to look.

It is generally supposed that Gibbs coined the term "Statistical Mechanics" for the title of this book. But in reading Gibbs' *Heterogeneous Equilibrium* we slowly developed a feeling, from the above handling of entropy and other incidents in it, that his thermodynamic formalism corresponds so closely to that of Statistical Mechanics – in particular, the grand canonical ensemble – that he must have known the main results of the latter already while writing the former.

This suspicion was confirmed in part by the discovery that the term "Statistical Mechanics" appears already in an Abstract of his dated 1884, of a paper read at a meeting but which, to the best of our knowledge, was never published and is lost. The abstract states that he will be concerned with Liouville's theorem and its applications to astronomy and thermodynamics; presumably, the latter is what appears in the first three Chapters of his *Statistical Mechanics*, 18 years later.

It seems likely, then, that Gibbs found the main results of his *Statistical Mechanics* very early; perhaps even as early as his attending Liouville's lectures in 1867. But he delayed finishing the book for many years, probably because of mysteries concerning specific heats that he kept hoping to resolve, but could not before the days of quantum mechanics; his remarks in the preface indicate how much this bothered him. The parts that seemed unsatisfactory would have been left unwritten in final form until the last possible moment; and those parts would be the ones where we are most likely to find small errors or incomplete statements.

Gibbs' concern about specific heats is no longer an issue for us today, but we want to understand why classical statistical mechanics appeared (at least to readers of the book) to fail badly in the matter of extensivity of entropy. He introduces the canonical ensemble in Chapter IV, as a probability density $P(p_1 \ldots q_n)$ in phase space, and writes (his Eq. 90; hereafter denoted by SM.90, etc.):

$$\eta = \frac{\psi - \epsilon}{\Theta} = \log P \qquad (SM.90)$$

where $\epsilon = \epsilon(p_1 \ldots q_n)$ is the Hamiltonian (same as the total energy, since he considers only conservative forces), Θ is the "modulus" of the distribution, and ψ is a normalization constant chosen so that $\int P dp_1, \ldots, dq_n = 1$. He notes that Θ has properties analogous to those of temperature, and strengthens the analogy by introducing externally variable coordinates a_1, a_2 with the meaning of volume, strain tensor components, height in a gravitational field, etc. with their conjugate forces $A_i = -\partial \epsilon / \partial a_i$. On an infinitesimal change (*i.e.*, comparing two slightly different canonical distributions) he finds the identity

$$d\bar{\epsilon} = -\Theta d\bar{\eta} - \sum \bar{A}_i \, da_i \qquad (SM.114)$$

where the bars denote canonical ensemble averages. He notes that this is identical in form with a thermodynamic equation if we neglect the bars and consider $(-\bar{\eta})$ as the analog of entropy, which he denotes, incredibly, by η.

Here Gibbs anticipates – and even surpasses – the confusion caused 46 years later when Claude Shannon used H for what Boltzmann had called $(-H)$. In addition, the prospective reader is warned that from p. 44 on, Gibbs uses the same symbol η to denote both the "index of probability" as in (SM.90), and the thermodynamic entropy, as in (SM.116); his meaning can be judged only from the context. In the following we deviate from Gibbs' notation by using Clausius' symbol S for thermodynamic entropy.

Note that Gibbs, writing before the introduction of Boltzmann's constant k, uses Θ, with the dimensions of energy, for what we should today call kT; consequently his thermodynamic analog of entropy is what we should today call S/k, and is dimensionless.*

The thermodynamic equation analogous to (SM.114) is then

$$d\epsilon = TdS - \sum A_i \, da_i \qquad\qquad (SM.116)$$

Now Gibbs notes that in the thermodynamic equation, the entropy S "is a quantity which is only defined by the equation itself, and incompletely defined in that the equation only determines its differential, and the constant of integration is arbitrary. On the other hand, the $\overline{\eta}$ in the statistical equation has been completely defined." Then interpreting $(-\overline{\eta})$ as entropy leads to the familiar conclusion, stressed by later writers, that entropy is not extensive in classical statistical mechanics.

Right here, we suggest, two fine points were missed, but the Pauli analysis conditions us to recognize them at once. The first is that normalization requires that P has the physical dimensions $(\text{action})^{-n}$, while the argument of a logarithm should be dimensionless. To obtain a truly dimensionless $\overline{\eta}$, we should rewrite (SM.90) as $\eta = \log(\xi^n P)$ where ξ is some quantity of the dimensions of action. This point has, of course, been noted before many times.

The second fine point is more serious and does not seem to have been noted before, even by Pauli; the question whether $-\overline{\eta}$ is the proper statistical analog of thermodynamic entropy cannot be answered merely by examination of (SM.114). Let us denote by σ the correct (dimensionless) statistical analog of entropy that Gibbs was seeking. The trouble is again that in (SM.114) we are varying only ϵ and the a_i; consequently it determines only how σ varies with ϵ and the a_i. As in (9), from Gibbs' (SM.114) we can infer only that the correct statistical analog of entropy must have the form

$$\sigma = -\overline{\eta} + g(N), \qquad\qquad (16)$$

where $N = n/3$ denotes as before the number of particles. Again, the "constant of integration" is not an arbitrary constant, but an arbitrary function $g(N)$. Clearly, no definite "statistical analog" of entropy has been defined until the function $g(N)$ is specified.

However, in defense of Gibbs, we should note that at this point he is not discussing extensivity of entropy at all, and so he could reply that he is considering only fixed values of N, and so is in fact concerned only with an arbitrary constant. Thus one can argue whether it was Gibbs or his readers who missed this fine point (it is only 160 pages later, in the final two paragraphs of the book, that Gibbs turns at last to the question of extensivity).

In any event, a point that has been missed for so long deserves to be stressed. For 60 years, all scientists have been taught that in the issue of extensivity of entropy we have a fundamental failure, not just of classical statistics, but of classical *mechanics*, and a triumph of quantum mechanics. The present writer was caught in this error just as badly as anybody else, throughout most of his teaching career. But now it is clear that the trouble was not in any failure of classical mechanics, but in our own failure to perceive all the freedom that Gibbs' (SM.114) allows. If we wish entropy to be extensive in classical

---

* In this respect Gibbs' notation is really neater formally – and more cogent physically – than ours; for Boltzmann's constant is only a correction factor necessitated by our custom of measuring temperature in arbitrary units derived from the properties of water. A really fundamental system of units would have $k \equiv 1$ by definition.

statistical mechanics, we have only to choose $g(N)$ accordingly, *as was necessary already in the phenomenological theory*. But curiously, nobody seems to have noticed, much less complained, that Clausius' definition $S \equiv \int dQ/T$ also failed in the matter of extensivity.

But exactly the same argument will apply in quantum statistical mechanics; in making the connection between the canonical ensemble and the phenomenological relations, if we follow Gibbs and use the Clausius definition of entropy as our sole guide in identifying the statistical analog of entropy, it will also allow an arbitrary additive function $h(N)$ and so it will not require entropy to be extensive any more than does classical theory. Therefore the whole question of in what way – or indeed, whether – classical mechanics failed in comparison with quantum mechanics in the matter of entropy, now seems to be re–opened.

Recognizing this, it is not surprising that entropy has been a matter of unceasing confusion and controversy from the day Clausius discovered it. Different people, looking at different aspects of it, continue to see different things because there is still unfinished business in the *fundamental definition of entropy*, in both the phenomenological and statistical theories.

In the case of the canonical ensemble, the oversight about extensivity is easily corrected, in a way exactly parallel to that indicated by Pauli. In the classical case, considering a gas defined by the thermodynamic variables $A_1 = p =$ pressure, $a_1 = V =$ volume, Gibbs' statistical analog equation (SM.114) may be written

$$d\bar{\epsilon} = -\Theta d\bar{\eta} - \bar{p}dV \tag{17}$$

The statistical analog of entropy must have the form (16); and if we want it to be extensive, it must also satisfy the scaling law (10). Now from the canonical ensemble for an ideal gas, with phase space probability density

$$P = \frac{1}{(2\pi m\Theta)^{3N/2}V^N} \exp\left[-\sum \frac{p_i^2}{2m\Theta}\right] \tag{18}$$

(which we note is dimensionally correct and normalized, although it does not contain Planck's constant), we have from the dimensionally corrected (SM.90),

$$-\bar{\eta} = N \log V + \frac{3N}{2}\left[\log(2\pi m\Theta) + 1\right] - 3N \log \xi \tag{19}$$

Then, substituting (19) and (16) into (10), we find that $g(N)$ must satisfy the same functional equation (11), with the same solution (13). The Gibbs statistical analog of entropy (16) is now

$$\sigma = N\left[\log \frac{V}{N} + \frac{3}{2}\log \frac{2\pi m\Theta}{\xi^2} + \frac{3}{2} + g(1)\right]. \tag{20}$$

The extensive property (9) now holds, and the constants $g(1)$ and ξ combine to form essentially the chemical constant. This is not determined by the above arguments, just as it was not determined in the phenomenological arguments.

Therefore it might appear that the shortcoming of classical statistical mechanics was not any failure to yield an extensive entropy function, but only its failure to determine the numerical value of the chemical constant. But even this criticism is not justified at present; the mere fact that it is believed to involve Planck's constant is not conclusive, since e and c

are classical quantities and Planck's constant is only a numerical multiple of e^2/c. We see no reason why the particular number 137.036 should be forbidden to appear in a classical calculation. Since the problem has not been looked at in this way before, and nobody has tried to determine that constant from classical physics, we see no grounds for confident claims either that it can or cannot be done.

But the apparent involvement of e and c suggests, as Gibbs himself noted in the preface to his *Statistical Mechanics*, that electromagnetic considerations may be important even in the thermodynamics of electrically neutral systems. Of course, from our modern knowledge of the electromagnetic structure of atoms, the origin of the van der Waals forces, *etc.*, such a conclusion is in no way surprising. Electromagnetic radiation is surely one of the mechanisms by which thermal equilibrium is maintained in any system whose molecules have dipole or quadrupole moments, rotational/vibrational spectrum lines, *etc.* and no system is in true thermal equilibrium until it is bathed in black–body radiation of the same temperature.

At this point the reader may wonder: Why do we not turn to the grand canonical ensemble, in which all possible values of N are represented simultaneously? Would this enable us to determine by physical principles how entropy varies with N? Unfortunately, it cannot do so as long as we try to set up entropy in statistical mechanics by finding a mere *analogy* with the Clausius definition of entropy, $S = \int dQ/T$. For the Clausius definition was itself logically incomplete in just this respect; the information needed is not in it.

The Pauli correction was an important step in the direction of getting "the bulk of things" right pragmatically; but it ignores the small deviations from extensivity that are essential for treatment of some effects; and in any event it is not a fundamental theoretical principle. A truly general and quantitatively accurate definition of entropy must appeal to a deeper principle which is hardly recognized in the current literature, although we think that a cogent special case of it is contained in some early work of Boltzmann, Einstein, and Planck.

10. Summary and Unfinished Business

We have shown here that the Clausius definition of entropy was incomplete already in the phenomenological theory; therefore it was inadequate to determine the proper analog of entropy in the statistical theory. The phenomenological theory, classical statistical mechanics, and quantum statistical mechanics, were all silent on the question of extensivity of entropy *as long as one tried to identify the theoretical entropy merely by analogy with the Clausius empirical definition.* The Pauli type analysis is a partial corrective, which applies equally well and is equally necessary in all three cases, if we wish entropy to be extensive.

It is curious that Gibbs, who surely recognized the incompleteness of Clausius' definition, did not undertake to give a better one in his *Heterogeneous Equilibrium*; he merely proceeded to use entropy in processes where N changes and never tried to interpret it in such terms as phase volume, as Boltzmann, Planck, and Einstein did later. In view of Gibbs' performance in other matters, we cannot suppose that this was a mere oversight or failure to understand the logic. More likely, it indicates some further deep insight on his part about the difficulty of such a definition.

A consequence of the above observations is that the question whether quantum theory really gave an extensive entropy function and determined the value of the chemical constant, also needs to be re–examined. Before a quantum theory analog of entropy has been defined, one must consider processes in which N changes. If we merely apply the scaling law as we did above, the quantum analog of entropy will have the form (20) in which ξ is set equal to Planck's constant; but a new arbitrary constant $\alpha \equiv f_q(1)$ is introduced that has not been considered heretofore in quantum statistics. If we continue to define entropy by the Clausius definition (7), quantum statistics does not determine α.

This suggests that, contrary to common belief, the value of the chemical constant is not determined by quantum statistics as currently taught, any better than it was by classical theory. It seems to be a fact of phenomenology that quantum statistics with $\alpha = 0$ accounts fairly well for a number of measurements; but it gives no theoretical reason why α should be zero. The Nernst Third Law of Thermodynamics does not answer this question; we are concerned rather with the experimental accuracy of such relations as the Sackur–Tetrode formula for vapor pressure. Experimental vapor pressures enable us to determine the difference $\alpha_{gas} - \alpha_{liquid}$. Therefore we wonder how good is the experimental evidence that α is not needed, and for how many substances we have such evidence.

For some 60 years this has not seemed an issue because one thought that it was all settled in favor of quantum theory; any small discrepancies were ascribed to experimental difficulties and held to be without significance. Now it appears that the issue is reopened; it may turn out that α is really zero for all systems; or it may be that giving it nonzero values may improve the accuracy of our predictions. In either case, further theoretical work will be needed before we can claim to understand entropy.

There is a conceivable simple resolution of this. Let us conjecture that the present common teaching is correct; i.e., that new precise experiments will confirm that present quantum statistics does, after all give the correct chemical constants for all systems. If this conjecture proves to be wrong, then some of the following speculations will be wrong also.

We should not really expect that a phenomenological theory, based necessarily on a finite number of observations that happened to feasible in the time of Clausius – or in our time – could provide a full definition of entropy in the greatest possible generality. The question is not an empirical one, but a conceptual one; and only a definite theoretical principle, that rises above all temporary empirical limitations, can answer it. In other words, it is a mistake to try to define entropy in statistical mechanics merely by analogy with the phenomenological theory. The only truly fundamental definition of entropy must be provided directly by the statistical theory itself; and comparison with observed phenomenology takes place only afterward.

This is how relativity theory was constituted: empirical results like the Michelson–Morley experiment might well suggest the principle of relativity; but the deduction of the theory from this principle was made without appeal to experiment. After the theory was developed, one could test its predictions on such matters as abberation, time dilatation, and the orbits of fast charged particles.

The answer is rather clear; for both Clausius and Gibbs the theoretical principles that were missing did not lie in quantum theory (that was only a quantitative detail). The fundamental principles were the principles of probability theory itself; how does one set up a probability distribution over microstates – classical or quantum – that represents correctly our information about a macrostate? If that information includes all the conditions needed in the laboratory to determine a reproducible result – equilibrium or nonequilibrium – then the theory should be able to predict that result.

We suggest that the answer is the following. For any system the entropy is a property of the macrostate (more precisely, it is a function of the macrovariables that we use to define the macrostate), and it is defined by a variational property: it is the upper bound of $-k\,\mathrm{Tr}(\rho \log \rho)$ over all density matrices that agree with those macrovariables. This will agree with the Clausius and Pauli prescriptions in those cases where they were valid; but it automatically provides the extra terms that Gibbs needed to analyze surface effects, if we apply it to the finite sized systems that actually exist in the laboratory.

When thermodynamic entropy is defined by this variational property, the long confusion about order and disorder (which still clutters up our textbooks) is replaced by a remarkable simplicity and generality. The conventional Second Law follows *a fortiori*: since entropy is defined as a constrained maximum, whenever a constraint is removed, the entropy will tend to increase, thus paralleling in our mathematics what is observed in the laboratory. But it provides generalizations far beyond that, to many nonequilibrium phenomena not yet analyzed by any theory.

And, just as the variational formalism of Gibbs' *Heterogeneous Equilibrium* could be used to derive useful rules of thumb like the Gibbs phase rule, which were easier to apply than the full formalism, so this generalization leads to simple rules of thumb like the phase volume interpretation $S = k \log W$ of Boltzmann, Einstein, and Planck, which are not limited to equilibrium conditions and can be applied directly for very simple applications like the calculation of muscle efficiency in biology (Jaynes, 1989).

REFERENCES

Callen, H. B. (1960), *Thermodynamics*, J. Wiley & Sons, Inc., New York.

Epstein, P. S. (1937), *Textbook of Thermodynamics*, J. Wiley & Sons, Inc., New York.

Gibbs, J. Willard (1875–78), "On the Equilibrium of Heterogeneous Substances", Connecticut Acad. Sci. Reprinted in *The Scientific Papers of J. Willard Gibbs*, by Dover Publications, Inc., New York (1961).

Gibbs, J. Willard (1902), *Elementary Principles in Statistical Mechanics*, Yale University Press, New Haven, Conn. Reprinted in *The Collected Works of J. Willard Gibbs*, Vol. 2, by Dover Publications, Inc., New York (1960).

Jaynes, E. T. (1965), "Gibbs vs Boltzmann Entropies", *Am. Jour. Phys.* **33**, 391–398.

Jaynes, E. T. (1967), "Foundations of Probability Theory and Statistical Mechanics", in *Delaware Seminar in Foundations of Physics*, M. Bunge, Editor, Springer–Verlag, Berlin. Reprinted in Jaynes (1983).

Jaynes, E. T. (1983), *Papers on Probability, Statistics, and Statistical Physics*, D. Reidel Publishing Co., Dordrecht, Holland, R. D. Rosenkrantz, Editor. Reprints of 13 papers.

Jaynes, E. T. (1986), "Macroscopic Prediction", in *Complex Systems – Operational Approaches*, H. Haken, Editor; Springer–Verlag, Berlin.

Jaynes, E. T. (1988), "The Evolution of Carnot's Principle", in *Maximum Entropy and Bayesian Methods in Science and Engineering*, Vol. 1, G. J. Erickson and C. R. Smith, Editors; Kluwer Academic Publishers, Dordrecht, Holland.

Jaynes, E. T. (1989), "Clearing up Mysteries; the Original Goal", in Proceedings of the 8'th International Workshop in Maximum Entropy and Bayesian Methods, Cambridge, England, August 1–5, 1988; J. Skilling, Editor; Kluwer Academic Publishers, Dordrecht, Holland.

Klein, M. J. (1973), "The Development of Boltzmann's Statistical Ideas", in E. C. D. Cohen & W. Thirring, eds., *The Boltzmann Equation*, Springer–Verlag, Berlin; pp. 53–106.

Pauli, W. (1973), *Thermodynamics and the Kinetic Theory of Gases*, (Pauli Lectures on Physics, Vol. 3); C. P. Enz, Editor; MIT Press, Cambridge MA.

Planck, M. (1926), *Treatise on Thermodynamics*, Dover Publications, Inc., N. Y.

Zemansky, M. W. (1943), *Heat and Thermodynamics*, McGraw–Hill Book Co., New York.

BAYESIAN SOLUTION OF ORDINARY DIFFERENTIAL EQUATIONS

John Skilling
Department of Applied Mathematics and Theoretical Physics
Silver Street
Cambridge CB3 7HJ
England

ABSTRACT. In the numerical solution of ordinary differential equations, a function $y(x)$ is to be reconstructed from knowledge of the functional form of its derivative: $dy/dx = f(x, y)$, together with an appropriate boundary condition. The derivative f is evaluated at a sequence of suitably chosen points (x_k, y_k), from which the form of $y(\cdot)$ is estimated. This is an inference problem, which can and perhaps should be treated by Bayesian techniques. As always, the inference appears as a probability distribution $\mathrm{prob}(y(\cdot))$, from which random samples show the probabilistic reliability of the results. Examples are given.

1. Introduction

The classic ordinary differential equation problem is the simplest initial value problem

$$\text{``Given } dy/dx = f(x, y) \text{ and } y(0) = 0, \text{ find } Y = y(X)\text{''}. \tag{1}$$

A variant on this is to require

$$y(x) \ \forall \ 0 \leq x \leq X \tag{2}$$

We are to solve equation (1) numerically and approximately, using as few evaluations of f as possible. The special case where f is independent of y reduces to simple integration, but when f depends on y as well, the problem is more awkward. The difficulty is that we never know the correct value of y at any selected abscissa x_k. When we estimate it, as y_k say, we will almost certainly be wrong. Hence our evaluations of data,

$$D_k = f(x_k, y_k) \tag{3}$$

will not measure dy/dx accurately. In other words, we remain uncertain of precisely what we are integrating. Clearly this is an inference problem of some subtlety. Probability calculus being the only calculus for consistently reasoning from incomplete information, we should nevertheless recommend it and try to use it in such problems.

This paper suggests ways of setting up a Bayesian treatment. Far from considering these suggestions to be the last word, the author considers them to be merely pointers to what could, and perhaps should, develop into an open field of research, complementing and if appropriate superseding the traditional algorithms.

23

C. R. Smith et al. (eds.), Maximum Entropy and Bayesian Methods, Seattle, 1991, 23–37.
© 1992 Kluwer Academic Publishers.

As always, Bayesian analysis requires
a) a hypothesis space $\{h\}$ which will here involve the set of admissible functions $y(\cdot)$, augmented by such extra parameters as may be required;
b) a prior probability distribution $\text{pr}(h)$ over these hypotheses;
c) the likelihood $\text{pr}(D|h)$, being the probability of acquiring any supplied data constraints D, given one of the hypotheses. Bayes' theorem, that

$$\begin{aligned} \text{pr}(h, D) &= \text{pr}(h)\,\text{pr}(D|h) \equiv \text{prior} \times \text{likelihood} \\ &= \text{pr}(D)\text{pr}(h|D) \equiv \text{evidence} \times \text{inference} \end{aligned} \tag{4}$$

(conditional on whatever background information $\mathcal{I}$ is currently assumed) allows us to draw probabilistic conclusions in the form of the overall evidence

$$\text{pr}(D) = \int dh\,\text{pr}(h, D) \tag{5}$$

and the posterior inference

$$\text{pr}(h|D) = \text{pr}(h, D)/\text{pr}(D) \tag{6}$$

The evidence, more properly written as $\text{pr}(D|\mathcal{I})$, allows us to assess any rival choices for $\mathcal{I}$.

It may at first seem optimistic to attempt this approach, which immediately involves integration over large spaces, and we may surmise this to be why numerical analysis did not historically take this path. However, we shall try to obey the Bayesian rules.

2. Spatial Correlation in the Hypothesis Space

Presumably f is reasonably smooth in x and y, otherwise its sampled values would offer inadequate guidance to construct $y(x)$, and the problem would be hopeless from the start. This is a fundamental property which must be part of our treatment. We assign

$$W = \text{length-scale with respect to } x, \tag{7}$$

representing the variability in f, which will ultimately be reflected in our inference about the solution $y(\cdot)$. Similarly, we assign a stiffness constant

$$c = \|\partial f/\partial y\| \tag{8}$$

to quantify the sensitivity of f to variations in y. Although c has dimensions formally inverse to W, it seems better to keep the two parameters independent.

Experience in other areas of scientific data analysis suggests that the natural way of imposing a preference for smoothness in x is to introduce an intrinsic correlation function (ICF) which explicitly smooths some sharper, hidden function by averaging it over the width W. Here, it is the differential y' which is assumed to be smooth, rather than y itself, so we take y' to be a smoothed version of some hidden function $h(x)$. Specifically, we shall suppose

$$y'(x) = \int_{-\infty}^{\infty} du\,R(x, u)\,h(u) \tag{9}$$

or, in shorter matrix notation,

$$y' = R h \tag{10}$$

where

$$R(x, u) = e^{-(x-u)^2/2W}/\sqrt{2\pi W} \tag{11}$$

Purely for convenience, this ICF R is given the form of a Gaussian convolution of width W, which incidentally requires h to be defined for negative as well as positive x. This approach to smoothness differs from the traditional approach of assuming some limit on some higher differential, typically $\|y''\|_\infty$.

There are reasons for preferring the ICF approach.

a) We seek the macroscopic structure of $y(\cdot)$, for which local irregularities in a high differential would scarcely be relevant, so that integration is a more natural representation of the smoothness property we need.

b) Explicit limits on high differentials are often unavailable, especially in non-trivial problems, where the algebraic form of a high differential may be excessively long.

c) Any prior probability distribution for a high differential would have to be supplemented by awkward subsidiary conditions on the corresponding constants of integration in order to reach a normalised prior for y itself.

d) Provided only that h is integrable ($h \in \mathcal{L}_1$), its smoothed form Rh shares all the continuity and differentiability properties of R, so that when R is Gaussian, Rh is automatically differentiable to all orders. On the other hand, the (realistic) error bars on the results from this approach lack the absolute force of the (pessimistic) worst case bounds which the traditional approach can give in favourable examples.

It follows from the assumed form of y' that y is yet smoother, effectively being h convolved with a smoothed step function instead of the smoothed delta function represented by the Gaussian function R. More precisely, we integrate (using $y(0) = 0$) to obtain

$$y(x) = \int_0^x dz\, y'(z) = \int_{-\infty}^\infty du\, Q(x, u)\, h(u) \tag{12}$$

or, in shorter matrix notation,

$$y = Q h \tag{13}$$

where

$$Q(x, u) = \int_0^x dz\, R(z, u) = \tfrac{1}{2}\left(\mathrm{erf}\left((x - u)/\sqrt{2}W\right) + \mathrm{erf}\left(u/\sqrt{2}W\right)\right) \tag{14}$$

Q may be viewed as the ICF for y itself, as opposed to y'.

As an alternative, we can assume that y is usefully twice differentiable, meaning that it is y'' which is the convolution with the original ICF R. In order to fix the second constant of integration, we then use the initial datum $D_0 = f(0, 0)$. This gives us a different hypothesis about the form of solution y to be expected. Indeed, we could go further, and let yet higher derivatives be the convolution with R (though the treatment of the extra constants of integration would become less clear). The point at issue is not whether or not the derivatives *exist*, because that will be determined by the differentiability of R, but rather whether or not the derivatives are *useful*. The "order" of these Bayesian algorithms can be defined as the order of differential which is most simply related to the hidden function.

Thus the *first order* Bayesian algorithm has $y' = Rh$ as in (9), the *second order* Bayesian algorithm has $y'' = Rh$ instead, and so on.

3. The Prior

Assuming, as above, that the only useful differential of y is the first, we need a quantified prior on the hidden function $h(\cdot)$. For tractability, we adopt a Gaussian with parameter α on a flat (unit) measure. In matrix notation, using a large number M of closely spaced values of x,

$$\mathrm{pr}(h)\, d^M h = \sqrt{\det(\alpha I/2\pi)}\, e^{-\alpha h^T h/2}\, d^M h \tag{15}$$

where

$$h^T h = \int du\, h(u)^2, \tag{16}$$

$$\det(\alpha I/2\pi) = (\alpha/2\pi)^M \tag{17}$$

I being the unit matrix. By placing h almost certainly in $\mathcal{L}_2$, it must almost certainly be in $\mathcal{L}_1$ as well, as required. As usual, the value of α needs to be discussed. Its dimensions are

$$[\alpha] = 1/[h^T h] = 1/[xh^2] = [x/y^2]. \tag{18}$$

Properly, α should be given a normalised prior biassed towards values of the order of

$$\alpha \sim \frac{(\text{expected scale in } x)}{(\text{expected scale in } y)^2} \tag{19}$$

In practice, once several data have been obtained the posterior distribution of α tends to be fairly sharply peaked around its maximum, so that it suffices to fix α at this maximising value. This completes the preliminaries. The hypothesis space has been set up, involving a hidden function h related to the required function y by an ICF of assumed Gaussian form and assumed width W. On this space, the prior on h has been defined, conditional upon the parameter α which can effectively be fixed *a posteriori*. The stiffness parameter c is also available to enter the analysis at a later stage.

4. The Likelihood

At any stage in the estimation of y, we have a list of coordinates (x_k, y_k) at which f has been evaluated as $D_k = f(x_k, y_k)$. Even at the very beginning, we may allow ourselves $D_0 = f(0,0)$. These data will (somehow) modify our estimates of the hidden function h, and thence of y and the ultimate required value $Y = y(X)$. However, we need to know how accurately the data represent the slope y' of the true curve $y(x)$.

Suppose that we can (somehow) use the data to estimate a mean $\hat{y}(x)$ and standard deviation $\delta\hat{y}(x)$, allowing us to infer

$$y(x_k) = \hat{y}_k \pm \delta\hat{y}_k \tag{20}$$

at each measurement abscissa x_k. Awkwardly, the measurement ordinate y_k will usually be different again. Although one hopes that it was chosen intelligently, it will necessarily have been selected on the basis of less data than are currently available. Hence we may write the positional error of y_k, relative to the true curve, as

$$\Delta y_k \equiv y_k - y(x_k) = (y_k - \hat{y}_k) + (\hat{y}_k - y(x_k)) \tag{21}$$

On the right, $y(x_k)$ is the only unknown. Its estimated mean and variance lead us to the variance

$$\langle (\Delta y_k)^2 \rangle = (y_k - \hat{y}_k)^2 + (\delta \hat{y}_k)^2 \tag{22}$$

describing our uncertainty about the true curve, relative to the measurement ordinate y_k. This uncertainty in position y can be rescaled through the stiffness constant to give a corresponding uncertainty in f:

$$\Delta f = c \Delta y \tag{23}$$

This represents the difference between the true slope $f(x_k, y(x_k))$ at abscissa x_k and the measured slope $D_k = f(x_k, y_k)$. Accordingly, D_k measures the true slope at abscissa x_k, with an error σ_k given by

$$\sigma_k^2 = c^2 \left((y_k - \hat{y}_k)^2 + (\delta \hat{y}_k)^2 \right) \tag{24}$$

Datum D_k becomes interpreted as a noisy measurement of the true slope, just as in ordinary data analysis. The main difference from ordinary data analysis is that the data uncertainties are not fixed in advance. The errors depend on the current inference through $\hat{y}$ and $\delta\hat{y}$, and these are likely to evolve as more data are acquired. However, we shall shortly see how to construct $\hat{y}$ and $\delta\hat{y}$ from a given dataset, from any given errors σ. A few iterations of this allows $\hat{y}$ and $\delta\hat{y}$ to be determined self-consistently along with σ: there seems no reason to expect any instability in this scheme.

Given a set of errors σ, albeit provisional, as well as the data D, we can finally construct the likelihood. In its usual Gaussian form,

$$\text{pr}(D|h) = \sqrt{\det(\sigma^{-2}/2\pi)} \, e^{\frac{1}{2}(Rh-D)^T \sigma^{-2}(Rh-D)} \tag{25}$$

where σ^{-2} is the $n \times n$ inverse covariance matrix of the errors. In this preliminary study, we take σ^{-2} to be diagonal, ignoring any possible correlation between the errors of different data: it is likely that this simplification detracts from the quality of the results.

5. Evidence and Inference

Both the prior $\text{pr}(h)$ and the likelihood $\text{pr}(D|h)$ are now available as Gaussians in h, albeit dependent on the iteratively-derived errors σ. Accordingly, the joint distribution is also Gaussian. Its integral gives the evidence

$$\text{pr}(D) = e^{-\frac{1}{2}D^T B^{-1} D} / \sqrt{\det(2\pi B)} \tag{26}$$

where

$$B = \sigma^2 + RR^T/\alpha \tag{27}$$

The parameter α is chosen to maximise the value of the evidence. Also, the mean and covariance give

$$\hat{h} = R^T B^{-1} D / \alpha \tag{28}$$

$$\langle \delta \hat{h} \delta \hat{h}^T \rangle = (I - R^T B^{-1} R/\alpha)/\alpha \qquad (29)$$

¿From h, we can estimate the result $y = Qh$ and its derivative $y' = Rh$ as

$$\hat{y} = Q R^T B^{-1} D/\alpha \qquad (30)$$

$$\langle \delta \hat{y} \delta \hat{y}^T \rangle = (QQ^T - QR^T B^{-1} RQ^T/\alpha)/\alpha \qquad (31)$$

$$\hat{y}' = R R^T B^{-1} D/\alpha \qquad (32)$$

$$\langle \delta \hat{y}' \delta \hat{y}'^T \rangle = (RR^T - RR^T B^{-1} RR^T/\alpha)/\alpha \qquad (33)$$

These results evaluated at the measurement abscissae define the errors σ self-consistently.

As a consequence of using Gaussian distributions, we have been able to write the "visible" results y and y' without using M-dimensional "hidden" space at all: the matrices Q and R which contain this dimension only appear in the combinations

$$RR^T(x, z) = \int_{-\infty}^{\infty} du\, R(x, u)\, R(z, u)$$
$$= e^{-(x-z)^2/4W^2}/2\sqrt{\pi}W \qquad (34)$$

$$QR^T(x, z) = \int_{-\infty}^{\infty} du\, Q(x, u)\, R(z, u)$$
$$= \tfrac{1}{2}\left(\mathrm{erf}((z - x)/2W) + \mathrm{erf}(x/2W)\right) \qquad (35)$$

$$QQ^T(x, z) = \int_{-\infty}^{\infty} du\, Q(x, u)\, Q(z, u)$$
$$= \tfrac{1}{2}\left(x\,\mathrm{erf}(x/2W) + z\,\mathrm{erf}(z/2W) - (x - z)\,\mathrm{erf}((x - z)/2W)\right)$$
$$+ \pi^{-\frac{1}{2}}W\left(e^{-x^2/4W^2} + e^{-z^2/4W^2} - e^{-(x-z)^2/4W^2} - 1\right) \qquad (36)$$

These combinations operate on the n-dimensional vector D and $n \times n$ matrix B. Thus the dimensionality of the matrix calculations is only n (the number of data), augmented perhaps by the number of abscissae at which results are required.

We can now estimate the desired result $Y = y(X)$, with its probabilistic uncertainty, as

$$\hat{Y} = \hat{y}(X) \pm \delta \hat{y}(X) \qquad (37)$$

If the uncertainty $\delta \hat{y}(X)$ is acceptably small, all well and good. But if the uncertainty is unacceptably large, we must acquire more data, by evaluating more values of $f(x, y)$. This raises the question of where such samples should be taken.

6. Sampling Strategy

In numerical analysis, as opposed to more examples of scientific data analysis, we often have freedom to place our data samples where we wish. In our ordinary differential equation problem, we can select our next sample, or samples, at any arbitrary coordinates x and y. Presumably these samples should be chosen sensibly, rather than just being purely random. Indeed, having selected a value ξ for x, then the selection $\hat{y}(\xi)$ seems appropriate for y, in order to minimise the error of σ of this new measurement, as expected on the basis of the currently available data. However, the choice of ξ is less obvious.

One could use some pre-assigned strategy. The simplest would be to set some fairly small step-length Δx, and steadily increase ξ by this amount. For extra safety, one could re-sample y a few times at each chosen x. One could also emulate Runge-Kutta by returning to the interior of the current step and re-sampling within. In fact any traditional strategy could be incorporated into the Bayesian formulation, because any strategy whatever will produce a dataset which the Bayesian can interpret.

However it is more interesting, and may ultimately prove more productive, to seek that sample abscissa ξ for which the new datum is expected to be *most informative* about our result Y. In this way, we may hope to gain information as rapidly as possible, and thus converge to an accurate result with fewest samples.

If we were to measure at a particular place ξ, we would refine our result from (say)

$$Y = \mu_0 \pm \sigma_0 \tag{38}$$

to (say)

$$Y = \mu_1 \pm \sigma_1 \tag{39}$$

both uncertanties being Gaussian. The natural measure of information gleaned in this refinement is the (positive) cross-entropy

$$
\begin{aligned}
S &= \int dY \; \mathrm{pr}_1(Y) \log\left(\mathrm{pr}_1(Y)\mathrm{pr}_0(Y)\right) \\
&= \left((\mu_1 - \mu_0)^2 + \sigma_1^2 - \sigma_0^2\right)/2\sigma_0^2 - \log(\sigma_1/\sigma_0)
\end{aligned}
\tag{40}
$$

As one would hope, this formula demonstrates a gain both for the reduction in variance σ^2 occasioned by acquiring more data, and for any change in the expectation value μ.

Suppose that we were to obtain a new datum $D^* = f(\xi, \hat{y}(\xi))$ at our selected abscissa ξ. Ignoring any iterative refinement of errors σ in the light of the new value D^*, we could already assign its uncertainty σ^* on the basis of the expected misfit $\delta\hat{y}(\xi)$ between our current estimate $\hat{y}(\xi)$ and the true curve $y(\xi)$. Accordingly, we could update the inference for Y, and obtain the corresponding information gain S^*. Until we perform the evaluation, we do not know the actual value of D^*, but we *can* predict what D^* measures, namely the slope $y'(\xi)$ which we expect to observe at ξ. Its mean and variance are the results (31) and (32) for y', evaluated at ξ. Averaging over this expected range of values for D^* gives us the expectation information $\langle S^*(\xi)\rangle$ which we *expect* to gain *if* we were to sample at ξ. The averaging can be carried out analytically, so that $\langle S^*(\xi)\rangle$ can be computed reasonably quickly for any ξ. It is then straightforward to evaluate it at sufficient trial values of ξ to locate its maximum.

We use this maximum as the new sample abscissa ξ, and proceed to the actual new measurement, augmenting the current dataset by

$$D_{n+1} = f(\xi, \hat{y}(\xi)) \tag{41}$$

(An alternative procedure would be to treat $S^*(\xi)$ as an additive distribution, and sample ξ randomly from it.) We now have a full, usable algorithm for solving the original problem (1). It involves evaluating f at successive points $(\xi, \hat{y}(\xi))$ which can be chosen to be optimal for the specific f in hand, and then repeating the process until the estimate of the final result $Y = y(X)$ is sufficiently accurate for its purpose.

7. Deletion of data

As the number n of data increases, the cost of their acquisition in terms of evaluations of f clearly increases in direct proportion. However, the overheads increase much faster, with the inversion of B being $\mathcal{O}(N^3)$ and the evaluations of $\langle S^* \rangle$ being $\mathcal{O}(N^2 n)$. If the cost of the overheads is not to overwhelm the cost of evaluating f, there must be a mechanism for deleting points from the dataset, as well as including new ones. Presumably, the point to be deleted should be that one which causes least damage to the result Y. Again, such degradation can be quantified in terms of the cross-entropy between the initial (with all data) and the degraded (with one datum deleted) probability distributions of Y. Given a choice, we shall choose to delete that datum for which the cross-entropy damage is *least*.

According to this scheme, the algorithm will proceed to loop indefinitely if the most recently added point turns out to be the least informative. When that happens, it seems that the algorithm has proceeded as far as it can with its restricted size of dataset. This gives a natural termination criterion.

8. Probabilistic displays

Finally, it is always interesting, and can be useful, to display samples from the current posterior probability distribution for y. Indeed, this seems to be the only way of readily visualising the current state of the algorithm.

The first step is to compute a sample of the hidden function h on a discrete grid of M abscissae x. Introduce vectors a and b whose M and n components (respectively) are each independently drawn from the unit normal distribution $\mathcal{N}(0,1)$, so that their statistics are

$$\langle a \rangle = 0, \quad \langle aa^T \rangle = I \tag{42}$$

$$\langle b \rangle = 0, \quad \langle bb^T \rangle = I \tag{43}$$

Then it can be straightforwardly verified that the M-dimensional vector

$$\tilde{h} = R^T B^{-1} D / \alpha + \alpha^{-\frac{1}{2}} (a - R^T B^{-1} (Ra + \alpha^{\frac{1}{2}} \sigma b) / \alpha) \tag{44}$$

has the correct statistics (27) and (28) to be a random sample of y. After this has been calculated, application of $M \times M$ versions of the matrices Q and R yield the required sample

$$\tilde{y} = Q\tilde{h} \tag{45}$$

and, should it be needed, the corresponding derivative

$$\tilde{y}' = R\tilde{h} \tag{46}$$

The accompanying figures show overlaid plots of a few such random samples $\tilde{y}_1, \tilde{y}_2, \tilde{y}_3, \ldots$ from the full result $\text{pr}(y(\cdot))$ in particular examples.

However, a more dynamic and visually effective presentation can be achieved on a computer screen. Instead of superposing the original samples, generate a related sequence

$$\tilde{y}_1^* = \tilde{y}_1, \quad \tilde{y}_{j+1}^* = \tilde{y}_j \cos\theta + \tilde{y}_{j+1} \sin\theta \tag{47}$$

where $0 < \theta < \pi/2$. Each $\tilde{y}^*(\cdot)$ is itself a random sample from $\text{pr}(y(\cdot))$, but successive samples are correlated according to $\cos\theta$. The movie, in which successive frames show successive $\tilde{y}^*(\cdot)$, shows a single sample from $\text{pr}(y(\cdot))$ undergoing Brownian diffusion through the probability cloud, with mean free path governed by $\sin\theta$. Both the structure and the variability of the results are clearly seen by watching $\tilde{y}^*$ in such movies.

9. Results

Table 1 shows three test problems, all with $X = 1$, and with $n = 20$ evaluations of f without deletion.

Problem	Differential $f(x,y)$	Solution $y(x)$	Result $Y = y(1)$
A	$(1+y)/(1+x^2)$	$(1+y_0)\exp(\tan^{-1}x) - 1$	1.1933
B	$2\pi\cos(2\pi x)$	$y_0 + \sin(2\pi x)$	0
C	$5(y + .08 - x^2)$	$x^2 + .4*x + y_0 e^{5x}$	1.4000

Table 1: Test Problems

Problem A is a preliminary easy test, with a stable, nearly straight solution. Problem B is a little harder, in that the solution is oscillatory. Problem C is stiff: although the desired solution is a simple quadratic, any error becomes amplified by up to $e^5 \simeq 150$ by the time x reaches 1.

Firstly, each problem was solved by the first order Bayesian algorithm, in which y' is obtained from the hidden function h by smoothing it as in equation (8). Problems A and B were both assigned width $W = 0.2$ and stiffness $c = 1$. The rationale for the choice of width was that the functional form of f seemed fairly smooth, so that W might be fairly large. On the other hand, the solution y should not be artificially forced to be too smooth, so that W should not be too large. The rationale for the choice of stiffness was simply that a number of order unity seemed appropriate for a problem in which all variables were of that order. Unusually in a Bayesian calculation, it is not possible to tune the parameter values W and c directly from the evidence $\text{pr}(D|W,c)$. The reason is that different choices of parameters lead the algorithm to select different sample locations (x_k, y_k). With the datasets being different, the relative evidence values are meaningless. It might be possible to tune the parameters as the algorithm proceeds, using the existing data to refine their values, rather as α and σ are already refined: that is a matter for future investigation. In the present study, parameter values were accepted if they gave consistent and reasonably accurate results, for which the result $\hat{Y} \pm \delta\hat{Y}$ converged sensibly as data were acquired. The original values $W = 0.2$ and $c = 1$ were acceptable for Problems A and B.

Problem C, though, was different, because of its stiffness. Using the original values for W and c, the results as more data were acquired became seriously inconsistent. Thus Y was estimated as -1.0735 ± 0.0664 after 5 evaluations of f, but $+1.8701 \pm 0.0120$ after 20 evaluations. Meanwhile, the value of α bounced by a factor of 100. These are symptoms of assumptions which are badly matched to the actual problem in hand. Stability returned when the stiffness constant c was increased to 25, after which an increase in W to 0.6 gave lower estimated errors. These latter parameter values were used for Problem C. The "y' Bayes" column of Table 2 gives the results, which can be compared with the correct answers, and with the results from 5 applications of standard 4th-order Runge-Kutta integration, which also involved 20 evaluations of f.

Secondly, each problem was solved using the alternative second order model in which it is the second derivative y'' which is obtained from the hidden function h by smoothing with R. One might expect this method to prove better for problems in which the solution is reasonably straight, and this expectation is borne out. The results are shown in the "y'' Bayes" column of Table 2. The results for A and C have their quoted errors reduced by factors of about 3. On the other hand, the result for the oscillatory problem B has its quoted error increased instead. One can also note that all the Bayesian quoted errors are reasonably in accord with the actual deviations from the true answers, 4 out of 6 being within one standard deviation and all being within two.

Problem	W	c	y' Bayes	y'' Bayes	Runge-Kutta	True Y
A	0.2	1.	1.1924±.0028	1.1923±.0008	1.1933	1.1933
B	0.2	1.	0.0054±.0195	-0.0147±.0550	0 [†]	0
C	0.6	25.	1.1680±.1437	1.3554±.0533	1.3294	1.4000

Table 2: Test Results. ([†] Errors at intermediate steps $\simeq 0.0009$)

Some insight into the operation of the Bayesian algorithms can be gained from the Figures, which plot ten random samples from the posterior inference $\text{pr}(y(\cdot))$ as data are acquired. Figure 1 shows the evolution of the inference for Problem A, according to the original first order "y'" algorithm. With a Gaussian prior underlying y', the solution y is expected *a priori* to be on average flat, but to random-walk away from the initial value as x increases. The initial datum D_0 modifies this within a width W or so of the origin, as seen in the first frame of Figure 1. Given this distribution of probable solutions, the most informative place to select the next sample is at (0.54,0.33). The next frame shows the effect on the posterior inference of this new datum. It turned out to have a slope rather more positive than would have been expected on average, so that the distribution of solutions was sloped (and consequently shifted) upwards. The subsequent sample, at (0.89, 0.81), behaves similarly, and completes a first, rough, sampling of the interval [0,1]. The most informative next sample interleaves the earlier ones, and appears at (0.30, 0.30). Again, the posterior distribution narrows, until after 20 samples it forms a narrow band barely distinguishable visually from the true analytic solution.

Figure 2 shows similar behaviour. With an oscillatory equation, though, more samples are needed to localise the posterior adequately. After two additional samples, the overall "up-down-up" behaviour of the final solution has already begun to appear, but both these points are a long way from the analytic solution. Ultimately being the least informative about the final solution, they would be the first samples to be deleted from the dataset if deletion were necessary.

Figure 3 shows the evolution of the posterior inference for Problem C. Here, the behaviour is rather different. Because the stiffness constant is relatively high, any sample which lies substantially away from the true solution will be almost useless as a measurement of the slope. Accordingly, the only safe strategy is to proceed in small, careful steps. Indeed, the strategy outlined above does this, incrementing x fairly steadily by about 0.03 each step, with only 2 out of the first 20 samples interleaving previous ones. After 20 samples, x has only been able to reach 0.46, so that the extrapolation to $x = 1$ relies purely on the relatively large ICF width W. Whether or not the extrapolation is accurate depends on whether or not the very precise linear combination Qh of error functions defined by the 20 closely spaced samples is sufficient to describe the solution y.

Figures 4, 5, 6 show the posterior inferences for Problems A, B, C respectively, using the alternative second order "y''''" algorithm. Whereas the first order algorithm favours solutions $y(\cdot)$ which random-walk away from some arbitrary constant, the second order algorithm favours solutions which random-walk away from an arbitrary straight line. This behaviour can clearly be seen in the Figures. The second order algorithm is better for Problems A and C, when the solution incorporates a linear trend, but worse for Problem B which does not have a net trend.

10. Conclusions

Solving differential equations numerically should be seen as an inference problem, in which results are to be estimated on the basis of incomplete data. Ideally, inference problems should be solved by Bayesian probabilistic methods. The advantage of this is that one obtains the probability distribution of solutions, including the uncertainties, whereas traditional algorithms compute a single solution only, with any error analysis grafted on. This paper shows that it is *possible* to compute with the full probability distribution, and it is possible to set up properly Bayesian algorithms, although the extra computational overheads from the matrix calculations are relatively severe in test problems.

There is a wide choice of Bayesian algorithms, and this paper is merely a preliminary foray. Indeed, there is an obvious inefficiency in the preliminary algorithms presented here, in that the sample points visibly cluster too closely in x. Presumably this is due to the practically convenient (but wrong) assumption that the errors σ_k on the data D_k are independent. If the correlations were allowed for, it could no longer be thought advantageous to place a new sample arbitrarily close to an old one, because duplicating an evaluation of f does not actually make it any more accurate. Thus the sampling strategy clearly needs improvement. There may be another inefficiency in the choice of a Gaussian for the form of the intrinsic correlation function. A Cauchy distribution, having larger wings, might well give better extrapolation, whereas some form of multiquadric might be more appropriate if y were to have isolated irregularities. The questions of the best Bayesian order, and of methods of choosing the width and stiffness constants, also remain open. Much work remains to be done, so that it is hardly surprising that the algorithms presented here are not yet competitive with traditional methods: Table 2 shows simple Runge-Kutta obtaining more accurate results in 5 out of 6 cases.

The examples shown here have been one-dimensional, with only one component of y. However, the extension to multi-dimensional y, with several components, is as straightforward here as it is with traditional methods. Moreover, the extension to second or higher order differential equations is entirely natural in these Bayesian schemes. An interesting test example for this would be the two-point boundary problem with the second-order Airy equation, $y'' = xy$ with $y(a)$ and $y(b)$ given.

Beyond ordinary differential equations lie partial differential equations. These too could be investigated from a Bayesian viewpoint. The future is wide open.

ACKNOWLEDGMENTS. This paper arose from long exposure to Laplace/Cox/Jaynes probabilistic reasoning, combined with the University of Cambridge's desire that the author teach some (traditional) numerical analysis. The rest is common sense. In this circumstance, the author hopes he will be forgiven for not making the usual explicit selection of formal references from the burgeoning literature. Simply, Bayesian ideas are "in the air", to the extent that the author judged any limited selection to be seriously invidious by omission.

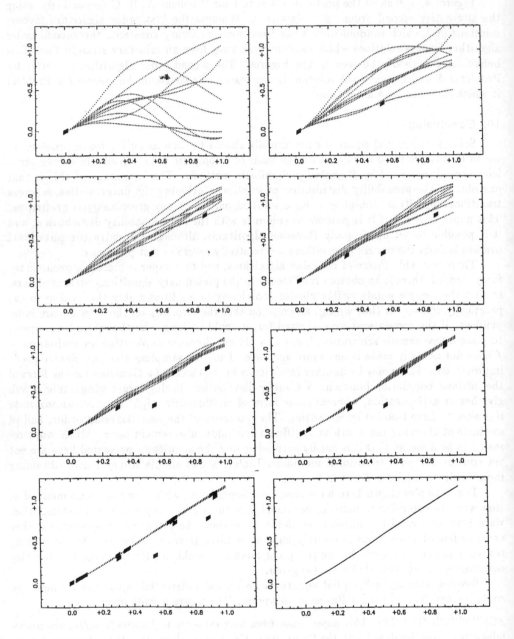

Fig. 1. Problem A, solved with the first order "y'" algorithm. The posterior distribution of $y(\cdot)$ after 1, 2, 3, 4, 5, 10, 20 samples, with the analytic solution.

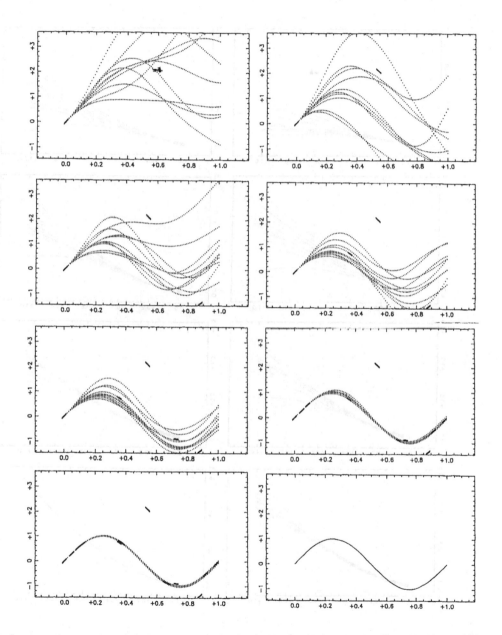

Fig. 2. Problem B, solved with the first order "y'" algorithm. The posterior distribution of $y(\cdot)$ after 1, 2, 3, 4, 5, 10, 20 samples, with the analytic solution.

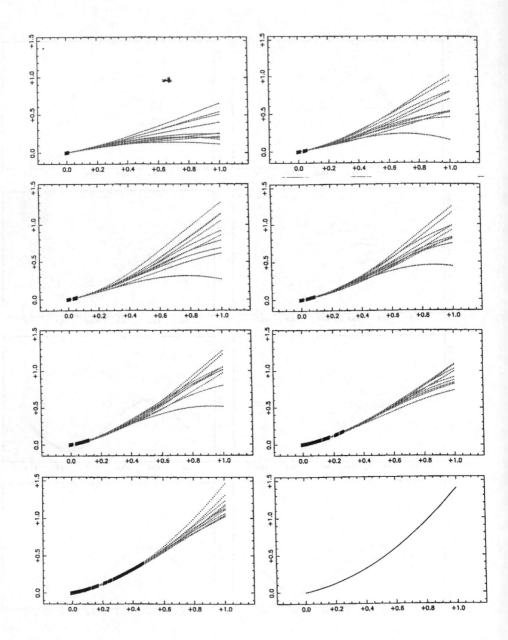

Fig. 3. Problem C, solved with the first order "y'" algorithm. The posterior distribution of $y(\cdot)$ after 1, 2, 3, 4, 5, 10, 20 samples, with the analytic solution.

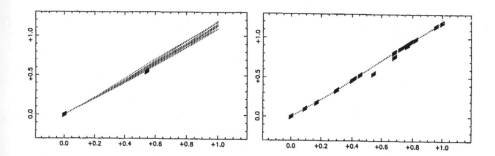

Fig. 4. Problem A, with the "y''" algorithm. The distribution of $y(\cdot)$ after 2 and 20 samples.

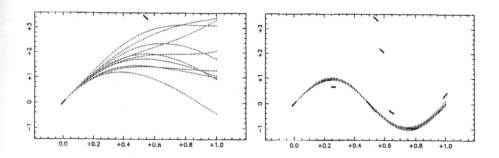

Fig. 5. Problem B, with the "y''" algorithm. The distribution of $y(\cdot)$ after 2 and 20 samples.

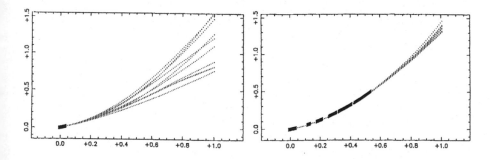

Fig. 6. Problem C, with the "y''" algorithm. The distribution of $y(\cdot)$ after 2 and 20 samples.

BAYESIAN INTERPOLATION

David J.C. MacKay[†]
Computation and Neural Systems
California Institute of Technology, 139-74
Pasadena, California 91125 USA
mackay@ras.phy.cam.ac.uk.

ABSTRACT. Although Bayesian analysis has been in use since Laplace, the Bayesian method of *model–comparison* has only recently been developed in depth. In this paper, the Bayesian approach to regularisation and model–comparison is demonstrated by studying the inference problem of interpolating noisy data. The concepts and methods described are quite general and can be applied to many other data modelling problems. Regularising constants are set by examining their posterior probability distribution. Alternative regularisers (priors) and alternative basis sets are objectively compared by evaluating the *evidence* for them. 'Occam's razor' is automatically embodied by this process. The way in which Bayes infers the values of regularising constants and noise levels has an elegant interpretation in terms of the effective number of parameters determined by the data set. This framework is due to Gull and Skilling.

1. Data modelling and Occam's razor

In science, a central task is to develop and compare models to account for the data that are gathered. In particular this is true in the problems of learning, pattern classification, interpolation and clustering. Two levels of **inference** are involved in the task of data modelling (figure 1). At the first level of inference, we assume that one of the models that we invented is true, and we fit that model to the data. Typically a model includes some free parameters; fitting the model to the data involves inferring what values those parameters should probably take, given the data. The results of this inference are often summarised by the most probable parameter values and error bars on those parameters. This is repeated for each model. The second level of inference is the task of model comparison. Here, we wish to compare the models in the light of the data, and assign some sort of preference or ranking to the alternatives.[1] For example, consider the task of interpolating a noisy data set. The data set could be interpolated using a splines model, using radial basis functions, using polynomials, or using feedforward neural networks. At the first level of inference,

---

[†] Current address: Cavendish laboratory, Madingley Road, Cambridge CB3 0HE, U.K.

[1] Note that both levels of *inference* are distinct from *decision theory*. The goal of inference is, given a defined hypothesis space and a particular data set, to assign probabilities to hypotheses. Decision theory typically chooses between alternative actions on the basis of these probabilities so as to minimise the expectation of a 'loss function.' This paper concerns inference alone and no loss functions or utilities are involved.

C. R. Smith et al. (eds.), Maximum Entropy and Bayesian Methods, Seattle, 1991, 39–66.

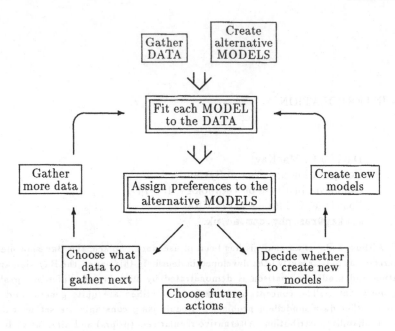

Fig. 1. Where Bayesian inference fits into the data modelling process.
This figure illustrates an abstraction of the part of the scientific process in which data are collected and modelled. In particular, this figure applies to pattern classification, learning, interpolation, etc.. The two double–framed boxes denote the two steps which involve *inference*. It is only in those two steps that Bayes' rule can be used. Bayes does not tell you how to invent models, for example.

The first box, 'fitting each model to the data', is the task of inferring what the model parameters might be given the model and the data. Bayes may be used to find the most probable parameter values, and error bars on those parameters. The result of applying Bayes to this problem is often little different from the answers given by orthodox statistics.

The second inference task, model comparison in the light of the data, is where Bayes is in a class of its own. This second inference problem requires a quantitative Occam's razor to penalise over–complex models. Bayes can assign objective preferences to the alternative models in a way that automatically embodies Occam's razor. From (MacKay, 1992a).

we take each model individually and find the best fit interpolant for that model. At the second level of inference we want to rank the alternative models and state for our particular data set that, for example, 'splines are probably the best interpolation model,' or 'if the interpolant is modelled as a polynomial, it should probably be a cubic.'

Bayesian methods are able consistently and quantitatively to solve both these inference tasks. There is a popular myth that states that Bayesian methods only differ from orthodox (also known as 'frequentist' or 'sampling theory') statistical methods by the inclusion of subjective priors which are arbitrary and difficult to assign, and usually don't make much difference to the conclusions. It is true that at the first level of inference, a Bayesian's results will often differ little from the outcome of an orthodox attack. What is not widely

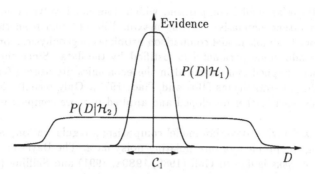

Fig. 2. Why Bayes embodies Occam's razor

This figure gives the basic intuition for why complex models are penalised. The horizontal axis represents the space of possible data sets D. Bayes rule rewards models in proportion to how much they *predicted* the data that occurred. These predictions are quantified by a normalised probability distribution on D. In this paper, this probability of the data given model $\mathcal{H}_i$, $P(D|\mathcal{H}_i)$, is called the evidence for $\mathcal{H}_i$.

A simple model $\mathcal{H}_1$ makes only a limited range of predictions, shown by $P(D|\mathcal{H}_1)$; a more powerful model $\mathcal{H}_2$, that has, for example, more free parameters than $\mathcal{H}_1$, is able to predict a greater variety of data sets. This means however that $\mathcal{H}_2$ does not predict the data sets in region $\mathcal{C}_1$ as strongly as $\mathcal{H}_1$. Assume that equal prior probabilities have been assigned to the two models. Then if the data set falls in region $\mathcal{C}_1$, the *less powerful* model $\mathcal{H}_1$ will be the *more probable* model. From (MacKay, 1992a).

appreciated is how Bayes performs the second level of inference. It is here that Bayesian methods are totally different from orthodox methods. Indeed, when regression and density estimation are discussed in most statistics texts, the task of model comparison is virtually ignored and no general orthodox method exists for solving this problem.

Model comparison is a difficult task because it is not possible simply to choose the model that fits the data best: more complex models can always fit the data better, so the maximum likelihood model choice would lead us inevitably to implausible over–parameterised models which generalise poorly. 'Occam's razor' is the principle that states that unnecessarily complex models should not be preferred to simpler ones. Bayesian meth-ods automatically and quantitatively embody Occam's razor (Gull, 1988, Jeffreys, 1939), without the introduction of ad hoc penalty terms. Complex models are automatically self–penalising under Bayes' rule. Figure 2 gives the basic intuition for why this should be expected. The rest of this paper will explore this property in depth.

Bayesian methods were first laid out in depth by the Cambridge geophysicist Sir Harold Jeffreys (1939). The logical basis for the Bayesian use of probabilities as measures of plau-sibility was subsequently established by Cox (1946), who proved that consistent inference in a closed hypothesis space can be mapped onto probabilities. For a general review of Bayesian philosophy the reader is encouraged to read the excellent papers by Jaynes (1986) and Loredo (1989), and the text by Box and Tiao (1973). Since Jeffreys the emphasis of most Bayesian probability theory has been 'to formally utilize prior information' (Berger, 1985), *i.e.* to perform inference in a way that makes explicit the prior knowledge and igno-rance that we have, which orthodox methods omit. However, Jeffreys' work also laid the foundation for Bayesian model comparison, which does not involve an emphasis on prior information, but rather emphasises getting maximal information from the data. Jeffreys applied this theory to simple model comparison problems in geophysics, for example testing whether a single additional parameter is justified by the data. Since the 1960s, Jeffreys' methods have been applied and extended in the economics literature (Zellner, 1984), and by a small number of statisticians (Box and Tiao, 1973). Only recently has this aspect of Bayesian analysis been further developed and applied to more complex problems in other fields.

This paper will review Bayesian model comparison, 'regularisation,' and noise estimation, by studying the problem of interpolating noisy data. The Bayesian framework I will describe for these tasks is due to Gull (1988, 1989a, 1991) and Skilling (1991), who have used Bayesian methods to achieve the state of the art in image reconstruction. The same approach to regularisation has also been developed in part by Szeliski (1989). Bayesian model comparison is also discussed by Bretthorst (1990), who has used Bayesian methods to push back the limits of NMR signal detection. The same Bayesian theory underlies the unsupervised classification system, AutoClass (Hanson et. al., 1991). The fact that Bayesian model comparison embodies Occam's razor has been rediscovered by Kashyap (1977) in the context of modelling time series; his paper includes a thorough discussion of how Bayesian model comparison is different from orthodox 'Hypothesis testing.' One of the earliest applications of these sophisticated Bayesian methods of model comparison to real data is by Patrick and Wallace (1982); in this fascinating paper, competing models accounting for megalithic stone circle geometry are compared within the description length framework, which is equivalent to Bayes.

As the quantities of data collected throughout science and engineering continue to increase, and the computational power and techniques available to model that data also multiply, I believe Bayesian methods will prove an ever more important tool for refining our modelling abilities. A companion paper (MacKay, 1992d) demonstrates how these techniques can be fruitfully applied to backpropagation neural networks. Separate publications show how this framework relates to the task of selecting where next to gather data so as to gain maximal information about our models (MacKay, 1992e) and demonstrate the application of Bayesian methods to adaptive classifiers (MacKay, 1992f).

2. The evidence and the Occam factor

Let us write down Bayes' rule for the two levels of inference described above, so as to see explicitly how Bayesian model comparison works. Each model $\mathcal{H}_i$ ($\mathcal{H}$ stands for 'hypothesis') is assumed to have a vector of parameters $\mathbf{w}$. A model is defined by its functional form and two probability distributions: a 'prior' distribution $P(\mathbf{w}|\mathcal{H}_i)$ which states what values the model's parameters might plausibly take; and the predictions $P(D|\mathbf{w},\mathcal{H}_i)$ that the model makes about the data D when its parameters have a particular value $\mathbf{w}$. Note that models with the same parameterisation but different priors over the parameters are therefore defined to be different models.

1. **Model fitting.** At the first level of inference, we assume that one model $\mathcal{H}_i$ is true, and we infer what the model's parameters $\mathbf{w}$ might be given the data D. Using Bayes' rule, the **posterior probability** of the parameters $\mathbf{w}$ is:

$$P(\mathbf{w}|D,\mathcal{H}_i) = \frac{P(D|\mathbf{w},\mathcal{H}_i)P(\mathbf{w}|\mathcal{H}_i)}{P(D|\mathcal{H}_i)} \tag{1}$$

In words:

$$\text{Posterior} = \frac{\text{Likelihood} \times \text{Prior}}{\text{Evidence}}$$

The normalising constant $P(D|\mathcal{H}_i)$ is commonly ignored, since it is irrelevant to the first level of inference, *i.e.* the choice of $\mathbf{w}$; but it will be important in the second level of inference, and we name it the **evidence** for $\mathcal{H}_i$. It is common to use gradient–based methods to find the maximum of the posterior, which defines the most probable value for the parameters, $\mathbf{w}_{\text{MP}}$; it is then common to summarise the posterior distribution by the value of $\mathbf{w}_{\text{MP}}$, and error bars on these best fit parameters. The error bars are obtained from the curvature of the posterior; writing the Hessian $\mathbf{A} = -\nabla\nabla \log P(\mathbf{w}|D, \mathcal{H}_i)$ and Taylor–expanding the log posterior with $\Delta\mathbf{w} = \mathbf{w} - \mathbf{w}_{\text{MP}}$,

$$P(\mathbf{w}|D,\mathcal{H}_i) \simeq P(\mathbf{w}_{\text{MP}}|D,\mathcal{H}_i)\exp\left(-\frac{1}{2}\Delta\mathbf{w}^{\text{T}}\mathbf{A}\Delta\mathbf{w}\right), \qquad (2)$$

we see that the posterior can be locally approximated as a gaussian with covariance matrix (error bars) $\mathbf{A}^{-1}$.[2]

2. **Model comparison.** At the second level of inference, we wish to infer which model is most plausible given the data. The posterior probability of each model is:

$$P(\mathcal{H}_i|D) \propto P(D|\mathcal{H}_i)P(\mathcal{H}_i) \qquad (3)$$

Notice that the data–dependent term $P(D|\mathcal{H}_i)$ is the evidence for $\mathcal{H}_i$, which appeared as the normalising constant in (1). The second term, $P(\mathcal{H}_i)$, is a 'subjective' prior over our hypothesis space which expresses how plausible we thought the alternative models were before the data arrived. We will see later that this subjective part of the inference will typically be overwhelmed by the objective term, the evidence. Assuming that we have no reason to assign strongly differing priors $P(\mathcal{H}_i)$ to the alternative models, **models $\mathcal{H}_i$ are ranked by evaluating the evidence.** Equation (3) has not been normalised because in the data modelling process we may develop new models after the data have arrived (figure 1), when an inadequacy of the first models is detected, for example. So we do not start with a completely defined hypothesis space. Inference is open–ended: we continually seek more probable models to account for the data we gather. New models are compared with previous models by evaluating the evidence for them.

The key concept of this paper is this: to assign a preference to alternative models $\mathcal{H}_i$, a Bayesian evaluates the evidence $P(D|\mathcal{H}_i)$. This concept is very general: the evidence can be evaluated for parametric and 'non–parametric' models alike; whether our data modelling

---

[2]Whether this approximation is a good one or not will depend on the problem we are solving. For the interpolation models discussed in this paper, there is only a single maximum in the posterior distribution, and the gaussian approximation is exact. For more general statistical models we still expect the posterior to be dominated by locally gaussian peaks on account of the central limit theorem (Walker, 1967). Multiple maxima which arise in more complex models complicate the analysis, but Bayesian methods can still successfully be applied (Hanson et. al., 1991, MacKay, 1992d, Neal, 1991).

task is a regression problem, a classification problem, or a density estimation problem, the evidence is the Bayesian's transportable quantity for comparing alternative models. In all these cases the evidence naturally embodies Occam's razor; we will examine how this works shortly.

Of course, the evidence is not the whole story if we have good reason to assign unequal priors to the alternative models $\mathcal{H}$. (To only use the evidence for model comparison is equivalent to using maximum likelihood for parameter estimation.) The classic example is the 'Sure Thing' hypothesis, © E.T Jaynes, which is the hypothesis that the data set will be D, the precise data set that actually occurred; the evidence for the Sure Thing hypothesis is huge. But Sure Thing belongs to an immense class of similar hypotheses which should all be assigned correspondingly tiny prior probabilities; so the posterior probability for Sure Thing is negligible alongside any sensible model. Models like Sure Thing are rarely seriously proposed in real life, but if such models are developed then clearly we need to think about precisely what priors are appropriate. Patrick and Wallace (1982), studying the geometry of ancient stone circles (for which some people have proposed extremely elaborate theories!), discuss a practical method of assigning relative prior probabilities to alternative models by evaluating the lengths of the computer programs that decode data previously encoded under each model. This procedure introduces a second sort of Occam's razor into the inference, namely a *prior* bias against complex models. However, this paper will not include such prior biases; we will address only the data's preference for the alternative models, *i.e.* the evidence, and the Occam's razor that it embodies. In the limit of large quantities of data this objective Occam's razor will always be the more important of the two.

A MODERN BAYESIAN APPROACH TO PRIORS

It should be pointed out that the emphasis of this modern Bayesian approach is not on the inclusion of priors into inference. There is not one significant 'subjective prior' in this entire paper. (For problems in which significant subjective priors do arise see (Gull, 1989b, Skilling, 1989b).) The emphasis is on the idea that consistent degrees of preference for alternative hypotheses are represented by probabilities, and relative preferences for models are assigned by evaluating those probabilities. Historically, Bayesian analysis has been accompanied by methods to work out the 'right' prior $P(\mathbf{w}|\mathcal{H})$ for a problem, for example, the principles of insufficient reason and maximum entropy. The modern Bayesian however does not take a fundamentalist attitude to assigning the 'right' priors — many different priors can be tried; each particular prior corresponds to a different hypothesis about the way the world is. We can compare these alternative hypotheses in the light of the data by evaluating the evidence. This is the way in which alternative regularisers are compared, for example. If we try one model and obtain awful predictions, we have *learnt* something. 'A failure of Bayesian prediction is an opportunity to learn' (Jaynes, 1986), and we are able to come back to the same data set with new models, using new priors for example.

EVALUATING THE EVIDENCE

Let us now explicitly study the evidence to gain insight into how the Bayesian Occam's razor works. The evidence is the normalising constant for equation (1):

$$P(D|\mathcal{H}_i) = \int P(D|\mathbf{w}, \mathcal{H}_i) P(\mathbf{w}|\mathcal{H}_i)\, d\mathbf{w} \tag{4}$$

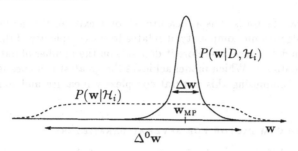

Fig. 3. The Occam factor

This figure shows the quantities that determine the Occam factor for a hypothesis $\mathcal{H}_i$ having a single parameter $\mathbf{w}$. The prior distribution (dotted line) for the parameter has width $\Delta^0\mathbf{w}$. The posterior distribution (solid line) has a single peak at $\mathbf{w}_{MP}$ with characteristic width $\Delta\mathbf{w}$. The Occam factor is $\frac{\Delta\mathbf{w}}{\Delta^0\mathbf{w}}$. From (MacKay, 1992a).

For many problems, including interpolation, it is common for the posterior $P(\mathbf{w}|D,\mathcal{H}_i) \propto P(D|\mathbf{w},\mathcal{H}_i)P(\mathbf{w}|\mathcal{H}_i)$ to have a strong peak at the most probable parameters $\mathbf{w}_{MP}$ (figure 3). Then the evidence can be approximated by the height of the peak of the integrand $P(D|\mathbf{w},\mathcal{H}_i)P(\mathbf{w}|\mathcal{H}_i)$ times its width, $\Delta\mathbf{w}$:

$$P(D|\mathcal{H}_i) \simeq \underbrace{P(D|\mathbf{w}_{MP},\mathcal{H}_i)}_{} \underbrace{P(\mathbf{w}_{MP}|\mathcal{H}_i)\,\Delta\mathbf{w}}_{} \qquad (5)$$

$$\text{Evidence} \simeq \text{Best fit likelihood} \quad \text{Occam factor}$$

Thus the evidence is found by taking the best fit likelihood that the model can achieve and multiplying it by an 'Occam factor' (Gull, 1988), which is a term with magnitude less than one that penalises $\mathcal{H}_i$ for having the parameter $\mathbf{w}$.

INTERPRETATION OF THE OCCAM FACTOR

The quantity $\Delta\mathbf{w}$ is the posterior uncertainty in $\mathbf{w}$. Imagine for simplicity that the prior $P(\mathbf{w}|\mathcal{H}_i)$ is uniform on some large interval $\Delta^0\mathbf{w}$, representing the range of values of $\mathbf{w}$ that $\mathcal{H}_i$ thought possible before the data arrived. Then $P(\mathbf{w}_{MP}|\mathcal{H}_i) = \frac{1}{\Delta^0\mathbf{w}}$, and

$$\text{Occam factor} = \frac{\Delta\mathbf{w}}{\Delta^0\mathbf{w}},$$

i.e. **the ratio of the posterior accessible volume of $\mathcal{H}_i$'s parameter space to the prior accessible volume,** or the factor by which $\mathcal{H}_i$'s hypothesis space collapses when the data arrive (Gull, 1988, Jeffreys, 1939). The model $\mathcal{H}_i$ can be viewed as being composed of a certain number of equivalent submodels, of which only one survives when the data arrive. The Occam factor is the inverse of that number. The log of the Occam factor can be interpreted as the amount of information we gain about the model when the data arrive.

Typically, a complex model with many parameters, each of which is free to vary over a large range $\Delta^0\mathbf{w}$, will be penalised with a larger Occam factor than a simpler model. The Occam factor also provides a penalty for models which have to be finely tuned to fit the data; the Occam factor promotes models for which the required precision of the parameters

Δw is coarse. The Occam factor is thus a measure of complexity of the model, but unlike the V–C dimension or algorithmic complexity, it relates to the complexity of the predictions that the model makes in data space; therefore it depends on the number of data points and other properties of the data set. Which model achieves the greatest evidence is determined by a trade–off between minimising this natural complexity measure and minimising the data misfit.

Occam factor for several parameters

If w is k-dimensional, and if the posterior is well approximated by a gaussian, the Occam factor is obtained from the determinant of the gaussian's covariance matrix:

$$
\underbrace{P(D\,|\mathcal{H}_i)}_{\text{Evidence}} \simeq \underbrace{P(D\,|\mathbf{w}_{\mathrm{MP}}, H_i)}_{\text{Best fit likelihood}} \underbrace{P(\mathbf{w}_{\mathrm{MP}}|\mathcal{H}_i)\,(2\pi)^{k/2}\mathrm{det}^{-\frac{1}{2}}\mathbf{A}}_{\text{Occam factor}} \tag{6}
$$

where $\mathbf{A} = -\nabla\nabla \log P(\mathbf{w}|D, \mathcal{H}_i)$, the Hessian which we already evaluated when we calculated the error bars on $\mathbf{w}_{\mathrm{MP}}$. As the amount of data collected, N, increases, this gaussian approximation is expected to become increasingly accurate on account of the central limit theorem (Walker, 1967). For the linear interpolation models discussed in this paper, this gaussian expression is exact for any N.

Comments

- Bayesian model selection is a simple extension of maximum likelihood model selection: **the evidence is obtained by multiplying the best fit likelihood by the Occam factor.** To evaluate the Occam factor all we need is the Hessian $\mathbf{A}$, if the gaussian approximation is good. Thus the Bayesian method of model comparison by evaluation of the evidence is computationally no more demanding than the task of finding for each model the best fit parameters and their error bars.
- It is common for there to be degeneracies in models with many parameters; *i.e.* several equivalent parameters could be relabelled without affecting the likelihood. In these cases, the right hand side of equation (6) should be multiplied by the degeneracy of $\mathbf{w}_{\mathrm{MP}}$ to give the correct estimate of the evidence.
- 'Minimum description length' (MDL) methods are closely related to this Bayesian framework (Rissanen, 1978, Wallace and Boulton, 1968, Wallace and Freeman, 1987). The log evidence $\log_2 P(D|\mathcal{H}_i)$ is the number of bits in the ideal shortest message that encodes the data D using model $\mathcal{H}_i$. Akaike's criterion (Akaike, 1970) can be viewed as an approximation to MDL (Schwarz, 1978, Zellner, 1984). Any implementation of MDL necessitates approximations in evaluating the length of the ideal shortest message. Although some of the earliest work on complex model comparison involved the MDL framework (Patrick and Wallace, 1982), MDL has no apparent advantages, and in my work I approximate the evidence directly.
- It should be emphasised that the Occam factor has nothing to do with how computationally complex it is to *use* a model. The evidence is a measure of *plausibility* of a model. How much CPU time it takes to use each model is certainly an interesting issue which might bias our decisions towards simpler models, but Bayes rule does not address that issue. Choosing between models on the basis of how many function calls

they need is an exercise in *decision theory*, which is not addressed in this paper. Once the probabilities described above have been inferred, optimal actions can be chosen using standard decision theory with a suitable utility function.

3. The noisy interpolation problem

Bayesian interpolation through *noise–free* data has been studied by Skilling and Sibisi (1991). In this paper I study the problem of interpolating through data where the dependent variables are assumed to be noisy (a task also known as 'regression,' 'curve–fitting,' 'signal estimation,' or, in the neural networks community, 'learning'). I am not examining the case where the independent variables are also noisy. This different and more difficult problem has been studied for the case of straight line–fitting by Gull (1989b).

Let us assume that the data set to be interpolated is a set of pairs $D = \{x_m, t_m\}$, where $m = 1 \ldots N$ is a label running over the pairs. For simplicity I will treat x and t as scalars, but the method generalises to the multidimensional case. To define a linear interpolation model, a set of k fixed basis functions[3] $\mathcal{A} = \{\phi_h(x)\}$ is chosen, and the interpolated function is assumed to have the form:

$$y(x) = \sum_{h=1}^{k} w_h \phi_h(x) \tag{7}$$

where the parameters w_h are to be inferred from the data. The data set is modelled as deviating from this mapping under some additive noise process $\mathcal{N}$:

$$t_m = y(x_m) + \nu_m \tag{8}$$

If ν is modelled as zero–mean gaussian noise with standard deviation σ_ν, then the probability of the data[4] given the parameters $\mathbf{w}$ is:

$$P(D \,|\, \mathbf{w}, \beta, \mathcal{A}) = \frac{\exp(-\beta E_D(D|\mathbf{w}, \mathcal{A}))}{Z_D(\beta)} \tag{9}$$

where $\beta = 1/\sigma_\nu^2$, $E_D = \sum_m \frac{1}{2}(y(x_m) - t_m)^2$, and $Z_D = (2\pi/\beta)^{N/2}$. $P(D \,|\, \mathbf{w}, \beta, \mathcal{A})$ is called the likelihood. It is well known that finding the maximum likelihood parameters $\mathbf{w}_{\mathrm{ML}}$ may be an 'ill–posed' problem. That is, the $\mathbf{w}$ that minimises E_D is underdetermined and/or depends sensitively on the details of the noise in the data; the maximum likelihood interpolant in such cases oscillates wildly so as to fit the noise. Thus it is clear that to complete an interpolation model we need a prior $\mathcal{R}$ that expresses the sort of smoothness we expect the interpolant $y(x)$ to have. A model may have a prior of the form

$$P(y|\mathcal{R}, \alpha) = \frac{\exp(-\alpha E_y(y|\mathcal{R}))}{Z_y(\alpha)} \tag{10}$$

---

[3] the case of *adaptive* basis functions, also known as feedforward neural networks, is examined in a companion paper.

[4] Strictly, this probability should be written $P(\{t_m\}|\{x_m\}, \mathbf{w}, \beta, \mathcal{A})$, since these interpolation models do not predict the distribution of input variables $\{x_m\}$; this liberty of notation will be taken throughout this paper and its companion.

where E_y might be for example the functional $E_y = \int y''(x)^2 dx$ (which is the regulariser for cubic spline interpolation[5]). The parameter α is a measure of how smooth $f(x)$ is expected to be. Such a prior can also be written as a prior on the parameters $\mathbf{w}$:

$$P(\mathbf{w}|\mathcal{A},\mathcal{R},\alpha) = \frac{\exp(-\alpha E_W(\mathbf{w}|\mathcal{A},\mathcal{R}))}{Z_W(\alpha)} \tag{11}$$

where $Z_W = \int d^k\mathbf{w}\,\exp(-\alpha E_W)$. E_W (or E_y) is commonly referred to as a regularising function.

The interpolation model $\mathcal{H}$ is now complete, consisting of a choice of basis functions $\mathcal{A}$, a noise model $\mathcal{N}$ with parameter β, and a prior (regulariser) $\mathcal{R}$, with regularising constant α.

THE FIRST LEVEL OF INFERENCE

If α and β are known, then the posterior probability of the parameters $\mathbf{w}$ is:[6]

$$P(\mathbf{w}|D,\alpha,\beta,\mathcal{A},\mathcal{R},\mathcal{N}) = \frac{P(D|\mathbf{w},\beta,\mathcal{A},\mathcal{N})P(\mathbf{w}|\alpha,\mathcal{A},\mathcal{R})}{P(D|\alpha,\beta,\mathcal{A},\mathcal{R},\mathcal{N})} \tag{12}$$

Writing[7]

$$M(\mathbf{w}) = \alpha E_W + \beta E_D,$$

the posterior is

$$P(\mathbf{w}|D,\alpha,\beta,\mathcal{A},\mathcal{R},\mathcal{N}) = \frac{\exp(-M(\mathbf{w}))}{Z_M(\alpha,\beta)} \tag{13}$$

where $Z_M(\alpha,\beta) = \int d^k\mathbf{w}\,\exp(-M)$. We see that minimising the combined objective function M corresponds to finding the *most probable interpolant*, $\mathbf{w}_{MP}$. Error bars on the best fit interpolant can be obtained from the Hessian of M, $\mathbf{A} = \nabla\nabla M$, evaluated at $\mathbf{w}_{MP}$. These error bars represent the uncertainty of the interpolant, and should not be confused with the typical scatter of noisy data points relative to the interpolant.

This is the well known Bayesian view of regularisation (Poggio *et. al.*, 1985, Titterington, 1985).

Bayesian methods provide far more than just an interpretation for regularisation. What has been described so far is just the first of three levels of inference. (The second level described in sections 1 and 2, 'model comparison,' splits into a second and a third level for this problem, because each interpolation model is made up of a continuum of sub–models with different values of α and β.) At the second level, Bayes allows us to objectively assign values to α and β, which are commonly unknown *a priori*. At the third, Bayes enables us to quantitatively rank alternative basis sets $\mathcal{A}$, alternative regularisers (priors)

---

[5]Strictly, this particular prior may be improper because a $y(x)$ of the form $w_1 x + w_0$ is not constrained by this prior.

[6]The regulariser $\alpha,\mathcal{R}$ has been omitted from the conditioning variables in the likelihood because the data distribution does not depend on the prior once $\mathbf{w}$ is known. Similarly the prior does not depend on $\beta,\mathcal{N}$.

[7]The name M stands for 'misfit'; it will be demonstrated later that M is the natural measure of misfit, rather than $\chi_D^2 = 2\beta E_D$.

$\mathcal{R}$, and, in principle, alternative noise models $\mathcal{N}$ (see Box and Tiao, 1973). Furthermore, we can quantitatively compare interpolation under any model $\mathcal{H} = \{\mathcal{A}, \mathcal{N}, \mathcal{R}\}$ with other interpolation and learning models such as neural networks, if a similar Bayesian approach is applied to them. Neither the second nor the third level of inference can be successfully executed without Occam's razor.

The Bayesian theory of the second and third levels of inference has only recently been worked out; this paper's goal is to review that framework. Section 4 will describe the Bayesian method of inferring α and β; section 5 will describe Bayesian model comparison for the interpolation problem. Both these inference problems are solved by evaluation of the appropriate evidence.

4. Selection of parameters α and β

Typically, α is not known a priori, and often β is also unknown. As α is varied, the properties of the best fit (most probable) interpolant vary. Assume that we are using a prior that encourages smoothness, and imagine that we interpolate at a very large value of α; then this will constrain the interpolant to be very smooth and flat, and it will not fit the data at all well (figure 4a). As α is decreased, the interpolant starts to fit the data better (figure 4b). If α is made even smaller, the interpolant oscillates wildly so as to overfit the noise in the data (figure 4c). The choice of the 'best' value of α is our first 'Occam's razor' problem: large values of α correspond to simple models which make constrained and precise predictions, saying 'the interpolant is expected to not have extreme curvature anywhere;' a tiny value of α corresponds to the more powerful and flexible model that says 'the interpolant could be anything at all, our prior belief in smoothness is very weak.' The task is to find a value of α which is small enough that the data are fitted but not so small that they are overfitted. For more severely ill–posed problems such as deconvolution, the precise value of the regularising parameter is increasingly important. Orthodox statistics has ways of assigning values to such parameters, based for example on misfit criteria, the use of test data, and cross–validation. Gull (1989a) has demonstrated why the popular use of misfit criteria is incorrect and how Bayes sets these parameters. The use of test data may be an unreliable technique unless large quantities of data are available. Cross–validation, the orthodox 'method of choice' (Eubank, 1988), will be discussed more in section 6 and (MacKay, 1992d). I will explain the Bayesian method of inferring α and β after first reviewing some statistics of misfit.

MISFIT, χ^2, AND THE EFFECT OF PARAMETER MEASUREMENTS

For N independent gaussian variables with mean μ and standard deviation σ, the statistic $\chi^2 = \sum(x-\mu)^2/\sigma^2$ is a measure of misfit. If μ is known a priori, χ^2 has expectation $N \pm \sqrt{N}$. However, if μ is fitted from the data by setting $\mu = \bar{x}$, we 'use up a degree of freedom,' and χ^2 has expectation $N - 1$. In the second case μ is a 'well-measured parameter.' When a parameter is determined by the data in this way it is unavoidable that the parameter fits some of the noise in the data as well. That is why the expectation of χ^2 is reduced by one. This is the basis of the distinction between the σ_N and σ_{N-1} buttons on your calculator. It is common for this distinction to be ignored, but in cases such as interpolation where the number of free parameters is similar to the number of data points, it is essential to find and make the analogous distinction. It will be demonstrated that

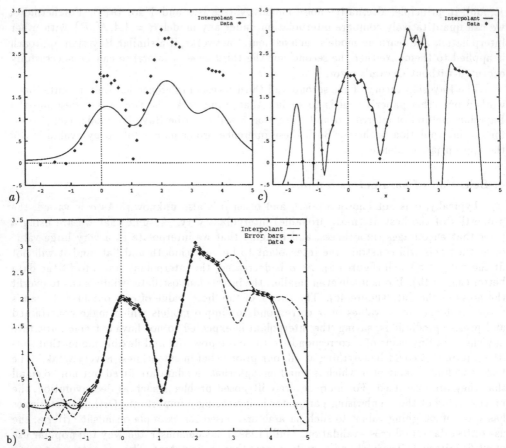

Fig. 4. How the best interpolant depends on α

These figures introduce a data set, 'X,' which is interpolated with a variety of models in this paper. Notice that the density of data points is not uniform on the x-axis. In the three figures the data set is interpolated using a radial basis function model with a basis of 60 equally spaced Cauchy functions, all with radius 0.2975. The regulariser is $E_W = \frac{1}{2} \sum w^2$, where w are the coefficients of the basis functions. Each figure shows the most probable interpolant for a different value of α: a) 6000; b) 2.5; c) 10^{-7}. Note at the extreme values how the data are oversmoothed and overfitted respectively. $\alpha = 2.5$ is the *most probable* value of α. In b), the most probable interpolant is displayed with its 1σ error bars, which represent how uncertain we are about the interpolant at each point, under the assumption that the interpolation model and the value of α are correct. Notice how the error bars increase in magnitude where the data are sparse. The error bars do not include the datapoint close to $(1,0)$, because the radial basis function model does not expect sharp discontinuities; the error bars are obtained assuming the model is correct, so that point is interpreted as an improbable outlier. From (MacKay, 1992a).

the Bayesian choices of both α and β are most simply expressed in terms of the effective number of well–measured parameters, γ, to be derived below.

Misfit criteria are 'principles' which set parameters like α and β by requiring that χ^2 should have a particular value. The discrepancy principle requires $\chi^2 = N$. Another principle requires $\chi^2 = N - k$, where k is the number of free parameters. We will find that an intuitive misfit criterion arises for the most probable value of β; on the other hand, the Bayesian choice of α will be unrelated to the value of the misfit.

BAYESIAN CHOICE OF α AND β

To infer from the data what value α and β should have, Bayesians evaluate the posterior probability distribution:

$$P(\alpha, \beta | D, \mathcal{H}) = \frac{P(D|\alpha, \beta, \mathcal{H})P(\alpha, \beta)}{P(D|\mathcal{H})} \tag{14}$$

The data dependent term $P(D|\alpha, \beta, \mathcal{H})$ has already appeared earlier as the normalising constant in equation (12), and it is called the evidence for α and β. Similarly the normalising constant of (15) is called the evidence for $\mathcal{H}$, and it will turn up later when we compare alternative models $\mathcal{H}$ in the light of the data.

If $P(\alpha, \beta)$ is a flat prior[8] (which corresponds to the statement that we don't know what value α and β should have), the evidence is the function that we use to assign a preference to alternative values of α and β. It is given in terms of the normalising constants defined earlier by

$$P(D|\alpha, \beta, \mathcal{H}) = \frac{Z_M(\alpha, \beta)}{Z_W(\alpha)Z_D(\beta)}, \tag{15}$$

Occam's razor is implicit in this formula: if α is small, the large freedom in the prior range of possible values of $\mathbf{w}$ is automatically penalised by the consequent large value of Z_W; models that fit the data well achieve a large value of Z_M. The optimum value of α achieves a compromise between fitting the data well and being a simple model.

Now to assign a preference to (α, β), our computational task is to evaluate the three integrals Z_M, Z_W and Z_D. We will come back to this task in a moment.

BUT THAT SOUNDS LIKE DETERMINING YOUR PRIOR AFTER THE DATA HAVE ARRIVED!

When I first heard the preceding explanation of Bayesian regularisation I was discontent because it seemed that the prior is being chosen from an ensemble of possible priors *after* the data have arrived. To be precise, as described above, the most probable value of α is selected; then the prior corresponding to that value of α alone is used to infer what the interpolant might be. This is not how Bayes would have us infer the interpolant. It is the combined ensemble of priors that define our prior, and we should integrate over this ensemble when we do inference.[9] Let us work out what happens if we follow this proper approach. The preceding method of using only the most probable prior will emerge as a good approximation.

---

[8]Since α and β are scale parameters, this prior should be understood as a flat prior over $\log \alpha$ and $\log \beta$.

[9]It is remarkable that Laplace almost got this right in 1774 (Stigler, 1986); when inferring the mean of a Laplacian distribution, he both inferred the posterior probability of a nuisance parameter like β in (15), and then attempted to integrate out the nuisance parameter as in equation (17).

The true posterior $P(\mathbf{w}|D,\mathcal{H})$ is obtained by integrating over α and β:

$$P(\mathbf{w}|D,\mathcal{H}) = \int P(\mathbf{w}|D,\alpha,\beta,\mathcal{H})P(\alpha,\beta|D,\mathcal{H})\,d\alpha\,d\beta \qquad (16)$$

In words, the posterior probability over $\mathbf{w}$ can be written as a linear combination of the posteriors for all values of α,β. Each posterior density is weighted by the probability of α,β given the data, which appeared in (15). This means that if $P(\alpha,\beta|D,\mathcal{H})$ has a single peak at $\hat{\alpha},\hat{\beta}$, then the true posterior $P(\mathbf{w}|D,\mathcal{H})$ will be dominated by the density $P(\mathbf{w}|D,\hat{\alpha},\hat{\beta},\mathcal{H})$. As long as the properties of the posterior $P(\mathbf{w}|D,\alpha,\beta,\mathcal{H})$ do not change rapidly with α,β near $\hat{\alpha},\hat{\beta}$ and the peak in $P(\alpha,\beta|D,\mathcal{H})$ is strong, we are justified in using the approximation:

$$P(\mathbf{w}|D,\mathcal{H}) \simeq P(\mathbf{w}|D,\hat{\alpha},\hat{\beta},\mathcal{H}) \qquad (17)$$

This approximation is valid under the same conditions as in footnote 10. It is a matter of ongoing research to develop computational methods for cases where this approximation is invalid (Sibisi and Skilling, personal communication, Neal, personal communication).

WHY NOT FIND THE JOINT OPTIMUM IN $\mathbf{w},\alpha,\beta$?

It is not generally satisfactory to simply maximise the likelihood simultaneously over $\mathbf{w}$, α and β; the likelihood has a skew peak such that the maximum likelihood value for the parameters is not in the same place as most of the posterior probability (Gull, 1989a). To get a feeling for this here is a more familiar problem: examine the posterior probability for the parameters of a gaussian (μ,σ) given N samples: the maximum likelihood value for σ is σ_N, but the most probable value for σ (found by integrating over μ) is σ_{N-1}. It should be emphasised that this distinction has nothing to do with the prior over the parameters, which is flat here. It is the process of marginalisation that corrects the bias of maximum likelihood and MAP estimators.

EVALUATING THE EVIDENCE

Let us return to our train of thought at equation (16). To evaluate the evidence for α,β, we want to find the integrals Z_M, Z_W and Z_D. Typically the most difficult integral to evaluate is Z_M.

$$Z_M(\alpha,\beta) = \int d^k\mathbf{w}\,\exp(-M(\mathbf{w},\alpha,\beta)).$$

If the regulariser $\mathcal{R}$ is a quadratic functional (and the favourites are), then E_D and E_W are quadratic functions of $\mathbf{w}$, and we can evaluate Z_M exactly. Letting $\nabla\nabla E_W = \mathbf{C}$ and $\nabla\nabla E_D = \mathbf{B}$ then using $\mathbf{A} = \alpha\mathbf{C} + \beta\mathbf{B}$, we have:

$$M = M(\mathbf{w}_{\mathrm{MP}}) + \frac{1}{2}(\mathbf{w} - \mathbf{w}_{\mathrm{MP}})^T\mathbf{A}(\mathbf{w} - \mathbf{w}_{\mathrm{MP}})$$

where $\mathbf{w}_{\mathrm{MP}} = \beta\mathbf{A}^{-1}\mathbf{B}\mathbf{w}_{\mathrm{ML}}$. This means that Z_M is the gaussian integral:

$$Z_M = e^{-M_{\mathrm{MP}}}(2\pi)^{k/2}\det^{-\frac{1}{2}}\mathbf{A}. \qquad (18)$$

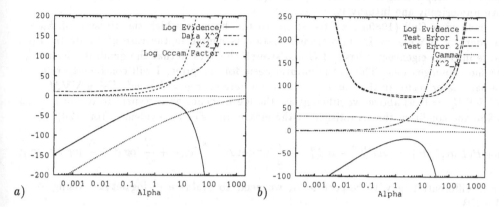

Fig. 5. Choosing α

a) **The evidence as a function of α:** Using the same radial basis function model as in figure 4, this graph shows the log evidence as a function of α, and shows the functions which make up the log evidence, namely the data misfit $\chi_D^2 = 2\beta E_D$, the weight penalty term $\chi_W^2 = 2\alpha E_W$, and the log of the volume ratio $(2\pi)^{k/2}\det^{-\frac{1}{2}}\mathbf{A}/Z_W(\alpha)$.

b) **Criteria for optimising α:** This graph shows the log evidence as a function of α, and the functions whose intersection locates the evidence maximum: the number of good parameter measurements γ, and χ_W^2. Also shown is the test error (rescaled) on two test sets; finding the test error minimum is an alternative criterion for setting α. Both test sets were more than twice as large in size as the interpolated data set. Note how the point at which $\chi_W^2 = \gamma$ is clear and unambiguous, which cannot be said for the minima of the test energies. The evidence gives α a 1σ confidence interval of $[1.3, 5.0]$. The test error minima are more widely distributed because of finite sample noise. From (MacKay, 1992a).

In many cases where the regulariser is not quadratic (for example, entropy–based), this gaussian approximation is still servicable (Gull, 1989a). Thus we can write the log evidence for α and β as:

$$\log P(D|\alpha,\beta,\mathcal{H}) = -\alpha E_W^{MP} - \beta E_D^{MP} - \frac{1}{2}\log\det\mathbf{A} - \log Z_W(\alpha) - \log Z_D(\beta) + \frac{k}{2}\log 2\pi \quad (19)$$

The term βE_D^{MP} represents the misfit of the interpolant to the data. The three terms $-\alpha E_W^{MP} - \frac{1}{2}\log\det\mathbf{A} - \log Z_W(\alpha)$ constitute the log of the 'Occam factor' penalising over-powerful values of α: the ratio $(2\pi)^{k/2}\det^{-\frac{1}{2}}\mathbf{A}/Z_W(\alpha)$ is the ratio of the posterior accessible volume in parameter space to the prior accessible volume, and the term αE_W^{MP} measures how far $\mathbf{w}_{MP}$ is from its null value. Figure 5a illustrates the behaviour of these various terms as a function of α for the same radial basis function model as illustrated in figure 4.

Now we could just proceed to evaluate the evidence numerically as a function of α and β, but a more deep and fruitful understanding of this problem is possible.

Properties of the evidence maximum

The maximum over α, β of $P(D|\alpha,\beta,\mathcal{H}) = \frac{Z_M(\alpha,\beta)}{Z_W(\alpha)Z_D(\beta)}$ has some remarkable properties which give deeper insight into this Bayesian approach. The results of this section are useful both numerically and intuitively.

Following Gull (1989a), we transform to the basis in which the Hessian of E_W is the identity, $\nabla\nabla E_W = \mathbf{I}$. This transformation is simple in the case of quadratic E_W: rotate into the eigenvector basis of $\mathbf{C}$ and stretch the axes so that the quadratic form E_W becomes homogeneous. This is the natural basis for the prior. I will continue to refer to the parameter vector in this basis as $\mathbf{w}$, so from here on $E_W = \frac{1}{2}\sum w_i^2$. Using $\nabla\nabla M = \mathbf{A}$ and $\nabla\nabla E_D = \mathbf{B}$ as above, we differentiate the log evidence with respect to α and β so as to find the condition that is satisfied at the maximum. The log evidence, from (20), is:

$$\log P(D|\alpha,\beta,\mathcal{H}) = -\alpha E_W^{MP} - \beta E_D^{MP} - \frac{1}{2}\log\det\mathbf{A} + \frac{k}{2}\log\alpha + \frac{N}{2}\log\beta - \frac{N}{2}\log 2\pi \quad (20)$$

First, differentiating with respect to α, we need to evaluate $\frac{d}{d\alpha}\log\det\mathbf{A}$. Using $\mathbf{A} = \alpha\mathbf{I} + \beta\mathbf{B}$,

$$\frac{d}{d\alpha}\log\det\mathbf{A} = \mathrm{Trace}\left(\mathbf{A}^{-1}\frac{d\mathbf{A}}{d\alpha}\right) = \mathrm{Trace}\left(\mathbf{A}^{-1}\mathbf{I}\right) = \mathrm{Trace}\mathbf{A}^{-1}$$

This result is exact if E_W and E_D are quadratic. Otherwise this result is an approximation, omitting terms in $\partial\mathbf{B}/\partial\alpha$. Now, differentiating (21) and setting the derivative to zero, we obtain the following condition for the most probable value of α:

$$2\alpha E_W^{MP} = k - \alpha\mathrm{Trace}\mathbf{A}^{-1} \quad (21)$$

The quantity on the left is the dimensionless measure of the amount of structure introduced into the parameters by the data, *i.e.* how much the fitted parameters differ from their null value. It can be interpreted as the χ^2 of the parameters, since it is equal to $\chi_W^2 = \sum w_i^2/\sigma_W^2$, with $\alpha = 1/\sigma_W^2$.

The quantity on the right of (22) is called the number of good parameter measurements, γ, and has value between 0 and k. It can be written in terms of the eigenvalues of $\beta\mathbf{B}$, λ_a, where the subscript a runs over the k eigenvectors. The eigenvalues of $\mathbf{A}$ are $\lambda_a + \alpha$, so we have:

$$\gamma = k - \alpha\,\mathrm{Trace}\mathbf{A}^{-1} = k - \sum_{a=1}^{k}\frac{\alpha}{\lambda_a + \alpha} = \sum_{a=1}^{k}\frac{\lambda_a}{\lambda_a + \alpha}$$

Each eigenvalue λ_a measures how strongly one parameter is determined by the data. The constant α measures how strongly the parameters are determined by the prior. The term $\lambda_a/(\lambda_a+\alpha)$ is a number between 0 and 1 which measures the strength of the data in direction a relative to the prior (figure 6): the components of $\mathbf{w}_{\mathrm{MP}}$ are given by $\mathbf{w}_{\mathrm{MP}\,a} = \gamma_a\mathbf{w}_{\mathrm{ML}\,a}$. A direction in parameter space for which λ_a is small compared to α does not contribute to the number of good parameter measurements. γ is thus a measure of the effective number of parameters which are well determined by the data. As $\alpha/\beta \to 0$, γ increases from 0 to k. The condition (22) for the most probable value of α can therefore be interpreted as an estimation of the variance σ_W^2 of the gaussian distribution from which the weights are drawn, based on γ effective samples from that distribution: $\sigma_W^2 = \sum w_i^2/\gamma$.

This concept is not only important for locating the optimum value of α: it is only the γ good parameter measurements which are expected to contribute to the reduction of the data misfit that occurs when a model is fitted to noisy data. In the process of fitting $\mathbf{w}$ to

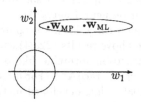

Fig. 6. Good and bad parameter measurements

Let w_1 and w_2 be the components in parameter space in two directions parallel to eigenvectors of the data matrix **B**. The circle represents the characteristic prior distribution for **w**. The ellipse represents a characteristic contour of the likelihood, centred on the maximum likelihood solution $\mathbf{w}_{ML}$. $\mathbf{w}_{MP}$ represents the most probable parameter vector. w_1 is a direction in which λ_1 is small compared to α, i.e. the data have no strong preference about the value of w_1; w_1 is a poorly measured parameter, and the term $\frac{\lambda_1}{\lambda_1 + \alpha}$ is close to zero. w_2 is a direction in which λ_1 is large; w_2 is well determined by the data, and the term $\frac{\lambda_2}{\lambda_2 + \alpha}$ is close to one. From (MacKay, 1992a).

the data, it is unavoidable that some fitting of the model to noise will occur, because some components of the noise are indistinguishable from real data. Typically, one unit (χ^2) of noise will be fitted for every well–determined parameter. Poorly determined parameters are determined by the regulariser only, so they do not reduce χ_D^2 in this way. We will now examine how this concept enters into the Bayesian choice of β.

Recall that the expectation of the χ^2 misfit between the true interpolant and the data is N. However we do not know the true interpolant, and the only misfit measure to which we have access is the χ^2 between the *inferred* interpolant and the data, $\chi_D^2 = 2\beta E_D$. The 'discrepancy principle' of orthodox statistics states that the model parameters should be adjusted so as to make $\chi_D^2 = N$. Work on un–regularised least–squares regression suggests that we should estimate the noise level so as to set $\chi_D^2 = N - k$, where k is the number of free parameters. Let us find out the opinion of Bayes' rule on this matter.

We differentiate the log evidence (21) with respect to β and obtain:

$$2\beta E_D = N - \gamma \tag{22}$$

Thus the most probable noise estimate, $\hat{\beta}$, does not satisfy $\chi_D^2 = N$ or $\chi_D^2 = N - k$; rather, $\chi_D^2 = N - \gamma$. This Bayesian estimate of noise level naturally takes into account the fact that the parameters which have been determined by the data inevitably suppress some of the noise in the data, while the poorly measured parameters do not. Note that the value of χ_D^2 only enters into the determination of β: misfit criteria have no role in the Bayesian choice of α (Gull, 1989a).

In summary, at the optimum value of α and β, $\chi_W^2 = \gamma$, $\chi_D^2 = N - \gamma$. Notice that this implies that the total misfit $M = \alpha E_W + \beta E_D$ satisfies the simple equation $2M = N$.

The interpolant resulting from the Bayesian choice of α is illustrated by figure 4b. Figure 5b illustrates the functions involved with the Bayesian choice of α, and compares them with the 'test error' approach. Demonstration of the Bayesian choice of β is omitted, since it is straightforward; β is fixed to its true value for the demonstrations in this paper.

These results generalise to the case where there are two or more separate regularisers with independent regularising constants $\{\alpha_c\}$ (Gull, 1989a). In this case, each regulariser has a number of good parameter measurements γ_c associated with it. Multiple regularisers will be used in the companion paper on neural networks.

Finding the evidence maximum with a head–on approach would involve evaluating $\det \mathbf{A}$ while searching over α, β; the above results (22,24) enable us to speed up this search (for example by the use of re–estimation formulae like $\alpha := \gamma/2E_W$) and replace the evaluation of $\det \mathbf{A}$ by the evaluation of $\mathrm{Trace}\mathbf{A}^{-1}$. For large–dimensional problems where this task is demanding, Skilling (1989b) has developed methods for estimating $\mathrm{Trace}\mathbf{A}^{-1}$ statistically in k^2 time.

5. Model comparison

To rank alternative basis sets $\mathcal{A}$, noise models $\mathcal{N}$ and regularisers (priors) $\mathcal{R}$ in the light of the data, we examine the posterior probabilities for alternative models $\mathcal{H} = \{\mathcal{A}, \mathcal{N}, \mathcal{R}\}$:

$$P(\mathcal{H}|D) \propto P(D|\mathcal{H})P(\mathcal{H}) \tag{23}$$

The data–dependent term, the evidence for $\mathcal{H}$, appeared earlier as the normalising constant in (15), and is evaluated by integrating the evidence for (α, β):

$$P(D|\mathcal{H}) = \int P(D|\mathcal{H}, \alpha, \beta)P(\alpha, \beta)\, d\alpha\, d\beta \tag{24}$$

Assuming that we have no reason to assign strongly differing priors $P(\mathcal{H})$, alternative $\mathcal{H}$ are ranked just by examining the evidence. The evidence can also be compared with the evidence found by an equivalent Bayesian analysis of other learning and interpolation models so as to allow the data to assign a preference to the alternative models. Notice as pointed out earlier that this modern Bayesian framework includes no emphasis on defining the 'right' prior $\mathcal{R}$ with which we ought to interpolate. Rather, we invent as many priors (regularisers) as we want, and allow the data to tell us which prior is *most probable*. Having said this, experience recommends that the 'maximum entropy principle' and other respected guides should be consulted when inventing these priors (see Gull, 1988, for example).

Evaluating the evidence for $\mathcal{H}$

As α and β vary, a single evidence maximum is obtained, at $\hat{\alpha}, \hat{\beta}$ (at least for quadratic E_D and E_W). The evidence maximum is often well approximated[10] by a separable gaussian, and differentiating (21) twice we obtain gaussian error bars for $\log \alpha$ and $\log \beta$:

$$(\Delta \log \alpha)^2 \simeq 2/\gamma$$
$$(\Delta \log \beta)^2 \simeq 2/(N - \gamma)$$

Putting these error bars into (26), we obtain the evidence.[11]

$$P(D|\mathcal{H}) \simeq P(D|\hat{\alpha}, \hat{\beta}, \mathcal{H})P(\hat{\alpha}, \hat{\beta}|\mathcal{H})\, 2\pi\, \Delta\log \alpha\, \Delta\log \beta \tag{25}$$

---

[10] This approximation is valid when, in the spectrum of eigenvalues of $\beta \mathbf{B}$, the number of eigenvalues within e-fold of $\hat{\alpha}$ is $O(1)$.

[11] There are analytic methods for performing such integrals over β (Bretthorst, 1990).

How is the prior $P(\hat{\alpha}, \hat{\beta}|\mathcal{H})$ assigned? This is the first time in this paper that we have met one of the infamous 'subjective priors' which are supposed to plague Bayesian methods. Here are some answers to this question. (a) Any other coherent method of assigning a preference to alternatives must implicitly assign such priors. Bayesians adopt the healthy attitude of not sweeping them under the carpet. (b) With some thought, reasonable values can usually be assigned to subjective priors, and the degree of reasonable subjectivity in these assignments can be quantified, and the sensitivity of our inferences to these priors can be quantified (Box and Tiao, 1973). For example, a reasonable prior on an unknown standard deviation states that σ is unknown over a range of (3 ± 2) orders of magnitude. This prior contributes a subjectivity of about ± 1 to the value of the log evidence. This degree of subjectivity is often negligible compared to the log evidence differences. (c) In the noisy interpolation example, all models considered include the free parameters α and β. So in this paper I do not need to assign a value to $P(\hat{\alpha}, \hat{\beta}|\mathcal{H})$; I assume that it is a flat prior (flat over $\log \alpha$ and $\log \beta$, since α and β are scale parameters) which cancels out when we compare alternative interpolation models.

6. A Demonstration

These demonstrations will use two one–dimensional data sets, in imitation of (Sibisi, 1991). The first data set, 'X,' has discontinuities in derivative (figure 4), and the second is a smoother data set, 'Y' (figure 8). In all the demonstrations, β was not left as a free parameter, but was fixed to its known true value.

The Bayesian method of setting α, assuming a single model is correct, has already been demonstrated, and quantified error bars have been placed on the most probable interpolant (figure 4). The method of evaluating the error bars is to use the posterior covariance matrix of the parameters w_h, $\mathbf{A}^{-1}$, to get the variance on $y(x)$, which for any x is a linear function of the parameters, $y(x) = \sum_h \phi_h(x) w_h$. The error bars at a single point x are given by $\text{var } y(x) = \phi^\mathrm{T} \mathbf{A}^{-1} \phi$. Actually we have access to the full covariance information for the entire interpolant, not just the pointwise error bars. It is possible to visualise the *joint* error bars on the interpolant by making typical samples from the posterior distribution, performing a random walk around the posterior 'bubble' in parameter space (Sibisi, 1991, Skilling *et. al.*, 1991). Figure 8 shows data set Y interpolated by three typical interpolants found by random sampling from the posterior distribution. These error bar properties are found under the assumption that the model is correct; so it is possible for the true interpolant to lie significantly outside the error bars of a poor model. (The pointwise error bars are directly related to the expected generalisation error at x, assuming that the model is true, evaluated in Levin *et. al.*, 1989, Tishby *et. al.*, 1989.)

In this section Bayesian model comparison will be demonstrated first with models differing only in the number of free parameters (for example polynomials of different degrees), then with comparisons between models as disparate as splines, radial basis functions and feedforward neural networks. The characters of some of these models are illustrated in figure 9, which shows a typical sample from each. For each individual model, the value of α is optimised, and the evidence is evaluated by integrating over α using the gaussian approximation. All logarithms are to base e.

LEGENDRE POLYNOMIALS: OCCAM'S RAZOR FOR THE NUMBER OF BASIS FUNCTIONS

Figure 7a shows the evidence for Legendre polynomials of different degrees for data set

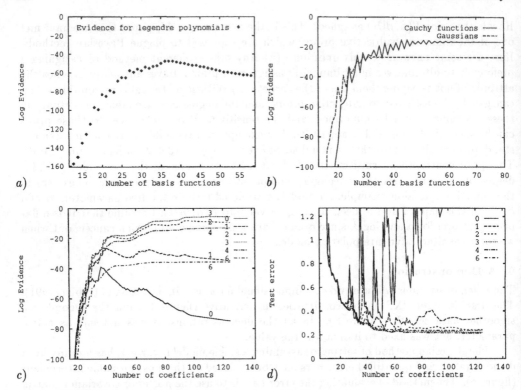

Fig. 7. The Evidence for data set X (see also table 1)

a) **Log Evidence for Legendre polynomials.** Notice the evidence maximum. The gentle slope to the right is due to the 'Occam factors' which penalise the increasing complexity of the model. b) **Log Evidence for radial basis function models.** Notice that there is no Occam penalty for the additional coefficients in these models, because increased density of radial basis functions does not make the model more powerful. The oscillations in the evidence are due to the details of the pixellation of the basis functions relative to the data points. c) **Log Evidence for splines.** The evidence is shown for the alternative splines regularisers $p = 0 \ldots 6$ (see text). In the representation used, each spline model is obtained in the limit of an infinite number of coefficients. For example, $p = 4$ yields the cubic splines model. d) **Test error for splines.** The number of data points in the test set was 90, c.f. number of data points in training set = 37. The y axis shows E_D; the value of E_D for the true interpolant has expectation 0.225 ± 0.02. From (MacKay, 1992a).

X. The basis functions were chosen to be orthonormal on an interval enclosing the data, and a regulariser of the form $E_W = \sum \frac{1}{2} w_h^2$ was used.

Notice that an evidence maximum is obtained: beyond a certain number of terms, the evidence starts to decrease. This is the Bayesian Occam's razor at work. The additional terms make the model more powerful, able to make more predictions. This flexibility is automatically penalised. Notice the characteristic shape of the 'Occam hill.' On the left, the hill is steep as the over–simple models fail to fit the data; the penalty for misfitting the data scales as N, the number of data measurements. The other side of the hill is

Table 1. Evidence for models interpolating data sets X and Y
All logs are natural. The evidence $P(D|\mathcal{H})$ is a density over D space, so the absolute value of the log evidence is arbitrary within an additive constant. Only differences in values of log evidences are relevant, relating directly to probability ratios.

Model	Data Set X		Data Set Y	
	Best parameter values	Log evidence	Best parameter values	Log evidence
Legendre polynomials	$k = 38$	-47	$k = 11$	23.8
Gaussian radial basis functions	$k > 40$, $r = .25$	-28.8 ± 1.0	$k > 50$, $r = .77$	27.1 ± 1.0
Cauchy radial basis functions	$k > 50$, $r = .27$	-18.9 ± 1.0	$k > 50$, $r = 1.1$	25.7 ± 1.0
Splines, $p = 2$	$k > 80$	-9.5	$k > 50$	8.2
Splines, $p = 3$	$k > 80$	-5.6	$k > 50$	19.8
Splines, $p = 4$	$k > 80$	-13.2	$k > 50$	22.1
Splines, $p = 5$	$k > 80$	-24.9	$k > 50$	21.8
Splines, $p = 6$	$k > 80$	-35.8	$k > 50$	20.4
Hermite functions	$k = 18$	-66	$k = 3$	42.2
Neural networks	8 neurons, $k = 25$	-12.6	6 neurons, $k = 19$	25.7

much less steep; the log Occam factors here only scale as $k \log N$, where k is the number of parameters. We note in table 1 the values of the maximum evidence achieved by these two models, and move on to alternative models.

The choice of orthonormal Legendre polynomials described above was motivated by a maximum entropy argument (Gull, 1988). Models using other polynomial basis sets have also been tried. For less well motivated basis sets such as Hermite polynomials, it was found that the Occam factors were far bigger and the evidence was substantially smaller. If the size of the Occam factor increases rapidly with over–parameterisation, it is generally a sign that the space of alternative models is poorly matched to the problem.

FIXED RADIAL BASIS FUNCTIONS

For a radial basis function or 'kernel' model, the basis functions are $\phi_h(x) = g((x - x_h)/r)$, with x_h equally spaced over the range of interest. I examine two choices of g: a gaussian and a Cauchy function, $1/1 + x^2$. We can quantitatively compare these alternative models of spatial correlation for any data set by evaluating the evidence. The regulariser is $E_W = \sum \frac{1}{2} w_h^2$. Note that this model includes one new free parameter, r; in these demonstrations this parameter has been set to its most probable value (i.e. the value which maximises the evidence). To penalise this free parameter an Occam factor is included, $\sqrt{2\pi} \Delta \log r P(\log r)$, where $\Delta \log r$ = posterior uncertainty in $\log r$, and $P(\log r)$ is the prior on $\log r$, which is subjective to a small degree (I used $P(\log r) = 1/(4 \pm 2)$). This radial basis function model is the same as the 'intrinsic correlation' model of Charter (1991), Gull (1989a), Skilling and Sibisi (1991).

Figure 7b shows the evidence as a function of the number of basis functions, k. Note that for these models there is *not* an increasing Occam penalty for large numbers of parameters. The reason for this is that these extra parameters do not make the model any more powerful (for fixed α and r). The increased density of basis functions does not enable the model to make any significant new predictions because the kernel g band–limits the possible interpolants.

Splines: Occam's Razor for the Choice of Regulariser

The splines model was implemented as follows: let the basis functions be a fourier set $\cos hx, \sin hx$, $h = 0, 1, 2, \ldots$. Use the regulariser $E_W = \sum \frac{1}{2} h^p w^2_{h(\cos)} + \sum \frac{1}{2} h^p w^2_{h(\sin)}$. If $p = 4$ then in the limit $k \to \infty$ we have the cubic splines regulariser $E_y^{(4)} = \int y''(x)^2 dx$; if $p = 2$ we have the regulariser $E_y^{(2)} = \int y'(x)^2 dx$, etc. Notice that the 'non–parametric' splines model can easily be put in an explicit parameterised representation. Note that none of these splines models include 'knots'.

Figure 7c shows the evidence for data set X as a function of the number of terms, for $p = 0, 1, 2, 3, 4, 6$. Notice that in terms of Occam's razor, both cases discussed above occur: for $p = 0, 1$, as k increases, the model becomes more powerful and there is an Occam penalty. For $p = 3, 4, 6$, increasing k gives rise to no penalty. The case $p = 2$ seems to be on the fence between the two.

As p increases, the regulariser becomes more opposed to strong curvature. Once we reach $p = 6$, the model becomes improbable because the data demand sharp discontinuities. The evidence can choose the order of our splines regulariser for us. For this data set, it turns out that $p = 3$ is the most probable value of p, by a few multiples of e.

In passing, the radial basis function models described above can be transformed into the Fourier representation of the splines models. If the radial basis function kernel is $g(x)$ then the regulariser in the splines representation is $E_W = \sum \frac{1}{2} (w^2_{h(\cos)} + w^2_{h(\sin)}) G_h^{-2}$, where G_h is the discrete Fourier transform of g.

Results for a Smoother Data Set

Figure 8 shows data set Y, which comes from a much smoother interpolant than data set X. Table 1 summarises the evidence for the alternative models. We can confirm that the evidence behaves in a reasonable manner by noting the following differences between data sets X and Y:

In the splines family, the most probable value of p has shifted upwards to the stiffer splines with $p = 4 - 5$, as we would intuitively expect.

Legendre polynomials: an observant reader may have noticed that when data set X was modelled with Legendre polynomials, the most probable number of coefficients $k = 38$ was suspiciously similar to the number of data points $N = 37$. For data set Y, however, the most probable number of coefficients is 11, which confirms that the evidence does not always prefer the polynomial with $k = N$! Data set X behaved in this way because it is very poorly modelled by polynomials.

Hermite functions, which were a poor model for data set X, are now the most probable, by a long way (over a million times more probable). The reason for this is that actually the data *were* generated from a Hermite function!

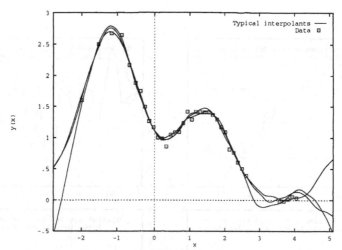

Fig. 8. Data set 'Y,' interpolated with splines, $p = 5$.

The data set is shown with three typical interpolants drawn from the posterior probability distribution. Contrast this with figure 4b, in which the most probable interpolant is shown with its pointwise error bars. From (MacKay, 1992a).

WHY BAYES CAN'T SYSTEMATICALLY REJECT THE TRUTH

Let us ask a sampling theory question: if one of the models we offer to Bayes is actually true, *i.e.* it is the model from which the data were generated, then is it possible for Bayes to systematically (over the ensemble of possible data sets) prefer a false model? Clearly under a worst case analysis, a Bayesian's posterior may favour a false model. Furthermore, Skilling (1991) demonstrated that with some data sets a free form (maximum entropy) model can have greater evidence than the truth; but is it possible for this to happen in the *typical* case, as Skilling seems to claim? I will show that the answer is no, the effect that Skilling demonstrated cannot be systematic. To be precise, the expectation over possible data sets of the log evidence for the true model is greater than the expectation of the log evidence for any other fixed model (Osteyee and Good, 1974).[12]

Proof. Suppose that the truth is actually $\mathcal{H}_1$. A single data set arrives and we compare the evidences for $\mathcal{H}_1$ and $\mathcal{H}_2$, a different fixed model. Both models may have free parameters, but this will be irrelevant to the argument. Intuitively we expect that the evidence for $\mathcal{H}_1$, $P(D|\mathcal{H}_1)$, should usually be greatest. Examine the difference in log evidence between $\mathcal{H}_1$ and $\mathcal{H}_2$. The expectation of this difference, given that $\mathcal{H}_1$ is true, is

$$\left\langle \log \frac{P(D|\mathcal{H}_1)}{P(D|\mathcal{H}_2)} \right\rangle = \int d^N D \, P(D|\mathcal{H}_1) \log \frac{P(D|\mathcal{H}_1)}{P(D|\mathcal{H}_2)}.$$

(Note that this integral implicitly integrates over all $\mathcal{H}_1$'s parameters according to their prior distribution under $\mathcal{H}_1$.) Now it is well known that for normalised p and q, $\int p \log \frac{p}{q}$

---

[12] Skilling's result presumably occurred because the particular parameter values of the true model that generated the data were not typical of the prior used when evaluating the evidence for that model. In such a case, the log evidence difference can show a transient bias against the true model; such biases are usually reversed by greater quantities of data.

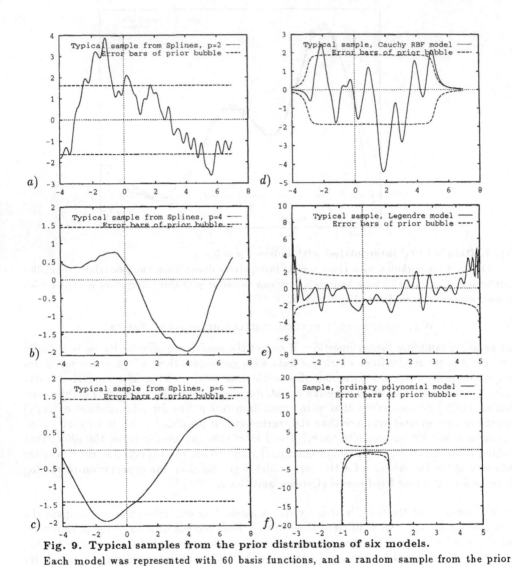

Fig. 9. Typical samples from the prior distributions of six models.

Each model was represented with 60 basis functions, and a random sample from the prior distribution is shown. a) Splines, $p = 2$. b) Splines, $p = 4$ (cubic splines). c) Splines, $p = 6$. The splines were represented with a Fourier set with period 12.0. Notice how the spikiness of the typical sample decreases as the order of the spline increases. d) Cauchy radial basis functions. The basis functions were equally spaced from -3.0 to 5.0, and had scale $r = 0.2975$. e) Legendre polynomials. The polynomials were stretched so that the interval [-3.0, 5.0] corresponds to the natural interval. Notice that the characteristic amplitude diverges at the boundaries, and the characteristic frequency of the typical sample also increases towards the boundaries. f) Ordinary polynomials. This figure illustrates what bad results can be obtained if a prior is carelessly assigned. A uniform prior over the coefficients of $y = \sum w_h x^h$ yields a highly non-uniform typical sample. From (MacKay, 1992a).

is minimised by setting $q = p$ (Gibbs' theorem). Therefore a distinct model $\mathcal{H}_2$ is never expected to *systematically* defeat the true model, for just the same reason that it is not wise to bet differently from the true odds.

This result has two important implications. First, it gives us frequentist confidence in the ability of Bayesian methods on the average to identify the true model. Secondly, it provides a severe test of any numerical implementation of Bayesian model comparison: imagine that we have written a program that evaluates the evidence for models $\mathcal{H}_1$ and $\mathcal{H}_2$; then we can generate mock data from sources simulating $\mathcal{H}_1$ and $\mathcal{H}_2$ and evaluate the evidences; if there is any systematic bias, averaged over several mock data sets, for the estimated evidence to favour the false model, then we can be sure that our numerical implementation is not evaluating the evidence correctly.

This issue is illustrated using data set Y. The 'truth' is that this data set was actually generated from a quadratic Hermite function, $1.1(1-x+2x^2)e^{-x^2/2}$. By the above argument the evidence ought probably to favour the model 'the interpolant is a 3–coefficient Hermite function' over our other models. Table 1 shows the evidence for the true Hermite function model, and for other models. Notice that the truth is indeed considerably more probable than the alternatives.

Having demonstrated that Bayes cannot systematically fail when one of the models is true, we now examine the way in which this framework can fail, if none of the models offered to Bayes is any good.

COMPARISON WITH 'GENERALISATION ERROR'

It is a popular and intuitive criterion for choosing between alternative interpolants (found using different models) to compare their errors on a test set that was not used to derive the interpolants. 'Cross–validation' is a more refined and more computationally expensive version of this same idea. How does this method relate to the evaluation of the evidence described in this paper?

Figure 7c displayed the evidence for the family of spline interpolants. Figure 7d shows the corresponding test error, measured on a test set with size over twice as big (90) as the 'training' data set (37) used to determine the interpolant. A similar comparison was made in figure 5b. Note that the overall trends shown by the evidence are matched by trends in the test error (if you flip one graph upside down). Also, for this particular problem, the ranks of the alternative spline models under the evidence are similar to their ranks under the test error. And in figure 5b, the evidence maximum over α was surrounded by the test error minima. Thus this suggests that the evidence might be a reliable predictor of generalisation ability. However, this is not necessarily the case. There are five reasons why the evidence and the test error might not be correlated.

First, the test error is a noisy quantity. It is necessary to devote large quantities of data to the test set to obtain a reasonable signal to noise ratio. In figure 5b more than twice as much data is in each test set but the difference in α between the two test error minima exceeds the size of the Bayesian confidence interval for α.

Second, the model with greatest evidence is not expected to be the best model all the time — Bayesian inferences are uncertain. The whole point of Bayes is that it quantifies precisely those uncertainties: the relative values of the evidence for alternative models express the plausibility of the models, given the data and the underlying assumptions.

Third, there is more to the evidence than there is to the generalisation error. For example, imagine that for two models, the most probable interpolants happen to be identical. In this case, the two solutions will have the same generalisation error, but the evidence will not in general be the same: typically, the model that was *a priori* more complex will suffer a larger Occam factor and will have a smaller evidence.

Fourth, the test error is a measure of performance only of the single most probable interpolant: the evidence is a measure of plausibility of the entire posterior ensemble around the best fit interpolant. Probably a stronger correlation between the evidence and the test statistic would be obtained if the test statistic used were the average of the test error over the posterior ensemble of solutions. This ensemble test error is not so easy to compute.

The fifth and most interesting reason why the evidence might not be correlated with the generalisation error is that there might be a flaw in the underlying assumptions such that the models being compared might all be poor models. If a poor regulariser is used, for example, one that is ill–matched to the statistics of the world, then the Bayesian choice of α will often not be the best in terms of generalisation error (Davies and Anderssen, 1986, Gull, 1989a, Haussler *et. al.*, 1991). Such a failure occurs in the companion paper on neural networks. What is our attitude to such a failure of Bayesian prediction? The failure of the evidence does not mean that we should discard Bayes' rule and use the generalisation error as our criterion for choosing α. A failure is an opportunity to learn; a healthy scientist actively searches for such failures, because they yield insights into the defects of the current model. The detection of such a failure (by evaluating the generalisation error for example) motivates the search for new models which do not fail in this way; for example alternative regularisers can be tried until a model is found that makes the data more probable.

If one only uses the generalisation error as a criterion for model comparison, one is denied this mechanism for learning. The development of image deconvolution was held up for decades because no–one used the Bayesian choice of α; once the Bayesian choice of α was used (Gull, 1989a), the results obtained were most dissatisfactory, making clear what a poor regulariser was being used; this motivated an immediate search for alternative priors; the new, more probable priors discovered by this search are now at the heart of the state of the art in image deconvolution (Weir, 1991).

ADMITTING NEURAL NETWORKS INTO THE CANON OF BAYESIAN INTERPOLATION MODELS

A second paper will discuss how to apply this Bayesian framework to feedforward neural networks. Preliminary results using these methods are included in table 1. Assuming that the approximations used were valid, it is interesting that the evidence for neural nets is actually good for both the spiky and the smooth data sets. Furthermore, neural nets, in spite of their arbitrariness, yield a relatively compact model, with fewer parameters needed than to specify the splines and radial basis function solutions.

7. Conclusions

The recently developed methods of Bayesian model comparison and regularisation have been presented. Models can be ranked by evaluating the evidence, a solely data–dependent measure which intuitively and consistently combines a model's ability to fit the data with its complexity. The precise posterior probabilities of the models also depend on the subjective priors that we assign to them, but these terms are typically overwhelmed by the evidence.

Regularising constants are set by maximising the evidence. For many regularisation problems, the theory of the number of well–measured parameters makes it possible to perform this optimisation on–line.

In the interpolation examples discussed, the evidence was used to set the number of basis functions k in a polynomial model; to set the characteristic size r in a radial basis function model; to choose the order p of the regulariser for a spline model; and to rank all these different models in the light of the data.

Further work is needed to formalise the relationship of this framework to the pragmatic model comparison technique of cross–validation. Using the two techniques in parallel, it is possible to detect flaws in the underlying assumptions implicit in the data models being used. Such failures direct our search for superior models, providing a powerful tool for human learning.

There are thousands of data modelling tasks waiting for the evidence to be evaluated. It will be exciting to see how much we can learn when this is done.

ACKNOWLEDGMENTS. I thank Mike Lewicki, Nick Weir and David R.T. Robinson for helpful conversations, and Andreas Herz for comments on the manuscript. I am grateful to Dr. R. Goodman and Dr. P. Smyth for funding my trip to Maxent 90. This work was supported by a Caltech Fellowship and a Studentship from SERC, UK.

REFERENCES

H. Akaike (1970). 'Statistical predictor identification', *Ann. Inst. Statist. Math.* **22**, 203–217.

J. Berger (1985). *Statistical decision theory and Bayesian analysis*, Springer.

G.E.P. Box and G.C. Tiao (1973). 'Bayesian inference in statistical analysis', Addison–Wesley.

G.L. Bretthorst (1990). 'Bayesian Analysis. I. Parameter Estimation Using Quadrature NMR Models. II. Signal Detection and Model Selection. III. Applications to NMR.', *J. Magnetic Resonance*, **88** 3, 533–595.

M.K. Charter (1991). 'Quantifying drug absorption', in Grandy and Schick, eds., 245–252.

R.T. Cox (1946). 'Probability, frequency, and reasonable expectation', *Am. J. Physics* **14**, 1–13

A.R. Davies and R.S. Anderssen (1986). 'Optimization in the regularization of ill–posed problems', *J. Austral. Mat. Soc. Ser. B*, **28**, 114–133.

R.L. Eubank (1988). 'Spline smoothing and non–parametric regression', Marcel Dekker.

W.T. Grandy, Jr. and L.H. Schick, eds. (1991). *Maximum Entropy and Bayesian Methods*, Laramie 1990, Kluwer.

S.F. Gull (1988). 'Bayesian inductive inference and maximum entropy', in *Maximum Entropy and Bayesian Methods in science and engineering, vol. 1: Foundations*, G.J. Erickson and C.R. Smith, eds., Kluwer.

S.F. Gull (1989). 'Developments in Maximum entropy data analysis', in J. Skilling, ed., 53–71.

S.F. Gull (1989). 'Bayesian data analysis: straight–line fitting', in J. Skilling, ed., 511–518.

S.F. Gull and J. Skilling (1991). *Quantified Maximum Entropy. MemSys5 User's manual.* M.E.D.C., 33 North End, Royston, SG8 6NR, England.

R. Hanson, J. Stutz and P.Cheeseman (1991). 'Bayesian classification theory', NASA Ames TR FIA–90–12–7–01.

D. Haussler, M. Kearns and R. Schapire (1991). 'Bounds on the sample complexity of Bayesian learning using information theory and the VC dimension', Preprint.

E.T. Jaynes (1986). 'Bayesian methods: general background', in *Maximum Entropy and Bayesian Methods in applied statistics*, ed. J.H. Justice, C.U.P.

H. Jeffreys (1939). *Theory of Probability*, Oxford Univ. Press.

R.L. Kashyap (1977). 'A Bayesian comparison of different classes of dynamic models using empirical data', *IEEE Transactions on Automatic Control*, **AC-22** 5, 715–727.

T.J. Loredo (1989). 'From Laplace to supernova SN 1987A: Bayesian inference in astrophysics', in *Maximum Entropy and Bayesian Methods*, ed. P. Fougere, Kluwer.

D.J.C. MacKay (1992a). 'Bayesian interpolation', *Neural computation*, **4** 3.

D.J.C. MacKay (1992d). 'The evidence for neural networks', this volume.

D.J.C. MacKay (1992e). 'Information–based objective functions for active data selection for optimal learning', to appear, *Neural computation*.

D.J.C. MacKay (1992f). 'The evidence framework applied to classification networks', to appear, *Neural computation*.

R.M. Neal (1991). 'Bayesian mixture modeling by Monte Carlo simulation', Preprint.

D.B. Osteyee and I.J. Good (1974). *Information, weight of evidence, the singularity between probability measures and signal detection*, Springer.

J.D. Patrick and C.S. Wallace (1982). 'Stone circle geometries: an information theory approach', in *Archaeoastronomy in the Old World*, D.C. Heggie, editor, Cambridge Univ. Press.

T. Poggio, V. Torre and C. Koch (1985). 'Computational vision and regularization theory', *Nature*, **317** 6035, 314–319.

J. Rissanen (1978). 'Modeling by shortest data description', *Automatica*, **14**, 465–471.

G. Schwarz (1978). 'Estimating the dimension of a model', *Ann. Stat.* **6** 2, 461–464.

S. Sibisi (1991). 'Bayesian interpolation', in Grandy and Schick, eds., 349–355.

J. Skilling, editor (1989). *Maximum Entropy and Bayesian Methods, Cambridge 1988*, Kluwer.

J. Skilling (1989). 'The eigenvalues of mega–dimensional matrices', in J. Skilling, ed., 455–466.

J. Skilling (1991). 'On parameter estimation and quantified MaxEnt', in Grandy and Schick, eds., 267–273.

J. Skilling, D.R.T. Robinson, and S.F. Gull (1991). 'Probabilistic displays', in Grandy and Schick, eds., 365–368.

S.M. Stigler (1986). 'Laplace's 1774 memoir on inverse probability', *Stat. Sci.*, **1** 3, 359–378.

R. Szeliski (1989). 'Bayesian modeling of uncertainty in low level vision', Kluwer.

N. Tishby, E. Levin and S.A. Solla (1989). 'Consistent inference of probabilities in layered networks: predictions and generalization', in *Proc. IJCNN*, Washington.

D. Titterington (1985). 'Common structure of smoothing techniques in statistics', *Int. Statist. Rev.*, **53**, 141–170.

A.M. Walker (1967). 'On the asymptotic behaviour of posterior distributions', *J. R. Stat. Soc. B*, **31**, 80–88.

C.S. Wallace and D.M. Boulton (1968). 'An information measure for classification', *Comput. J.*, **11** 2, 185–194.

C.S. Wallace and P.R. Freeman (1987). 'Estimation and Inference by Compact Coding', *J. R. Statist. Soc. B*, **49** 3, 240-265.

N. Weir (1991). 'Applications of maximum entropy techniques to HST data', *Proceedings of the ESO/ST–ECF Data Analysis Workshop*, April 1991.

A. Zellner (1984). *Basic issues in econometrics*, Chicago.

ESTIMATING THE RATIO OF TWO AMPLITUDES IN NUCLEAR MAGNETIC RESONANCE DATA

G. Larry Bretthorst
Washington University
Department of Chemistry
Campus Box 1134
St. Louis, Missouri 63130-4899.

ABSTRACT. Probability theory is applied to the problem of estimating the ratio of the amplitudes of two sinusoids in nuclear magnetic resonance data. The posterior probability-density for the amplitude ratio is derived independent of the phase, frequencies, decay-rate constants, variance of the noise, and the amplitude of the other sinusoid. This probability-density function is then applied in an illustrative example, and the results are contrasted with those obtained by traditional analysis.

1. Introduction

Investigating the molecular structure of a compound in a nondestructive manner is difficult. If the nuclei of the compound have a magnetic moment, then one way to investigate the structure is to place the compound in a high magnetic field, excite the system with radio-frequency energy, and listen to it "ring." This type of experiment is called a nuclear magnetic resonance (NMR) experiment. The "ringing" is called a free induction decay (FID). When nuclei of the same type, for example protons, are in different electronic environments (as they are when they are bound to different nuclei) they resonate at slightly different frequencies. These frequencies provide information about the local environments, while the intensities are be related to the relative concentrations of the nuclei.

Traditionally, FID data have been analyzed using the fast Fourier transform after placing zeroes on the end of the FID (i.e., zero-padding the total number of complex points to a power of two). The frequencies are estimated from the real part of the Fourier transform by peak picking, and the amplitudes by integration. After estimating the amplitudes, the ratio is computed. The amplitude ratio is important, because a spectrometer only measures a relative amplitude; not an absolute amplitude.

Bayesian probability theory has recently been applied to the frequency estimation problem [1–4] in NMR, and more recently to amplitude estimation [5,6]. In this paper, probability theory is applied to the problem of determining the amplitudes ratio of the sinusoids. Specifically, the problem is: given that the data contain two exponentially decaying sinusoids with unknown frequencies, decay-rate constants, amplitudes, phases, and variance of the noise, derive the "best" estimate of the ratio of the amplitudes given the data and the prior information. In Bayesian probability theory, all of the information

67

C. R. Smith et al. (eds.), Maximum Entropy and Bayesian Methods, Seattle, 1991, 67–77.
© *1992 Kluwer Academic Publishers.*

relevant to this question is contained in a probability-density function. This function is independent of the unknown parameters appearing in the model function. Such a probability density function is called a *marginal* probability density function. This marginal posterior probability-density for the amplitude ratio is derived here. The calculation will be for the amplitude ratio of the first sinusoid to the second, but which sinusoid is first and which is second, is a matter of convention; at the end of the calculation the labels on the frequencies and decay-rate constants may be exchanged to obtain the probability-density function for the amplitude ratio of the second sinusoid to the first.

2. The Posterior Probability For The Amplitude Ratio

The problem addressed is: given a quadrature-detected FID containing two in phase exponentially decaying sinusoids with different amplitudes, frequencies, and decay-rate constants, calculate the posterior probability for the amplitude ratio of the sinusoids, independent of all other parameters. In quadrature detected data there are two data sets: the real data (0° phase), and the imaginary data (90° phase). The real data is assumed to be the sum of a signal plus noise:

$$d_R(t_i) = G_R(t_i) + e_i \tag{1}$$

where $d_R(t_i)$ denotes a real data item sampled at time ft_i ($1 \leq i \leq N$), and $D_R \equiv \{d_R(t_1), \ldots, d_R(t_N)\}$ will denote all of the real data.

The model signal, $G_R(t_i)$, is defined as

$$G_R(t_i) \equiv A_1 \cos(\omega_1 t_i + \theta)e^{-\alpha_1 t_i} + A_2 \cos(\omega_2 t_i + \theta)e^{-\alpha_2 t_i} \tag{2}$$

where A_1 is the amplitude of the first sinusoid, A_2 is the amplitude of the second sinusoid, θ is the common phase of the sinusoids, α_1 and α_2 are the decay-rate constants of the sinusoids, and e_i represents noise at time t_i. Note that in phase sinusoids are the rule rather than the exception in NMR FID data.

In addition to the real data, the imaginary data contain the same signal, shifted by 90°:

$$d_I(t_i) = G_I(t_i) + e_i \tag{3}$$

where $d_I(t_i)$ denotes an imaginary data item sampled at time t_i, and all of the imaginary data is represented by $D_I \equiv \{d_I(t_1), \ldots, d_I(t_N)\}$, and D will represent both the real and imaginary data, $D \equiv \{D_R, D_I\}$. The model signal in the imaginary channel is given by

$$G_I(t_i) \equiv -A_1 \sin(\omega_1 t_i + \theta)e^{-\alpha_1 t_i} - A_2 \sin(\omega_2 t_i + \theta)e^{-\alpha_2 t_i}. \tag{4}$$

The actual noise, e_i, realized in the imaginary channel is assumed to be different from the noise realized in the real channel. However, the calculation will be simplified by assuming the variance of the noise is the same in both channels.

To proceed, the ratio of the amplitudes must be introduced into the model. Here the ratio of the amplitude of the first sinusoid to the second sinusoid is the quantity of interest:

$$r \equiv \frac{A_2}{A_1}. \tag{5}$$

Introducing the change of variables $rA = A_2$, and $A = A_1$ the model equations are transformed into

$$G_R(t_i) \equiv A[\cos(\omega_1 t_i + \theta)e^{-\alpha_1 t_i} + r \cos(\omega_2 t_i + \theta)e^{-\alpha_2 t_i}] \tag{6}$$

for the real channel and

$$G_I(t_i) \equiv -A[\sin(\omega_1 t_i + \theta)e^{-\alpha_1 t_i} + r\sin(\omega_2 t_i + \theta)e^{-\alpha_2 t_i}] \qquad (7)$$

for the imaginary channel.

In previous work [4,5], it was assumed that a sample of the noise was known. When this sample was present, it placed a scale in the problem against which probability theory could measure small effects. When this sample was not present, the generalized results reduced to those found when no noise sample was gathered [1]. The same strategy will be used here and the more general results will be derived directly. Thus, a noise sample is assumed to be available. The real noise sample is denoted as $D_{\sigma R}$ with $D_{\sigma R} \equiv \{d_{\sigma R}(t_{\sigma 1}), \cdots, d_{\sigma R}(t_{\sigma N_\sigma})\}$. The imaginary noise sample is denoted as $D_{\sigma I}$ with $D_{\sigma I} \equiv \{d_{\sigma I}(t_{\sigma 1}), \cdots, d_{\sigma I}(t_{\sigma N_\sigma})\}$, and $D_\sigma \equiv \{D_{\sigma R}, D_{\sigma I}\}$ denotes both samples. The discrete times $t_{\sigma i} \equiv \{t_{\sigma 1}, \cdots, t_{\sigma N_\sigma}\}$ are assumed distinct from the sampling times for the data D.

The posterior probability for the amplitude ratio will be denoted as $P(r|D_\sigma, D, I)$. According to Bayes' theorem [7], this is given by

$$P(r|D_\sigma, D, I) = \frac{P(r|I)P(D_\sigma, D|r, I)}{P(D_\sigma, D|I)} \qquad (8)$$

where $P(r|I)$ is the prior probability for the amplitude ratio, $P(D_\sigma, D|r, I)$ is the joint marginal probability for the data and the noise sample, and $P(D_\sigma, D|I)$ is a normalization constant.

The symbol "I" being carried in these probability functions represents all of the assumptions that go into the calculation; explicitly it represents "everything known about the problem." At present, these include the quadrature nature of the data, the separation of the data into a signal plus additive noise, and that the data is composed of two exponentially decaying sinusoids. These assumptions are hypotheses just like any others appearing inside a probability symbol, and could be tested using the rules of probability theory. However, in this calculation they are assumed known.

3. Assigning Probabilities

Throughout this paper uninformative priors will be used. Normalization will be done at the end of the calculation, and priors ranges for uniform priors will be absorbed into this normalization constant. The first prior to be assigned, $P(r|I)$, will be taken to be a uniform prior $0 \le r \le max$ where max is the maximum detectable dynamic range of the digitizer.

Parameter estimation problems using uninformative priors reduce to finding the joint marginal probability for the data and the noise sample (when it is available):

$$P(r|D_\sigma, D, I) \propto P(D_\sigma, D|r, I). \qquad (9)$$

The joint marginal probability for the data and the noise sample can be computed from the joint marginal probability for the data, the noise sample, and the parameters:

$$P(r|D_\sigma, D, I) \propto \int A\,dA\,d\theta\,d\omega_1\,d\omega_2\,d\alpha_1\,d\alpha_2\,P(D_\sigma, D, A, \theta, \omega_1, \omega_2, \alpha_1, \alpha_2|r, I) \qquad (10)$$

where the extra factor, A, is the volume element is due to working in polar coordinates.

The product rule may be used to factor the right-hand-side of this equation into a joint prior for the parameters and a direct probability given the parameters. Assuming the joint prior factors, assigning uniform priors, and dropping all of the prior ranges, one obtains

$$P(r|D_\sigma, D, I) \propto \int AdAd\theta d\omega_1 d\omega_2 d\alpha_1 d\alpha_2 P(D_\sigma, D|r, A, \theta, \omega_1, \omega_2, \alpha_1, \alpha_2, I) \qquad (11)$$

as the posterior probability for the amplitudes ratio. Applying the product rule a second time one obtains

$$P(r|D_\sigma, D, I) \propto \int AdAd\theta d\omega_1 d\omega_2 d\alpha_1 d\alpha_2 P(D_\sigma|I) \\ \times P(D|r, A, \theta, \omega_1, \omega_2, \alpha_1, \alpha_2, I) \qquad (12)$$

where it has been assumed that the noise sample does not depend on the signal parameters: $P(D_\sigma|r, A, \theta, \omega_1, \omega_2, \alpha_1, \alpha_2, I) = P(D_\sigma|I)$ and it was assumed that the probability for the data sample did not depend on the presence of the noise sample.

The data, D, and the noise sample, D_σ, are actually joint hypotheses. The data, D, stands for both the real, and imaginary data and the noise sample, D_σ, stands for the real and imaginary noise samples. Making this substitution on the right-hand-side of equation (12) one has

$$P(r|D_\sigma, D, I) \propto \int AdAd\theta d\omega_1 d\omega_2 d\alpha_1 d\alpha_2 P(D_{\sigma R}, D_{\sigma I}|I) \\ \times P(D_R, D_I|r, A, \theta, \omega_1, \omega_2, \alpha_1, \alpha_2, I) \qquad (13)$$

as the posterior probability for the amplitude ratio.

If the product rule is applied to $P(D_{\sigma R}, D_{\sigma I}|I)$, one obtains

$$P(D_{\sigma R}, D_{\sigma I}|I) = P(D_{\sigma R}|I)P(D_{\sigma I}|D_{\sigma R}, I). \qquad (14)$$

The two channels of an NMR spectrometer are specifically designed to give projections onto orthogonal functions; the noise in the two channels should be uncorrelated. Having no reason to assume dependence, it will be assumed that $P(D_{\sigma I}|D_{\sigma R}, I) = P(D_{\sigma I}|I)$ and the posterior probability for the amplitude ratio becomes

$$P(r|D_\sigma, D, I) \propto \int AdAd\theta d\omega_1 d\omega_2 d\alpha_1 d\alpha_2 P(D_{\sigma R}|I)P(D_{\sigma I}|I) \\ \times P(D_R, D_I|r, A, \theta, \omega_1, \omega_2, \alpha_1, \alpha_2, I). \qquad (15)$$

Applying the product rule to $P(D_R, D_I|r, A, \theta, \omega_1, \omega_2, \alpha_1, \alpha_2, I)$, one obtains

$$P(D_R, D_I|r, A, \theta, \omega_1, \omega_2, \alpha_1, \alpha_2, I) = P(D_R|r, A, \theta, \omega_1, \omega_2, \alpha_1, \alpha_2, I) \\ \times P(D_I|D_R, r, A, \theta, \omega_1, \omega_2, \alpha_1, \alpha_2, I). \qquad (16)$$

Again the NMR spectrometer is designed so that the two channels give independent uncorrelated measurements of the signal. Because the measurements are independent, the probability for the imaginary data should reflect this information and be independent of the real data and vice versa. With these assumptions the joint probability of the data factors and

$$P(D_R, D_I | r, A, \theta, \omega_1, \omega_2, \alpha_1, \alpha_2, I) = P(D_R | r, A, \theta, \omega_1, \omega_2, \alpha_1, \alpha_2, I)$$
$$\times P(D_I | r, A, \theta, \omega_1, \omega_2, \alpha_1, \alpha_2, I). \tag{17}$$

The posterior probability for the amplitude ratio is given by

$$P(r | D_\sigma, D, I) \propto \int A dA d\theta d\omega_1 d\omega_2 d\alpha_1 d\alpha_2 \, P(D_{\sigma R} | I) P(D_{\sigma I} | I)$$
$$\times P(D_I | r, A, \theta, \omega_1, \omega_2, \alpha_1, \alpha_2, I) P(D_R | r, A, \theta, \omega_1, \omega_2, \alpha_1, \alpha_2, I) \tag{18}$$

where $P(D_R | r, A, \theta, \omega_1, \omega_2, \alpha_1, \alpha_2, I)$ is the direct probability for the real data, and the direct probability for the imaginary data is $P(D_I | r, A, \theta, \omega_1, \omega_2, \alpha_1, \alpha_2, I)$.

The posterior probability for the amplitude ratio has now been sufficiently simplified to permit assignment of the various terms. To do this assignment, note that equations (1) and (3) constitute a definition of what is meant by noise in this calculation. The direct probability for the data given the parameters *is* the noise prior probability given the parameters. Before these probability-density functions can be assigned, one must assign a prior probability for the noise. To do this the assumptions made about the noise must be explicitly stated. As in previous works [1–4,8] it will be assumed that the noise carries a finite, but unknown total power. Using this assumption in a maximum entropy calculation [9] results in the assignment of a Gaussian for the noise prior probability-density:

$$P(e_1, \cdots, e_N | \sigma, I) = (2\pi\sigma^2)^{-\frac{N}{2}} \exp\left\{ -\sum_{i=1}^{N} \frac{e_i^2}{2\sigma^2} \right\}. \tag{19}$$

The direct probability for obtaining the real data is given by

$$P(D_R | \sigma, r, A, \theta, \omega_1, \omega_2, \alpha_1, \alpha_2, I) = (2\pi\sigma^2)^{-\frac{N}{2}} \exp\left\{ -\sum_{i=1}^{N} \frac{[d_R(t_i) - G_R(t_i)]^2}{2\sigma^2} \right\} \tag{20}$$

where the notation has been adjusted as follows: first, $e_1, \cdots, e_N$, were replaced by D_R to indicate that it is the direct probability for the real data; and second σ, the standard deviation for the noise, was added to the probability-density function to indicate that it is a known quantity. Later, the product rule and sum rules of probability theory will be used to remove the dependence on the standard deviation σ. The direct probability for obtaining the imaginary data is given by

$$P(D_I | \sigma, r, A, \theta, \omega_1, \omega_2, \alpha_1, \alpha_2, I) = (2\pi\sigma^2)^{-\frac{N}{2}} \exp\left\{ -\sum_{i=1}^{N} \frac{[d_I(t_i) - G_I(t_i)]^2}{2\sigma^2} \right\}. \tag{21}$$

The direct probability for the real noise sample is given by

$$P(D_{\sigma R} | \sigma, I) = (2\pi\sigma^2)^{-\frac{N_\sigma}{2}} \exp\left\{ -\sum_{i=1}^{N_\sigma} \frac{d_R(t_i)^2}{2\sigma^2} \right\}. \tag{22}$$

Last, the direct probability for the imaginary noise sample is given by

$$P(D_{\sigma I}|\sigma, I) = (2\pi\sigma^2)^{-\frac{N_\sigma}{2}} \exp\left\{-\sum_{i=1}^{N_\sigma} \frac{d_I(t_i)^2}{2\sigma^2}\right\}. \tag{23}$$

Combining these four terms, the posterior probability for the amplitude ratio is given by

$$P(r|\sigma, D_\sigma, D, I) \propto \int dA dA d\theta d\omega_1 d\omega_2 d\alpha_1 d\alpha_2 \sigma^{-2N-2N_\sigma} \exp\left\{-\frac{2N\overline{d^2} + 2N_\sigma \overline{d_\sigma^2}}{2\sigma^2}\right\}$$

$$\times \exp\left\{\frac{2A[R_1\cos(\theta) - I_1\sin(\theta) + rR_2\cos(\theta) - rI_2\sin(\theta)]}{2\sigma^2}\right\} \tag{24}$$

$$\times \exp\left\{-\frac{A^2[C_{11} + 2rC_{12} + r^2C_{22}]}{2\sigma^2}\right\}$$

where

$$R_x \equiv R(\omega_x, \alpha_x) \qquad \text{with } x = 1 \text{ or } 2, \tag{25}$$

and

$$C_{jk} \equiv C(\omega_j - \omega_k, \alpha_j + \alpha_k) \qquad \text{with } j \text{ and } k = 1 \text{ or } 2 \tag{26}$$

and similarly for I_x and S_{jk}.

In obtaining the above, the identity $\sin^2(x) + \cos^2(x) = 1$, and the trigonometric relations for the sum of two angles were used. The mean-square of the data value, $\overline{d^2}$, is defined as

$$\overline{d^2} \equiv \frac{1}{2N} \sum_{i=1}^{N} d_R(t_i)^2 + d_I(t_i)^2. \tag{27}$$

The mean-square of noise value, $\overline{d_\sigma^2}$, is defined as

$$\overline{d_\sigma^2} \equiv \frac{1}{2N_\sigma} \sum_{i=1}^{N_\sigma} d_{\sigma R}(t_i)^2 + d_{\sigma I}(t_i)^2. \tag{28}$$

And the notation "$\cdot$" means the two functions of discrete times are to be multiplied and a sum over discrete times performed, for example:

$$d_R \cdot \cos(\omega t + \theta)e^{-\alpha t} \equiv \sum_{i=1}^{N} d_R(t_i)\cos(\omega t_i + \theta)e^{-\alpha t_i}. \tag{29}$$

The functions $R(\omega, \alpha)$ and $I(\omega, \alpha)$ are defined as

$$R(\omega, \alpha) \equiv d_R \cdot \cos(\omega t)e^{-\alpha t} - d_I \cdot \sin(\omega t)e^{-\alpha t} \tag{30}$$

and

$$I(\omega, \alpha) \equiv d_R \cdot \sin(\omega t)e^{-\alpha t} + d_I \cdot \cos(\omega t)e^{-\alpha t}. \tag{31}$$

When the data are uniformly sampled and ω is taken on a discrete grid, $\omega_i = 2\pi i/N$ ($i = 0, 1, \cdots, N - 1$), the functions $R(\omega_i, \alpha)$ and $I(\omega_i, \alpha)$ are the real and imaginary parts of the fast Fourier transform of the complex FID data when the data have been multiplied by a decaying exponential of decay-rate α. The function $C(\omega, \alpha)$ is defined as

$$C(\omega, \alpha) \equiv \sum_{i=1}^{N} \cos(\omega t_i) e^{-\alpha t_i}. \tag{32}$$

If uniform sampling is used, then $C(\omega, \alpha)$ may be expressed in closed form. Taking the sampling times to be $t_i = \{0, 1, \cdots, N - 1\}$, then the frequencies and decay-rate constants are measured in radians and the sum, appearing in $C(\omega, \alpha)$, may be done explicitly to obtain

$$C(\omega, \alpha) = \frac{1 - \cos(\omega)e^{-\alpha} - \cos(N\omega)e^{-N\alpha} + \cos[(N-1)\omega]e^{-(N+1)\alpha}}{1 - 2\cos(\omega)e^{-\alpha} + e^{-2\alpha}}. \tag{33}$$

If nonuniform sampling is being used, $C(\omega, \alpha)$, $R(\omega, \alpha)$, and $I(\omega, \alpha)$ must be computed from their definitions.

4. Removing Nuisance Parameters

There are seven nuisance parameters: two frequencies ω_1 and ω_2, two decay-rate constants, α_1 and α_2, one phase θ, amplitudes, A, and the variance of the noise, σ. The two frequencies and decay-rate constants appear in the posterior in a nonlinear way and the integrals over these parameters cannot be done explicitly; approximations for the integrals will be used. The remaining integrals may be done in closed form.

To evaluate the integral over the phase, one uses the relationship

$$a\cos(x) + b\sin(x) = \sqrt{a^2 + b^2}\cos(x + \chi) \quad \text{where} \quad \chi = \tan^{-1}(b/a). \tag{34}$$

This relationship transforms the θ integral into an integral representation of the Bessel function of a real argument I_0:

$$P(r|\sigma, D_\sigma, D, I) \propto \int A dA d\omega_1 d\omega_2 d\alpha_1 d\alpha_2 \sigma^{-2N-2N_o+2}$$

$$\times \exp\left\{-\frac{2N\overline{d^2} + 2N_o\overline{d_\sigma^2} + A^2X}{2\sigma^2}\right\} I_0\left(\frac{AY}{\sigma^2}\right) \tag{35}$$

where

$$X \equiv C_{11} + 2rC_{12} + r^2C_{22}, \tag{36}$$

and

$$Y \equiv \sqrt{[R_1 + rR_2]^2 + [I_1 + rI_2]^2}. \tag{37}$$

The amplitude integral may be evaluated by using the following integral relation

$$\int_0^\infty dx\, x e^{-ax^2} I_0(bx) = \frac{1}{2a}\exp\left\{-\frac{b^2}{4a}\right\}. \tag{38}$$

Evaluating this integral one obtains:

$$P(r|\sigma, D_\sigma, D, I) \propto \int d\omega_1 d\omega_2 d\alpha_1 d\alpha_2 \frac{\sigma^{-2N-2N_\sigma+2}}{X} \exp\left\{-\frac{2N\overline{d^2} + 2N_\sigma \overline{d_\sigma^2} - X^2/Y}{2\sigma^2}\right\} \quad (39)$$

as the posterior probability for the amplitude ratio.

The standard deviation, σ, is the only remaining parameters that may be removed in closed form. Before doing so, note that the frequencies appear in the integral in the form of $[R(\omega_1, \alpha_1) + rR(\omega_2, \alpha_2)]^2 [I(\omega_x, \alpha_x) + rI(\omega_2, \alpha_2)]^2$. If one considers this quantity as a function of ω_1, and holds the other parameters constant, then it is varying like a power spectrum. Power spectra often change by many orders of magnitude over a small frequency range and it is the *exponential* of this quantity that is being computed. Assuming even moderate signal-to-noise (frequency domain peak signal-to-noise RMS noise ratio of at least 2 or 3), this frequency integral is well approximated by a delta function. The same argument applies equally well to the integral over ω_2.

The integrals over the decay-rate constants are similar. The function $R(\omega_1, \alpha_1)$, $I(\omega_1, \alpha_1)$ are the the real and imaginary parts of the discrete Fourier transform of the complex FID data when an exponential filter is applied. When this exponential is decaying at the same rate as the first sinusoid, the filter is "matched" and small changes in the decay rate constant will make large changes in the height of the posterior probability. Similarly for $R(\omega_2, \alpha_2)$ and $I(\omega_2, \alpha_2)$. At these matched values, small changes in exponential decay rate constants will cause corresponding large changes in the height of the posterior, and again the integrals may be approximated by delta functions. Designating $\hat{\omega}_1$, $\hat{\alpha}_1$, $\hat{\omega}_2$, and $\hat{\alpha}_2$ as the values that maximize the joint posterior probability for the frequencies and decay-rate constants (the matched values), the posterior probability for the amplitude ratio is approximately given by

$$P(r|\sigma, D_\sigma, D, I) \approx \int d\sigma \frac{\sigma^{-2N-2N_\sigma+2}}{X} \exp\left\{-\frac{2N\overline{d^2} + 2N_\sigma \overline{d_\sigma^2} - X^2/Y}{2\sigma^2}\right\}\Bigg|_{\hat{\omega}_1 \hat{\omega}_2 \hat{\alpha}_1 \hat{\alpha}_2} \quad (40)$$

The last to remove the standard deviation of the noise, note that

$$P(r|D_\sigma, D, I) = \int d\sigma P(r, \sigma|D_\sigma, D, I) = \int P(\sigma|I)P(r|\sigma, D_\sigma, D, I). \quad (41)$$

Using a Jeffreys prior [9,10] for the standard deviation $(1/\sigma)$ and evaluating the integral over σ one obtains

$$P(r|D_\sigma, D, I) \propto \int \frac{d\omega_1 d\omega_2 d\alpha_1 d\alpha_2}{X} \left[1 - \frac{X^2}{Y(2N\overline{d^2} + 2N_\sigma \overline{d_\sigma^2})}\right]^{1-N-N_\sigma} \quad (42)$$

Substituting the definitions X and Y Eq. (36,37) one obtains

$$P(r|D_\sigma, D, I) \propto X^{-1}\left[1 - \frac{(R_1 + rR_2)^2 + (I_1 + rI_2)^2}{(C_{11} + 2rC_{12}r^2C_{22})(2N\overline{d^2} + 2N_\sigma \overline{d_\sigma^2})}\right]^{3-N-N_\sigma}\Bigg|_{\hat{\omega}_1 \hat{\omega}_2 \hat{\alpha}_1 \hat{\alpha}_2} \quad (43)$$

as the posterior probability for the amplitude ratio.

5. Example

To illustrate the use of the previous calculation, consider Fig. 1A. This is a plot of the real part of the fast Fourier transform for computer-generated FID containing two exponentially decaying sinusoids. The data for the real channel $(0°)$ were generated from

$$d_R(T_i) = 100 \cos(\omega_1 t_i)e^{-\alpha_1 t_i} + 200 \cos(\omega_2 t_i)e^{-\alpha_2 t_i} + e_i \qquad (44)$$

where e_i represents a random Gaussian noise component of standard deviation one. The imaginary data $(90°)$ were generated from the same equation, except the sinusoids are shifted by 90°. For data taken every millisecond for 2.048 seconds, there are $N = 2048$ points in the real and imaginary channel. The frequencies and decay-rate constants are given by

$$\omega_1 = 47.7 \quad \text{Hz and} \quad \alpha_1 = 1.6 \, \text{Hz} \qquad (45)$$

$$\omega_2 = 55.7 \quad \text{Hz and} \quad \alpha_2 = 16 \, \text{Hz.} \qquad (46)$$

The peak amplitude to RMS signal-to-noise ratio is large; nonetheless, because of overlap, traditional methods can not be estimate the amplitudes, so no estimate of the amplitude ratio is available. Using traditional methods, the amplitudes are estimated by integrating the areas under the peaks in the absorption spectrum; unless these peaks are well separated, the integral estimates the combined area.

To apply the previous calculation, one must first locate the maximum of the posterior probability for the frequencies and decay-rate constants. This is done using the procedures described in [3]. After locating these values, one can estimate the amplitudes ratio by computing the posterior probability for the amplitudes ratio using Eq. (43). The resulting probability-density functions are shown in Figs. 1B and C. Notice that both amplitude ratios have been well determined using Bayesian methods.

For the ratio of the amplitude of the sinusoid at frequency 47.7 Hertz to the sinusoid at 55.7 Hertz the ratio is estimated to be 2.003 ± 0.006 at one standard deviation, where by standard deviation it is meant the smallest area that encloses 63% of the total probability. The reciprocal ratio, i.e., the ratio of the amplitude of the sinusoid at 55.7 Hz divided by the amplitude of the sinusoid at 47.7 Hz is 0.499 ± 0.017 at one standard deviation. The true values are 2 and one half respectively.

6. Summary And Conclusions

Probability theory has been successfully applied to the problem of estimating the amplitude ratio in NMR FID data when the data containing two sinusoids of different amplitudes, frequencies, decay-rate constants and the same phase. The posterior probability for the ratio of the amplitudes, independent of the values of all the other parameters was computed. An example illustrating that probability theory can easily estimate amplitude ratios under conditions where traditional discrete Fourier transform methods fail was given.

ACKNOWLEDGMENTS.

This work was supported by a gift from the Monsanto Company. The encouragement of Professor J. J. H. Ackerman is greatly appreciated as are the editorial comments of Dr. C. Ray Smith and extensive conversations with Professor E. T. Jaynes.

Figure 1: Estimation of the Amplitude Ratio

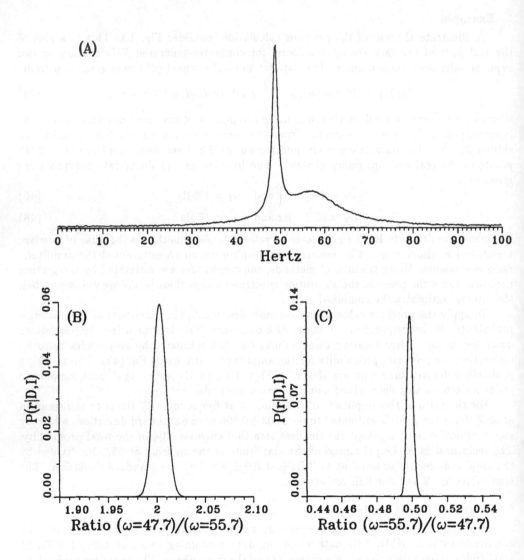

Fig. 1A is the absorption spectrum (the real part of the Fourier transform) of the computer simulated FID data. Traditional methods cannot estimate either the frequencies or the amplitudes of the sinusoids due to the overlap exhibited by the two NMR resonances. Panel B is the posterior probability for ratio of the amplitudes of the sinusoids at frequency 47.7 Hz to the sinusoid at 55.7 Hz. Panel C is the posterior probability for the reciprocal. The true ratios are 2 and one half respectively.

REFERENCES

Bretthorst, G. Larry, "Bayesian Analysis. I. Parameter Estimation Using Quadrature NMR Models," *J. Magn. Reson.*, **88**, pp. 533-551 (1990).

Bretthorst, G. Larry, "Bayesian Analysis. II. Model Selection," *J. Magn. Reson.*, **88**, pp. 552-570 (1990).

Bretthorst, G. Larry, "Bayesian Analysis. III. Applications to NMR Signal Detection, Model Selection and Parameter Estimation," *J. Magn. Reson.*, **88**, pp. 571-595 (1990).

Bretthorst, G. Larry, "Bayesian Analysis. IV. Noise and Computing Time Considerations," *J. Magn. Reson.*, June 15, (1991).

Bretthorst, G. Larry, "Bayesian Analysis. V. Amplitude Estimation, The One Frequency Model," submitted *J. Magn. Reson.*,.

Bretthorst, G. Larry, "Amplitude Estimation In Nuclear Magnetic Resonance Data," in the Proceedings of the Third Valencia Conference on Bayesian Statistics (J. Bernardo *ed*), in press.

Rev. T. Bayes, *Philos. Trans. R. Soc. London* **53**, 370 (1763); reprinted in *Biometrika* **45**, 293 (1958), and "Facsimiles of Two Papers by Bayes," with commentary by W. Edwards Deming Hafner, New York, 1963.

Bretthorst, G. Larry, *Bayesian Spectrum Analysis and Parameter Estimation,* in "Lecture Notes in Statistics" **48**, Springer-Verlag, New York, New York, 1988.

Jaynes, E. T., " Papers on Probability, Statistics and Statistical Physics" R. D. Rosenkrantz, *ed.*, D. Reidel, Dordrecht, The Netherlands, 1983.

Jeffreys, H., "Theory of Probability," Oxford University Press, London, 1939; Later editions, 1948, 1961.

REFERENCES

A BAYESIAN METHOD FOR THE DETECTION OF A PERIODIC SIGNAL OF UNKNOWN SHAPE AND PERIOD

P.C. Gregory
Department of Physics
University of British Columbia
6224 Agricultural Road
Vancouver, British Columbia V6T 1Z1

T.J. Loredo
Department of Astronomy
Space Sciences Building
Cornell University
Ithaca, New York 14853

ABSTRACT.   We present a new method for the detection and measurement of a periodic signal in a data set when we have no prior knowledge of the existence of such a signal or of its characteristics. It is applicable to data consisting of the locations or times of individual events. To address the detection problem, we use Bayes' theorem to compare a constant rate model for the signal to models with periodic structure. The periodic models describe the signal plus background rate as a stepwise distribution in m bins per period, for various values of m. The Bayesian posterior probability for a periodic model contains a term which quantifies Ockham's razor, penalizing successively more complicated periodic models for their greater complexity even though they are assigned equal prior probabilities. The calculation thus balances model simplicity with goodness-of-fit, allowing us to determine both whether there is evidence for a periodic signal, and the optimum number of bins for describing the structure in the data. Unlike the results of traditional "frequentist" calculations, the outcome of the Bayesian calculation does not depend on the number of periods examined, but only on the range examined. Once a signal is detected, we again use Bayes' theorem to estimate the frequency of the signal. The probability density for the frequency is inversely proportional to the multiplicity of the binned events and is thus maximized for the frequency leading to the binned event distribution with minimum combinatorial entropy. The method is capable of handling gaps in the data due to intermittent observing or dead time.

1. Introduction

A frequent problem that arises in science, engineering and economics is to determine if there is a periodic signal present in some data. In some cases it is known that a periodic signal of known shape and possibly known period is present but difficult to find because of additive noise. For an excellent Bayesian treatment of this problem in the case of additive Gaussian noise see Bretthorst (1988). In this paper we are concerned with the problem of detecting a periodic signal for which we have no prior knowledge of its period or shape.

79

C. R. Smith et al. (eds.), Maximum Entropy and Bayesian Methods, Seattle, 1991, 79–103.
© 1992 Kluwer Academic Publishers.

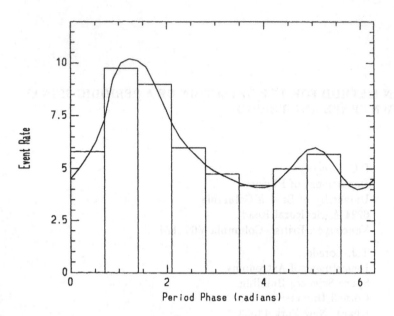

Figure 1. The periodic model, M_m, assumes the periodic signal plus background can be modeled by a stepwise distribution in m bins as illustrated here.

Usually the experimenter has some reason to suspect the presence of a periodic signal but in other cases the search for a periodicity represents an unexplored avenue with important consequences if successful. Given some data we seek the optimum approach to answering the question, "Is there evidence for a periodic signal, and if so, what is the best estimate of its period and shape?"

There are two special cases of general interest, depending on whether the appropriate sampling distribution is Gaussian or Poisson. This particular study is motivated by astronomical data in which the objective was the detection of X-ray pulsars. The data are the arrival times of individual X-ray photons, some or all of which are background events, and the appropriate sampling distribution is the Poisson distribution. Our method is directly applicable to data consisting of any event locations (*e.g.* spatial locations or redshifts), not just time series.

Our method detects a signal by comparing a constant model for the signal to members of a class of models with periodic structure, and if a periodic signal is detected, estimates the signal frequency and shape. The periodic models describe the signal plus background with a stepwise function with m phase bins resembling a histogram (Figure 1). Such a model is capable of approximating a lightcurve of essentially arbitrary shape.

Our calculations use Bayes' theorem both to estimate signal parameters and to compare rival signal models. The use of Bayes' theorem to estimate parameters in a model is probably familiar to many readers. Given some proposition, M, specifying a model with parameters denoted collectively by θ, and a proposition, D, specifying data relevant to the model, one calculates the posterior distribution for the parameters, $p(\theta \mid D, M)$, according to,

$$p(\theta \mid D, M) = p(\theta \mid M) \frac{p(D \mid \theta, M)}{p(D \mid M)}. \tag{1}$$

In words, the posterior distribution is found by multiplying the prior distribution, $p(\theta \mid M)$, by the likelihood function, $p(D \mid \theta, M)$, and dividing by the global likelihood, $p(D \mid M)$. The global likelihood is independent of θ and merely plays the role of a normalization constant whose value is given by integrating the product of the prior and the likelihood:

$$p(D \mid M) = \int d\theta \, p(\theta \mid M) \, p(D \mid \theta, M). \tag{2}$$

Readers unfamiliar with Bayesian parameter estimation will find useful reviews, derivations, and references in Bretthorst (1990) and Loredo (1990).

To compare rival models, we again use Bayes' theorem. This use of Bayes' theorem is probably less familiar to most readers, though it is exactly analogous to use of Bayes' theorem for parameter estimation. We begin by specifying a set of competing models (this corresponds to specifying the parameter space of a single model in parameter estimation). We use the symbol M_i to denote a proposition asserting that model i describes the data, and the symbol I to denote a proposition asserting that one of the models being considered describes the data (I = "M_1 or M_2 or ..."). Then we use Bayes' theorem to calculate the probability for model M_i,

$$p(M_i \mid D, I) = p(M_i \mid I) \frac{p(D \mid M_i, I)}{p(D \mid I)}. \tag{3}$$

This is very much like equation (1), with M_i now playing the role of the parameter, and I now playing the role of the model. The term $p(M_i \mid I)$ is the prior probability for model M_i. The proposition (M_i, I) ("M_i and I") is true if and only if model M_i is true, that is, it is equivalent to the proposition M_i itself. Thus $p(D \mid M_i, I) = p(D \mid M_i)$, the global likelihood for model M_i, which is calculate according to equation (2). The global likelihood for a model, which plays the uninteresting role of a normalization constant in parameter estimation, plays a key role in model comparison. Just as the posterior distribution for a model's parameters is proportional to the product of a prior and a likelihood, the posterior probability for a model as a whole is proportional to the product of a prior for the model and the model's global likelihood. The global likelihood is called the *evidence* for the model in some works (MacKay 1992).

It is sometimes more convenient to work with ratios of model probabilities, particularly when there is a special "default" model (the constant, nonperiodic model in our case). Ratios of probabilities are called odds ratios; the odds ratio in favor of model M_i over model M_1 is,

$$
\begin{aligned}
O_{m1} &= \frac{p(M_m \mid D, I)}{p(M_1 \mid D, I)} \\
&= \frac{p(M_m \mid I)}{p(M_1 \mid I)} \frac{p(D \mid M_m, I)}{p(D \mid M_1, I)} \\
&\equiv \frac{p(M_m \mid I)}{p(M_1 \mid I)} B_{m1},
\end{aligned}
\tag{4}
$$

where the first factor is the prior odds ratio, and the second factor is called the *Bayes factor*. The Bayes factor is simply the ratio of the global likelihoods of the models. Note that the normalization constant in equation (3), $p(D \mid I)$, drops out of the odds ratio. We can go back and forth between odds ratios and probabilities by inverting equation (4):

$$p(M_i \mid I) = \frac{O_{i1}}{\sum_{j=1}^{N_{\text{mod}}} O_{j1}}, \tag{5}$$

where N_{mod} is the number of models being considered, and of course $O_{11} = 1$.

If two models have equal prior probabilities, $O_{m1} = B_{m1}$, and the models are compared simply by comparing their global likelihoods. We shall see in this work that the calculation of global likelihoods implements an automatic and objective posterior "Ockham's Razor", leading one to prefer simpler models unless the data provide substantial evidence in favor of a more complicated alternative, even when the rival models are assigned *equal* prior probabilities. Interested readers can learn more about Bayesian model comparison by consulting works in earlier *Maximum Entropy and Bayesian Methods* proceedings (Gull 1988; Loredo 1990) or the fine review of MacKay in these proceedings (MacKay 1992).

In this work, we are careful to distinguish between the problem of detecting a signal and the problem of estimating the value of parameters describing a detected signal. To detect a signal, we use Bayesian model comparison calculations to compare a nonperiodic class of models (the class of constant models) to a periodic class of models (stepwise models with varying numbers of bins). If a signal is detected, we then use Bayesian parameter estimation calculations to estimate the frequency and shape of the periodic signal. In our model comparison calculations, a Bayesian "Ockham's Razor" works on two levels. First, it leads us to prefer nonperiodic models to periodic ones, even though we will assign the nonperiodic and periodic model classes equal prior probabilities. Second, if the periodic class is favored, it will lead us to prefer periodic models with smaller numbers of bins, even though we will assign models with varying bin numbers equal prior probabilities.

The remainder of this paper is organized as follows. In §2 we derive a general likelihood function for arrival time data. Then in §3 we specialize this likelihood function to the constant and stepwise models used in this work, and we assign the priors needed to use Bayes' theorem. We show that the likelihood function for the stepwise model leads one to count data falling in various phase bins in a manner analogous to the "epoch folding" method commonly used in astrophysics (Leahy *et al.* 1983).

In §4 we perform the calculations needed to compare periodic and constant models using Bayes' theorem. In §5 we describe how to use Bayes' theorem to estimate the frequency of a signal if the model comparison calculations show a signal is present.

We illustrate our method with an application to a simulated data set in §6. In §7 we elaborate on the relationship between our method and two other better-known statistics: configurational entropy and the χ^2 statistic used in the epoch folding method. A concluding section summarizes the results. A more extensive description of the method is available in Gregory and Loredo (1992).

2. Likelihood Function for Arrival Time Data

In this section we derive a general likelihood function for arrival time data with an arbitrary time-dependent rate, $r(t)$. We then specialize to the models of interest in this work in the following section.

The data are the arrival times for each of N events, $D = \{t_i\}$, $i = 1$ to N, over some observing interval of duration T. The probability for D given some specified rate function—the likelihood function—can be calculated as follows.

We divide the observation period into small time intervals, Δt, each containing either no event or one event. From the Poisson distribution, the probability of seeing n events in an interval Δt about time t is,

$$p_n(t) = \frac{[r(t)\Delta t]^n e^{-r(t)\Delta t}}{n!}. \tag{6}$$

We have assumed that the rate does not vary substantially within Δt, so the average rate in the interval is approximately equal to the rate at any time within the interval. If N and Q are the number of time intervals in which one event and no event are detected, respectively, then the likelihood function is given by

$$p(D \mid r, I) = \prod_{i=1}^{N} p_1(t_i) \prod_{k=1}^{Q} p_0(t_k). \tag{7}$$

From equation (6), with $n = 0$ and 1,

$$p_0(t) = e^{-r(t)\Delta t}, \tag{8}$$

and

$$p_1(t) = r(t)\Delta t e^{-r(t)\Delta t}, \tag{9}$$

so the likelihood function takes the form,

$$
\begin{aligned}
p(D \mid r, I) &= \Delta t^N \left[\prod_{i=1}^{N} r(t_i) \right] \exp\left[-\sum_{k=1}^{N+Q} r(t_k)\Delta t \right] \\
&= \Delta t^N \left[\prod_{i=1}^{N} r(t_i) \right] \exp\left[-\int_T dt\, r(t) \right].
\end{aligned} \tag{10}
$$

In the last equation we have identified the sum of the rates in all intervals with the integral of the rate over the entire observing interval, with the range of integration $T = (N+Q)\Delta t$.

Equation (10) is the general likelihood function we use throughout this work. The intervals, Δt, could correspond to the precision of the clock recording the arrival times. But when used in Bayes' theorem, the Δt^N factor in the likelihood will cancel with the same factor in the global likelihood, so inferences will not depend on the size of Δt, and are well-behaved even in the limit in which the time intervals become infinitesimal.

Note that in equation (10) the time T is the total duration of the intervals in which it is known that either no event or one event was detected. These intervals need not be contiguous; there can be gaps during which there are no data.

3. Constant and Periodic Models

In this section we specialize equation (10) to the models considered in this paper. In addition, we assign prior probabilities for the parameters of the models, and for the models themselves, to prepare us to use Bayes' theorem to perform model comparison and parameter estimation in later sections.

CONSTANT MODEL

The simplest model for the data is the constant model, which has only one parameter, the constant event rate, A. We denote the information specifying this one-parameter model by the symbol M_1. Setting $r(t_i) = A$ in equation (10), the likelihood function for this model is,

$$p(D \mid A, M_1) = \Delta t^N A^N e^{-AT}. \tag{11}$$

We will assume that the information M_1 leads to a prior density for A that is constant from $A = 0$ to some upper limit, A_{max}. Thus the normalized prior for A is,

$$p(A \mid M_1) = \frac{1}{A_{max}}. \tag{12}$$

Our results are insensitive to the form of this prior, and are well-behaved in the limit where $A_{max} \to \infty$.

PERIODIC STEPWISE LIKELIHOOD

As noted in the introduction, our model for a periodic signal is a stepwise function with a constant rate in each of m bins per period ($m \geq 2$). This is not a single model, but a class of models, one for each choice of m. We denote the information specifying each such model by the symbol M_m. Model M_m has $(m + 2)$ parameters: an angular frequency ω (or alternatively a period, P, with $\omega = 2\pi/P$); a phase, ϕ, specifying the location of the bin boundaries; and m values, r_j, specifying the rate in each phase bin, with $j = 1$ to m. The value of the subscript j corresponding to any particular time t is given by,

$$j(t) = \text{int}[1 + m\{(\omega t + \phi) \bmod 2\pi\}/2\pi]. \tag{13}$$

We sometimes denote the full set of m values of r_j by the symbol $\vec{r}$.

To facilitate comparison with the constant model, it is convenient to re-express the r_j parameters. We write the rate as the time-averaged rate, A, times a normalized stepwise function that describes the shape of the periodic lightcurve. The average rate is,

$$A = \frac{1}{m} \sum_{j=1}^{m} r_j. \tag{14}$$

The lightcurve shape is completely described by the fraction of the total rate per period that is in each phase bin. These fractions are,

$$f_j = \frac{r_j}{\sum_{k=1}^{m} r_k}$$
$$= \frac{r_j}{mA}. \tag{15}$$

Equations (14) and (15) let us write r_j in terms of A and f_j:

$$r_j = mAf_j. \tag{16}$$

In this way the m values of r_j are replaced by A and the m values of f_j. There are still only m degrees of freedom associated with the rate because by definition the f_j must satisfy the constraint,

$$\sum_{j=1}^{m} f_j = 1, \tag{17}$$

so only $(m-1)$ of them are free. We sometimes denote the full set of m values of f_j by the symbol $\vec{f}$.

With this form for the r_j parameters, the likelihood function for model M_m can be calculated using equation (10), giving

$$P(D \mid \omega, \phi, A, \vec{f}, M_m) = \Delta t^N \left[\prod_{i=1}^{N} m A f_{j(t_i)} \right] \exp \left[-\sum_{k=1}^{N+Q} m A f_{j(t_k)} \Delta t \right]$$

$$= \Delta t^N (mA)^N \left[\prod_{j=1}^{m} f_j^{n_j} \right] \exp \left[-mA \sum_{j=1}^{m} f_j \tau_j(\omega, \phi) \right], \tag{18}$$

where $j(t_k)$ is given by equation (13), n_j is the number of events that fall into phase bin j, and $\tau_j(\omega, \phi)$ is the total integration time for bin j. We have implicitly assumed that the intervals Δt are small compared to the width, P/m, of the r_j bins, so that each event has an unambiguous bin assignment for given ω and ϕ.

The $\tau_j(\omega, \phi)$ depend on ω and ϕ (and m), thus the exponential term depends on all of the model parameters. But in many cases this term will be essentially independent of all parameters except A, as we now show.

Unless the "off" times are concentrated in particular bins, we expect the integration time per bin to be approximately T/m. Thus we write the sum in the exponential as follows;

$$mA \sum_{j=1}^{m} f_j \tau_j = AT \sum_{j=1}^{m} f_j \frac{\tau_j}{T/m}$$

$$= AT \sum_{j=1}^{m} f_j s_j, \tag{19}$$

where we have defined the bin time factors $s_j(\omega, \phi)$ by

$$s_j(\omega, \phi) = \frac{\tau_j(\omega, \phi)}{T/m}. \tag{20}$$

Like the τ_j, the s_j are not new parameters; they are numbers that are determined by the data and the model parameters ω, ϕ, and m.

If the observing interval, T, is a contiguous interval containing an integral number of periods, then $\tau_j = T/m$, so $s_j = 1$ for all j. More generally, the observing interval will not be an integral number of periods, and may have gaps, so the s_j will differ from unity. But as long as the number of periods in the observing interval is large, and as long as the gaps are not somehow concentrated in certain bins, the s_j will be very close to unity, and to a good approximation the sum in equation (19) will be equal to $AT \sum_j f_j = AT$. Then the likelihood function is well approximated by,

$$P(D \mid \omega, \phi, A, \vec{f}, M_m) = \Delta t^N (mA)^N e^{-AT} \left[\prod_{j=1}^{m} f_j^{n_j} \right]. \tag{21}$$

We use this likelihood function in the remainder of this work. If the observing duration does not contain a large number of periods, or if there is significant dead time, the s_j may depart significantly from unity, complicating the analysis. For a discussion of this case see Gregory and Loredo (1992).

Note that when $m = 1$, so that the single value of the shape parameter $f_1 = 1$, equation (21) takes the same form as equation (11). Thus model M_1 is the $m = 1$ case of model M_m, as suggested by our notation.

PRIORS FOR PERIODIC MODEL PARAMETERS

We will assume that we do not have any prior information linking the frequency, phase, and lightcurve shape, so that the priors for ω, ϕ, and $\vec{r}$ are all independent of one another. We will further assume that there is no prior information linking the shape of the lightcurve to its average value. When reparametrized in terms of A and $\vec{f}$, the joint prior will thus be of the form,

$$p(\omega, \phi, A, \vec{f} \mid M_m) = p(\omega \mid M_m)\, p(\phi \mid M_m)\, p(A \mid M_m)\, p(\vec{f} \mid M_m). \tag{22}$$

We will assign priors assuming that little is known about the model parameters *a priori*, aside from their physical significance.

The prior density for ϕ we take to be uniform over the interval $[0, 2\pi]$,

$$p(\phi \mid M_m) = \frac{1}{2\pi}. \tag{23}$$

Formally, this prior can be derived from an invariance argument requiring investigators with different origins of time to reach the same conclusions. A similar invariance argument, this time requiring observers with different units of time to reach the same conclusions (Bretthorst 1988), leads to a prior density for ω of the form,

$$p(\omega \mid M_m) = \frac{1}{\omega \ln \frac{\omega_{hi}}{\omega_{lo}}}, \tag{24}$$

where $[\omega_{lo}, \omega_{hi}]$ is a prior range for ω (which we might set from Nyquist arguments) and the $\ln \frac{\omega_{hi}}{\omega_{lo}}$ factor is a normalization constant ensuring that the integral of $p(\omega \mid M_m)$ over the prior range is equal to 1. This density is uniform in the logarithm of ω, and is form-invariant with respect to reparameterization in terms of period, P.

The average rate, A, is not the same quantity as the constant rate that appears in model M_1, so technically we should use a different symbol for it. If we knew that the rate was constant, so that $f_j = 1/m$ for all j, we expect our inferences about A under model M_m to be identical to those under model M_1. Combined with our assumption of independent priors for the shape and average rate, this requires that we use the same prior for the average rate under M_m and the constant rate under M_1. Accordingly, our prior is,

$$p(A \mid M_m) = \frac{1}{A_{\max}}. \tag{25}$$

Since the A parameters for all of the models enter their respective likelihood functions in the same way, and have the same priors, no confusion is caused by using the same symbol for them.

Finally, we must assign a prior joint density for $\vec{f}$. We simply assume that all values between 0 and 1 are equally probable subject to the constraint that $\sum f_j = 1$. Thus we write,

$$p(\vec{f} \mid I) = K_m \delta \left(1 - \sum_{j=1}^{m} f_j \right), \tag{26}$$

where $\delta(x)$ denotes the Dirac Delta-function, and K_m is a normalization constant that depends on the value of m.

We can evaluate K_m from the requirement that $\int d\vec{f}\, p(\vec{f} \mid I) = 1$. The required integral is a special case of the generalized Beta integral that we will require later:

$$\int_0^\infty dx_1 \ldots \int_0^\infty dx_m\, x_1^{k_1-1} \ldots x_m^{k_m-1} \delta \left(a - \sum_{j=1}^{m} x_j \right) = \frac{\Gamma(k_1) \ldots \Gamma(k_m)}{\Gamma(k)}\, a^k, \tag{27}$$

where $k = \sum_j k_j$, and $\Gamma(x)$ is the Gamma function, with $\Gamma(n) = (n-1)!$ when n is a positive integer. To evaluate K_m, we consider the case with $a = 1$ and all $k_j = 1$, so $k = m$. Thus $\int d\vec{f}\, p(\vec{f} \mid I) = K_m/(m-1)!$, so

$$K_m = (m-1)!. \tag{28}$$

Our specification of model M_m is now complete.

PRIORS FOR MODEL COMPARISON

Finally, we need to assign a prior probability to each model as a whole in order to perform model comparison calculations. As noted in the introduction, we will consider the hypotheses of the presence and absence of a periodic signal to be equally probable a priori. Thus we assign equal prior probabilities of 1/2 to the class of nonperiodic models and to the class of periodic models. Since the nonperiodic class contains only the constant ($m = 1$) model, we have

$$p(M_1 \mid I) = \frac{1}{2}, \tag{29}$$

where we use the symbol I to denote the information specifying the classes of models we are comparing, as in the Introduction. The periodic class will consist of some finite number of stepwise models with m varying from $m = 2$ to $m = m_{\max}$. We consider each member of this class to be equally probable a priori, so that the probability of 1/2 assigned to the periodic class is spread equally among the $\nu = m_{\max} - 1$ members of this class. Thus,

$$p(M_m \mid I) = \frac{1}{2\nu}. \tag{30}$$

Alternatively, we could view our stepwise models as a single model, and m as a discrete parameter in this model; the $1/\nu$ factor in equation (30) then plays the role of a flat prior distribution for the parameter m.

We now have all the probabilities we need. We note that either model class could be expanded to more comprehensively cover the possible forms of periodic or nonperiodic signals. For example, we could enlarge the periodic class to contain sinusoidal or other simple functional models (Loredo 1992b). Or we could enlarge the nonperiodic class to include varying but nonperiodic models, such as the stepwise nonperiodic model discussed in Appendix C of Gregory and Loredo (1992), or a simple polynomial variation with time. In either case, we would further spread the prior probability of $1/2$ over the additional members of the model class. Alternatively, if we knew the precise shape of a possible signal *a priori*, we could shrink the periodic model class, assigning its full prior probability of $1/2$ to a single model.

4. Odds Ratios for Signal Detection

In this section we use the likelihoods and priors from the previous section to test the hypothesis that the signal is periodic. We do this by comparing members of the class of stepwise periodic models to the constant model that comprises the nonperiodic class using the model comparison methods briefly described in the Introduction. We calculate the global likelihoods for all models, and use them to find the odds ratios, O_{m1}, in favor of each periodic model over the constant model. The probability for a model can be calculated from the odds ratios using equation (5). In particular, the probability for the nonperiodic (constant) model is,

$$p(M_1 \mid D, I) = \frac{1}{1 + \sum_{m=2}^{m_{max}} O_{m1}}, \tag{31}$$

and the probability that the signal is periodic is just the sum of the probabilities of the ν periodic models,

$$p(m > 1 \mid D, I) = \frac{\sum_{m=2}^{m_{max}} O_{m1}}{1 + \sum_{m=2}^{m_{max}} O_{m1}}, \tag{32}$$

This is simply $[1 - p(M_1 \mid D, I)]$. The ratio of equation (32) to equation (31) is the odds ratio, O_{per}, in favor of the hypothesis that the signal is periodic,

$$O_{per} = \sum_{m=2}^{m_{max}} O_{m1}. \tag{33}$$

When O_{per} is greater than unity ($p(m > 1 \mid D, I) > 1/2$), there is evidence for a periodic signal, the magnitude of O indicating the strength of this evidence.

We discuss three cases: the case when the period and phase of the signal is known, that when only the period is known, and that when neither the period nor the phase is known. In all cases the shape of the lightcurve will be considered unknown.

Period and Phase Known

Our first case is of little practical interest: if we do not know the shape of the lightcurve, it is not meaningful to say we know the phase. However, if the phase is fixed, the needed calculations can be done analytically, and the result can be readily interpreted, facilitating our understanding of the results for more complicated cases.

As explained in the Introduction, we need the global likelihoods of the models to perform model comparison. These are calculated by integrating the product of the prior and the likelihood for each model as illustrated in equation (2). From equations (11) and (12), the global likelihood for the constant model is,

$$
\begin{aligned}
p(D \mid M_1) &= \int_0^{A_{max}} dA\, p(A \mid M_1)\, p(D \mid M_1, A, I) \\
&= \frac{\Delta t^N}{A_{max}} \int_0^{A_{max}} dA\, A^N e^{-AT} \\
&= \frac{\Delta t^N \gamma(N+1, A_{max} T)}{A_{max} T^{N+1}}.
\end{aligned}
\tag{34}
$$

Here $\gamma(n, x)$ denotes the incomplete Gamma function defined by,

$$
\gamma(n, x) = \int_0^x y^{n-1} e^{-y} dy,
\tag{35}
$$

where the usual Gamma function $\Gamma(n) = \gamma(n, \infty)$.

With ω and ϕ known, the global likelihood for a periodic model is similarly calculated by integrating the product of equations (21), (25), and (26) over A and $\vec{f}$. The integral over A is the same as that just performed for the constant model, and the constrained integral over $\vec{f}$ can be performed using the generalized Beta integral of equation (27). The result is,

$$
\begin{aligned}
p(D \mid \omega, \phi, M_m) &= \int d\vec{f} \int_0^{A_{max}} dA\, p(A \mid M_m)\, p(\vec{f} \mid M_m)\, p(D \mid \omega, \phi, A, \vec{f}, M_m) \\
&= \frac{\Delta t^N m^N (m-1)!}{A_{max}} \int_0^{A_{max}} dA\, A^N e^{-AT} \int d\vec{f} \prod_{j=1}^m f_j^{n_j} \delta\left(1 - \sum_{j=1}^m f_j\right) \\
&= \frac{\Delta t^N (m-1)!\, N!\, \gamma(N+1, A_{max} T)}{A_{max} T^{N+1} (N+m-1)!} \frac{m^N}{W_m(\omega, \phi)},
\end{aligned}
\tag{36}
$$

where $W_m(\omega, \phi)$ is the multiplicity of the binned distribution of events, the number of ways the particular set of n_j values can be made with N events,

$$
W_m(\omega, \phi) = \frac{N!}{n_1!\, n_2! \, \ldots \, n_m!}.
\tag{37}
$$

The multiplicity is a function of ω and ϕ because the n_j are. Note that m^N is the total number of possible arrangements of N events in m bins, so equation (36) is inversely proportional to the ratio of the number of ways the observed n_j can be made to the total number of ways N events can be placed in m bins.

The probability of a model is proportional to its global likelihood. Thus from equation (36) we find the intuitively appealing result that the probability of a periodic model is inversely proportional to the multiplicity of its resulting binned distribution. Crudely, if the number of ways the binned distribution could have arisen "by chance" is large, the probability that the distribution is due to a genuinely periodic signal is small.

The many factors common to the global likelihoods in equations (34) and (36) cancel when we calculate the odds ratio comparing one of the periodic models to the constant model. Using the prior probabilities for the models given by equations (29) and (30), the odds ratio in favor of an m-bin periodic model over the constant model, conditional on ω and ϕ, is,

$$O_{m1}(\omega, \phi) = \frac{1}{\nu} \binom{N + m - 1}{N}^{-1} \frac{m^N}{W_m(\omega, \phi)}. \tag{38}$$

Note that $A_{\max}$ has canceled out of the odds ratio, so that the result of comparing the models is independent of the prior range for A. This is generally the case in model comparison calculations when a parameter is common to all models being considered, and its value is independent of the values of the other parameters: only the ranges associated with the additional parameters affect the outcome.

Values of $O_{m1}(\omega, \phi) > 1$ indicate that model M_m is more probable than the constant rate model for the frequency and phase considered. From the conditional odds ratios we can easily calculate the odds ratio, $O_{\mathrm{per}}(\omega, \phi)$, in favor of the periodic class of models over the constant model,

$$O_{\mathrm{per}}(\omega, \phi) = \sum_{m=2}^{m_{\max}} O_{m1}(\omega, \phi). \tag{39}$$

This is the conditional counterpart to equation (33). Note that $O_{\mathrm{per}}(\omega, \phi)$ could exceed unity even if no single periodic model is more probable than the constant model. This can arise in practice when there is a weak periodic signal present whose shape is not well-modelled by a single stepwise curve, so that probability is spread over several stepwise models.

OCKHAM FACTORS

The odds ratios $O_{m1}(\omega, \phi)$ contain factors that penalize models with larger numbers of bins. We can see this by writing the global likelihood for each model as the product of its maximum likelihood and a remaining factor, the Ockham factor. In traditional "frequentist" statistics, ratios of maximum likelihoods are commonly used to compare models. However, more complicated models almost always have higher likelihoods than simpler competitors, so more complicated models are only accepted if the maximum likelihood ratio in their favor exceeds some subjectively specified critical amount, expressing a subjective prior preference for simplicity. But Bayesian methods compare *global* likelihoods, not maximum likelihoods, and tend to favor simpler models even when simple and complicated models are assigned equal priors. By factoring the global likelihood into the product of the maximum likelihood and an additional *Ockham factor*, we can better understand the nature of the Bayesian posterior preference for simplicity.

We thus implicitly define the Ockham factor, Ω_θ, associated with the parameters, θ, of a model, M, by writing $p(D \mid M) \equiv \mathcal{L}_{\max}\Omega_\theta$, where $\mathcal{L}_{\max}$ is the maximum value of the likelihood function, $\mathcal{L}(\theta) \equiv p(D \mid \theta, M)$. Recalling equation (2) for the global likelihood, this implies

$$\Omega_\theta = \frac{1}{\mathcal{L}_{\max}} \int d\theta \, p(\theta \mid M) \, \mathcal{L}(\theta). \tag{40}$$

Assuming, as is generally the case, that the prior varies slowly compared to the likelihood, the integral in this equation is approximately equal to $p(\hat{\theta} \mid M) \int d\theta \mathcal{L}(\theta)$, where $\hat{\theta}$ is the maximum likelihood value of θ. If we write the integral of the likelihood function as the maximum likelihood value times a characteristic width of the likelihood, $\delta\theta$, we find that,

$$\Omega_\theta \approx p(\hat{\theta} \mid M) \, \delta\theta. \tag{41}$$

When the prior is constant, with width $\Delta\theta$, we find that $\Omega_\theta \approx \delta\theta/\Delta\theta$, the ratio of the posterior range of the parameter to its prior range. This number will be less than one, and in this manner the Ockham factor penalizes the maximum likelihood, the penalty generally growing with the number of parameters. We now proceed to find the Ockham factors associated with some of the parameters in our models.

The constant model has an Ockham factor associated with its single parameter, the rate, A. To find it, we differentiate equation (11) with respect to A, and choose A so the derivative vanishes, leading to a maximum likelihood value for A of N/T, as we might guess. Thus the maximum value of the likelihood itself is,

$$\mathcal{L}_{\text{max},1} = \Delta t^N N^N T^{-N} e^{-N}. \tag{42}$$

From this result and equation (34), the global likelihood for M_1 can be written,

$$p(D \mid M_1) = \mathcal{L}_{\text{max},1} \frac{e^N N^{-N} \gamma(N+1, A_{\text{max}}T)}{A_{\text{max}}T}$$

$$= \mathcal{L}_{\text{max},1} \, \Omega_A, \tag{43}$$

where we have identified the Ockham factor associated with the parameter A,

$$\Omega_A = \frac{e^N N^{-N} \gamma(N+1, A_{\text{max}}T)}{A_{\text{max}}T}. \tag{44}$$

When $A_{\text{max}}T \gg N$ (the prior upper limit for A is larger than the observed rate), as will usually be the case, the incomplete Gamma function is equal to $N!$ to a very good approximation. Using Stirling's approximation, $N! \approx \sqrt{2\pi N} \, N^N e^{-N}$, we find that,

$$\Omega_A \approx \sqrt{2\pi} \, \frac{N^{1/2}/T}{A_{\text{max}}}. \tag{45}$$

It is easy to show that the standard deviation of the posterior distribution for A is $N^{1/2}/T$ (the familiar "root-N" result; see Loredo 1992a); thus the Ockham factor associated with A is approximately $\sqrt{2\pi}$ times the posterior standard deviation divided by the prior range for A. That is, it has the form of the ratio of a posterior range to a prior range, as noted above following equation (41).

The Ockham factors associated with the A and $\vec{f}$ parameters of a periodic model with known ω and ϕ can be found in a similar manner. As with the constant model, the maximum likelihood value of A is N/T. We show below that maximization of equation (21) with respect to the f_j leads to maximum likelihood values of $f_j = n_j/N$, as one might expect. Thus the maximum value of the likelihood for a known ω and ϕ is,

$$\mathcal{L}_{\text{max},m}(\omega, \phi) = \left[\Delta t^N N^N T^{-N} e^{-N}\right] \left[m^N N^{-N} \prod_{j=1}^{m} n_j^{n_j}\right]. \tag{46}$$

We have grouped terms to facilitate comparison with the calculations for the constant model. We can now write the global likelihood in equation (36) as,

$$p(D \mid \omega, \phi, M_m) = \mathcal{L}_{\mathrm{max},m}(\omega, \phi) \left[\frac{e^N N^{-N} \gamma(N+1, A_{\mathrm{max}}T)}{A_{\mathrm{max}}T} \right] \times$$

$$\binom{N+m-1}{N}^{-1} \frac{N^N \prod_j n_j!}{N! \prod_j n_j^{n_j}}, \tag{47}$$

where we have identified a combination of factorials that are equal to a binomial coefficient,

$$\binom{N+m-1}{N} = \frac{(N+m-1)!}{N! \, (m-1)!}. \tag{48}$$

Comparing equation (47) with equation (44), we see that we can write the global likelihood for M_m as,

$$p(D \mid \omega, \phi, M_m) = \mathcal{L}_{\mathrm{max},m}(\omega, \phi) \, \Omega_A \, \Omega_m, \tag{49}$$

where Ω_A is given by equation (44), and the Ockham factor associated with the m values of f_j is,

$$\Omega_m = \binom{N+m-1}{N}^{-1} \frac{N^N \prod_j n_j!}{N! \prod_j n_j^{n_j}}. \tag{50}$$

Using Stirling's approximation for the various factorials, one can show that

$$\Omega_m \approx \sqrt{\frac{N}{2\pi}} \left(1 + \frac{m-1}{N} \right)^{-(m-\frac{1}{2})} \left[(m-1)! \prod_{j=1}^{m} \frac{\sqrt{2\pi n_j}}{N} \right]. \tag{51}$$

We show below that the posterior standard deviation for f_j is approximately $\sqrt{n_j}/N$, thus the term in brackets is the product of the prior and $\sqrt{2\pi}$ times of the posterior standard deviations, as in the approximation given by equation (41) for a one parameter model. The remaining terms arise to give the exact, multivariate result for this model.

Using equations (43) and (49), we can write the odds ratio in equation (38) as the product of a prior odds ratio $(1/\nu)$, a maximum likelihood ratio, and and Ockham factor:

$$O_{m1}(\omega, \phi) = \frac{1}{\nu} R_{m1} \, \Omega_m, \tag{52}$$

where the ratio of the maximum values of the likelihoods of the models is,

$$R_{m1} \equiv \frac{\mathcal{L}_{\mathrm{max},m}(\omega, \phi)}{\mathcal{L}_{\mathrm{max},1}}$$

$$= \left(\frac{m}{N} \right)^N \prod_{j=1}^{m} n_j^{n_j}. \tag{53}$$

Note that Ω_A has cancelled out of the odds ratio. Our remark above regarding the cancellation of A_{max} in the odds ratio applies to Ockham factors as well: when a parameter

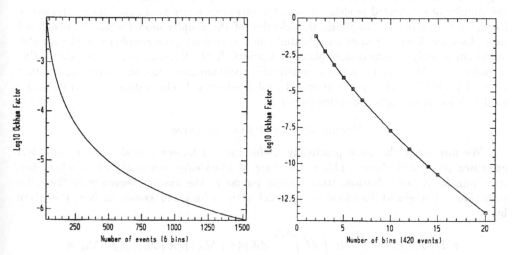

Figure 2. Logarithm of the Ockham factor, Ω_m, for N uniformly distributed events. (a) Ω_m versus N for $m = 6$ bins. (b) Ω_m versus m, for $N = 420$ events. Only points with N/m an integer are plotted.

is common to all models being compared, and its inferred value is not correlated with the values of the other parameters, only the Ockham factors associated with the additional parameters affect the outcome of model comparison calculations.

To illustrate the effect of the Ockham factor, Ω_m, assume that we only had one model, with fixed m, in the periodic model class (so $\nu = 1$). Then the prior odds for this model over the constant model would be unity. However, the Ockham factor in equation (52) implements an "Ockham's razor," strongly favoring the simpler constant rate model, M_1, unless the data justify the model M_m with its larger number of free parameters. The effect of the Ockham factor can be appreciated most readily for the special case where the total number of events, N, is an integer multiple of the number of bins, m, and the events fall uniformly in the bins, so that $n_j = N/m$, an integer. In this case the likelihood ratio $R_{m1} = 1$, as we expect, so the odds ratio is equal to the Ockham factor, which takes the value,

$$\Omega_m = \binom{N + m - 1}{N}^{-1} \frac{m^N}{N!} \left(\frac{N}{m}!\right)^m . \qquad (54)$$

Figure 2a shows a plot of $\log_{10} \Omega_m$ versus the total number of events, N, while the number of bins is held constant at $m = 6$. For $N \approx 500$ events, $\Omega_m \approx 10^{-5}$. Figure 2b shows $\log_{10} \Omega_m$ versus m for N fixed at 420 events. In this case for 5 bins ($m = 5$), $\Omega_5 = 9.3 \times 10^{-5}$, and for 10 bins, $\Omega_{10} = 2.0 \times 10^{-8}$. In all cases, the Ockham factor strongly penalizes the complicated models, and the odds ratio favors the constant model, even though the likelihood ratio itself does not favor one model over another for this hypothetical flat data set.

These results may be surprising to those familiar with more traditional frequentist approaches to model comparison based on best-fit likelihood ratios or their logarithms, (e.g., differences in χ^2, such as are used in the F-test). Likelihood ratios can never favor the simpler of two nested models; at best, the ratio can be unity. In such tests "Ockham's Razor" must be invoked to justify the selection of the simpler model when the likelihood ratio does not favor the more complicated one too strongly, the simplicity of the simpler model supposedly making it more plausible *a priori*. In the Bayesian analysis a quantitative *a posteriori* Ockham factor arises as a derivable consequence of the basic sum and product rules of probability theory. Thus model probabilities and odds ratios can favor simpler models even when likelihood ratios do not.

PERIOD KNOWN, PHASE UNKNOWN

We now treat the more practically useful case of known period or frequency, but unknown phase and shape. This is the state of knowledge one might be in when, say, searching for X-ray pulsations from a radio pulsar at the known frequency of the radio pulsations. The global likelihood for a model with m bins and known frequency is given by,

$$p(D \mid \omega, M_m) = \int_0^{2\pi} d\phi \int d\vec{f} \int_0^{A_{\max}} dA \, p(\phi \mid M_m) \, p(A \mid M_m) \, p(\vec{f} \mid M_m) \times$$
$$p(D \mid \omega, \phi, A, \vec{f}, M_m). \tag{55}$$

This is simply equation (36), multiplied by the prior for ϕ, and integrated over ϕ. Thus,

$$p(D \mid \omega, M_m) = \frac{\Delta t^N (m-1)! \, N! \, \gamma(N+1, A_{\max}T)}{2\pi A_{\max} T^{N+1} (N+m-1)!} \int_0^{2\pi} d\phi \, \frac{m^N}{W_m(\omega, \phi)}. \tag{56}$$

The integral must be evaluated numerically. Numerical integrals over ϕ are most efficiently done by integrating over one bin width ($\phi = 0$ to $2\pi/m$), and multiplying the result by m, since phase shifts larger than one bin width simply correspond to cyclically permuting the f_j, which does not change the value of the multiplicity.

From equations (34) and (56), the odds ratio in favor of an m-bin model with known frequency is,

$$O_{m1}(\omega) = \frac{1}{2\pi\nu} \binom{N+m-1}{N}^{-1} \int_0^{2\pi} d\phi \, \frac{m^N}{W_m(\omega, \phi)}, \tag{57}$$

that is, it is the odds ratio for the case with known period and phase, equation (38), averaged over phase. This is smaller than $O_{m1}(\omega, \phi)$, maximized with respect to ϕ, by an additional Ockham factor that penalizes the model for its unknown phase (the $1/2\pi$ factor is just the part of Ω_ϕ arising from the prior for ϕ).

As in equation (39), the odds ratios given by equation (57) can be used to calculate the odds ratio in favor of the periodic class, $O_{per}(\omega)$, according to,

$$O_{per}(\omega) = \sum_{m=2}^{m_{\max}} O_{m1}(\omega). \tag{58}$$

This odds ratio, or equivalently, the probability of the periodic class given by equation (32), is the quantity one would use to detect a signal at a frequency which is known *a priori* (say, from measurements at another wavelength), when the shape of the lightcurve is unknown.

PERIOD AND PHASE UNKNOWN

Finally, if the frequency or period is also unknown, we can find the global likelihood for an m-bin model by multiplying equation (56) by the prior for ω and integrating over ω;

$$p(D \mid M_m) = \frac{\Delta t^N (m-1)! \, N! \, \gamma(N+1, A_{max}T)}{2\pi A_{max}(N+m-1)! \, T^{N+1} \ln \frac{\omega_{hi}}{\omega_{lo}}} \int_{\omega_{lo}}^{\omega_{hi}} \frac{d\omega}{\omega} \int_0^{2\pi} d\phi \, \frac{m^N}{W_m(\omega, \phi)}. \tag{59}$$

Using this global likelihood, the odds ratio in favor of M_m with unknown period and phase is,

$$O_{m1} = \frac{1}{2\pi\nu \ln \frac{\omega_{hi}}{\omega_{lo}}} \binom{N+m-1}{N}^{-1} \int_{\omega_{lo}}^{\omega_{hi}} \frac{d\omega}{\omega} \int_0^{2\pi} d\phi \, \frac{m^N}{W_m(\omega, \phi)}. \tag{60}$$

As before, this expression must be evaluated numerically. If there is significant evidence for a periodic signal, the integral is usually completely dominated by a single peak in the inverse multiplicity, even for moderately small numbers of events. The integration introduces an additional Ockham factor penalizing the model for its unknown frequency (the $1/\ln \frac{\omega_{hi}}{\omega_{lo}}$ factor is the part of Ω_ω arising from the normalization constant for the prior for ω). This Ockham factor, which arises from marginalizing ω, is the Bayesian counterpart to the adjustment of the significance of a frequentist period search for the number of independent periods searched. We elaborate on the distinction between these methods below.

The odds ratios given by equation (60) can be used to calculate the odds ratio in favor of the hypothesis that a periodic signal of unknown frequency and unknown shape is present by calculating O_{per} using equation (33).

If the results of the signal detection calculations just described lead us to conclude that a periodic signal is present, then the problem becomes one of estimating the frequency and shape of the lightcurve. Estimation of the frequency is treated in the following section, and estimation of the lightcurve shape is treated in Gregory and Loredo (1992).

5. Estimation of the Frequency

Our signal detection calculations have focused on the global likelihoods for various models. For parameter estimation, we will focus instead on the posterior distribution for model parameters, in which the global likelihood merely plays the role of a normalization constant.

Assuming the truth of a particular model, M_m, Bayes' theorem for the posterior distribution for the frequency reads,

$$p(\omega \mid D, M_m) = p(\omega \mid M_m) \frac{p(D \mid \omega, M_m)}{p(D \mid M_m)}. \tag{61}$$

The probabilities on the right hand side are by given by equations (24), (56), and (59). The result is,

$$p(\omega \mid D, M_m) = \frac{C}{\omega} \int d\phi \, \frac{1}{W_m(\omega, \phi)}. \tag{62}$$

where C is a normalization constant,

$$C = \left[\int_{\omega_{lo}}^{\omega_{hi}} \frac{d\omega}{\omega} \int_0^{2\pi} d\phi \, \frac{1}{W_m(\omega, \phi)} \right]^{-1}. \tag{63}$$

If we wish to estimate the phase and frequency jointly, a similar calculation, using equations (23), (24), (36), and (59), gives,

$$p(\omega, \phi \mid D, M_m) = \frac{C}{\omega} \frac{1}{W_m(\omega, \phi)}. \tag{64}$$

These marginal distributions show that the multiplicity, $W_m(\omega, \phi)$, of the binned arrival times contains all the information provided by the data about the frequency and phase (it is a "sufficient statistic" for the frequency and phase) when we are not interested in the exact shape of the lightcurve.

These calculations are all conditional on the choice of a particular model, M_m. But the Bayesian model comparison calculations of the previous section do not isolate a single model; rather, they assign a probability to each possible model (just as Bayesian parameter estimation does not produce a single point in parameter space, but rather a posterior "bubble"—a region of high posterior probability). Our estimate of the frequency should include the information provided by all of the models, essentially marginalizing (summing) the joint distribution for ω and m over m, which we can consider a discrete "nuisance parameter." We can show this formally as follows.

Let $(m > 1)$ stand for the proposition, "the signal is periodic, not constant." Then a complete description of our knowledge of the frequency of the periodic signal can be calculated according to,

$$p(\omega \mid m > 1, D, I) = \sum_{m=2}^{m_{\max}} p(M_m, \omega \mid m > 1, D, I)$$

$$= \sum_{m=2}^{m_{\max}} p(M_m \mid D, I)\, p(\omega \mid D, M_m), \tag{65}$$

where $p(M_m \mid D, I)$ can be calculated from the odds ratios in equation (60) using equation (5), and $p(\omega \mid D, M_m)$ is given by equation (62). Equation (65) is a weighted sum of the marginal posteriors for all the periodic models being considered, from those with $m = 2$ to $m = m_{\max}$.

Though equation (65) is the full Bayesian estimate for the frequency, we find in practice that it is often adequate simply to compute equation (62), conditional on the most probable choice of m (that with the largest O_{m1}).

6. Application to Simulated Data

In this section we illustrate the use of our Bayesian method by applying it to data simulated from a stepwise lightcurve with 7 bins. The period of the signal was $P = 2.05633$ s ($\omega = 3.05553$ s^{-1}), and we simulated data over an observing interval of $T = 60$ s (about 29 periods). The average signal rate was 4.6 s^{-1} (≈ 9.5 events per period). Approximately 76% of this rate was in a constant background; 16% (≈ 1.5 events per period) was in a 1-bin pulse in bin 2, and 8% ($\approx .75$ events per period) was in a 1-bin pulse in bin 4. A total of 276 events were simulated. The simulated data were analyzed "blind," though by construction the signal period was chosen to be in the Nyquist range we standardly search, and the number of bins was chosen to be in the range 2 to 12 (we typically use $m_{\max} = 12$ to 15, a computationally convenient number that is capable of representing the simulated and real lightcurves we have analyzed).

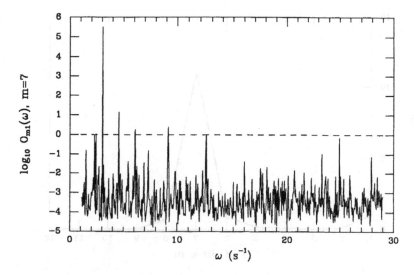

Figure 3. Logarithm of the frequency dependent odds ratio $O_{m1}(\omega)$ versus ω for $m = 7$ bins, for data simulated from a stepwise rate function with 7 bins and $\omega = 3.05553$.

In Figure 3 we have plotted $O_{m1}(\omega)$ (eqn. (57)) in the range $\omega_{lo} = 20\pi/T$ to $\omega_{hi} = 2\pi N/T$ (twice effective Nyquist frequency), for $m = 7$ and $\nu = 1$. This is the quantity that one would calculate were the frequency and number of bins known, and one would then evaluate it only at the known frequency. We plot it for illustrative purposes. Were the frequency and number of bins known, Figure 3 shows that the odds ratio in favor of a periodic signal would be greater than 2.5×10^5 to 1. But if the frequency and number of bins are not known, the evidence for a periodic signal decreases due to the Ockham factors associated with m and ω. This is illustrated by Figure 4, which shows O_{m1} versus m, as given by equation (60). Here the odds ratio in favor of the $m = 7$ model has dropped from 2.5×10^5 to 10.9. The odds ratio in favor of a periodic signal is the sum of the odds ratios for the models considered; here we find $O_{per} = 13.03$, corresponding to a 93% probability that the signal is periodic, given the class of models assumed.

Since this is strong evidence for a signal, we considered a signal to be present, and estimated its frequency and shape. Figure 5 shows the marginal posterior density for ω, for $m = 7$. The large-scale plot shows a strong and extremely narrow peak near the true frequency. The inset details this peak, which is indeed at the true frequency; the frequency is measured with an accuracy of approximately 0.6 mHz (for comparison, the "Nyquist frequency interval" $1/T = 17$ mHz).

In Gregory and Loredo (1992), we show how to estimate the shape of the lightcurve. We use Bayesian methods to find the mean and standard deviation for the shape, $r(t)/A$, at a set of times, t. The calculation averages over all model parameters, including the number of bins, effectively estimating the shape by superposing stepwise functions with various frequencies, phases, and numbers of bins. The result is an estimate that can be significantly smoother than a stepwise curve when the underlying signal is smooth, but

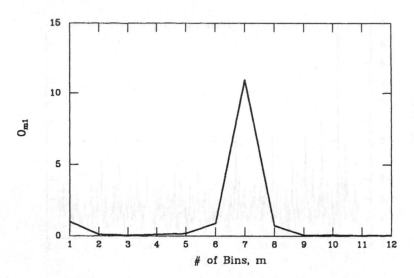

Figure 4. O_{m1} versus m for data simulated from a stepwise rate function with 7 bins.

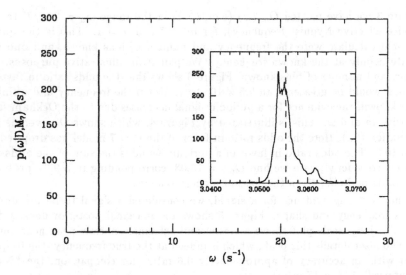

Figure 5. Marginal posterior density for ω, for $m = 7$ bins, for data simulated from a stepwise rate function with 7 bins and $\omega = 3.05553$. Inset shows a detail of the narrow peak; the dashed vertical line indicates the true frequency.

which is capable of detecting sharp features. We applied the method to the simulated data to produce Figure 6. The solid lines show the $\pm 1\sigma$ region for $r(t)/A$, and the dotted curve

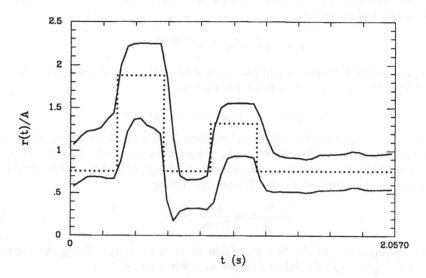

Figure 6. Shape estimate for data simulated from a stepwise rate function with 7 bins and 2.05633 s period. Solid curves show ±1 standard deviation estimates; dotted curve shows true shape.

shows the true shape that generated the data. The shape estimate satisfactorily reproduces the true shape to within the uncertainties.

In Gregory and Loredo (1992) we additionally apply our methods to data simulated from a smooth, sinusoidal signal, demonstrating that our model can adequately model smooth shapes as well as "spiky" shapes.

7. Relation to Existing Statistics

Our Bayesian calculation bears a close relationship to other statistical methods already in use. In particular, the multiplicity that plays a key role in our analysis is related to entropy and to the χ^2 statististic used in the epoch folding method. In this section we elucidate these relationships, and highlight some distinctions between our Bayesian calculation and more familiar frequentist methods. More extensive discussion is available in Gregory and Loredo (1992).

MULTIPLICITY AND ENTROPY

Using Stirling's approximation, $N! \approx \sqrt{2\pi N}\, N^N e^{-N}$, for the factorials appearing in equation (37), we can approximate the logarithm of the multiplicity as follows:

$$-\ln W(\omega, \phi) \approx \ln \frac{C}{\sqrt{N}} + \frac{1}{2} \sum_{j=1}^{m} \ln n_j + N \sum_{j=1}^{m} \frac{n_j}{N} \ln \frac{n_j}{N}, \tag{66}$$

where $C = (2\pi)^{m-1}$. The last, leading order term is proportional to the combinatorial entropy, $H(\omega, \phi)$, of the folded lightcurve (binned data) which is defined as,

$$H(\omega, \phi) = -\sum_{j=1}^{m} \frac{n_j}{N} \ln \frac{n_j}{N}. \tag{67}$$

Thus the joint posterior for the frequency and phase is approximately inversely proportional to the exponential of the entropy,

$$p(\omega, \phi \mid D, M_m) \propto e^{-NH(\omega, \phi)}. \tag{68}$$

Thus the most probable frequency and phase, for a given choice of the number of bins, are those leading to the lightcurve with minimum entropy.

MULTIPLICITY AND "CHI-SQUARED"

We can connect our results with the χ^2 statistic by making a further approximation. Consider evaluating the multiplicity when the binned data are nearly uniform, with $n_j \approx N/m$. This would be the case if there was no periodic signal, or if there was a signal but ω was not close to a harmonic of the true frequency. Writing,

$$\ln \frac{n_j}{N} = \ln \frac{n_j}{N/m} - \ln m \tag{69},$$

we expect the argument of the first logarithm to be near unity. Using the expansion $\ln x \approx (x - 1) - \frac{1}{2}(x - 1)^2$, the entropy can be approximated by,

$$H(\omega, \phi) = +\frac{1}{N} \sum_{j=1}^{m} \left[n_j \log m + n_j \log \frac{N/m}{n_j} \right]$$

$$\approx \log m + \frac{1}{N} \sum_{j=1}^{m} n_j \left(\frac{n_j - N/m}{n_j} \right) - \frac{1}{2N} \sum_{j=1}^{m} n_j \left(\frac{n_j - N/m}{n_j} \right)^2. \tag{70}$$

The first sum is just $\sum_j n_j - \sum_j (N/m) = N - N = 0$, so we find,

$$H(\omega, \phi) \approx \log m - \frac{1}{2N} \sum_{j=1}^{m} \frac{(n_j - N/m)^2}{n_j}$$

$$= \log m - \frac{1}{2N} \chi^2(\omega, \phi), \tag{71}$$

where $\chi^2(\omega, \phi)$ is the same χ^2 statistic used in the "epoch folding method" (Leahy et al. 1983). Using this approximation in equation (68), we have shown that, roughly,

$$p(\omega, \phi \mid D, M_m) \propto e^{\chi^2/2}. \tag{72}$$

Note, however, that the approximation breaks down just in the interesting case where the binned data are nonuniform so there is evidence for a period. Note also that, were we interested in the frequency alone, marginalizing the phase corresponds to averaging the *exponential* of $\chi^2/2$ over phase, not averaging χ^2 itself, as has been advocated in the literature (Collura, et al. 1987).

It is interesting to compare how the multiplicity is used in Bayesian calculations with how the popular epoch folding (EF) method uses the χ^2 statistic, an approximation of the logarithm of the multiplicity. In EF (see, e.g., Leahy et al. 1983) it is assumed that there is indirect evidence for a periodic signal if we can reject the null hypothesis that the data is consistent with a constant rate model by examining values of the χ^2 misfit between the binned phases of the events for various periods and a flat distribution. The percent confidence level, C, associated with the test is found by identifying the maximum observed χ^2 value, χ^2_{obs}, and calculating

$$C = 100[1 - N_{per}Q_{m-1}(\chi^2_{obs})], \tag{73}$$

where $Q_{m-1}(\chi^2_{obs})$ is the area above χ^2_{obs} in the χ^2 distribution for $m-1$ degrees of freedom, and N_{per} is the number of independent frequencies searched. One problem is that there is no general formula for N_{per}, it depends on the number of events, their detailed spacing, and knowledge of the true shape of the signal, and must be estimated by numerical simulations (e.g. Horne and Baliunas 1986). Also, the probability $C/100$ is not a probability for the observed data set. For example, $C = 95\%$ does not mean that there is only a 5% chance that we would obtain a χ^2 as large as that observed in our single sample if the null hypothesis is correct. Rather, it means that if we were to repeatedly draw samples of the same size from the population we would expect only 5% to have a value of χ^2 *greater than or equal to* the value obtained for the actual sample recorded. So how does one interpret any particular confidence level with regard to the presence of a periodic signal in the one observed data set? Of course what we really want is the probability that the data were generated by a periodic signal, not the probability of an unobserved class of hypothetical data.

In contrast the odds ratio can be used to determine directly the desired probability without having to estimate the number of independent frequencies searched; only the prior range for the frequency is required. Also the Bayesian calculation automatically incorporates Ockham's razor and permits the determination of the optimum number of bins, features which are unique to the Bayesian approach. On the question of computational speed, we have found that the Bayesian method can be made at least as fast as the EF method (with phase-averaged χ^2) by pre-computing an array of natural logarithms of factorials that are repeatedly required in the evaluation of the multiplicity.

8. Conclusions

Bayesian probability theory has yielded a solution to the important problem of the detection of a periodic signal in a data set consisting of the locations or times of individual events, when we have no *a priori* knowledge about the nature of the periodic signal. This has been accomplished by using Bayes' theorem to compare a model with a constant rate to members of a class of periodic models. The odds ratio is used to decide if a periodic signal is present in a specified frequency range. At first sight it would appear to be an overwhelming computational problem to allow for all possible signal shapes in the calculation of the odds ratio. In the Bayesian calculation this is accomplished by marginalizing (integrating) out the shape parameters. Our choice of models allows us to perform the needed marginalizations analytically, leading to an algorithm with computational speed comparable to that of the popular epoch folding method based on χ^2.

The odds ratio contains a factor—the Ockham factor—that is a quantitative expression of Ockham's razor. It penalizes the periodic models for their greater complexity in a manner determined by the data and the structure of the models, and not by subjective criteria. In traditional statistical tests, Ockham's razor is invoked to justify considering simpler models to be more plausible *a priori*. In the Bayesian analysis the Ockham factor is a derivable consequence of the basic sum and product rules of probability theory and arises *a posteriori*; our calculations assume the models to have *equal* prior probabilities. The calculation thus balances model simplicity with goodness-of-fit, allowing us to determine both whether there is evidence for a periodic signal, and the optimum number of bins for describing the structure in the data.

When the odds ratio is greater than unity, indicating the presence of a periodic signal, the problem becomes one of estimating the parameters describing the lightcurve, such as the signal frequency. With our models, the posterior probability density for the signal frequency is shown to be inversely proportional to the multiplicity or combinatorial entropy of the binned events. The probability is a maximum for the period corresponding to minimum entropy.

When there are gaps in the observing period (*e.g.*, from dead time, or intermittent observing) our method can be applied without modification. However, to simplify the calculations, we have assumed that the gaps affect the phase bins nearly uniformly. This may not be the case if a strong periodic signal is observed with significant dead time, since the dead time will accumulate in highly populated bins.

The ability of Bayesian methods to straightforwardly handle gaps, determine the optimum number of bins, and quantify Ockham's razor, are features which are unique to the Bayesian approach. We believe these features commend the use of our method. For a more complete discussion of the method, including, (a) a computationally efficient modification to deal with significant data gaps, and (b) a full Bayesian estimate of the shape of the lightcurve and its uncertainty, see Gregory and Loredo (1992).

ACKNOWLEDGMENTS. We thank David MacKay for helping us better understand Bayesian model comparison and Ockham factors. We also thank the editors, Gary Erickson and Ray Smith, for organizing a splendid conference, and for their extreme patience regarding this manuscript! This research was supported in part by a grant from the Canadian Natural Sciences and Engineering Research Council at the University of British Columbia, and by a NASA GRO fellowship, NASA grant NAGW–666, and NSF grants AST–87–14475 and AST–89–13112 at Cornell University.

REFERENCES

Bretthorst, G.L. (1988) *Bayesian Spectrum Analysis and Parameter Estimation*, Springer-Verlag, New York.

Bretthorst, G.L. (1990) 'An Introduction to Parameter Estimation Using Bayesian Probability Theory', in P.F. Fougere (ed.), *Maximum Entropy and Bayesian Methods*, Kluwer Academic Publishers, Dordrecht, 53.

Broadhurst, T.J., R.S. Ellis, D.C. Koo, and A.S. Szalay (1990) *Nature* **343**, 726.

Cleveland , T. (1983) *Nuc. Instr. and Meth.* **214**, 451.

Collura, A., A. Maggio, S. Sciortino, S. Serio, G.S. Vaiana, and R. Rosner (1987) *Ap. J.* **315**, 340.

Gregory, P.C., and T.J. Loredo (1992) 'A New Method for the Detection of a Periodic Signal of Unknown Shape and Period', submitted to *Ap. J.*

Gull, S.F. (1988) 'Bayesian Inductive Inference and Maximum Entropy', in G.J. Erickson and C.R. Smith (eds.), *Maximum-Entropy and Bayesian Methods in Science and Engineering, Vol. 1*, Kluwer Academic Publishers, Dordrecht, p. 53.

Horne, J.H. and S.L. Baliunas (1986) *Ap. J.* **302**, 757.

de Jager, O.C., J.W.H. Swanepoel, and B.C. Raubenheimer (1986) *Astron. Ap.* **170**, 187.

Leahy, D.A., W. Darbro, R.F. Elsner, M.C. Weisskopf, P.G. Sutherland, S. Kahn, and J.E. Grindley (1983) *Ap. J.* **266**, 160.

Loredo, T.J. (1990) in P.F. Fougere (ed.), *Maximum Entropy and Bayesian Methods*, Kluwer Academic Publishers, Dordrecht, 81.

Loredo, T.J. (1992a) 'The Promise of Bayesian Inference for Astrophysics', in E. Feigelson and G. Babu (eds.), *Statistical Challenges in Modern Astronomy*, Springer-Verlag, New York, in press.

Loredo, T.J. (1992b) 'Bayesian Inference With the Poisson Distrubution', in preparation.

MacKay, D. (1992) 'Bayesian Interpolation', in these proceedings; also to appear in *Neural Computation*.

LINKING THE PLAUSIBLE AND DEMONSTRATIVE INFERENCES

Vicente Solana
Unidad de Matemáticas y Sistemas
National Research Council of Spain
Serrano 123, E-28006 Madrid, Spain

ABSTRACT.

A set of patterns of plausible reasoning are available which agree with probability rules. But only some plausible patterns and probability rules have been related to demonstrative patterns of classic logic. The connection between all the plausible inference patterns and the demonstrative patterns is examined in a sentential plausible logic language. A common–sense approach is presented as a formalized system from only two elementary monotonicity patterns of plausible reasoning, given by logic equations. Two elementary patterns of demonstrative reasoning are also stated and formalized in this language. Syllogism forms derived from one of these patterns are identified as the "modus ponens" in this language. Uniqueness of elementary plausible and demonstrative patterns is shown. The plausible and demonstrative elementary patterns are expressed by logic equations which are exactly complementary. This complementarity condition determines the linking form between all the plausible and demonstrative patterns derived from the elementary patterns.

1. INTRODUCTION

To find the relation between the demonstrative reasoning and the plausible ways of reasoning usually recognized as common-sense, is an important and old-age question in science and phylosophy.

Demonstrative or deductive ways of reasoning were large and successfully researched from more than two millenia, and now they are well-established and formulated as different kind of logics. Non–demónstrative or plausible ways of reasoning, on the other hand, were also studied ever since, but the major research progress must be associated, since Laplace, with the development of logical probability theory.

Polya (1954) identified a set of patterns of plausible reasoning which agree with probability theory rules. They are the fundamental two–premises inference patterns examining a consequence and a possible ground, the "shaded" inference patterns, the patterns of conflicting conjectures and the pattern of inference from analogy.

This author remarked that some syllogism forms of the fundamental inference patterns may be simply related to demonstrative syllogisms as the "weakness" process of a major premise, from the logical equivalence to a logical implication, or the inverse process in which the logical equivalence becomes the "limiting" form of successive logical implications.

C. R. Smith et al. (eds.), Maximum Entropy and Bayesian Methods, Seattle, 1991, 105–120.

Polya inquired himself then, *How all forms of plausible reasoning could be linked to the demonstrative patterns of classic logic?*.

Jaynes (1957-88) showed that the limiting forms of the sum and multiplicative rules of probability correspond to demonstrative inference patterns. In particular, the two limiting forms of the Bayes theorem are the "modus ponens" and "modus tollens" of classical logic. Hence, this author affirmed that: *"The relation is simply: the Aristotelian deductive logic is the limiting form of our rules for plausible reasoning when one becomes more and more certain of its conclusions"*.

The problem of formalizing common–sense by using only first order logic as a language has been also investigated for more than three decades in the artificial intelligence field (Mc.Carthy 1957, Hayes 1978). However, negative conclusions were drawn about the progress of this logicist research by Mc. Dermont (1987). The reason is that the argument that a lot of common–sense reasoning can be analyzed as deductive or approximately deductive, is erroneous.

Any attempt to formalize common–sense has to consider logical probabilities, i.e., the probabilities given as plausibilities or certainty degrees. Then, a plausible logic language which contain not only the elements of a logic language system, but also the necessary elements to express plausibilities, is required. Such a system was formalized by Carnap (1950), who regarded logical probabilities as degrees of confirmation. This paper follows a simplified version of this system, which is here named the sentential plausible logic language. This language that uses plausibilities for inferred sentences on given evidences (data), is here summarized is section 2.

A new approach to common–sense, in the sentential plausible logic language, has been developed as a system formalized from only two elementary monotonicity patterns of plausible reasoning (Solana, 1991). In this system, all inference patterns established in Polya (1954) has been obtained from the new elementary patterns. Moreover, the strict monotonicity of the Cox's functions, which is necessary for axiomatic derivation of probabilities, has been also demonstrated from the new patterns.

This paper first raises the question of formalizing the demonstrative patterns of classical logic in the sentential plausible logic language. Next, it examines the connection between the plausible and demonstrative ways of reasoning in the new formal system of common–sense.

The elementary patterns of plausible reasoning and their logical equations are presented in section 3. They are identified as the monotonicity and data-monotonicity elementary plausible patterns. Analysis of patterns is made in section 4, where uniqueness as well as completeness of the two elementary plausible patterns are shown.

Two new elementary patterns of demonstrative reasoning are stated and formulated by logic equations in section 5. They are the monotonicity and data–monotonicity elementary demonstrative patterns. Uniqueness of the elementary demostrative patterns is also obtained in section 6, where the syllogism forms derived from these patterns are analyzed.

Finally, the paper establishes in section 7 that plausible and demonstrative elementary patterns are exactly complementary. It determines the reason for which all forms of demonstrative and plausible reasoning derived from elementary patterns may be linked each other.

2. PLAUSIBLE LOGIC SYSTEMS

A formal language using plausibilities is required to formulate the new elementary patterns of plausible and demonstrative reasoning to be established in sections 3 and 5.

The sentential plausible logical systems in this paper are an extension of the truth–values sentential logical systems that assign two values, i.e. true or false. These two–values logical systems employ a truth–value semantic that has to be specified in terms of true–sets. The propositions are atomic sentences and the true–sets are the ranges of each sentence (Carnap, 1950). Then, the range of any sentence makes true this sentence.

The extension of the truth semantic to plausible logical systems provides the sentential plausible logic. The formal language of each plausible logic system contains on the one hand, the formal language for sentential logic and, on the other hand, some additional elements related to plausibilities.

Let L be a formal language of sentential plausible logic systems. First, the general contents of the sentential language are:

- S is the set of all sentences of L.
- $A, B, C,...,$ are atomic sentences of S.
- T_a, T_b,..., are the ranges of the sentences $A, B, ...$ The range T_a of an atomic sentence A is the subset of all sentences of S that holds A, Carnap (1950). Hence if T_a is a true-set then A is true.
- The L–logic implication denoted by the symbols $\Rightarrow$ is so defined, Carnap (1950)

$$(A \Rightarrow B) \leftrightarrow (T_a \subset T_b).$$

- The L–logic equivalence denoted by the symbols $\Leftrightarrow$ is defined, (Carnap 1950) by

$$(A \Leftrightarrow B) \leftrightarrow (T_a \equiv T_b).$$

Next, the specific elements related to plausibilities are:

- The truth-content of the range T_a of one sentence into the range T_b of another sentence, represented by the intersection $T_a \cap T_b$.
- The amount of the truth-content $T_a \cap T_b$ on the range T_b denoted by $T_a|T_b$.
- The plausibility $A|B$ defined by the amount of the true–value of the sentence A given that B is true, such that the truth–content amount $T_a|T_b$ makes true the true–value amount $A|B$, as the logic equation

$$T_a|T_b \leftrightarrow A|B.$$

All logic equations in the paper, see for instance eqs. (1) and (2), contain two symbols for implications. The symbol $\Rightarrow$ represents the logical implication in L, whereas the symbols $\rightarrow$ and $\leftrightarrow$ respectively, are the simple and double conditional implications "if then" and "if and only if", from any antecedent to a consequence.

The formal language L also contains all connectives other than logical implications and logical equivalence, and all symbols used in Boolean algebra.

3. ELEMENTARY PATTERNS OF PLAUSIBLE REASONING

Two elementary patterns of plausible reasoning, namely the monotonicity and the data-monotonicity patterns were identified (Solana 1991).

These patterns are *elementary* in the sense that they logically imply other patterns but they cannot be implied from any pattern. The elementary patterns, on the other hand, correspond to the simple case of inductive inference in which plausibilities can be ordered. This is the case of the inferences among three different sentences when two of them are logically implied. Then, the plausibilities refer to any sentence considered either as an inferred sentence or an evidence, with respect to every sentence of a logic implication.

In this section, the statements of the patterns are given and its logical equations are formulated in the plausible logic language of section 2. Alternative logical equations are also given for each pattern.

Monotonicity elementary Pattern

This pattern has a unique mode for sentential logic systems which may be equivalently formulated as the logic equations (1) and (2). It refers to the monotonicity of the plausibility function for successively implied inferred sentences on a common evidence.

The monotonicity pattern states that:

"Given two inferred sentences A_1 and A_2 on any given evidence B, such that this evidence does not imply the first sentence A_1 nor the denial of the second A_2, the plausibility for the first inferred sentence is lower than that for the second, if and only if the first sentence implies the second (Solana 1991)".

It is formalized in L as the following logic equation

$$\{\forall B \not\Rightarrow A_1, B \not\Rightarrow \overline{A}_2 \cdot (A_1|B < A_2|B)\} \leftrightarrow \{A_1 \Rightarrow A_2\}. \tag{1}$$

The pattern may be also expressed by an analogous statement related to inferred denials of sentences instead of inferred sentences. It is also written by the logic equation

$$\{\forall B \not\Rightarrow A_1, B \not\Rightarrow \overline{A}_2 \cdot (\overline{A}_1|B > \overline{A}_2|B)\} \leftrightarrow \{A_1 \Rightarrow A_2\}. \tag{2}$$

Logic equations (1) and (2) respectively mean the strictly increasing monotony of the plausibility fuction for succesively implied inferred sentences, and the strictly decreasing monotony of the plausibility function for inferred denials of sentences. The expression $\forall B \not\Rightarrow A_1, B \not\Rightarrow \overline{A}_2$ is the quantifier into the first terms of both equations.

The logical equivalence of expressions (1) and (2) is shown as follows. Consider the logical equivalences:

$$(A_1 \Rightarrow A_2) \leftrightarrow (\overline{A}_1 \Leftarrow \overline{A}_2),$$

$$(B \not\Rightarrow A_1) \leftrightarrow (\overline{B} \not\Leftarrow \overline{A}_1),$$

$$(B \not\Rightarrow A_2) \leftrightarrow (\overline{B} \not\Leftarrow \overline{A}_2). \tag{3}$$

Then, the substitution of the equivalent terms of (3) into the equation (1) and the change of sentences A_1 for $\overline{A}_2$ and A_2 for $\overline{A}_1$ yield a new logic equation which is logically equivalent according with (3) to the logic equation (2).

DATA-MONOTONICITY ELEMENTARY PATTERN

This pattern has two modes for sentential logic systems. They refer to the monotonicity of the plausibility function on different evidences for a common inferred sentence in mode I, and for a common inferred denial of a sentence in mode II.

Mode I

The first mode of the data-monotonicity pattern states that

"*For any inferred sentence A on two given evidences B_1 and B_2, such that the inferred sentence implies the first evidence B_1 but not the denial of the second $\overline{B}_2$, the plausibility for the inferred sentence on the first evidence is greater than that on the second evidence, if and only if the first evidence implies the second one (Solana 1991)*".

This mode has been formalized as the following logic equation

$$\{\forall A \Rightarrow B_1, A \not\Rightarrow \overline{B}_2 \cdot (A|B_1 > A|B_2)\} \leftrightarrow \{B_1 \Rightarrow B_2\}. \tag{4}$$

This mode may be also expressed by an analogus statement related to the plausibilities for a common inferred sentence on the denials of different evidences. It is written by the logic equation

$$\{\forall A \not\Rightarrow B_1, A \Rightarrow \overline{B}_2 \cdot (A|\overline{B}_1 < A|\overline{B}_2)\} \leftrightarrow \{B_1 \Rightarrow B_2\}. \tag{5}$$

Mode II

The second mode of the data-monotonicity pattern states that:

"*For any inferred sentence A on two given evidences B_1 and B_2, such that the inferred sentence implies the first cvidence B_1, but not the denial of the second $\overline{B}_2$, the plausibility of the denial of the inferred sentence on the first evidence is lower than that of the same denial on the second evidence, if and only if the first evidence implies the second one (Solana 1991)*".

This mode has been formalized using the same quantifier of the equation (4), as the following logic equation

$$\{\forall A \Rightarrow B_1, A \not\Rightarrow \overline{B}_2 \cdot (\overline{A}|B_1 < \overline{A}|B_2)\} \leftrightarrow \{B_1 \Rightarrow B_2\}. \tag{6}$$

This mode may be equivalently expressed by an analogous statement related to the plausibilities for the inferred denial of a common sentence on the denials of different evidences. It is written by the logic equation

$$\{\forall A \not\Rightarrow B_1, A \Rightarrow \overline{B}_2 \cdot (\overline{A}|\overline{B}_1 > \overline{A}|\overline{B}_2)\} \leftrightarrow \{B_1 \Rightarrow B_2\}. \tag{7}$$

Logic equivalences between the expressions (4) and (5), and expressions (6) and (7) are shown as follows. Consider the logical equivalences

$$(A \Rightarrow B_1) \leftrightarrow (\overline{A} \Leftarrow \overline{B}_1),$$
$$(A \not\Rightarrow \overline{B}_2) \leftrightarrow (\overline{A} \not\Leftarrow B_2),$$
$$(B_1 \Rightarrow B_2) \leftrightarrow (\overline{B}_1 \Leftarrow \overline{B}_2). \tag{8}$$

Then, the substitutions of logically-equivalent terms of equation (8) into the logical equations (4) and (6), and the changes of sentences B_1 for $\overline{B}_2$ and B_2 for $\overline{B}_1$, yield two new logic equations which are equivalent by (8) to the logic equations (5) and (7).

Logical equations (4) and (5) of the mode I respectively mean the strictly decreasing monotony of the plausibility function for an inferred proposition on successively implied evidences, and the strictly increasing monotony of the plausibility function on denials of successively implied evidences.

On the other hand, the logical equations (6) and (7) of the mode II respectively mean the strictly increasing monotony of the plausibility function for the inferred denial of a sentence on successively implied evidences, and the strictly decreasing monotony of the plausibility fuction on the denials of successively implied evidences.

4. ANALYSIS OF THE ELEMENTARY PLAUSIBLE PATTERNS

Logical expressions for elementary patterns given in section 3 have been formed in the metalanguage where A and A_i are inferred sentences and B and B_j are evidences. This is a suitable way to order the plausibilities $A_i|B$ and $A|B_j$ of the individual elementary patterns that correspond to a common evidence and a common inferred sentence. However, it makes difficult the analysis of the patterns.

Since any sentence of L may be considered either an inferred proposition A or an evidence B, the joint analysis of elementary patterns requires a more general metalanguage in which they can be formulated making no distinction of the symbols used for inferred sentences and evidences. In this section the metalanguage includes the symbols A and B as well as A_i and B_j, but all of them denote only different sentences.

Now consider the simple case of the plausible inferences among three different sentences in which the elementary patterns were formalized. Let A_1 and A_2 be two given sentences, such that $A_1 \Rightarrow A_2$, and let B denote any sentence of the logic system L, such that B is not logically–equivalent to A_1, A_2 $\overline{A}_1$, or $\overline{A}_2$.

In the metalanguage of this section, the monotonicity pattern applies directly to this case according to the same logic equations (1) and (2), whereas the data–monotonicity pattern is reformalized as new logic equations by changing the symbols A for B and B_i for A_i into the logical equations (4) to (7). In this way, the new logic equations of the mode I of data–monotonicity pattern result in

$$\{\forall B \Rightarrow A_1, B \not\Rightarrow \overline{A}_2 \cdot (B|A_1 < B|A_2)\} \leftrightarrow \{A_1 \Rightarrow A_2\}, \tag{9}$$

$$\{\forall B \Rightarrow A_1, B \Rightarrow \overline{A}_2 \cdot (B|\overline{A}_1 < B|\overline{A}_2)\} \leftrightarrow \{A_1 \Rightarrow A_2\}; \tag{10}$$

and the logical equations of the mode II of data–monotonicity pattern are:

$$\{\forall B \Rightarrow A_1, B \not\Rightarrow \overline{A}_2 \cdot (\overline{B}|A_1 < \overline{B}|A_2)\} \leftrightarrow \{A_1 \Rightarrow A_2\}, \tag{11}$$

$$\{\forall B \not\Rightarrow A_1, B \Rightarrow \overline{A}_2 \cdot (\overline{B}|\overline{A}_1 > \overline{B}|\overline{A}_2)\} \leftrightarrow \{A_1 \Rightarrow A_2\}; \tag{12}$$

such that, equations (9) and (10) as well as equations (11) and (12) are logically-equivalent.

Hence, the elementary patterns of plausible reasoning may be summarized by the set of logic equations (1), (9) and (12) or other equivalent sets that result when changing one or more of these equations for the logically–equivalent equations (2), (10) and (11). Here, the set of equations (1), (9) and (12), having distinct quantifiers, is selected for pattern analysis.

Uniqueness of the Elementary Plausible Patterns.

Now consider the set of all possible quantifiers of B defined by the logic product of two sentences such that every sentence is one of the logic implications $B \Rightarrow A_1$, $B \Rightarrow A_2$, $B \Rightarrow \overline{A}_1$ and $B \Rightarrow \overline{A}_2$ and the reverse implications, or the denial of any of these implications. This set contains the quantifiers $(\forall B \not\Rightarrow A_1, B \not\Rightarrow \overline{A}_2)$, $(\forall B \Rightarrow A_1, B \not\Rightarrow \overline{A}_2)$ and $(\forall B \not\Rightarrow A_1, B \Rightarrow \overline{A}_2)$ of the logic equations (1), (9) and (12) of elementary patterns.

The analysis reduces to the set of quantifiers formed only by logic implications and denials, and the quantifiers having only one reverse logic implication or denial related to A_2 and $\overline{A}_2$. Any other quantifier is equivalent to one of this set when changing B for $\overline{B}$.

The quantifiers $(\forall B \Rightarrow A_1, B \Rightarrow \overline{A}_2)$, $(\forall B \Rightarrow A_1, B \not\Rightarrow A_2)$, $(\forall B \not\Rightarrow \overline{A}_1, B \Rightarrow \overline{A}_2)$, $(\forall B \Rightarrow A_1, B \Leftarrow A_2)$, $(\forall B \Rightarrow A_1, B \Leftarrow \overline{A}_2)$, $(\forall B \Rightarrow \overline{A}_1, B \Leftarrow A_2)$, and $(\forall B \Rightarrow \overline{A}_1, B \Leftarrow \overline{A}_2)$ of this reduced set are incompatible with the initial logic implication $A_1 \Rightarrow A_2$, in the sense that no sentence B exists that holds all implications. But there are in the set other quantifiers compatible with $A_1 \Rightarrow A_2$, that could be candidates to form logic expressions for another elementary pattern of plausible reasoning.

Uniqueness of the elementary patterns in shown by examining the feasibility of new patterns based on candidate quantifiers.

The candidate quantifiers may be classified in three groups:
- First, the quantifiers $(\forall B \Rightarrow A_1, B \Rightarrow A_2)$, $(\forall B \Rightarrow \overline{A}_1, B \Rightarrow \overline{A}_2)$, $(\forall B \not\Rightarrow A_1, B \Leftarrow A_2)$, $(\forall B \not\Rightarrow A_1, B \Leftarrow \overline{A}_2)$, $(\forall B \not\Rightarrow \overline{A}_1, B \Leftarrow A_2)$, and $(\forall B \not\Rightarrow \overline{A}_1, B \Leftarrow \overline{A}_2)$ which are equivalent to the quantifiers of logic equations (9) and (12), in the sense that they describe the same sentences.
- Second, the quantifiers $(\forall B \not\Rightarrow A_1, B \Rightarrow A_2)$, $(\forall B \Rightarrow \overline{A}_1, B \Rightarrow \overline{A}_2)$, $(\forall B \not\Rightarrow \overline{A}_1, B \not\Rightarrow A_2)$, $(\forall B \Rightarrow \overline{A}_1, B \Rightarrow A_2)$, $(\forall B \not\Rightarrow A_1, B \Leftarrow \overline{A}_2)$, $(\forall B \not\Rightarrow A_1, B \Leftarrow A_2)$, $(\forall B \not\Rightarrow \overline{A}_1, B \Leftarrow \overline{A}_2)$, and $(\forall B \not\Rightarrow \overline{A}_1, B \Leftarrow A_2)$ which describe several subsets of the sentences described by the quantifier of the logic equation (1).
- Third, the remained quantifiers $(\forall B \not\Rightarrow A_1, B \not\Rightarrow A_2)$ $(\forall B \not\Rightarrow A_1, B \not\Rightarrow \overline{A}_2)$, $(\forall B \not\Rightarrow \overline{A}_1, B \Rightarrow \overline{A}_2)$, $(\forall B \Rightarrow \overline{A}_1, B \Rightarrow A_2)$, $(\forall B \not\Rightarrow A_1, B \not\Leftarrow \overline{A}_2)$, $(\forall B \not\Rightarrow A_1, B \not\Leftarrow A_2)$, $(\forall B \not\Rightarrow \overline{A}_1, B \not\Leftarrow A_2)$, and $(\forall B \not\Rightarrow \overline{A}_1, B \not\Leftarrow \overline{A}_2)$ which describe several joint subsets of the sentences described by two of the quantifiers of equations (1), (9) and (12).

All inference patterns with logic equations formed using the second group of quantifiers should be contained by the logic equation (1) of the monotonicity pattern. However, they cannot be said elementary, in the sense that the logic equation (1) is not implied by the logical equations of these inference patterns.

On the other hand, the inference patterns based on the third group of quantifiers should be formed by the logic sum of two of the plausibility inequalities in the logic equations (1) and (9) or (12). However, application of each inequality is not specified by the quantifier. Hence, the patterns are not elementary, in the sense that logic equations (1), (9) or (12) do not derive from them.

Therefore, there are no other elementary pattern of plausible reasoning in this case, but the elementary patterns given in section 3, which were formulated by logic equations that cannot be implied individually by logic equations of any other pattern, and the equivalent patterns formed by the first group of quantifiers.

Likewise, the set of monotonicity and data–monotonicity elementary patterns is *complete* in the sense that its logic equations contain the logic equations of any other inference pattern applicable to this case, including the pattern equations of the second and third group of quantifiers, and the one way logic equations and syllogism forms.

5. ELEMENTARY DEMONSTRATIVE PATTERNS

This section studies the simple case of demonstrative inference in an analogous way to the plausible inference case examined in sections 3 and 4.

In this case the plausibilities refer to an inferred sentence, or an evidence, which is logically–equivalent to any of the sentences or denials of two successively implied sentences.

The quantifiers correspond in this case to the logic product of two sentences such that one of them is a logical equivalence and the other is either a logical implication or the denial of a logical implication.

Two elementary patterns of demonstrative reasoning, namely the monotonicity and the data–monotonicity demonstrative patterns, are stated. The patterns are formalized by logic equations. Syllogism forms including the classical demonstrative patterns are derived from these logic equations.

MONOTONICITY ELEMENTARY DEMONSTRATIVE PATTERN.

This pattern is written in a parallel way to the monotonicity pattern of plausible reasoning given in section 3. It states that:

For two inferred sentences A_1 and A_2 on a given evidence B such that this evidence is logically–equivalent to the first inferred sentence A_1 and it does not imply the second A_2, every inferred sentence is logically-true given that the evidence is true, if and only if the first sentence implies the second.

The pattern is formalized in L using plausibilities by the following logic equation

$$\{\forall B \Leftrightarrow A_1, B \not\Rightarrow \overline{A_2} \cdot (A_1|B = A_2|B)\} \leftrightarrow \{A_1 \Rightarrow A_2\}. \tag{13}$$

It may be also expressed by an analogous statement related to inferred denials of sentences instead of the inferred sentences. Thus, the pattern is alternatively formulated by the logic equation

$$\{\forall B \not\Rightarrow A_1, B \Leftrightarrow \overline{A_2} \cdot (\overline{A_1}|B = \overline{A_2}|B)\} \leftrightarrow \{A_1 \Rightarrow A_2\}. \tag{14}$$

The equivalence of equations (13) and (14) is easily shown in a similar way to the case of equations (1) and (2), taking the logical equivalences (3) and the expression $\{(B \Leftrightarrow \overline{A_2}) \leftrightarrow (\overline{B} \Leftrightarrow A_2)\}$ into account.

DATA–MONOTONICITY ELEMENTARY DEMONSTRATIVE PATTERN

This pattern is also expressed in a parallel way to the data-monotonicity pattern of plausible reasoning given in section 3. However, in this case the pattern exhibits a unique mode. This pattern states that:

For an inferred sentence A on two evidences B_1 and B_2, such that the inferred sentence A is logically-equivalent to the second evidence B_2 and it does not imply the first evidence B_1, the inferred sentence is logically-true given that every evidence is true, if and only if the first evidence implies the second.

It is formalized by the following logic equation

$$\{\forall A \not\Rightarrow B_1, A \Leftrightarrow B_2 \cdot (A|B_1 = A|B_2)\} \leftrightarrow \{B_1 \Rightarrow B_2\}. \tag{15}$$

The data–monotonicity pattern may be also written in an analogous statement related to the evidences given as denials of sentences instead of sentences. It is alternatively formulated by the next logic equation

$$\{\forall A \Leftrightarrow \overline{B}_1, A \not\Rightarrow \overline{B}_2 \cdot (A|\overline{B}_1 = A|\overline{B}_2)\} \leftrightarrow \{B_1 \Rightarrow B_2\}. \tag{16}$$

The logic equivalence of (15) and (16) is also shown, first by substituting the equivalent terms of (8) and the expression $\{(A \Leftrightarrow B_2) \leftrightarrow (\overline{A} \Leftrightarrow \overline{B}_2)\}$ into the equation (15) and changing B_1 for $\overline{B}_2$ and B_2 for $\overline{B}_1$, and next, by restituting the above equivalent terms.

Monotonicity Demonstrative Pattern and Syllogism Forms

Logic equations (13) and (14) of the monotonicity elementary demonstrative pattern contain the following one way logic equations

$$\forall B \Leftrightarrow A_1, B \not\Rightarrow \overline{A}_2 \cdot \{(A_1 \Rightarrow A_2) \rightarrow (A_1|B = A_2|B)\}, \tag{17}$$

$$\forall B \not\Rightarrow A_1, B \Leftrightarrow \overline{A}_2 \cdot \{(A_1 \Rightarrow A_2) \rightarrow (\overline{A}_1|B = \overline{A}_2|B)\}. \tag{18}$$

These equivalent equations determine a new demonstrative pattern derived from the monotonicity elementary pattern. This pattern may be expressed as two syllogism forms according to the meaning of plausibility in the plausible logic language L.

The first syllogism form of the demonstrative pattern given by (17) may be expressed as follows

$$B \Leftrightarrow A_1$$

$$B \not\Rightarrow \overline{A}_2$$

$$A_1 \Rightarrow A_2$$

$$A_1 \text{ is } L - \text{true given } B \text{ is true}$$

$$A_2 \text{ is } L - \text{true given } B \text{ is true,}$$

where two first premises describe the quantifier of (17) and the inferred sentences A_1 and A_2 are logically–true with regard to a given true evidence B. The last premise which is missing in equation (17), is a consequence of $B \Leftrightarrow A_1$ and "B is true". Indeed, the inferred sentence A_1 is logically-true with regard to any true sentence B being logically-equivalent to A_1.

Analogically the second syllogism form of the demonstrative pattern given by equation (18) is expressed as follows

$$B \Leftrightarrow \overline{A}_2$$

$$B \not\Rightarrow A_1$$

$$A_1 \Rightarrow A_2$$

$$\overline{A}_2 \ is \ L - true \ given \ B \ is \ true$$

$$\overline{A}_1 \ is \ L - true \ given \ B \ is \ true,$$

where the two first premises describe the quantifier of equation (18), the inferred denials of sentences A_1 and A_2 are logically–true with regard to a given true evidence B, and the last premise is a consequence of $B \Leftrightarrow \overline{A}_2$ and "B is true".

CLASSICAL DEMONSTRATIVE PATTERN

As the general metalanguage of the section 4 making no distinction of symbols for inferred sentences and evidences is employed, the above syllogisms simplify as two new forms having only two premises.

Thus, the first syllogism of the monotonicity demonstrative pattern given by equation (17) is written by changing A_1 for B, as follows

$$B \Rightarrow A_2$$

$$B \ is \ L - true \ given \ B \ is \ true$$

$$A_2 \ is \ L - true \ given \ B \ is \ true,$$

where the inferred sentence A_2 is logically–true with regard to a given true evidence B, which is also logically–true with regard to itself.

Similarly, the second syllogism of the monotonicity demostrative pattern given by equation (18) is written, changing B for $\overline{A}_2$, by

$$A_1 \Rightarrow A_2$$

$$\overline{A}_2 \ is \ L - true \ given \ \overline{A}_2 \ is \ true$$

$$\overline{A}_1 \ is \ L - true \ given \ \overline{A}_2 \ is \ true,$$

where the inferred denial $\overline{A}_1$ is logically-true with regard to a given true denial evidence $\overline{A}_2$, which is also logically-true regarding itself true.

These syllogism are close to the "modus ponens" and "modus tollens" patterns of classical logic which are respectively expresed, in this case, by

$$B \Rightarrow A_2$$
$$B \text{ is true}$$

$$\overline{}$$

$$A_2 \text{ is true,}$$

and

$$A_1 \Rightarrow A_2$$
$$\overline{A_2} \text{ is true}$$

$$\overline{}$$

$$\overline{A_1} \text{ is true.}$$

The syllogism forms formulated in this subsection not only explain the classical demonstrative patterns, by using relative truth concepts, but themselves are the new "modus ponens" and "modus tollens" in the plausible logic language systems. The first syllogism form of the elementary demonstrative pattern, or "modus ponens", states that:

Any inferred sentence is logically-true with regard to a given true sentence that logically-implies it.

Likewise, the second syllogism form of the monotonicity demostrative pattern, or "modus tollens", states that:

The inferred denial of any sentence is logically-true with regard to a given true denial of a sentence that is logically-implied by the first sentence.

such that, both statements are equivalent according to the equivalence of logic equations (13) and (14).

DATA-MONOTONICITY DEMONSTRATIVE PATTERN AND SYLLOGISM FORMS

Logic equations (15) and (16) of the data-monotonicity elementary pattern contain the next one way logic equations

$$\forall A \not\Rightarrow B_1, A \Leftrightarrow B_2 \cdot \{(B_1 \Rightarrow B_2) \rightarrow (A|B_1 = A|B_2)\}, \tag{19}$$

$$\forall A \Leftrightarrow \overline{B}_1, A \not\Rightarrow \overline{B}_2 \cdot \{(B_1 \Rightarrow B_2) \rightarrow (A|\overline{B}_1 = A|\overline{B}_2)\}. \tag{20}$$

These equivalent equations determine another new demonstrative pattern derived from the data–monotonicity elementary pattern. This pattern may be also expressed as two syllogism forms.

The first syllogism form of the demonstrative pattern given by (19) is expressed by

$$A \not\Leftrightarrow B_1$$

$$A \Leftrightarrow B_2$$

$$B_1 \Rightarrow B_2$$

$$A \ is \ L-true \ given \ B_2 \ is \ true$$

$$A \ is \ L-true \ given \ B_1 \ is \ true,$$

where the two first premises are the quantifier of equation (19), and the inferred sentence A is logically–true with regard to each of the given true evidences B_1 and B_2. Here, the last premise which is missing in equation (19) is a consequence of $A \Leftrightarrow B_2$ and "B_2 *is true*".

The second syllogism form of the demonstrative pattern given by (20) is analogously expressed by

$$A \Leftrightarrow \overline{B}_1$$

$$A \not\Leftrightarrow \overline{B}_2$$

$$B_1 \Rightarrow B_2$$

$$A \ is \ L-true \ given \ \overline{B}_1 \ is \ true$$

$$A \ is \ L-true \ given \ \overline{B}_2 \ is \ true,$$

such that the two first premises describe the quantifier of (20), and the inferred sentence A is logically–true with regard to each of the given true denials of sentences $\overline{B}_1$ and $\overline{B}_2$.

As the general metalanguage of section 5 is employed the above syllogisms simplify as two new forms having only two premises. The syllogism forms of the data–monotonicity demonstrative pattern obtained from equations (19) and (20) are respectively given by

$$B_1 \Rightarrow A$$

$$A \ is \ L-true \ given \ A \ is \ true$$

$$A \ is \ L-true \ given \ B_1 \ is \ true,$$

and

$$B_1 \Rightarrow B_2$$

$$\overline{B}_1 \ is \ L-true \ given \ \overline{B}_1 \ is \ true$$

$$\overline{B}_1 \ is \ L-true \ given \ \overline{B}_2 \ is \ true.$$

Since the second premise and the conclusion of each syllogism are here referred to different given true evidences, they cannot be expressed by simple true sentences like the classic patterns do. Consequently these syllogism forms are not identified with any other demonstrative patterns of classic logic. The new syllogisms in this subsection can be only formulated in the sentential plausible logic language.

The first syllogism form of the data–monotonicity demonstrative pattern states that:

A given inferred sentence is logically-true with regard to any true sentence that logically–implies it;

and, the second syllogism of the same pattern states that:

The given inferred denial of a sentence is logically–true with regard to the true denial of any sentence that is logically–implied by the first sentence.

Both statements are equivalent, since they derive from the equivalent equations (15) and (16) of the elementary data–monotonicity demonstrative pattern.

6. UNIQUENESS OF ELEMENTARY DEMONSTRATIVE PATTERNS

Analysis of elementary demonstrative patterns is made in this section using a general metalanguage in which the symbols A and A_i, and B and B_i denote either inferred sentences or evidences.

In this metalanguage the monotonicity demonstrative pattern is formulated as the same logic equations (13) and (14), but the data–monotonicity demonstrative pattern is formulated, by changing the symbols A for B and B_i for A_i into the logic equations (15) and (16) as the following new logic equations

$$\{\forall B \not\Rightarrow A_1, B \Leftrightarrow A_2 \cdot (B|A_2 = B|A_1)\} \leftrightarrow \{A_1 \Rightarrow A_2\}, \tag{21}$$

$$\{\forall B \Leftrightarrow \overline{A}_1, B \not\Rightarrow \overline{A}_2 \cdot (B|\overline{A}_1 = B|\overline{A}_2)\} \leftrightarrow \{A_1 \Rightarrow A_2\}, \tag{22}$$

which are also logically–equivalent.

The analysis here refers to the simple case of demonstrative inferences in which the elementary patterns have been formulated. Let A_1 and A_2 be two given sentences such that $A_1 \Rightarrow A_2$ and B denote any sentence of the logic system being logically–equivalent to one of the sentences A_1, A_2, $\overline{A}_1$ and $\overline{A}_2$.

Consider in this case the set of all possible quantifiers of B defined by the logic product of two sentences, such that one of the sentences is any of the logic equivalences $B \Leftrightarrow A_1$, $B \Leftrightarrow A_2$, $B \Leftrightarrow \overline{A}_1$ and $B \Leftrightarrow \overline{A}_2$, and the other sentence is any compatible sentence of the logic implications $B \Rightarrow A_1$, $B \Rightarrow A_2$, $B \Rightarrow \overline{A}_1$, $B \Rightarrow \overline{A}_2$ and reverse implications, or the denial of any of these implications. The possibility of other quantifiers formed by two logic equivalences has been disregarded because they should be contradictory with the basic implication $A \Rightarrow A_2$ to form elementary patterns.

This set of all possible quantifiers of B includes the quantifiers $(\forall B \Leftrightarrow A_1, B \not\Rightarrow \overline{A}_2)$ and $(\forall B \not\Rightarrow A_1, B \Leftrightarrow \overline{A}_2)$ of equations (13) and (14), and the quantifiers $(\forall B \not\Rightarrow A_1, B \Leftrightarrow A_2)$ and $(\forall B \Leftrightarrow \overline{A}_1, B \not\Rightarrow \overline{A}_2)$ of equations (21) and (22), that respectively correspond to the monotonicity and data–monotonicity elementary demonstrative patterns.

Likewise, the set contains other quantifiers which are equivalent to the quantifiers of logic equations (13), (14), (21) and (22). They are, first, the quantifiers $(\forall B \Leftrightarrow A_1, B \Rightarrow A_2)$, $(\forall B \Leftrightarrow A_1, B \not\Leftarrow \overline{A}_2)$, $(\forall B \Leftrightarrow A_1, B \Leftarrow A_2)$ and the quantifiers $(\forall B \Rightarrow \overline{A}_1, B \Leftrightarrow \overline{A}_2)$, $(\forall B \not\Leftarrow A_1, B \Leftrightarrow \overline{A}_2)$ and $(\forall B \not\Leftarrow A_1, B \Leftrightarrow A_2)$, respectively equivalent to those of the logic equations (13) and (14); second, the quantifiers $(\forall B \not\Rightarrow \overline{A}_1, B \Leftrightarrow A_2)$, $(\forall B \not\Leftarrow A_1, B \Leftrightarrow A_2)$, $(\forall B \not\Leftarrow \overline{A}_1, B \Leftrightarrow A_2)$ and the quantifiers $(\forall B \Leftrightarrow \overline{A}_1, B \not\Rightarrow A_2)$,$(\forall B \Leftrightarrow \overline{A}_1, B \Leftarrow A_2)$ and $(\forall B \Leftrightarrow \overline{A}_1, B \not\Leftarrow \overline{A}_2)$, respectively equivalent to the quantifiers of equations (21) and (22). Therefore, all these quantifiers can be used to form expressions of demonstrative patterns which must be exactly equivalent to the elementary demonstrative patterns given by logic equations (13), (14), (21) and (22).

All other quantifiers of the set of candidate quantifiers of B are incompatible with the initial logic implication $A_1 \Rightarrow A_2$, in the sense that no one sentence B exists that holds these implications. Therefore they cannot be used to build logic equations for demonstrative patterns.

Consequently, there are only two possible elementary demonstrative patterns which are the monotonicity and data-monotonicity elementary patterns stated in section 5. Uniqueness of elementary patterns means here that no elementary demonstrative pattern exists but the elementary demonstrative patterns given by logic equations equivalent to (13), (14), (21) and (22).

7. COMPLEMENTARITY OF ELEMENTARY PATTERNS

Two simple cases have been examined in this paper for all plausible inferences among three sentences denoted A_1, A_2 and B, when two of them are related to by the logical implication $A_1 \Rightarrow A_2$.

The first simple case deals with the inference patterns of *plausible reasoning* presented in sections 3 and 4. It concerns with any possible sentence B of a logic system that is not logically–equivalent to $A_1, A_2, \overline{A}_1$ and $\overline{A}_2$.

Here all possible connections of B with both sentences of the implication $A_1 \Rightarrow A_2$, can be formulated through the quantifiers of B determined by the logic products of two sentences taken from the logic implications of B regarding $A_1, A_2, \overline{A}_1$ and $\overline{A}_2$, the reverse implications or the denials of these implications. All these logical products describe, in different ways, the quantifier for any B of the logic system with exception of the sentences being logically– equivalent to $A_1, A_2, \overline{A}_1$ and $\overline{A}_2$.

In this case, examination of these quantifiers showed that there are only two elementary patterns of plausible reasoning: the monotonicity pattern and the data–monotonicity pattern which has two independent modes. Logic equations of these patterns were summarized in a general metalanguage by the set of equations (1), (9) and (12), or any other with equations logically–equivalent to this set.

The second simple case corresponds to the inference patterns of *demonstrative reasoning* presented in sections 5 and 6. It refers to any possible sentence of a logic system being logically–equivalent to $A_1, A_2, \overline{A}_1$ and $\overline{A}_2$. Therefore, it is just complementary to the first simple case in the sense that they describe together the universal quantifier of B in a logic system. This second case considers all possible connections of B with every sentence of the logic implication $A_1 \Rightarrow A_2$, formulated as the quantifiers including both one logic equivalence of B regarding $A_1, A_2, \overline{A}_1$ and A_2, and any of the logic implications considered in the first case.

Hence, examination of all possible quantifiers showed that there are only two elementary patterns of demonstrative reasoning: the monotonicity pattern and the data-monotonicity pattern having only one mode. Logical equations of these patterns are summarized in a general metalanguage by the set of equations (21) and (22), or another set of equivalent equations.

Complementarity of the quantifiers independently examined in both simple cases determines that the elementary patterns stated in each case are also complementary, in the sense that no other elementary patterns are possible but the demonstrative and plausible patterns.

Moreover, this complementarity means that the elementary patterns of plausible and demonstrative reasoning cannot be derived from each other. Indeed, the logical equivalences in the quantifiers used for logic equations of demonstrative patterns cannot be derived from the logical implications in the quantifiers of logic equations of plausible reasoning patterns.

This complementarity of elementary patterns constitutes the essential linking form of the plausible and demonstrative reasoning.

Distinct examples connecting the plausible and demonstrative ways of reasoning as "limiting forms" , derive from the complementarity link betwen elementary patterns. Thus the "limiting forms" in Polya (1954) and Jaynes (1957, 91) could be interpretated as substitutions of any symbol of logical implication or denial, by another symbol of logical equivalence, what define complementary patterns and syllogism forms.

8. CONCLUSIONS

(1) A formalized system of common–sense can be developed from only two elementary patterns of plausible reasoning, namely the monotonicity and the data–monotonicity patterns of plausible inference. These patterns were formulated in the formal language of sentential plausible logic, by logical equations using quantifiers. All Polya's patterns and other forms of plausible reasoning were derived from the elementary patterns.

(2) The elementary monotoniciy and data–monotonicity plausible patterns are the only possible elementary patterns of plausible reasoning, in the sense that any other plausible pattern either derives from them, or can be formulated by logical equations which are logically–equivalent to the logic equations of these elementary patterns.

(3) Two new elementary patterns of demonstrative reasoning, namely the monotonicity and data–monotonicity demonstrative patterns, are stated and formulated in the plausible logic language, by logic equations using quantifiers.

The elementary monotonicity demonstrative pattern contains logic equations which may be expressed by syllogism forms similar to the "modus ponens" and the "modus tollens" of classic logic. These syllogism forms are the "modus ponens" and "modus tollens" in the plausible logic language .

The elementary data–monotonicity demonstrative pattern, however, contains logic equations and syllogism forms of demonstrative inference that cannot be identified as similar forms in classic logic. These syllogisms only appear in the plausible logic language. They are complementary to the syllogisms of the monotonicity demonstrative inference pattern.

(4) The elementary monotonicity and data–monotonicity demonstrative patterns are the only possible elementary demonstrative patterns, in the sense that any other demonstrative pattern either derives from them, or can be formulated by logical equations

which are logically–equivalent to the equations of these elementary patterns.

(5) Elementary patterns of plausible and demonstrative inferences are complementary in the sense that no other elementary patterns are possible. The set of elementary patterns is therefore irreducible.

This complementarity condition constitutes the essential link between the plausible and demonstrative ways of reasoning. It determines the way in which all possible inference patterns of plausible and demonstrative reasoning are related each other, as inquired in Polya (1954).

ACKNOWLEDGMENTS. The work reported herein received finantial support from DGYCIT of Spain as a part of the research project PB 89-0077.

REFERENCES

Carnap, R.: 1950, *Logical Foundations of Probability*, University of Chicago Press.

Cox, R.T.: 1961, *The Algebra of Probable Inference*, John Hopkins University Press, Baltimore MD.

Cheesemen, P.: 1988, 1990, "An Inquiry into Computer Understanding" *Computational Intelligence*, Vol. 4, 57-67, Discussion and Defense in Vol. 4, 67-141 and Vol. 6, 179-192.

Jaynes, E.T.: 1957, *"How Does the Brain Do Plausible Reasoning"* Microwave Laboraty Report, No. 421. University of Stanford; reprinted in G.J. Ericson and C.R. Smith (eds), *Maximum Entropy and Bayesian Methods in Science and Engineering*, Vol.I, 1-24, Kluwer Academic Publishers, 1988.

Jaynes, E.T.: 1991, *Probability Theory. The Logic of Science*, to be published.

Mc.Dermott, D.: 1987, "A critique of pure reason" *Computational Intelligence*, Vol. 3, 151-160.

Polya, G.: 1954, *Mathematics and Plausible Reasoning*, Vol. 2, Princenton University Press.

Pearl, J.: 1988, *Probabilistic Reasoning in Intelligent Systems: Networks of Plausible Inference*, Morgan Kauffmann Publishers, San Mateo, California.

Smith, C.R. and Erickson, G.: 1990, "Probability Theory and the Associativity Equation", in Paul Fouguere (ed), *Maximum Entropy and Bayesian Methods*, 17-30, Kluwer Academic Publishers.

Solana, V.: 1991, *"Monotonicity Patterns of Plausible Reasoning and Logical Probabilities"*, Research Report, Unidad Matemticas y Sistemas, National Research Council of Spain, Madrid.

DIMENSIONAL ANALYSIS IN DATA MODELLING

G. A. Vignaux
Institute of Statistics and Operations Research
Victoria University
PO Box 600, Wellington, New Zealand
email: vignaux@isor.vuw.ac.nz

ABSTRACT. Dimensional Analysis can make a contribution to data modelling when some of the variables in the problem are physical. The analysis constructs the set of independent dimensionless factors that should be used as the major variables of the model in place of the original measurements. There are fewer of these than the originals and they may have a more appropriate interpretation. The technique is described briefly and its proposed role in data analysis and regression illustrated with an example.

When setting up a problem we should, to paraphrase Einstein, "use all the information we have but no more." To use all our information suggests a Bayesian approach; to use no more is in the spirit of Maximum Entropy. Despite this sage advice we often ignore information carried by the physical characteristics of the variables involved. Dimensional Analysis, written as DA throughout the rest of this paper, ensures that this physical information is always used.

Experienced statisticians have long argued that there is more to data analysis than throwing all the numbers into a computer program and reporting the results of the resulting regression. They prefer to conduct an exploratory data analysis, inspecting graphs and examining residuals. Based on the observed curvatures they then manipulate and transform the data until approximately linear relationships are obtained.

DA can be used in this preliminary analysis to assist, and perhaps even replace, the stage of data transformation. It has been a standard working tool for many of the greatest mathematical physicists, such as Rayleigh (1915). It should be familiar to readers but it is apparently new to many statisticians and operations research analysts. Or perhaps it is not new, for many will have learned it at school or college. What is novel is that it might actually be useful.

I propose to demonstrate its application to a well-known data analysis problem where at least some of the variables involved are physical quantities. For those to whom it is unfamiliar, DA is introduced briefly in the next section.

2. What is Dimensional Analysis?

DA is a technique for restructuring the original dimensional variables of a problem into a set of dimensionless products using constraints imposed upon them by their dimensions (Huntley, 1967; Vignaux, 1986, 1988). It is ultimately based on the simple requirement for

121

C. R. Smith et al. (eds.), Maximum Entropy and Bayesian Methods, Seattle, 1991, 121–126.

dimensional homogeneity in any relationship between the variables or, alternatively, the knowledge that the structure of physical laws cannot depend on the choice of units.

Buckingham (1914) showed in his Pi theorem that if the original unknown relationship is represented by $f(x_1, x_2, \ldots, x_n) = 0$, where the x_i are the variables, it can be transformed into a new relation $\phi(\pi_1, \pi_2, \ldots, \pi_{n-m}) = 0$ with $n - m$ independent dimensionless products π_j of the original x_i variables. m is the number of fundamental dimensions that the dimensions of the original physical variables are constructed from. In the physical sciences these are length, mass, and time $[L, M, T]$. In operations research we would add quantity and cost $[Q, \$]$.

Depending on the problem the x_i might also include some dimensional constants such as viscosity or resistance. Each of these corresponds to the statement of a relevant physical relationship such as Stokes' or Ohm's law.

A set of dimensionless πs can be constructed in a fairly mechanical manner. One needs a basis of m of the original variables to satisfy the m constraints in the exponents corresponding to each of the fundamental dimensions. It should include only variables with linearly independent exponent vectors (for example, one cannot have two velocities in the basis). Any basis can be used but some bases may be better than others as we will see later.

The number of variables to take part in any regression has now been reduced from the original n xs to the $n - m$ πs. This clearly makes the problem simpler and may make it trivial. It may even make a regression unnecessary.

For example, a popular demonstration of DA involves the derivation of the formula for the period of a pendulum (Taylor, 1974). What starts out as a 4-variable problem is reduced to finding the value of a single dimensionless constant. A similar collapse to a single dimensionless constant is seen in the derivation of the optimum lot-size in simple deterministic inventory problems (Naddor, 1966; Vignaux, 1986, 1988).

Reminded again of the reaction to Maximum Entropy methods, it appears at first sight that in using DA we are getting something for nothing. In fact we are only applying at an early stage those constraints that we know must be satisfied by the final solution.

3. An Example - Hocking's Automobile Data

Hocking's classical problem is to fit a regression model to predict the gas consumption m (miles per gallon) from other characteristics of the automobile. The collection of gas mileage data from the 1974 *Motor Trend* magazine was used as a test case by Hocking (1976) to investigate automatic methods of regression. Henderson and Velleman (1982) examined the same data (together with a second set, collected from *Consumer Reports*), and argued for an alternative philosophy of computer–assisted data analysis - a collaboration between the analyst and the computer in an interactive mode in accord with the philosophy of 'data analysis' rather than blind regression.

To these approaches we add a preliminary Dimensional Analysis, combined, as it should be, with some physical insight. We introduce additional dimensional constants and convert the set of variables to a smaller number of dimensionless πs. The regression is then carried out on the dimensionless products.

EXAMINING THE VARIABLES

The data, reproduced in (Henderson, 1982), contains 11 variables, of which 6 are already dimensionless, being, in the previous authors' notation, either ratios (such as $DRAT$, the final drive ratio) or numbers (CYL, the number of cylinders). The four variables that are clearly physical, m, the variable of interest and h, w and q are listed in Table 1. with their dimensions. The remaining variable, $DISP$, the cylinder displacement, has an intermediate nature in that it has dimensions (cubic inches in this case) but is really a surrogate for engine size. m is plotted against w in Figure 1.

Table. 1. The variables and their dimensions.

Variable	Symbol	Description	Dimensions
m	MPG	miles per gallon	LG^{-1}
h	HP	horsepower	ML^2T^{-3}
w	WT	weight (mass)	M
q	$QSEC$	quarter mile time	T

The dimensionless variables and ratios are appropriate candidates for direct inclusion in the data analysis or regression; they are already of the correct dimensionless product form and may carry information about the behaviour of the system. The DA must now construct dimensionless products from the remaining four.

FURTHER CONSIDERATION

m $[LG^{-1}]$ clearly cannot immediately be combined with the other variables to make a dimensionless product because of the $[G]$ dimension which does not appear in any of the other three. (DA experts will notice that rather than replace dimension $[G]$ with a standard volume measurement $[L^3]$ it has been kept separate as it is a measure of the amount of fuel, not travel distance or vehicle size.) This suggests that we could with advantage introduce a new dimensional constant E $[ML^2T^{-2}G^{-1}]$ giving the energy content $[ML^2T^{-2}]$ in each gallon to reflect the physical law that 1 gallon contains a fixed amount of energy. This has the disadvantage of adding another dimensional constant, E, to our set of 4 variables but has gained us a sensible link between m and the others. We naturally expect E to remain constant, though with a more extensive data set it might change its value with different types of fuel.

The variable q $[T]$, quarter mile time, is the time to accelerate over a distance of a quarter of a mile. At least we should add to the set of problem variables this distance, d $[L]$, though this will, of course, again be constant.

We may wish to consider further physical effects. These might include the effect of rolling or frictional resistance or air resistance. These will have physical relationships that connect the variables to each other and to new dimensional constants. Since rolling or frictional resistance is likely to depend on the weight of the vehicle, I have added the gravitational acceleration g $[LT^{-2}]$ to the set of variables.

DIMENSIONAL ANALYSIS

Including the energy coefficient of gas E, the gravitational acceleration g, and the distance measurement, d, we have 7 dimensional variables and constants. With 4 fundamental dimensions $[M, L, T, G]$ we clearly need $7 - 4 = 3$ dimensionless πs to connect them.

Though problems of this size can easily be analysed by hand (Huntley, 1967; Taylor, 1974) a program in the statistical programming language, S, was written to carry out the analysis for a given basis, generate a list of the expressions for the πs, calculate the numerical values of the πs, and can then go on to fit a regression line.

One dimensional analysis with a particular choice of basis yields the following three dimensionless products. Normally one scales the products to have integer powers of the variables, if possible. Here the major variable (m, h, and q) in each product has been constrained to have power 1.

$$\frac{mwg}{E}, \quad \frac{h}{wd^{0.5}g^{1.5}}, \quad \frac{qg^{0.5}}{d^{0.5}} \tag{1}$$

The first product is a measure of energy efficiency, the second is a relationship between horsepower and weight, and the third is the ratio of q to the quarter mile time that would be achieved with an acceleration of one g.

REGRESSION WITH THE NEW DIMENSIONLESS PRODUCTS

DA by itself cannot get us further. In the absence of further physical or engineering input, we must leave the final determination of the relationship between these factors and the other dimensionless factors of the original problem to the usual methods of fitting. In terms of the products we have:

$$\frac{mwg}{E} = \phi\left(\frac{h}{wd^{0.5}g^{1.5}}, \frac{qg^{0.5}}{d^{0.5}}, \cdots\right) \tag{2}$$

where the ... represent the list of dimensionless variables such as $DRAT$ and CYL in the observations.

If we are interested in a forecast of the miles per gallon, m, we would expect a relation of the form:

$$m = \frac{E}{wg}\phi\left(\frac{h}{wd^{0.5}g^{1.5}}, \frac{qg^{0.5}}{d^{0.5}}, \cdots\right) \tag{3}$$

The simplest form of (2) would be

$$\frac{mwg}{E} = const \tag{4}$$

This was fitted to Hocking's data and the corresponding graph of m against w is plotted in Figure 1.

Henderson and Velleman (1982) noted after examining the data that it would be better to use w^{-1} as the dependent variable to get a linear relationship with m. Our dimensional analysis, and the graph in Figure 1, confirms this observation.

Direct regression with the πs will incorporate the natural nonlinearities of the problem but it will also bring in an alternative error structure. Minimising the sum of squared deviations about mwg/E is not the same as the (assumed) original problem of deviations about m. Of course, any other transformation of the variables (as might be applied by a statistician) would have a corresponding effect.

4. Conclusion

The DA literature is replete with warnings about the inappropriate use of the technique. First one should not make the mistake of applying it to situations in which even the physical

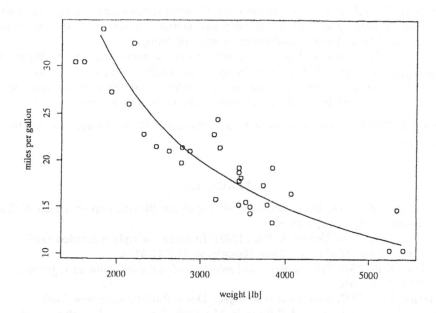

Fig. 1 Automobile Miles per Gallon, m, versus Weight, w. The curve is given by the simplest dimensionless model.

laws involved are unknown. Taylor (1974) remarks that such attempts "give the subject the reputation of being a black art, which in these circumstances it is indeed." Though the conversion from the original variables to a set of πs can be automatic, careful consideration of the problem is important. DA can only deal with the variables it is presented with and cannot invent them.

Once the variables are selected, our choice of the basis, and hence the particular set of πs, is also important. Though there is a fixed number of them, the set of πs can be manipulated by multiplying them together or dividing one by another to produce new, but still dimensionless, factors.

We therefore have the additional flexibility of choosing the best set of πs. One criterion is essentially an aesthetic one - good products are those that have a physical interpretation. For example, one product may represent the balance of two important effects. Many specially named dimensionless products, such as Froude's and Reynold's numbers, are of this kind.

Alternatively, in an appeal to Ockham's razor, we can choose a set that forms the simplest model once the regression is finished. We might be lucky in that such a model would also have a clear physical meaning.

Nor are we limited to linear regression in the πs. We would, however, hope that a good fitted model would also be simple, either linear or of simple polynomial form. Even here there is some freedom of choice for the analyst and more sophisticated but still simple models might be tested. Kasprzak et al, (1990, p.28) give an example of a search for such a simple model.

Using another criterion, we might use a non-linear programming algorithm to minimise the residuals not only by fitting the regression parameters but also, by changing the basis, the πs to be used in it. Work is continuing on ways of doing this.

While there is often a tendency to regard the data as mere numbers, to throw them into a computer package and then try to interpret the results, good statisticians recognise the need for careful preliminary investigations. I believe that Dimensional Analysis will prove to be a useful addition to the set of tools available for this purpose.

ACKNOWLEDGMENTS. I wish to thank John L. Scott and Ms M. Vignaux for assistance in preparing this paper.

REFERENCES

Buckingham,E.: 1914, 'On physically similar systems: Illustrations of the use of dimensional equations', *Phys.Rev.*, 4:345–76.

Henderson, H.V. and Velleman, P.E.: 1982, 'Building multiple regression models interactively', *Biometrics*, 37:391–411, June. (discussion 38, 511-516).

Hocking, R.R.: 1976, 'The analysis and selection of variables in linear regression', *Biometrics*, 32:1–49, March.

Huntley, H.E.: 1967, *Dimensional Analysis*. Dover Publications, New York.

Kasprzak, W., Lysik, B, and Rybaczuk, M.: 1990, *Dimensional Analysis in the Identification of Mathematical Models*. World Scientific Press.

Naddor, E.: 1966, 'Dimensions in operations research', *Operations Research*, 14:508–514.

Rayleigh, J.W.S.: 1915, 'The principle of similitude', *Nature*, 95(66):591 and 644.

Taylor, E.S.: 1974, *Dimensional Analysis for Engineers*. Clarendon Press, Oxford.

Vignaux, G.A.: 1986, 'Dimensional analysis in operations research', *New Zealand Operational Research*, 14:81–92.

Vignaux, G.A. and Jain, S.: 1988, 'An approximate inventory model based on dimensional analysis', *Asia-Pacific Journal of Operational Research*, 5(2):117–123.

ENTROPIES OF LIKELIHOOD FUNCTIONS

Michael Hardy
School of Statistics
University of Minnesota
Minneapolis, Minnesota 55455 USA

ABSTRACT. We show that the normalized likelihood function needed to get from a prior proba-bility vector to the posterior that results from the minimum cross-entropy inference process has the highest entropy among all probability vectors satisfying an appropriate set of linear constraints. We regard the domains of the entropy and cross-entropy functions as groups. The relationship between the group structure and the entropy and cross-entropy functions is used to derive the result on entropies of likelihood functions.

1. Minimum Cross-Entropy Inference

The *entropy* of a probability vector $(p_1, \ldots, p_n)$ is $H(p_1, \ldots, p_n) = -\Sigma_i p_i \log p_i$. The *cross-entropy* of $(p_1, \ldots, p_n)$ relative to $(q_1, \ldots, q_n)$ is $H((p_1, \ldots, p_n)/(q_1, \ldots, q_n)) = \Sigma_i p_i \log p_i/q_i$. There is no minus sign in the definition of cross-entropy because we are following the conventions of Shore and Johnson (1980). In the minimum cross-entropy inference pro-cess we start with a prior probability vector $(q_1, \ldots, q_n) = (P(K_1 \mid I), \ldots, P(K_n \mid I))$. We then receive new information J that says the intrinsic frequencies or population proportions $q_1^*, \ldots, q_n^*$ of $K_1, \ldots, K_n$ satisfy $(q_1^*, \ldots, q_n^*)A = (b_1, \ldots, b_k)$, where A is $n \times k$ and $k \leq n - 2$. The vector of posterior probabilities $(p_1, \ldots, p_n) = (P(K_1 \mid I, J), \ldots, P(K_n \mid I, J))$ is then the unique probability vector that minimizes cross-entropy relative to $(q_1, \ldots, q_n)$ subject to the constraint J. The maximum entropy inference process is the special case in which the prior is uniform. In that case the entropy is a decreasing function of the cross-entropy.

2. Likelihood Functions

Is there a normalized likelihood function $(r_1, \ldots, r_n)/(r_1 + \cdots + r_n) = (P(J \mid I, K_1), \ldots, P(J \mid I, K_n))/(\text{normalizing constant})$ that takes us from the prior $(q_1, \ldots, q_n)$ to the cross-entropy minimizing posterior $(p_1, \ldots, p_n)$ when Bayes' theorem is applied? If D is any event for which $(P(D \mid I, K_1), \ldots, P(D \mid I, K_n))/(\text{normalizing constant}) = (r_1, \ldots, r_n) = (p_1/q_1, \ldots, p_n/q_n)/((p_1/q_1) + \cdots + (p_n/q_n))$ then clearly this will suffice. Let $(d_1, \ldots, d_k) = (r_1, \ldots, r_n)A$. We claim that among all probability vectors $(s_1, \ldots, s_n)$ satisfying $(d_1, \ldots, d_k) = (s_1, \ldots, s_n)A$ the one with the highest entropy $H(s_1, \ldots, s_n)$ is $(r_1, \ldots, r_n)$. This will be a corollary to our main result below.

This appears to be an intimate connection between Bayes' theorem and the minimum cross-entropy inference process. Another such connection was discovered by Arnold Zellner

C. R. Smith et al. (eds.), Maximum Entropy and Bayesian Methods, Seattle, 1991, 127–130.
© 1992 Kluwer Academic Publishers.

and described in Zellner (1988). Zellner used a concept of entropy of a likelihood function according to which the entropy depends on the posterior probabilities, in contrast to the definition above.

Several questions arise immediately. Does it really make sense to say that $(r_1, \ldots, r_n)$ is the likelihood function $(P(J \mid I, K_1), \ldots, P(J \mid I, K_n))/(\text{normalizing constant})$ where the new information J is the constraint $(q_1^*, \ldots, q_n^*)A = (b_1, \ldots, b_k)$? What is meant by the weighted averages $(r_1, \ldots, r_n)A$ when the vector $(r_1, \ldots, r_n)$ is a likelihood function rather than a probability measure? Is $H(r_1, \ldots, r_n)$ the amount of information in the likelihood function? (I think the answer to this last question is probably "no".) I cannot answer these questions now, and so am uncertain exactly what the significance of this result is, but (pardon the heretical frequentist language) it seems unlikely that this proposition would have been true under the null hypothesis that it has no such import. (A member of the audience said that the results of this talk are too pretty to not to be of some import. I believe this sort of "pure" mathematicians' intuition is a much better predictor of this sort of thing than a few skeptics here seem to think.)

Having said I am unsure of the import of this proposition, I will claim below that it is relevant to numerical computations. Here I am treading on thin ice, because I am not familier with entropy maximizing and cross-entropy minimizing algorithms, and the foremost expert on this field is in the audience. Perhaps I will be redoing in somewhat different language something very familiar to those who work with such algorithms.

3. Bayes groups

We define the nth Bayes group $\mathcal{B}_n$ to be the set of all n-tuples $(p_1, \ldots p_n)$ of strictly positive real numbers such that $p_1 + \ldots + p_n = 1$ with the operation $(p_1, \ldots, p_n) \circ (q_1, \ldots, q_n) = (p_1 q_1, \ldots, p_n q_n)/(p_1 q_1 + \ldots + p_n q_n)$. This is an abelian group. The unit element is $\mathbf{e} = (1/n, \ldots, 1/n)$. The inverse of $(p_1, \ldots, p_n)$ is $(1/p_1, \ldots, 1/p_n)/((1/p_1) + \ldots + (1/p_n))$. For $\mathbf{p} \in \mathcal{B}_n$ and t real we define $\mathbf{p}^t = (p_1^t, \ldots, p_n^t)/(p_1^t + \ldots + p_n^t)$. This is a group homomorphism from the additive group of real numbers into $\mathcal{B}_n$. In particular, it satisfies the familiar laws of exponents: $\mathbf{p}^{t+s} = \mathbf{p}^t \circ \mathbf{p}^s$, $\mathbf{p}^0 = \mathbf{e}$, $\mathbf{p}^1 = \mathbf{p}$. Here is one interpretation of this group operation. If $\mathbf{q} = $ prior probabilities and $\mathbf{r} = $ normalized likelihood of new data, then $\mathbf{p} \circ \mathbf{r} = $ posterior probabilities. Despite the disturbing asymmetry of this interpretation, it will be of some interest to us below. Here is another interpretation that is more symmetric. If $\mathbf{p} = (P(D \mid I, K_1), \ldots, P(D \mid I, K_n))/(\text{normalizing constant})$ and $\mathbf{r} = (P(E \mid I, K_1), \ldots, P(E \mid I, K_n))/(\text{normalizing constant})$ and D and E are conditionally independent given $K_1, \ldots, K_n$ then $\mathbf{p} \circ \mathbf{r} = (P(DE \mid I, K_1), \ldots, P(DE \mid I, K_n))/(\text{normalizing constant})=$composite likelihood function.

(Apparently the fact that Bayes' theorem defines a group operation has rarely been noticed. James Dickey has pointed out that he mentioned it in Dickey (1968).)

4. The Main Theorem

For $\mathbf{p} \in \mathcal{B}_n$ let $f(\mathbf{p}) = \mathbf{p}A$, where A is $n \times k$ and $k \leq n - 2$. Let $\mathbf{q} \in \mathcal{B}_n$. Define an *entropy maximizing point* of f to be a point with higher entropy than any other point in the same level set of f, and a *cross-entropy minimizing point relative to* $\mathbf{q}$ of f to be a point with lower cross-entropy relative to $\mathbf{q}$ than any other point in the same level set of f. If the matrix A is of rank k and a column of 1's is linearly independent of the columns of A then the

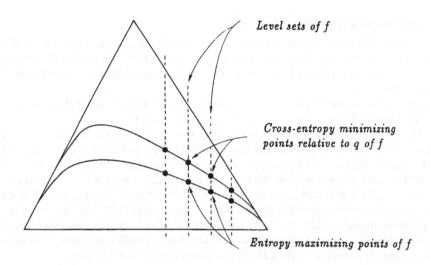

Level sets of f

Cross-entropy minimizing points relative to q of f

Entropy maximizing points of f

set of entropy maximizing points is a k-dimensional submanifold of the $(n-1)$-dimensional manifold B_n.

Main Theorem: (1) The set of entropy maximizing points of f is a subgroup (called the entropy maximizing subgroup associated with f). (2) The set of cross-entropy minimizing points relative to q of f is the coset containing q, of the entropy maximizing subgroup.

A proof is sketched in the next section.

Corollary: Let $q \sim p$ mean p is a cross-entropy minimizing point relative to q of f. Then $\sim$ is an equivalence relation, i.e., $\sim$ is reflexive, symmetric, and transitive. This follows from the fact that q and p belong to the same coset of the entropy maximizing subgroup.

Corollary: Suppose A is $n \times 1$, so that f is scalar-valued. Let the constraint be an inequality $pA \geq c$. Then the transitivity of $\sim$ implies consistency of updating in the minimum cross-entropy inference process when later information consists of a stronger inequality.

Corollary: The likelihood r needed to get from the prior q to the posterior p in minimum cross-entropy inference is an entropy maximizing point. This follows immediately from the first corollary above.

Now we consider an application to computation. Suppose A is $n \times 1$. Let r be the entropy maximizing point in any level set of f except the one containing e. Then every cross-entropy minimizing point relative to q is of the form $q \circ r^t$ for some real number t. Thus we can simply start at $t = 0$ and move in the right direction until $f(q \circ r^t)$ is where we want it. If we have, e.g., twenty problems with different priors and different constraints on the value of f, we need not run a cross-entropy minimizing algorithm twenty times. We need only run an entropy maximizing algorithm once.

If A is $n \times k$ where $k > 1$ then we would need k different entropy maximizing points $r_1, \ldots, r_k$ and k different scalars $t_1, \ldots, t_k$. Instead of $q \circ r^t$ we would use $q \circ r_1^{t1} \circ \cdots \circ r_k^{tk}$. Unfortunately now we have infinitely many directions to choose from, rather than only two.

But at least we have reduced an n-dimensional problem to a k-dimensional problem.

5. Sketch Proof of the Main Theorem

The proof can of course be summarized in two words: Lagrange multipliers. However, rather than just leave it at that I want to point out that the Lagrange multipliers give us an explicit group isomorphism from the entropy maximizing subgroup to euclidean k-space with addition.

Suppose $\mathbf{p}$ is an entropy maximizing point of f. Then there are scalars $\mu_0, \mu_1, \ldots, \mu_k$ such that $-(\mathrm{grad}\ H)(\mathbf{p}) = \mu_0(1, \ldots, 1) + \mu_1(a_{11}, \ldots, a_{n1}) + \ldots + \mu_k(a_{1k}, \ldots, a_{nk})$. (The vector $(1, \ldots, 1)$ comes from the constraint that says $\mathbf{p}$ is a probability vector.) Likewise if $\mathbf{q}$ is another entropy maximizing point then there are scalars $\nu_0, \nu_1, \ldots, \nu_k$ such that $-(\mathrm{grad}\ H)(\mathbf{p}) = \nu_0(1, \ldots, 1) + \nu_1(a_{11}, \ldots, a_{n1}) + \cdots + \nu_k(a_{1k}, \ldots, a_{nk})$. It is an exercise to check that $-(\mathrm{grad}\ H)(\mathbf{p} \circ \mathbf{q}^{-1}) = \lambda_0(1, \ldots, 1) + \lambda_1(a_{11}, \ldots, a_{n1}) + \cdots + \lambda_k(a_{1k}, \ldots, a_{nk})$ if $\lambda_i = \mu_i - \nu_i$ for $i \neq 0$ and $\lambda_0 = \mu_0 - \nu_0 - 1 + \log(p_1/q_1 + \cdots + p_n/q_n)$. The equality $\lambda_i = \mu_i - \nu_i$ for $i \neq 0$ implies that the map $\mathbf{p} \mapsto (\mu_1, \ldots, \mu_k)$ (note that μ_0 is not here!) is a group homomorphism. The strict upward concavity of the entropy function implies there can be no point other than $\mathbf{e}$ where all of the Lagrange multiplier are zero. This means the homomorphism is one-to-one. Because of the dimensions it is surjective.

REFERENCES

Dickey, James: 1968, Three Multidimensional-Integral Identities with Bayesian Applications, *Annals of Mathematical Statistics* **39**, 1615-1627.

Paris, J.B., and Vencovská, A.: 1990, A Note on the Inevitability of Maximum Entropy, *International Journal of Approximate Reasoning* 4, 183–223.

Shore, John E. and Johnson, Rodney W.: 1980, Axiomatic Derivation of the Principle of Maximum Entropy and the Principle of Minimum Cross-Entropy, *IEEE Transactions on Information Theory* **I.T.-26**, 26–37.

Zellner, Arnold: 1988, Optimal Information Processing and Bayes's Theorem, *The American Statistician* **42**, 278–284.

Maximum Likelihood Estimation of the Lagrange Parameters of the Maximum Entropy Distributions

Ali Mohammad–Djafari
Laboratoire des Signaux et Systèmes (CNRS–ESE–UPS)
École Supérieure d'Électricité
Plateau de Moulon, 91192 Gif–sur–Yvette Cédex, France

ABSTRACT. The classical maximum entropy (ME) problem consists of determining a probability distribution function from a finite set of expectations $\mu_n = \mathrm{E}\{\phi_n(x)\}$ of known functions $\phi_n(x), 0, \ldots, N$. The solution depends on $N + 1$ Lagrange multipliers which are determined by solving the set of nonlinear equations formed by the N data constraints and the normalization constraint. The problem we address here is different. It consists of estimating these Lagrange multipliers when the available data are the M samples $\{x_1, \ldots, x_M\}$ of the random variable X in place of the expectations μ_n. We propose and compare two methods: the maximum likelihood (ML) one and the method of moments (MM). We show also an interesting relation between the classical ME method and the ML method for this problem. Finally, we show the interest of these developpements in determining the prior law of an image in a Bayesian approach to solving the inverse problems of image restoration and reconstruction.

1. Introduction

Maximum Entropy (ME) and Maximum Likelihood (ML) are two different approaches to make inference about the probability law of a random variable (rv) X. The objectif of both approaches is to choose a probability distribution function (pdf) for the rv X which best represents the observed data. When the data are the expectations of some known functions, the ME approach is used and when the data are the samples of X, the ML approach is used. Let's resume these two approaches.

Maximum Entropy (ME)
Suppose that there exists a *pdf* $p(x) \in \mathcal{P}$ defined by:

$$\mathcal{P} = \left\{ p(x) \mid \mathrm{E}\{\phi_n(x)\} = \mu_n, n = 0, \ldots, N \right\} \tag{1}$$

where $\mu_0 = 1$, $\phi_0(x) = 1$ and $\phi_n(x), n = 1, \ldots, N$ are N known functions. If the only data available on the rv X are $\mu_n, n = 0, \ldots, N$ then the ME approach suggest us to choose $p(x)$ as

$$\hat{p}(x) = \arg \max_{p \in \mathcal{P}} \left\{ - \int p(x) \ln p(x) \, \mathrm{d}x \right\} \tag{2}$$

131

C. R. Smith et al. (eds.), Maximum Entropy and Bayesian Methods, Seattle, 1991, 131–139.
© 1992 Kluwer Academic Publishers.

or, equivalently

$$\text{maximize} \quad H = -\int p(x) \ln p(x) \, dx,$$

$$\text{subject to} \quad E\{\phi_n(x)\} = \int \phi_n(x) p(x) \, dx = \mu_n \quad n = 0, \ldots, N. \tag{3}$$

The classical solution of this problem is given by

$$p(x) = \exp\left[-\sum_{n=0}^{N} \lambda_n \phi_n(x)\right]. \tag{4}$$

The $(N + 1)$ Lagrange parameters $\lambda = \{\lambda_0, \ldots, \lambda_N\}$ are obtained by solving the following set of $(N + 1)$ nonlinear equations:

$$G_n(\lambda) = \int \phi_n(x) \exp\left[-\sum_{n=0}^{N} \lambda_n \phi_n(x)\right] \, dx = \mu_n, \quad n = 0, \ldots, N. \tag{5}$$

Maximum likelihood (ML)

Suppose that the *pdf* $p(x)$ of the *rv* X is known to belong to the parametric family

$$\mathcal{P} = \{p(x) \mid p(x) = p(x; \lambda), \lambda \in \Lambda\}. \tag{6}$$

Now, if the data available on X are the M independent samples $\{x_1, \ldots, x_M\}$, the ML approach suggests us to choose $p(x)$ as

$$\hat{p}(x) = \arg\max_{p \in \mathcal{P}} \left\{ p(x; \lambda) = \prod_{i=1}^{M} p(x_i; \lambda) \right\} \tag{7}$$

or equivalently:

$$\hat{\lambda} = \arg\max_{\lambda \in \Lambda} \left\{ L(\lambda) = \ln \prod_{i=1}^{M} p(x_i; \lambda) \right\} \tag{8}$$

The problem we address here is the following: We do not have directly the values $\mu_n, n = 0, \ldots, N$ to apply the ME approach. We have M independent samples $x = \{x_1, \ldots, x_M\}$ of the *rv* X and we make the hypothesis that the *pdf* of X is known to belong to the ME family (4). This hypothesis fixes the form and the number of its parameters. The problem now becomes a classical parameter estimation one. In what follows, we describe first, two methods, the maximum likelihood (ML) method and the method of the moments (MM) to solve this problem. We show then the equivalency between the two following algorithms.

• Determine $\mu_n, n = 1, \ldots, N$ by their empirical estimates:

$$\mu_n = \frac{1}{M} \sum_{i=1}^{M} \phi_n(x_i), \quad n = 1, \ldots, N \tag{9}$$

and use the ME approach to calculate the corresponding parameters λ.

• Suppose that X has a ME *pdf* as in (4) and use the ML approach to estimate its parameters λ.

We will see also that, in this last case, the two methods ML and MM lead to the same results only when $\phi_n(x) = x^n, n = 0, \ldots, N$. We will give then the details of the numerical implementation of these methods and compare the results obtained in the special case of the Gamma distribution where $\phi_1(x) = x$, and $\phi_2(x) = \ln(x)$.

Finally, we show the interest of these developpements in determining the prior law of an image in a Bayesian approach to solving the inverse problems of image restoration and reconstruction. Noting that the parameters of the prior law play the role of the regularization parameter, we will see how these developments can propose us a method to determine these parameters.

2. Standard Maximum Entropy Distribution

We have seen that the solution of the standard ME problem is given by (4) in which the Lagrange multipliers λ are obtained by solving the nonlinear equations (5). We give here the details of the numerical method that we implemented. When developing the $G_n(\lambda)$ in the equations (5) in first order Taylor's series around the trial λ^0, the resulting linear equations are given by

$$G_n(\lambda) = G_n(\lambda^0) + (\lambda - \lambda^0)^t \left[\mathbf{grad}\, G_n(\lambda)\right]_{(\lambda=\lambda^0)} = \mu_n, \quad n = 0, \ldots, N. \quad (10)$$

Noting the vectors $\boldsymbol{\delta}$ and $\boldsymbol{v}$ by

$$\boldsymbol{\delta} = \lambda - \lambda^0,$$

$$\boldsymbol{v} = \left[\mu_0 - G_0(\lambda^0), \ldots, \mu_N - G_N(\lambda^0)\right]^t,$$

and the matrix $\boldsymbol{G}$ by

$$\boldsymbol{G} = \left(g_{nk}\right) = \left(\frac{\partial G_n(\lambda)}{\partial \lambda_k}\right)_{(\lambda=\lambda^0)} \quad n, k = 0, \ldots, N, \quad (11)$$

then equation (10) becomes

$$\boldsymbol{G}\,\boldsymbol{\delta} = \boldsymbol{v}. \quad (12)$$

This system is solved for $\boldsymbol{\delta}$ from which we drive $\lambda = \lambda^0 + \boldsymbol{\delta}$, which becomes our new vector of trial λ^0 and the iterations continue until $\boldsymbol{\delta}$ becomes appropriately small. Note that the matrix $\boldsymbol{G}$ is a symmetric one and we have

$$g_{nk} = g_{kn} = -\int \phi_n(x)\,\phi_k(x)\,\exp\left[-\sum_{n=0}^{N} \lambda_n\,\phi_n(x)\right]\,dx \quad n, k = 0, \ldots, N. \quad (13)$$

Note that in each iteration we have to calculate the $N(N-1)/2$ integrals in equation (13). These are calculated numerically using the standard univariate Simpson's rule. The procedure has been programmed in MATLAB and given in the annex of reference [5].

3. Maximum Likelihood Estimation

In the standard ME problem we are directly given the values $\mu_n, n = 0, \ldots, N$. The problem we address here is different. We do not have these data. What we have is the M samples $x = \{x_1, \ldots, x_M\}$ of the random variable X. But we suppose that the random variable X has a probability density function $p(x)$ which has the following form

$$p(x) = \frac{1}{Z(\lambda)} \exp\left[-\sum_{n=1}^{N} \lambda_n \phi_n(x)\right] \tag{14}$$

depending on the N parameters $\lambda = \{\lambda_1, \ldots, \lambda_N\}$. $Z(\lambda)$ is the normalization constant which is called also the partition function. What we propose here is to use the ML approach to determine λ. If the M samples $x = \{x_1, \ldots, x_M\}$ are obtained independently. The likelihood function $l(\lambda)$ is given by

$$l(\lambda) = \prod_{i=1}^{M} p(x_i; \lambda) = \prod_{i=1}^{M} \frac{1}{Z(\lambda)} \exp\left[-\sum_{n=1}^{N} \lambda_n \phi_n(x_i)\right]. \tag{15}$$

The log-likelihood $L(\lambda)$ is then given by:

$$L(\lambda) = -M \ln Z(\lambda) - \sum_{i=1}^{M} \sum_{n=1}^{N} \lambda_n \phi_n(x_i). \tag{16}$$

The ML estimate $\hat{\lambda}$ of λ is then obtained by solving the following set of equations:

$$\frac{\partial \ln Z(\lambda)}{\partial \lambda_n} = \frac{1}{M} \sum_{i=1}^{M} \phi_n(x_i), \quad n = 1, \ldots, N, \tag{17}$$

where

$$Z(\lambda) = \int \exp\left[-\sum_{n=1}^{N} \lambda_n \phi_n(x)\right] dx, \tag{18}$$

or equivalently by:

$$\int \phi_n(x) \exp\left[-\sum_{n=1}^{N} \lambda_n \phi_n(x)\right] dx = \frac{1}{M} \sum_{i=1}^{M} \phi_n(x_i), \quad n = 1, \ldots, N. \tag{19}$$

Comparing this set of equations with the set of equations in (5), we see that $\mu_n, n = 0, \ldots, N$ are replaced by their empirical estimates

$$\mu_n = \frac{1}{M} \sum_{i=1}^{M} \phi_n(x_i), \quad n = 0, \ldots, N \tag{20}$$

from the samples $x_i, i = 1, \ldots, M$. So, the same numerical method used in the precedent section can be used to determine the parameters λ. The only difference is that, in a first step we must calculate the empirical estimate μ_n using (20) from the samples $x = \{x_1, \ldots, x_M\}$ and use them as the input of the standard ME program.

4. Method of Moments

The method of moments to estimate the parameters $\lambda = \{\lambda_0, \ldots, \lambda_N\}$ of a probability distribution function $p(x; \lambda)$ consists in relating the $(N + 1)$ moments of the *pdf* to the $(N+1)$ parameters of the *pdf* and try to invert them to find an estimate of these parameters. Let us define

$$G_n(\lambda) = \int x^n\, p(x; \lambda)\, dx, \quad n = 0, \ldots, N. \tag{21}$$

The method of moments to estimate λ consists of trying to solve the set of of $(N + 1)$ equations

$$G_n(\lambda) = \int x^n \exp\left[-\sum_{n=1}^{N} \lambda_n\, \phi_n(x)\right] dx = \frac{1}{M} \sum_{i=1}^{M} x_i^n, \quad n = 0, \ldots, N. \tag{22}$$

Comparing this equation with (19), we see that they become the same when $\phi_n(x) = x^n$, $n = 0, \ldots, N$. Note that in this last case, when using the numerical procedure of Section 2, we note that

$$g_{nk} = g_{kn} = -G_{n+k}(\lambda) \quad n, k = 0, \ldots, N \tag{23}$$

This means that the $[(N + 1) \times (N + 1)]$ matrix G becomes a symmetric Hankel matrix which is entirely defined by $2N - 1$ values $G_n(\lambda), n = 0, \ldots, 2N$. But in the general case this method is not equivalent to the ML method. To clarify the differences between the ME, ML and MM methods we summerize these three methods in the following.

- **ME**

  Problem: Given $\phi_n(x), \mu_n, n = 0, \ldots, N$, determine the ME probability distribution function of X.

  Solution: $p(x; \lambda)$ is the form

  $$p(x; \lambda) = \exp\left[-\sum_{n=0}^{N} \lambda_n\, \phi_n(x)\right], \tag{24}$$

  where the Lagrange multipliers λ are determined by solving

  $$G_n(\lambda) = \int \phi_n(x) \exp\left[-\sum_{n=0}^{N} \lambda_n\, \phi_n(x)\right] dx = \mu_n, \quad n = 0, \ldots, N. \tag{25}$$

- **ML**

  Problem: Given $\phi_n(x), n = 0, \ldots, N$, the M samples $x = \{x_1, \ldots, x_M\}$ and the knowledge that $p(x; \lambda)$ is the form (24), determine the parameters λ by the ML method.

  Solution: parameters λ are determined by solving

  $$G_n(\lambda) = \int \phi_n(x) \exp\left[-\sum_{n=0}^{N} \lambda_n\, \phi_n(x)\right] dx = \frac{1}{M} \sum_{i=1}^{M} \phi_n(x_i), \quad n = 0, \ldots, N. \tag{26}$$

- **MM**

    Problem: Given $\phi_n(x), n = 0, \ldots, N$, the M samples $x = \{x_1, \ldots, x_M\}$ and the knowledge that $p(x; \lambda)$ is the form (24), determine the parameters λ by the MM method.

    Solution: parameters λ are determined by solving

$$G_n(\lambda) = \int x^n \exp\left[-\sum_{n=0}^{N} \lambda_n \phi_n(x) \right] dx = \frac{1}{M} \sum_{i=1}^{M} x_i^n, \quad n = 0, \ldots, N. \tag{27}$$

Comparing (25) and (26), we see that the only difference is that the μ_n in (25) are replaced by their empirical estimates in (26). Comparing (26) and (27), we see that they will become equivalent only if $\phi_n(x) = x^n$.

5. Numerical Experiments

We consider the case of the Gamma distribution

$$p(x; \alpha, \beta) = \frac{\beta^{(1-\alpha)}}{\Gamma(1-\alpha)} x^\alpha \exp(-\beta x), \quad x > 0, \alpha < 1, \beta > 0. \tag{28}$$

This distribution can be considered as a ME distribution when the constraints are

$$\int p(x; \alpha, \beta) \, dx = 1 \quad \text{normalization} \quad \phi_0(x) = 1,$$

$$\mathrm{E}\left\{x\right\} = \int x p(x; \alpha, \beta) \, dx = \mu_1 \qquad\qquad \phi_1(x) = x, \tag{29}$$

$$\mathrm{E}\left\{\ln x\right\} = \int \ln(x) p(x; \alpha, \beta) dx = \mu_2 \qquad\qquad \phi_2(x) = \ln(x).$$

Note that we have an analytical relation between (α, β) and $\left(m = \mathrm{E}\left\{x\right\}, \sigma^2 = \mathrm{E}\left\{(x - m)^2\right\}\right)$ which is

$$\begin{cases} m = (1-\alpha)/\beta \\ \sigma^2 = (1-\alpha)/\beta^2 \end{cases} \text{ or inversely } \begin{cases} \alpha = (\sigma^2 - m^2)/\sigma^2 \\ \beta = m/\sigma^2 \end{cases}. \tag{30}$$

Now consider the following problems

1. Given μ_1 and μ_2 determine α and β and derive the mean m and the standard deviation σ. This is the standard ME method. Given μ_1 and μ_2, we can calculate α and β by the standard ME method. We can then use (30) to determine m and σ. Table (1) gives some numerical results.

Table 1.

μ_1	μ_2	α	β	m	σ^2
0.2000	-3.0000	0.2156	-3.0962	0.2533	0.0818
0.2000	-2.0000	-0.4124	-6.9968	0.2019	0.0289
0.3000	-1.5000	-0.6969	-5.3493	0.3172	0.0593

2. Given M samples $x_i, i = 1, \ldots, M$ determine (α, β) by either the ML method or the method of moments and derive the mean m and the standard deviation σ of the samples.

• In the ML method, first we estimate $\mu_1 = E\{x\}$ and $\mu_2 = E\{\ln x\}$ by their empirical estimates

$$\mu_1 = \frac{1}{M} \sum_{i=1}^{M} x_i \quad \text{and} \quad \mu_1 = \frac{1}{M} \sum_{i=1}^{M} \ln x_i, \tag{31}$$

then using these values in (26) will give us the parameters α and β from which we can derive m and σ using (30).

• In the MM method, first we estimate $\mu_1 = E\{x\}$ and $\mu_2 = E\{x^2\}$ by their empirical estimates

$$\mu_1 = \frac{1}{M} \sum_{i=1}^{M} x_i \quad \text{and} \quad \mu_1 = \frac{1}{M} \sum_{i=1}^{M} x_i^2, \tag{32}$$

then using these values in (27) will give us the parameters α and β from which we can deduce m and σ using (30).

Tables (2) and (3) give the results obtained, respectively, by the ML method and by the moments method, for the 3 different sample size, $M = 4096, 1024$ and 512. These samples are driven from a Gamma distribution with theoretical parameters $\alpha = -.2$ and $\beta = .3$.

Table 2.: Maximum Likelihood

Sample size	α	β	m	σ^2
$M = 4096$	-0.1872	0.2988	3.9732	13.2967
$M = 512$	-0.2029	0.2895	4.1547	14.3502
$M = 256$	-0.1412	0.2894	3.9433	13.6255

Table 3.: Method of Moments

Sample size	α	β	m	σ^2
$M = 4096$	-0.1782	0.2966	3.9718	13.3888
$M = 512$	-0.0880	0.2620	4.1521	15.8454
$M = 256$	-0.0787	0.2737	3.9417	14.4036

What we see from these two tables is that, in all cases, ML method has given better estimates of α and β than MM method.

6. Application to Bayesian ME image reconstruction

In many image restoration and reconstruction problems we have to solve the following problem

Determine the image $x \in \mathbf{R}_+^n$ from the data $y \in \mathbf{R}^m$

$$y = Ax + b, \tag{33}$$

where x represents the image, y the measured data, b the error and the noise and A is a known matrix which is, in general, either singular or ill-conditioned.

In a bayesian approach to solve this problem we have first to assign the probability distributions $p(x)$ and $p(y \mid x)$ and then use the Baye's rule to obtain the posterior law

$$p(x \mid y) \propto p(y \mid x) p(x). \tag{34}$$

Invoking the ME principle we have shown [5,6] the following
- When the noise b is a priori assumed to be white, with a covariance matrix

$$W = \mathrm{diag}\left[\frac{1}{\sigma_1^2}, \ldots, \frac{1}{\sigma_m^2}\right], \tag{35}$$

the ME principle will give us

$$p(y \mid x) \propto \exp\left[-Q(x)\right] \quad \text{with} \quad Q(x) = \left[y - Ax\right]^t W \left[y - Ax\right]. \tag{36}$$

- When our prior information about the image x is in the form

$$\begin{cases} \mathrm{E}\{\phi_1(x)\} = \mu_1 \\ \mathrm{E}\{\phi_2(x)\} = \mu_2 \end{cases} \quad \text{with} \quad \begin{cases} \phi_1(x) = \sum_{i=1}^n H(x_i) \\ \phi_2(x) = \sum_{i=1}^n S(x_i) \end{cases}, \tag{37}$$

the ME principle will give us

$$p(x) = \sum_{i=1}^n p(x_i) \propto \exp\left[-\lambda_1 \phi_1(x) - \lambda_2 \phi_2(x)\right]. \tag{38}$$

- An argument of scale invariance of the $p(x)$ yields a restricted set of admissible forms for $H(x)$ and $S(x)$:

$$\left\{(S(x), H(x))\right\} = \left\{(x^{r_1}, x^{r_2}), (x^{r_1}, \ln x), (x^{r_1}, x^{r_1} \ln x), (\ln x, \ln^2 x)\right\}. \tag{39}$$

Now, replacing (38) and (36) in (34) we obtain

$$p(x \mid y) \propto \exp\left\{-Q(x) - \lambda_1 \phi_1(x) - \lambda_2 \phi_2(x)\right\}, \tag{40}$$

and if we choose, as the solution, the maximum a posteriori (MAP), we have

$$\hat{x} = \arg\max_{x>0}\left\{p(x \mid y)\right\} = \arg\min_{x>0}\left\{J(x) = Q(x) + \lambda_1 \phi_1(x) + \lambda_2 \phi_2(x)\right\}. \tag{41}$$

At this point we can give a regularization interpretation to this result: $\phi_1(x)$ and $\phi_2(x)$ are the regularization functionals and $\{\lambda_1, \lambda_2\}$ are the regularization parameters. A well-posed Bayesiam ME problem is then the following

Given; $\{\sigma_1, \ldots, \sigma_M\}$, (prior about the noise b); $\{\mu_1, \mu_2\}$, (prior about the image x); y, (measured data) and the model A; estimate the parameters $\{\lambda_1, \lambda_2\}$ and the image x. This is a well-posed problem because $\{\lambda_1, \lambda_2\}$ can be determined from $\{\mu_1, \mu_2\}$ using the ME approach and then the MAP estimate $\hat{x}$ for the image x is determined by minimizing $J(x)$ in (41).

In practice, two problems must still be faced

- In fact, the parameters $\{\sigma_1,\ldots,\sigma_M\}$ and $\{\mu_1,\mu_2\}$ or equivalently $\{\lambda_1,\lambda_2\}$ are not known, so they must be estimated along with the image x.
- Practical computation of the true MAP solution is far from obvious. In general, implementable algorithms only yield suboptimal solutions, i.e., local minima of $J(x)$.

What we propose is an approximate and iterative solution to the following problem
Given; $\{\sigma_1,\ldots,\sigma_M\}$; y, (measured data) and the model A; estimate the parameters $\{\lambda_1,\lambda_2\}$ and the image x.

The proposed algorithm is
1. Start with a good initial estimate $x^{(0)}$.
2. From the current estimate $x^{(k)}$ infer about the parameters $\{\lambda_1,\lambda_2\}^{(k)}$ using either the (ME and ML) or (ME and MM) approaches described above.
3. Use a conjugate gradient method to reach a local minimum of the criterion $J(x)$.
4. Then iterate the procedure from step 2, until the variation of the estimated parameters $\{\lambda_1,\lambda_2\}$ is no longer significant, and adopt the final estimate as $\hat{x}$.

7. Conclusions

In this paper we addressed first the class of the ME distributions when the data available are a finite set of expectations $\mu_n = \mathrm{E}\{\phi_n(x)\}$ of some known functions $\phi_n(x), n = 0,\ldots,N$. This is the classical ME problem. Then we proposed two methods (ML and MM) to estimate the parameters of these distributions when we dispose of M samples of a random variable in place of expectations $\mu_n, n = 0,\ldots,N$. We have seen also that the MM and the ML methods will become the same in the special case where $\phi_n(x) = x^n, n = 0,\ldots,N$. We gave also some numerical results in the case of the Gamma distribution which can be considered as a special case of a ME distributions with $\phi_1(x) = x$ and $\phi_2(x) = \ln x$. Finally, we discussed the interest of these developements in a Bayesian approach with the maximum entropy priors to solve the inverse problem in image reconstruction.

REFERENCES

[1] A. Zellnerr and R. Highfiled, "Calculation of Maximum Entropy Distributions and Approximation of Marginal Posterior Distributions", *Journal of Econometrics* 37, 1988, 195–209, North Holland.

[2] Jaynes, "Papers on probability, statistics and statistical physics", Reidel Publishing Company, Dordrecht, Holland, 1983.

[3] Matz, "Maximum Likelihood parameter estimation for the quartic exponential distributions", *Technometrics*, 20, 475–484, 1978.

[4] Mohammad-Djafari A. et Demoment G., "Estimating Priors in Maximum Entropy Image Processing," *Proc. of ICASSP 1990*, pp: 2069-2072

[5] Mohammad-Djafari A. et Idier J., "Maximum entropy prior laws of images and estimation of their parameters," *Proc. of The 10th Int. MaxEnt Workshop, Laramie, Wyoming*, published in Maximum-entropy and Bayesian methods, T.W. Grandy ed., 1990.

[6] Mohammad-Djafari, "A Matlab Program to Calculate the Maximum Entropy Distributions," *Proc. of The 11th Int. MaxEnt Workshop, Seattle, USA*, 1991.

ENTROPY OF FORM AND HIERARCHIC ORGANIZATION

J. Wagensberg† and R. Pastor‡
† Museu de la Ciència de la
Fundació "la Caixa"
Teodor Roviralta, 55
08022 Barcelona (Spain).

‡ Departament de Física Fonamental
Universitat de Barcelona
Diagonal, 647
08028 Barcelona (Spain).

ABSTRACT. The use of the Theory of Information as a mathematical tool for the understanding of complex systems allows the application of the MaxEnt Principle as a criterion for the prediction of the final stationary states. The processes of *growth and differentiation* are subject to a similar treatment by means of a new entropic magnitude. The iterative construction of fractals suggests a mathematical model of the hierarchic processes of growth and differentiation whose final stationary states are defined by the MaxEnt Principle. This can be interpreted in a more general context concerning the concept of Adaptation.

1. Introduction

Entropic magnitudes and the MaxEnt Principle are time-honored tools for the study of complex systems, especially in situations when we lack information, i.e. the applications in the prediction of average magnitudes in statistical mechanics (in a situation where there is an extreme lack of information), the reconstruction of images hidden in noises, the prediction of ecological stationary states, etc. In a recent paper (Wagensberg et al., 1990) a model was suggested for the study of a rather general range of complex systems. This model can be represented by a flow network. Given a network of fluxes of some extensive magnitude, we define the *structural probabilities* (describing the network's internal structure). Following the Mathematical Theory of Information, these probabilities determine, in their turn, an Entropy of Connections and an Information Transfer. The Entropy of Connections is the central magnitude of a principle of adaptation in the context of the so-called MaxEnt philosophy, that is to say: stable networks ("adapted" to their environment, which is defined by a particular set of constraints) are those that maximize the Entropy of Connections.

This is a good starting point for the study of the adapatation of certain complex systems. However, many natural and fundamental complex systems cannot easily be represented by a flow-network description. This is the case of the so-called hierarchical systems, that is to say, a system exhibiting a structure with different levels, each one of which is

141

C. R. Smith et al. (eds.), Maximum Entropy and Bayesian Methods, Seattle, 1991, 141–151.
© 1992 *Kluwer Academic Publishers.*

ruled by different particular dynamic laws. For exemple, let us think of any biological organization. Living matter is characterized by different scales representing barriers between different levels of organization: biomolecules, cells, organs, living beings, populations, etc. Each level is ruled by laws that belong to that particular scale. It is obvious that hierarchic organizations of this kind can be considered (at least to a certain extent) as the result of an adaptation in which the different hierarchic levels "compete" among one another (given the conditions imposed by the environment) in order to satisfy some performance: to grow, to develop the complexity of the system, etc. It is quite clear that macroscopic restrictions might have an effect on microscopic levels, although the prevailing philosophy in physics only considers the other way around: the understanding of the higher levels starting from more elemental or microscopic interactions. Our central suggestion here is that these hierarchic situations can be explained from an adaptive perspective by maximizing an entropic magnitude that takes into account the contribution to the variety of the whole of the different hierarchic levels. This suggestion is visualized with a mathematical model based on the so-called *Iterated Function System*. In this model a new magnitude called *The Entropy of Form* can be used as an evolutionary potential.

2. The MaxEnt as a principle of adaptation

Many complex natural systems can be represented by a mathematical graph. Here nodes are some previously assigned compartments among which an exchange of a certain extensive magnitude takes place. In a previous paper (Wagensberg *et al.*, 1990) we suggested a more general study of a weighted and directed network, that is to say, a graph made up of n nodes and a certain number of arcs to which the intensity of flow of a certain extensive magnitude has been associated. Three sets of structural probabilities for each node are then easily defined: the probabilities of emission of energy by the generic node i, $p(x_i)$ the probabilities of reception of energy by the node j, $p(x_j)$; and a set of n^2 conditional probabilities, being the probabilities of emission of energy of node i, that is captured by node j, $p(x_i/y_j)$, namely, the ratio between the energy emitted by node i that reaches j and the total energy emitted by node i, for $i,j = 1,\ldots,n$. In this view, the system is characterized by its total entropy, the Entropy of Connections or connectivity:

$$H(X,Y) = -\sum_i \sum_j p(x_i, y_j) \log p(x_i, y_j) \qquad (1)$$

and by its Information Transfer:

$$I(X,Y) = H(X) + H(Y) - H(X,Y). \qquad (2)$$

Both magnitudes characterize a flow network structure of complexity Γ:

$$\Gamma(X,Y) = (H(X,Y), I(X,Y)). \qquad (3)$$

The possibility for a system of complexity Γ to exist is measured by some probability of such a system to belong to, let us say, some universe of real entities (Wagensberg *et al.*, 1990). Here we apply the celebrated *Principle of Maximum Entropy* in the context of the adaptive processes (Jaynes, 1957):

The least biased probability assignment (our structural probabilities) is that which maximizes the entropy (our connectivity H) subject to imposed constraints (our criterion of reality).

In this view, concepts such as adaptation and evolution are quantitatively introduced. The basic property of entropic magnitudes, such as the defined connectivity, is that there are more systems with a high entropy value than with a low one. An entropy function measures the degree of adaptation of a given system in a given real world. Adaptation simply means the adaptation to some given constraints through the MaxEnt Principle. On the other hand, evolution (innovation) implies a change in these constraints. This suggests that a principle of adaptation can always be proposed in relation to any definition of an entropy-like function. This is, in fact, a whole research method devoted to adaptive processes: an entropy-like function, a universe of realities and the MaxEnt Principle. We next use this method in order to start with a key question in both physics and mathematics: the *Hierarchic Organization*.

3. Entropy of Form

Let us start with an elementary mathematical model of a process of growth and differentiation, based on the so-called *deterministic fractals* (Barnsley, 1988).

These processes, especially significant for biological phenomena, are traditionally studied in the context of Non Equilibrium Thermodynamics. This is done by means of the entropy balance equation:

$$\frac{1}{m}\frac{dS}{dt} = \left\{\frac{1}{m}\frac{dm}{dt}\right\}\frac{S}{m} + \frac{d}{dt}\left\{\frac{S}{m}\right\} \tag{4}$$

This is the entropy balance representing the production of new mass together with the differentiation of the existing mass (Bermúdez *et al.*, 1986).

Starting from equation (4), we consider the whole process of growth and differentiation as a process of *adaptation* within our conceptual scheme. We can indeed consider growth and differentiation as a general process by means of which a system captures mass (or another extensive magnitude) from the rest of the universe and adds it on to its own structure. In fact, a growing and differentiating system shows two competing effects, namely:

- Acquisition of a great deal of mass that arranges itself according to certain given rules in a simple structure.
- Acquisition of little mass that arranges itself in a complicated structure.

The whole process can then be considered as the trend to a final state in which both behaviors are compatible with the maximum stability. Let us consider the most simple case: the production of a fractal object (Mandelbrot, 1982) according to a deterministic algorithm. The easiest way to introduce this concept is through the well-known Koch curve on the plane R^2. We start its generation using two simple forms, an *initiator* and a *generator*. The latter is a broken line made up of N segments equal in length r. Every iteration starts with a broken line, and it consists of replacing every rectilinear interval by a reduced copy of the generator shifted in such a way that it has the same extremes as the interval being replaced. By iterating the process n times, we will obtain a collection of N^n segments of length proportional to r^n, which will make up a curve of a total length :

$$L(n) \approx (Nr)^n \tag{5}$$

and which we characterize for r constant by a *mass*:

$$M(n) \stackrel{\text{def}}{=} N^n \tag{6}$$

which measures the number of copies of the initial segment that make up the last one, after n iterations.

　　We can now understand this process in a new way. In general, a real process of growth takes place in scales of time that we consider continuous; however, it is also reasonable to consider processes in which prevail scales of time large enough to be observed only in discrete values, that is to say, when noticeable macroscopic changes occur. We consider the iteration order n as a *discrete measure of time* and a state of growth characterized by a mass $M(n)$ and a length $L(n)$. The following state $n + 1$ is obtained by means of a relative mass increment:

$$\frac{\Delta M(n)}{M(n)} \equiv \frac{M(n+1) - M(n)}{M(n)} = N - 1 \tag{7}$$

This increment is "reorganized" over the original form to give a length $L(n + 1)$ in accordance with a pre-established pattern given by the iteration algorithm. The algorithm itself describes differentiation. This is, in fact, our model for growth and differentiation: a fractal structure whose number of points (its length in the case of the curve) increases with time and that arranges itself in space according to a specified deterministic algorithm. It should be pointed out that the geometrical algorithmic construction does not actually imply an acquisition of mass but a growing succession of points in space (as we shall see better further on).

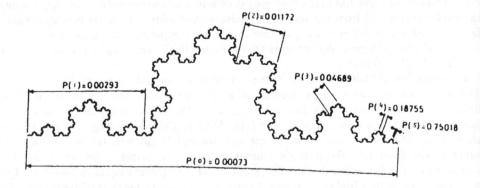

Fig. 1 Five iterations of the Koch's curve. The virtual segments are labelled with the corresponding probabilities.

We now introduce a new magnitude, the *Entropy of Form*, a measure of diversity which will take into account, in each state of iteration n, the whole of the preceding states which it has been through, that is, the whole process of growth. In order to do so, we start with a straight line of length 1 and we apply n times the particular iterative process. The final curve after n steps (the *n-fractal curve*) is a broken line made up of N^n intervals of length r^n. Nevertheless, we can also identify (Fig. 1) N^{n-1} intervals of length r^{n-1}, and in general, N^i intervals of length r^i, for $i = 0, 1, \ldots, n$. The global form of the n-fractal will, of course, depend on all this population of intervals of $n + 1$ different kinds of lengths, and its final form cannot be explained other than through the preceding forms. Some structural probabilities $p_n(i)$ are now easy to define:

$$p_n(i) = \frac{N^i}{M} \tag{8}$$

for $N > 1$, $i = 0, \ldots, n$, where M is a normalization factor:

$$M = \sum_{i=0}^{n} N^i = \frac{N^{n+1} - 1}{N - 1}. \tag{9}$$

This is a "good" (in the sense of Kolmogorov) distribution of probability that defines the diversity of the general form of order n, an actual Shannon entropy:

$$S(N, n) = S\{p_n(0), \ldots, p_n(n)\} = -\sum_{i=0}^{n} p_n(i) \log p_n(i) \tag{10}$$

bits per individual. Introducing the relations (8), (9) into the equation (10) we obtain:

$$S(N, n) = \frac{N}{N - 1} \log N - \log(N - 1) - \\ - \frac{N^{n+1}}{N^{n+1} - 1} \log(N^{n+1}) + \log(N^{n+1} - 1) \tag{11}$$

Our model of growth and differentiation provides then an entropic magnitude that measures the diversity of the different states the process goes through. The next step, of course, is to extend the Entropy of Form to more general constructions.

4. Generalization of the Entropy of Form

The most general way of considering the construction of fractal objects by iterative algorithms is by the so-called *Iterated Function Systems* (IFS) (Barnsley, 1988). A function:

$$\omega : R^d \longrightarrow R^d \tag{12}$$

is a *contraction* if there is a real $s \in R, 0 < s < 1$, called contractivity factor, such that:

$$\| \omega(x) - \omega(y) \| \leq s \| x - y \|, \quad \forall x, y \in R^d, \tag{13}$$

where the double bar shows the Euclid distance. An IFS is just a set of N contractions, $\mathcal{F} = \{\omega_\alpha;\ \alpha = 1, \ldots, N\}$, that applies on the space of all R^d compact sets:

$$A_0 \subset R^d, \quad \mathcal{F}(A_0) \overset{\text{def}}{=} \bigcup_\alpha \omega_\alpha(A_0) = \bigcup_\alpha \{\omega_\alpha(x), \forall x \in A_0\}. \tag{14}$$

A deterministic fractal is defined as an IFS attractor, through the limit of the following iterative process:

$$\begin{aligned} A_0 &\subset R^d \\ A_1 &= \mathcal{F}(A_0) \\ A_n &= \mathcal{F}(A_{n-1}) = \mathcal{F}^{(n)}(A_0) \\ A &\equiv \lim_{n \to \infty} A_n \quad \text{definition of attractor.} \end{aligned} \tag{15}$$

This attractor A can also be defined as being the only fixed point of the IFS, satisfying:

$$A = \mathcal{F}(A) \equiv \bigcup_\alpha \omega_\alpha(A) \tag{16}$$

We call the set A_n an *n-fractal*; it depends on the initial set but these dependencies vanishes in the limit (15).

Now, the Entropy of Form of the Koch curve can easily be generalized for the process producing some n-fractal A_n. We start with an arbitrary compact set A_0 (the *virtual image of order* 0). To this set we apply the IFS and we obtain A_1, a *virtual image of order* 1, a new image made up of N images of A_0. After n iterations A_n is obtained. It is the result of the union of N^n images of A_0 through all possible variations of the functions ω_α: these are *virtual images of order n*. Therefore, equation (8) also expresses, in this case, the probability of a virtual image to be of order i. This means that the same equation (11) also defines the Entropy of Form for a n-iterated IFS. The material building up the real object is distributed among the real images, but all the preceding virtual images also intervene in the diversity of form of the resulting n-fractal object.

The Entropy of Form (11) has the following fundamental properties:

P–1 The Entropy of Form is a well-defined function for any value of n and N, with $N \geq 1$; that is to say, we can associate a measure of diversity to any deterministic n-fractal.

P–2

$$S(N, n+1) > S(N, n) \tag{17}$$

P–3 The limit n tending to infinity leads to a finite entropy:

$$S(N) \equiv \lim_{n \to \infty} S(N, n) = \frac{N}{N-1} \log N - \log(N-1) \tag{18}$$

In the limit of infinite iteration we obtain a real fractal object which we can characterize by means of a well defined entropy value, $S(N)$.

P–4 The following inequalities are satisfied:

$$\begin{aligned} 0 &\leq S(N, n) < S(N) \\ 0 &\leq S(N) \leq S(2) \end{aligned}$$

The Entropy of Form is bounded for any value of n and increases with this variable. Both parameters appearing in the Entropy of Form $S(N, n)$ have a clear adaptive interpretation in our view. The parameter N, which we could call *order of complexity*, definitely represents the complexity of the structure growing in an algorithmic sense: the more functions in the IFS we need to specify a structure, the more complex it happens to be. In a certain sense, we could thus associate with N a quantitative measure of the necessary "genetic" contents to fully describe the process of growth and differentiation (the IFS). On the other hand, the parameter n, which we could call order of iteration, provides an idea of the level that the growth has reached. In a certain sense, we could give n the value of the "age" of the process of growth and differentiation.

Two competing aspects appear in this model of growth: the complexity with which the acquired mass (N) organizes itself, and the gain of new mass (n). The prevalence of either one factor or the other determines the dominant character of the whole process. But in any case, and this is our central working hypothesis, the MaxEnt Principle (the maximization of the Entropy of Form) determines the most stable final state. However, maximization of this function leads to trite results, as it lacks the local maximums and minimums for the rank of physical values of N and n. The suggested model of growth for a single hierarchic level is too simple an example. In the next section we extend these concepts to non-trivial cases.

5. The Extended Entropy of Form

Nature exhibits hierarchic organizations whose different levels are characterized by different values of the fractal dimension (Mandelbrot, 1982): coral reefs (Bradbury *et al.*, 1984), deciduous forest patterns (Krümmel *et al.*, 1987), etc. Fractal hierarchies with a constant fractal dimension are therefore observable for small-scale changes. On the other hand, large-scale changes show clear variations in the fractal dimension reporting a transition between hierarchical levels. In this sense one could affirm that hierarchic structures (i.e. fractal structures) are the result of some processes of growth and differentiation in which the patterns of organization of the acquired mass undergo discontinuous variations for certain values of the total mass. At a certain level, the aggregation of elementary particles leads to the formation of a set that turns out to be the elementary particle for the construction of the immediately higher level. We now extend our mathematical description in order to include this fundamental idea.

Hierarchic processes can be described by means of new constructions that we call *Meta Iterated Function Systems* (MIFS). Let us consider μ IFS's given by the sets of contractions on R^d:

$$\mathcal{F}_1 = \{\omega_{\alpha_1}; \; \alpha_1 = 1, \ldots, N_1\}$$

$$\vdots \tag{19}$$

$$\mathcal{F}_\mu = \{\omega_{\alpha_\mu}; \; \alpha_\mu = 1, \ldots, N_\mu\}$$

We start with a compact A_0 of R^d to which we apply n_μ iterations using the set of functions $\mathcal{F}_\mu$. We now apply $n_{\mu-1}$ iterations to the resulting set using the set of functions $\mathcal{F}_{\mu-1}$, and so on and so forth, up to $\mathcal{F}_1$. The resulting final set is:

$$A \equiv \mathcal{F}_1^{(n_1)} \circ \cdots \circ \mathcal{F}_\mu^{(n_\mu)}(A_0) \equiv \bigcup_{\alpha_1} \cdots \bigcup_{\alpha_\mu} \omega_{\alpha_1}^{(n_1)} \circ \cdots \circ \omega_{\alpha_\mu}^{(n_\mu)}(A_0) \tag{20}$$

where the indexes between brackets show the composition of functions. This object (the *n-fractal hierarchic* object), is characterized by a succession of fractal dimension values $D_1, D_2, \ldots, D_\mu$ (corresponding to the different sets $\mathcal{F}_j$) at intervals of decreasing characteristic lengths. We now look for the corresponding entropic magnitude.

Let us consider the process MIFS as the construction of the geometrical object A (20), starting from the compact A_0 and going through the distribution of virtual images:

$$A_{i_1,\ldots,i_\mu} \equiv \mathcal{F}_1^{(i_1)} \circ \cdots \circ \mathcal{F}_\mu^{(i_\mu)}(A_0) \tag{21}$$
$$i_j = 1, \ldots, N_j$$

The populations of these distributions of virtual images are given by:

$$N_1^{i_1} \cdots N_\mu^{i_\mu} \tag{22}$$

from which we obtain some well-defined *structural probabilities* which determine the distribution of images:

$$p(i_1, \ldots, i_\mu) = \frac{N_1^{i_1} \cdots N_\mu^{i_\mu}}{\sum_{j_1=0}^{n_1} \cdots \sum_{j_\mu=0}^{n_\mu} N_1^{j_1} \cdots N_\mu^{j_\mu}} \tag{23}$$
$$= p_{n_1}(i_1) \cdots p_{n_\mu}(i_\mu)$$

From the probabilistic point of view, the MIFS process coincides with a process of the combination of independent events (IFS), so that its entropy will be the addition of the entropies of the different independent events, namely:

$$S(A) \equiv \sum_{j=0}^{\mu} S(N_j, n_j) \tag{24}$$

We shall call this function the *Extended Entropy of Form*. It corresponds, of course, to a particular MIFS. As can be seen, this is a non-trivial function that accepts relative maximums for certain imposed external constraints.

Let us now analyze this model of hierarchic growth and differentiation from our adaptive perspective. The different patterns of organization $\{\mathcal{F}_j\}$, iterated a number of times $\{n_j\}$, compete among each other for the capturing of mass, which takes place at every level. Let us also consider that a series of constraints (that we know as C) influences the model. A natural constraint in any process of growth is a limitation on the maximum quantity of matter available. (This is related to the availability of resources from the environment in real systems.) In our model we shall express these constraints through the non-holonomic relation:

$$M = N_1^{n_1} \cdots N_\mu^{n_\mu} \le M_0; \tag{25}$$

This relation fixes a mass measured by the number of points of the final object A, being A_0 of finite number of points and N in (25) the relative increase of mass (eq. (6)). The process of growth takes place in accordance with the schemes defined by the MIFS $\{\mathcal{F}_j\}$. The final state of the growth is defined giving a set of values $\{n_j\}$; given the constraints C, the possible states compatible with them will be given by $\{n_j\}^C$. We understand the process of adaptation as competition, among the different hierarchic levels, to agglutinate mass among all the possible states $\{n_j\}^C$ to acquire that of the maximum stability. This is, again, the situation maximizing the Extended Entropy of Form (24) under the set of constraints C.

6. An Exemple

In order to illustrate the model, we consider the case of a computer simulation of a two-level hierarchy, the one that makes up the *leaf-tree system*, a familiar example in biology. In most cases it is not to be expected that both turn out to be fractal objects. However, it is easy to see examples of both cases, which enable us to consider this case as reproducible by the MIFS. As models of leaf and tree, we have chosen the IFS codes (Barnsley, 1982), given by the affinity transformations, detailed in Tables 1 and 2 respectively. The codes are expressed as follows:

$$\omega(\vec{x}) = \begin{pmatrix} a & b \\ c & d \end{pmatrix} \vec{x} + \begin{pmatrix} e \\ f \end{pmatrix}$$

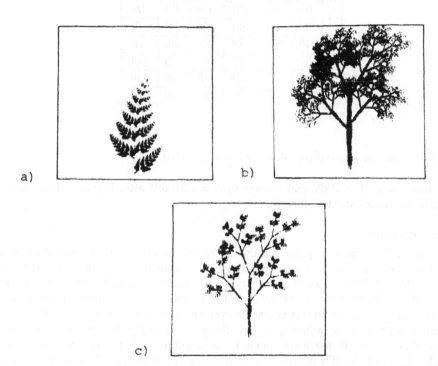

Fig. 2 a) A fractal leaf. b) A fractal tree. c) The leaf-tree compatibility predicted by MaxEnt.

We consider the maximum available mass $M_0 = 10^5$ as the only constraint imposed on the system. We next apply the MaxEnt Principle by maximizing by numerical methods the Extended Entropy of Form. The result obtained is that the most probable structure is given by the values of the iterations $n_2 = 4$ (leaf) and $n_1 = 3$ (tree), which correspond

	a	b	c	d	e	f
1	.00	.00	.00	.16	.00	.00
2	.85	.04	-.04	.85	.00	.26
3	.20	-.26	.22	.22	.00	.26
4	-.15	.28	.24	.24	.00	.07

Table 1 IFS code for a fractal fern, from (Barnsley, 1988). The parameters of the affine transformation has been slightly changed in order to implement our computer graphic algorithm.

	a	b	c	d	e	f
1	.05	.00	.00	.60	.00	.00
2	.05	.00	.00	-.50	.00	1.00
3	.46	-.32	.38	.38	.00	.60
4	.47	-.15	.17	.42	.00	1.00
5	.43	.27	-.25	.48	.00	1.00
6	.42	.25	-.35	.31	.00	.70

Table 2 IFS code for a fractal tree, from (Frame *et al.* 1990).

with a total mass $M = 55296$ and an entropy $S = 1.276395$ nits. Figure 2 shows what the corresponding leaves and tree look like.

7. Final comment

Our approach is a starting point for the study of one of the more interesting aspects of complex systems: the hierarchic organization. Living matter is of course the central paradigm. Indeed, Nature exhibits many systems with an essential fractal structure (Feder *et al.*, 1990). Many of them can be considered to be the result of a growth process (Vicsek, 1989) that maintains the fractal character over an interval of the growth. But the first criticism is related to the difficulty of describing a real object by means of an attractor of a set of functions. In particular, there is not a unique way to carry out such a thing, since the choice of the IFS functions is to some extent arbitrary. Nevertheless, there is, in principle, some hope concerning this indetermination. The key is the so-called *Collage Theorem* (Barnsley, 1988). This theorem states, roughly speaking, that any subset of R^d can produce an approximation (as the attractor of some IFS) of a real object, the error of that approximation (measured according to Haussdorf's metrics on compacts) being as small as required. Further "economic" criteria, which emerge from the so called *Algorithmic Information Theory* (Chaitin, 1975) could perhaps finally help to solve the problem.

     To sum up, our approach and our example are devoted to mathematical objects. Nevertheless, many real hierarchic organizations suggest that some entropy of form can be defined in each case (the virtual images are clearly observable in particular real structures). Our belief is that the whole conceptual scheme should remain useful in order to provide predictions concerning the compatibility of consecutive hierarchic levels.

REFERENCES

Barnsley, M.F.: 1988, *Fractals Everywhere*, Academic Press Inc., San Diego.

Bermúdez, J. and Wagensberg, J.: 1986, 'On the Entropy Production in Microbiological Stationary States', *J. Theor. Biol.* **122**, 347.

Bradbury, R. H., Reichtel, R. E. and Green, D. G.: 1984, 'Fractals in Ecology—Methods and Interpretation', *Mar. Ecol.Prog. Series*, **14** ,295.

Chaitin, G.: 1975, 'Randomness and Mathematical Proof', *Scientific American*, **232** (5), 47.

Feder, J. and Aharony, A. (eds).: 1990, *Fractals in Physics; Essays in Honour of B. B. Mandelbrot*, North-Holland, Netherlands.

Frame, M. and Erdman, L.: 1990, 'Coloring Schemes and the Dynamical Structure of Iterated Function Systems', *Computer in Physics*, **4** ,500.

Jaynes, E. T.:1957, 'Information Theory and Statistical Mechanics', *Phys. Rev.*, **106**, 620.

Krummel, J. R., Gardner, R. H., Sugihara, G., O'Neill, R. V. and Coleman, P. R.: 1987, 'Landscape Patterns in Disturbed Environment', *OIKOS*, **48**, 321.

Mandelbrot, B. B.: 1982, *The Fractal Geometry of Nature*, W. H. Freeman & Co., New York.

Vicsek, T.: 1989, *Fractal Growth Phenomena*, World Sientific, Singapore.

Wagensberg, J.,García, A. and Solé, R. V.: 1990, 'Energy Flow-Networks and the Maximum Entropy Formalism', in *Maximum Entropy and Bayesian Methods*, Kluwer, Dordrecht.

A BAYESIAN LOOKS AT THE ANTHROPIC PRINCIPLE

A.J.M. Garrett
Department of Physics and Astronomy
University of Glasgow
Glasgow G12 8QQ
Scotland, U.K.

ABSTRACT. The anthropic principle — that there is a correlation between key properties of the universe, and our presence within it — is analysed from a Bayesian point of view.

"Don't let me catch anyone talking about the Universe in my department."

Lord Rutherford [1]

1. Introduction

The anthropic principle refers to the idea that there exists a connection between material properties of the universe and the presence of human life (Greek: $\alpha\nu\theta\rho\omega\pi\sigma\varsigma$ = human). Therefore the observed fact of human life tells us something — nontrivial, we hope — about the universe.

Few ideas so spectacularly polarise a roomful of physicists. Reactions range, even among well-known physicists, from "the key to quantum cosmology", to "mere tautology". The idea is prominent in the (distorted) lay expositions of fundamental physics associated with the New Age movement. Nevertheless, if one reaches an anomalous answer by correct reasoning, one can criticise only the question. The correct format for reasoning is now known to be Bayesian. Let us examine the logical structure of the anthropic principle in this light.

Some preliminary work must be done, and some misconceptions dispelled. This paper was to have contained a paragraph on the distinction between correlation and causation, suggesting that while one particular universe might cause life to arise within it, such life could hardly be said to *cause* that disposition of the universe. Correlation, by contrast, is symmetrical and presents no difficulty. Correlation is easy to define, but causation has no invariant definition: it depends on which variables are taken as independent and which as dependent, and this is an arbitrary choice. By convention we take whatever variable we control as independent, but this differs in different experiments. For example, we can test the pressure-volume relation for a fixed mass of gas at constant temperature by moving (and holding) a piston along a scale, waiting for the temperature to settle down, and reading the pressure off a manometer; or we can arrange for the piston to move freely in a vertical direction, load it with standard weights to select the pressure, and read the volume from the scale. The inescapable conclusion is that theories, and their equations, deal exclusively with *correlations* between variables. (Of course, there is another use of the word

153

C. R. Smith et al. (eds.), Maximum Entropy and Bayesian Methods, Seattle, 1991, 153–164.

causation, denoted *time-ordering*; but this is separate.) Confusion also surrounds the word 'explain', which is reserved for supposedly causative connections; but is now meaningful in the correlative sense. It is up to us to state whether we are satisfied with an explanation; if not, further correlations must continue to be sought.

The central principle of logic involved in the anthropic principle is Bayes' theorem: if A, B, C denote propositions which may be true or false, then, in standard probability notation,

$$p(A|BC) = \frac{p(A|C)p(B|AC)}{p(B|C)}. \tag{1}$$

We begin by reviewing Bayesian probability theory.

2. Bayesian Probability

The probability $p(A|C)$ that A is true, assuming the truth of proposition C, represents the *degree of belief* which it is consistent to hold in the truth of A, given C. This is a number between 0 (false) and 1 (true) and is unique: precise algorithms exist for constructing it, and the same result is obtained when these are correctly implemented whether the hardware be one person's brain, another person's brain, an insect's brain or an electronic chip. Misunderstandings arise because different persons generally possess different information, and thereby assign different probabilities to the same proposition; in general, $p(A|C) \neq p(A|C')$. This is at the root of the misconception that Bayesian probability is flawed through being 'subjective', and it motivates the *frequentist* theory, in which probability is declared an observable quantity by identifying it with proportion. Observable quantities can, of course, be measured 'objectively'; but the probability of a proposition, A (say) is not observable and should not be, since room is denied for $p(A|C)$ and $p(A|C')$ to differ. A different probability is assigned to rain tomorrow if one happens to overhear an atrocious weather forecast. Probabilities are *always* conditional on other assumed truths: there is no such thing as 'absolute' or 'unconditional' probability.

Probability is not the same as proportion, numerically (though it may be equal) or conceptually, since it is as applicable to one-off events — the weather — as it is to coin-tossing. It applies equally to past and future: Given the travails of historians, how likely is it that Richard III had the Princes murdered in the Tower? Probability always applies when there is insufficient information for deductive certainty, and deductive logic is a special case; probability *is* inductive logic. Further, there is no such thing as a 'random' process; only an unknown one.

Let us state the rules governing the assignment and manipulation of probabilities. For assignment there is the principle of maximum entropy, which we need not discuss here (see [2,3]), and for manipulation, there are the familiar product and sum rules

$$p(AB|C) = p(A|BC)p(B|C), \tag{2}$$

$$p(A|C) + p(\bar{A}|C) = 1. \tag{3}$$

Since probability is a theory of logic, and not of physical science, these rules can be derived only by logical, rather than observational, means. In particular, the theory must pass tests of internal consistency, and this necessary condition in fact suffices to derive (2) and (3). For example, suppose the rule for decomposing the logical product is

$$p(XY|Z) = F\big(p(X|YZ), p(Y|Z)\big). \tag{4}$$

Dependence of $p(XY|Z)$ on all of $p(X|YZ)$, $p(Y|XZ)$, $p(X|Z)$, $p(Y|Z)$ can be whittled down by a combination of logical argument and symmetry to (4). We now use this relation to decompose the probability of the logical product of three propositions in two different ways: With $X = AB$, $Y = C$, $Z = D$, (4) gives

$$p(ABC|D) = F\big(p(AB|CD), p(C|D)\big). \tag{5}$$

A further decomposition of $p(AB|CD)$ in the RHS gives, putting now $X = A$, $Y = B$, $Z = CD$, the result

$$p(ABC|D) = F\big(F\big(p(A|BCD), p(B|CD)\big), p(C|D)\big). \tag{6}$$

A different decomposition is

$$p(ABC|D) = F\big(p(A|BCD), p(BC|D)\big) \tag{7}$$
$$= F\big(p(A|BCD), F\big(p(B|CD), p(C|D)\big)\big). \tag{8}$$

Since the logical product is commutative, these decompositions must be equal. With $u \equiv p(A|BCD)$, $v \equiv p(B|CD)$, $w \equiv p(C|D)$, this means that

$$F\big(F(u,v), w\big) = F\big(u, F(v,w)\big). \tag{9}$$

We quote the solution of this functional equation from the original 1946 paper of R. T. Cox [4]:

$$F(\alpha, \beta) = \phi^{-1}\big(\phi(\alpha)\phi(\beta)\big) \tag{10}$$

where ϕ is an arbitrary function, which must be single-valued for our theory to make sense; ϕ^{-1} is its inverse. The product rule (4) is therefore

$$\phi(p(XY|Z)) = \phi(p(X|YZ))\phi(p(Y|Z)), \tag{11}$$

and the $\phi(\bullet)$ can be absorbed into the $p(\bullet)$ without loss of generality, giving the familiar form of the product rule. Certainty is represented by $p = 1$: suppose that A is certain, given C. Then $p(AB|C) = p(B|C)$ and its probability cancels in (2), leaving $p(A|BC) = 1$ if A is certain, given C. It also follows that falsehood corresponds to $p = 0$: for suppose now that A is impossible, given C. Then AB is also impossible, given C, and $p(AB|C) = p(A|C)$, so that (2) becomes $p(FALSE) = p(FALSE)p(B|C)$ for arbitrary $p(B|C)$, whence $p(FALSE) = 0$.

The sum rule is derived by taking

$$p(\bar{X}|Z) = G\big(p(X|Z)\big). \tag{12}$$

Since $\bar{\bar{X}} = X$, we require $G^2 = I$ (in operational terms). This rule is used to express conjugates of propositions in terms of their affirmations. A relation of Boolean algebra is that $A + B = \overline{\bar{A}\bar{B}}$, and so the logical sum of A and B has probability

$$p(A + B|C) = p(\overline{\bar{A}\bar{B}}|C). \tag{13}$$

We use the product rule and (12) to express this in terms of $p(A|C)$, $p(B|C)$, $p(AB|C)$; interchanging A and B in the result and equating the two expressions yields, with $G(0) = 1$, a functional equation for G [4]:

$$uG\left(\frac{G(v)}{u}\right) = vG\left(\frac{G(u)}{v}\right). \tag{14}$$

This has solution [4]

$$G(\alpha) = \left(1 - \alpha^{1/n}\right)^n, \tag{15}$$

so that (12) becomes

$$p(X|Z)^n + p(\bar{X}|Z)^n = 1. \tag{16}$$

We may take $n = 1$ without loss of generality in the same way we absorbed ϕ into p at (11). The sum rule is the result.

If A and B are exclusive, so that AB is $FALSE$ and has probability zero, then (16) reduces to $p(A+B|C) = p(A|C)+p(B|C)$; more generally, given an exhaustive and exclusive set $A_1, A_2, ...$, then

$$p\left(\sum_i A_i|C\right) = \sum_i p(A_i|C) = 1. \tag{17}$$

Propositions can take the form "the value of the variable is x", which is often abbreviated to"x", as in "$p(x|C)$"; this is really a *function* of x. Should x be continuous, a probability density is defined.

Bayes' theorem (1) is a consequence of the product rule and commutativity of the logical product. A further consequence of the sum and product rules is the *marginalising rule*

$$p(A|C) = p(AB|C) + p(A\bar{B}|C). \tag{18}$$

This is used (with B, A interchanged) as

$$p(A|BC) = Kp(A|C)p(B|AC); \tag{19a}$$

$$K^{-1} = p(B|C) = p(A|C)p(B|AC) + p(\bar{A}|C)p(B|\bar{A}C). \tag{19b}$$

In this form of Bayes' theorem, K obviously acts as a normalising factor.

All of the relations we have derived for probabilities apply also to proportions, where they correspond to trivial algebraic identitites on the Venn diagram. We stress, however, that these applications are conceptually distinct.

More importantly, any deviation from the sum and product rules comprises a logical inconsistency in the strict sense that different numbers can be generated for the same probability conditioned on the same information; we have seen how the product rule was designed to prevent this. Many of the techniques of sampling theory, designed for data analysis of repeated samples, are inequivalent to (2) and (3) and can be made to look arbitrarily illogical by suitable datasets; Jaynes [5] gives examples.

In order to deal with unique events, like the weather, frequentists introduce the idea of the *ensemble*, a collection of imaginary possibilities. This enables them to maintain the notion that probability *is* a proportion, within this collection. Provided that calculations are properly executed, numerical results coincide with the Bayesian; but frequentists often talk as if the ensemble is real and this terminology is seriously misleading, for there is only one, imperfectly known, system. (A similar confusion afflicts the 'many worlds' interpretation of quantum mechanics, which has been discussed in anthropic contexts [6]. In order to allow for the possibility of improvement one instead posits hidden variables; ignorance of their values leads, by marginalisation, to the present, probabilistic mode of prediction in quantum measurements.)

Two propositions $A, \bar{A}$ and $B, \bar{B}$ are correlated, given C, if

$$S_A + S_B - S_{AB} > 0 \tag{20}$$

where

$$
\begin{aligned}
S_{AB} \equiv & - p(AB|C)\log p(AB|C) - p(A\bar{B}|C)\log p(A\bar{B}|C) \\
& - p(\bar{A}B|C)\log p(\bar{A}B|C) - p(\bar{A}\bar{B}|C)\log p(\bar{A}\bar{B}|C)
\end{aligned} \tag{21}
$$

and

$$S_A = -p(A|C)\log p(A|C) - p(\bar{A}|C)\log p(\bar{A}|C); \tag{22}$$

S_B is defined in a similar manner to S_A. (The base of logarithms exceeds unity.) The LHS of (20) is non-negative, and is zero only if $p(A|BC) = p(A|C)$ [in which case it is also equal to $p(A|\bar{B}C)$], so that the truth value of B does not affect the truth value of A, and $p(AB|C) = p(A|C)p(B|C)$: precisely what is meant by *uncorrelated*.

Before returning to the anthropic principle, let us illustrate the meaning of Bayes' theorem, (1). It is a principle of *inverse reasoning*: if A and B are correlated in such a way that the truth of A enhances the probability of truth of B (given C) so that $p(B|AC)/p(B|C) > 1$, then the converse holds: truth of B enhances the probability of truth of A, since this probability ratio equals $p(A|BC)/p(A|C)$ also. A theory, which states that what you saw was very likely, is preferentially boosted over a theory which asserts it is unlikely. You tend to believe that someone who runs an expensive car has a large salary, since people with large salaries are more likely to run expensive cars. This is the basic intuition which concerns us here and which is encapsulated in Bayes' theorem, though the theorem goes further: it also shows that the more likely is a proposition to begin with, the more likely it remains after the truth of another proposition has been incorporated. From (18), the posterior $p(A|BC)$ increases with the prior $p(A|C)$. In this parlance, $p(B|AC)$ is called the *likelihood*.

3. Anthropic Principles

Denote by U the proposition that the values of various parameters $1, 2, \ldots$ in our model M of the universe take values $\{U_1, U_2, \ldots\}$. Denote by L the proposition that life evolves. Then

$$p(U|LM) = Kp(U|M)p(L|UM) \tag{23}$$

where

$$K^{-1} = \sum_U p(U|M)p(L|UM). \tag{24}$$

The dependence of $p(U|LM)$ on $p(L|UM)$ demonstrates that the values of U for which $p(L|UM)$ is larger correspond to larger values of $p(U|LM)$: Since life *is* observed, those parameter values which make life more probable are preferred. This is the 'car-salary' principle in action, and the fact that the observation of human life and the calculation are both performed by human beings is irrelevant.

To go further we need to specify both the likelihood $p(L|UM)$ and the prior $p(U|M)$. The assignment of prior probabilities is not yet developed enough to deal with this (though it has made a good start [2]). We can be confident, though, that the prior will be broad, not sharply peaked. Then, if the likelihood $p(L|UM)$ is sharply peaked in U, this peak transfers directly to the posterior $p(U|LM)$; by Taylor-expanding the prior about the peak in the likelihood, we have

$$p(U|LM) \approx K p(L|UM) \quad ; \quad K^{-1} = \sum_U p(L|UM). \tag{25}$$

Much, therefore, rests on the likelihood $p(L|UM)$.

The likelihood $p(L|UM)$ is assigned from our knowledge of how stars form and shine, according to the values U; how planets are created in orbit around stars; what conditions are needed for life on a planet; how evolution might find the survival strategy called intelligence (if L is taken to mean 'intelligent life'); and so on. There are huge gaps in our knowledge here; nevertheless, if only one step in the chain is sharply peaked in U, that sharp peak transfers to $p(L|UM)$ as the intermediate steps are marginalised over:

$$p(L|UM) = \sum_{a=A,\bar{A}} \sum_{b=B,\bar{B}} \dots \quad p(L|ab\dots UM)p(a|b\dots UM)p(b|\dots UM)\dots \tag{26}$$

There are well-known physical arguments why stars would not work if things were only a little different [7]. There are also sound arguments why intelligent life must at least begin by being carbon-based, and so evolve on a planet. We therefore have a sharp peak with respect to U in the likelihood $p(L|UM)$, and so in $p(U|LM)$.

Our observations are that life exists, and that U has the value $G \approx 6.67 \times 10^{-11}$ N m^2 kg^{-2}, $e \approx 1.60 \times 10^{-19}$ C, m_p (proton) $\approx 1.67 \times 10^{-27}$ kg , and so on; and we have a theory indicating a correlation between the presence of life and these values. Theory and practice are in accord; Bayesian logic is operating; all is well. For example, the hydrogen atom is unstable in four or more dimensions, and in one and two dimensions there is insufficient connectivity to allow intelligent networks of cells to develop; so intelligent life demands dimension $D = 3$ — in accord with direct observation. It is merely an accident of timing that the idea has not been used to find more accurately the values of fundamental constants than by direct measurement, and such arguments may play a useful role in the future; for example in estimating the values of new constants in a sub-quantum 'hidden' variables theory, which we know must be nonlocal [8,9] and therefore experimentally difficult. Furthermore, an argument based on the observation of life *has* been used, by Richard Dicke [7], to explain the striking observation that the dimensionless number $Gm_p^2/\hbar c \approx 0.6 \times 10^{-38}$ is roughly the inverse square root of the number of nucleons N in the universe. This exemplary argument is worth summarising: the number of nucleons is approximately the ratio of the mass of the universe to m_p, so that if ρ is the density of the universe, and ct its radius (expressed in light-years t), then

$$N \approx \frac{4\pi}{3} \rho \frac{(ct)^3}{m_p}. \tag{27}$$

Cosmology tells us that $\rho \approx (Gt^2)^{-1}$, so that $N \approx c^3 t / Gm_p$. Now the heavier elements found in our bodies (and elsewhere) were produced in supernovae, and the time, from the Big Bang, to reach an era in which supernovae have occurred but other stars continue to shine onto planets can be estimated from the mass and luminosity of a typical star. (Luminosity is related to the mass-energy loss rate.) The result is approximately $(Gm_p^2/\hbar c)^{-1}(\hbar/m_p c^2) \approx 10^{10}$ years, which gives

$$N \approx \frac{c^3}{Gm_p} \left(\frac{Gm_p^2}{\hbar c} \right)^{-1} \frac{\hbar}{m_p c^2} \tag{28}$$

$$\approx \left(\frac{Gm_p^2}{\hbar c} \right)^{-2}. \tag{29}$$

This and other, similar arguments are collected and sharpened (with extensive cultural and philosophical discourse) in Reference [6], particularly Chapters 5, 6, 8. Clearly the idea has explanatory power.

Let us further illustrate the logic using analogy. In arid regions, seeds of many plants germinate only after persistent rain. Those seeds should not be surprised (granting them the faculty of reason) to find, on emerging, that it is wet. They can deduce it is wet directly, by observation; or by logical argument, coupled with the observation of their own germination.

Let us rephrase the argument in order to intersect with other intuitions. We can ask: How *surprised* are we that the constants of nature take their observed values? Again this is a probability-based argument: The smaller the probability of the observed values based on our prior knowledge, the more surprised we are at finding them. But surprise must be bounded in some sense: We have no surprise that the constants fall *somewhere*. In fact, surprise at an event is usefully defined as the logarithm of the reciprocal of its probability, following which the sum rule expresses the bounding principle. To be surprised both at truth and at falsehood of a proposition is to reason inconsistently.

Arguments of the present sort were named the *weak anthropic principle* by Brandon Carter [10], as a reaction against the 'Copernican principle' that humanity occupies no privileged niche in the universe. (The Gaia hypothesis is another.) Carter tells us this in a further paper [11] in which he also points out the Bayesian character of the argument — though we do not find his exposition of Bayesianism convincing.

Carter [10,11] also discusses a more speculative *strong anthropic principle*, that the universe *must* have those properties which allow life to develop within it; that (for example) dimension D *must* equal three. This argument has a different logical status from the previous ones, which were simply applications of Bayesian reasoning; it is a proposition, something whose probability might be assigned *using* Bayesian analysis.

Given — since life exists — that the universe *does* have those properties conducive to life, what does it mean to say that it *must*? The strong anthropic principle (SAP) asserts that, with $\mathcal{U}_L, \mathcal{U}_{\bar{L}} \equiv \overline{\mathcal{U}_L}$ denoting the sets of values of U respectively conducive and non-conducive to life,

$$p(U \in \mathcal{U}_L | \text{SAP}, I) = 1 \quad , \quad p(U \in \mathcal{U}_{\bar{L}} | \text{SAP}, I) = 0, \tag{30}$$

where I denotes any prior information other than the existence of life L. Now, trivially,

$$p(U \in \mathcal{U}_L | L, I) = 1 \quad , \quad p(U \in \mathcal{U}_{\bar{L}} | L, I) = 0. \tag{31}$$

The similarity of (30) to (31) does not imply the truth of SAP from the existence of life. We can state only, from the marginalising rule (18), that

$$p(\text{SAP}|L, I) = p(\text{SAP}, U \in \mathcal{U}_L|L, I) + p(SAP, U \in \mathcal{U}_L|L, I) \tag{32}$$
$$= p(\text{SAP}|U \in \mathcal{U}_L, L, I)p(U \in \mathcal{U}_L|L, I)$$
$$+ p(\text{SAP}|U \in \mathcal{U}_L, L, I)p(U \in \mathcal{U}_L|L, I) \tag{33}$$
$$= p(\text{SAP}|U \in \mathcal{U}_L, L, I) \tag{34}$$

on using (31) in (33). Clearly this is no help. The logic of (30) – (34) is that two observations which separately lead to the same conclusion need not imply each other: During daylight hours, both rain, and the absence of direct sunlight, imply cloud; but the absence of direct sunlight does not imply rain.

For the purposes of logic it makes no matter what motivates the SAP: The role of probability theory is to work with given propositions and hypotheses, not to generate them or query their origins. Nothing prevents us from tossing into play, at any stage, a deeper theory. However, the SAP is unsatisfactory: It does not tell us quantitatively how to redistribute the prior probability of those values of U non-conducive to life among the values conducive to life. (It is a separate issue that these boundaries are blurred.) More details are needed of the reasoning leading to the SAP; only then can we gain a usefully informed assignment of its probability, and its predictions. The SAP is not usefully predictive; the example given by Carter [10] is, on closer examination, of weak anthropic type, while SAP predictions based on quantum branching into an ensemble of universes [6,10] involves the frequentist fallacy which we have already discussed.

We expect that any reasoning leading to the SAP will be of two types. Type 1 is based on fundamental physics, and would imply, for example, that if the fundamental constants were otherwise, then an inconsistency arises; superstring theory attempts this with respect to dimension. This idea makes no reference to life. Type 2 is a 'design', or teleological theory, uniting physics and biology. Since physics and biology are complementary, operating at different levels of description, such theories must possess quite extraordinary features. The alleged role of consciousness in collapsing the wavefunction of an observed system is one example of a 'mingling of the levels', but we have argued against this elsewhere [9]. We are sceptical that quantum mechanics is derivable from the existence of conscious observers measuring the rest of the universe, as Patton and Wheeler have suggested [12].

A different Type 2 theory, perhaps held at a subconscious level, is to see the idea as a facet of God, since the theological axiom $p(L|\text{GOD}) = 1$ provides, through the inverse reasoning of Bayes' theorem, evidence for the existence of God. However, this argument advances neither science nor theology. It echoes the conjectural 'final anthropic principle' that intelligent life *must* emerge [6]. This has the same logical status as the strong anthropic principle — a hypothesis of no value.

Let us close this Section with a final demonstration of the power of the weak anthropic principle and of Bayesian reasoning. This is Carter's ingenious use of the similarity between the length of time life has existed on Earth, $\tau_L \approx 0.4 \times 10^{10}$ years, and the duration the Earth *could* have supported life, $\tau_E \approx 10^{10}$ years [11]. Of necessity, $\tau_E > \tau_L$ — but not, proportionally, by much, and there is no *a priori* reason why the time scales of a biological process and a physical process should coincide. Suppose that the probability density $p_t(t\,|\bar{t}, B)$ for life attaining its present state in time t from its inception, assuming uniformly supportive conditions and given our present knowledge of evolutionary biochemistry B, is clearly peaked about a time parameter $t = \bar{t}$, which we choose to write as part of the conditioning information. How is $\bar{t}$, the typical time interval for life to reach its present state from its start, related to τ_E, the time available? The probability density p_L for τ_L is simply p_t, truncated at τ_E and re-normalised:

$$p_L(\tau_L|\tau_E, \bar{t}, B) = \frac{p_t(\tau_L|\bar{t}, B)H(\tau_E - \tau_L)}{\int_0^{\tau_E} dt' p_t(t'|\bar{t}, B)}, \qquad (35)$$

where H denotes the step function. The probability density for $\bar{t}$, from Bayes' theorem, is

$$p_{\bar{t}}(\bar{t}|\tau_L, \tau_E, B) = Kp(\bar{t}|\tau_E, B)p_L(\tau_L|\tau_E, \bar{t}, B), \qquad (36)$$

where K is determined by normalisation *a posteriori*, and $p(\bar{t}|\tau_E, B)$ denotes the prior density for $\bar{t}$; this is independent of τ_E, which can be dropped in this distribution from the conditioning information. In (36), $p_L(\tau_L|\tau_E, \bar{t}, B)$ is the likelihood, favouring values of $\bar{t}$ at which it is large. We can see from plots of p_t what its truncation p_L looks like as τ_E and $\bar{t}$ vary: see Figure 1. Clearly τ_L is more likely to be close to τ_E if $\bar{t} \gg \tau_E$, reflecting the notion that, if life usually takes far longer than τ_E to evolve, then it is more likely to have needed most of that time given that it *has* evolved. The observation that $\tau_L \approx \tau_E$ is understood as evidence for this.

If, for simplicity, we model $p_t(t|\bar{t}, B)$ by a Gaussian

$$p_t(t|\bar{t}, B) = \frac{1}{\sqrt{2\pi\sigma^2}} e^{-(t-\bar{t})^2/2\sigma^2}, \qquad (37)$$

where $\bar{t} \gg \sigma$ (so that the restriction $t \not< 0$ introduces negligible error), we have from (35)

$$p_L(\tau_L|\tau_E, \bar{t}, B) = \frac{1}{\sqrt{2\pi\sigma^2}} \frac{e^{-(\tau_L - \bar{t})^2/2\sigma^2}}{\frac{1}{2}\text{erfc}(\frac{\bar{t}-\tau_E}{\sqrt{2}\sigma})} H(\tau_E - \tau_L), \qquad (38)$$

which introduces a factor

$$\frac{e^{-(\tau_L - \bar{t})^2/2\sigma^2}}{\text{erfc}(\frac{\bar{t}-\tau_E}{\sqrt{2}\sigma})} \qquad (39)$$

into the posterior probability distribution (36) for $\bar{t}$. Since $\text{erfc}(z) \sim \sqrt{2/\pi}e^{-z^2}/z$ for large z, this factor is of order

$$(\bar{t} - \tau_E)e^{-(\tau_E - \tau_L)\bar{t}/\sigma^2} \qquad (40)$$

except where $(\bar{t} - \tau_E)$ is $O(\sigma)$. This expression initially *increases* as $\bar{t}$ increases, maintaining that increase for longer as τ_L comes closer to τ_E.

Since we have learned to expect $\bar{t} \gg \tau_E$, we are surprised that intelligent life has evolved at all. Usually it takes much longer. But, though it is reasonable to be surprised life has evolved *on Earth*, it would be less surprising if we were to learn it evolved only on one out of a large number of habitable planets; just as you would be surprised to be in a car crash, but expect crashes to take place every day in a city. And, though we are very

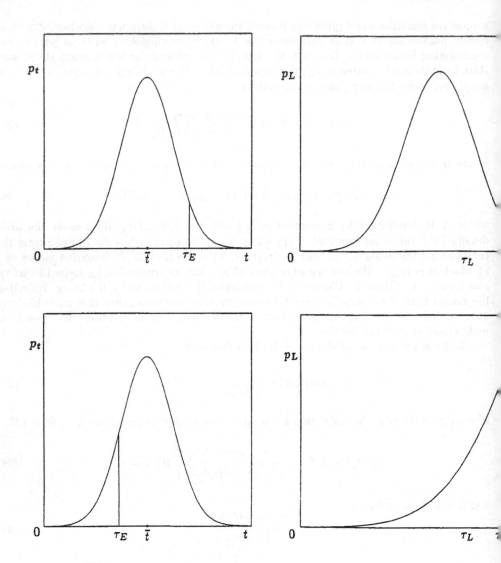

Fig. 1. The density p_t and its truncation p_L.

uncertain as to the number of habitable planets, no non-terrestrial life has yet been found. Perhaps we should not be so surprised at the way things are on this scale.

The fortuitous accidents leading to intelligent life are etched in fossils, in the Burgess shale. Stephen Jay Gould has explained [13] that these indicate the evolutionary process has extremely sensitive dependence on initial conditions: A tiny difference in our ancestors, or in the conditions under which they competed, is amplified enormously over time. Many species have died out. Intelligence need not have arisen as a survival strategy.

If a punctuated equilibrium model of evolution is adopted, a different form for $p_t(t\,|\bar{t},B)$

results, from which Bayesian analysis can estimate the number of steps in the chain [11].

4. Conclusions

This tutorial demonstrates once more that Bayesian expression of an idea such as the anthropic principle serves to separate the wheat from the chaff. The surviving arguments are most clearly formulated in Bayesian terms.

Specifically, observation of carbon-based intelligent life, coupled to some ingenious physical arguments, allow us to make predictions in good accord with observation and to explain, in a natural way, some otherwise remarkable numerical coincidences. All of this is the province of Bayesian reasoning using observed data, and these applications are collectively called the weak anthropic principle. Finally, the logical relationship between ourselves and the universe is reciprocal: To say that the universe constrains us is as meaningless as saying that we constrain the universe. There are only correlations; patterns. Science is no more or less than their manipulation.

ACKNOWLEDGMENTS. The author acknowledges the support of a Royal Society of Edinburgh Personal Research Fellowship.

REFERENCES

1. Rutherford, E., in A. L. Mackay (ed.), *The Harvest of a Quiet Eye: A Selection of Scientific Quotations*, Institute of Physics, London, UK, 1977, p131.

2. Jaynes, E. T.: 1983, *E. T. Jaynes: Papers on Probability, Statistics and Statistical Physics*, R. D. Rosenkrantz (ed.), Synthese Series **158**, Reidel, Dordrecht, Netherlands.

3. Garrett, A. J. M.: 1991, 'Macroirreversibility and Microreversibility Reconciled: The Second Law', in B. Buck and V. A. Macaulay (eds.), *Maximum Entropy in Action*, Oxford University Press, Oxford, UK, pp. 139-170.

4. Cox, R. T.: 1946, 'Probability, Frequency and Reasonable Expectation', *Am. J. Phys.* **14**, 1-13.

5. Jaynes, E. T.: 1976, 'Confidence Intervals vs Bayesian Intervals', in W. L. Harper and C. A. Hooker (eds.), *Foundations of Probability Theory, Statistical Inference and Statistical Theories of Science*, Reidel, Dordrecht, Netherlands, pp. 175-257. Largely reprinted as Chapter 9 of Reference [2].

6. Barrow, J. D. and Tipler, F. J.: 1986, *The Anthropic Cosmological Principle*, Clarendon Press, Oxford, UK.

7. Dicke, R. H.: 1961, 'Dirac's Cosmology and Mach's Principle', *Nature* **192**, 440-441.

8. Bell, J. S.: 1964, 'On the Einstein Podolsky Rosen Paradox', *Physics* **1**, 195-200.

9. Garrett, A. J. M.: 1990, 'Bell's Theorem and Bayes' Theorem', *Found. Phys.* **20**, 1475-1512 and **21**, 753-755.

10. Carter, B.: 1974, 'Large Number Coincidences and the Anthropic Principle in Cosmology', in M. S. Longair (ed.), *Confrontation of Cosmological Theories with Observational Data*, IAU Symposium **63**, Reidel, Dordrecht, Netherlands, pp. 291-298.

11. Carter, B.: 1983, 'The Anthropic Principle and its Implications for Biological Evolution', *Phil. Trans. Roy. Soc. Lond. A* **310**, 347-363.

12. Patton, C. M. and Wheeler, J. A.: 1975, 'Is Physics Legislated by Cosmogony?', in C. J. Isham, R. Penrose and D. W. Sciama (eds.), *Quantum Gravity: an Oxford Symposium*, Clarendon Press, Oxford, UK, pp. 538-605.

13. Gould, S. J.: 1989, *Wonderful Life: The Burgess Shale and the Nature of History*, Hutchinson Radius, London, UK.

THE EVIDENCE FOR NEURAL NETWORKS

David J.C. MacKay[†]
Computation and Neural Systems
California Institute of Technology, 139-74
Pasadena, California 91125 USA
mackay@ras.phy.cam.ac.uk.

ABSTRACT. A quantitative and practical Bayesian framework is described for learning of mappings in feedforward networks. The framework makes possible: (1) objective comparisons between solutions using alternative network architectures; (2) objective stopping rules for network pruning or growing procedures; (3) objective choice of magnitude and type of weight decay terms or additive regularisers (for penalising large weights, etc.); (4) a measure of the effective number of well–determined parameters in a model; (5) quantified estimates of the error bars on network parameters and on network output; (6) objective comparisons with alternative learning and interpolation models such as splines and radial basis functions. The Bayesian 'evidence' automatically embodies 'Occam's razor,' penalising over-flexible and over-complex models. The Bayesian approach helps detect poor underlying assumptions in learning models. For learning models well matched to a problem, a good correlation between generalisation ability and the Bayesian evidence is obtained.

PREREQUISITE

This paper makes use of the Bayesian framework for regularisation and model comparison described in the companion paper 'Bayesian interpolation' (MacKay, 1992c). This framework is due to Gull and Skilling (1989).

1. The gaps in backprop

There are many knobs on the black box of 'backprop' (learning by back–propagation of errors (Rumelhart *et. al.*, 1986)). Generally these knobs are set by rules of thumb, trial and error, and the use of reserved test data to assess generalisation ability (or more sophisticated cross–validation). The knobs fall into two classes: (1) parameters which change the effective learning model, for example, number of hidden units, and weight decay terms; and (2) parameters concerned with function optimisation technique, for example, 'momentum' terms. This paper is concerned with making objective the choice of the parameters in the first class, and with ranking alternative solutions to a learning problem in a way which makes full use of all the available data. Bayesian techniques will be described which are both theoretically well–founded and practically implementable.

[†] Current address: Cavendish laboratory, Madingley Road, Cambridge CB3 0HE, U.K.

C. R. Smith et al. (eds.), *Maximum Entropy and Bayesian Methods, Seattle, 1991*, 165–183.
© 1992 *Kluwer Academic Publishers.*

Let us review the basic framework for learning in networks, then discuss the points at which objective techniques are needed. The training set for the mapping to be learned is a set of input–target pairs $D = \{\mathbf{x}^m, \mathbf{t}^m\}$, where m is a label running over the pairs. A neural network architecture $\mathcal{A}$ is invented, consisting of a specification of the number of layers, the number of units in each layer, the type of activation function performed by each unit, and the available connections between the units. If a set of values $\mathbf{w}$ is assigned to the connections in the network, the network defines a mapping $\mathbf{y}(\mathbf{x}; \mathbf{w}, \mathcal{A})$ from the input activities $\mathbf{x}$ to the output activities $\mathbf{y}$.[1] The distance of this mapping to the training set is measured by some error function; for example the error for the entire data set is commonly taken to be

$$E_D(D\,|\mathbf{w}, \mathcal{A}) = \sum_m \frac{1}{2}\Big(\mathbf{y}(\mathbf{x}^m; \mathbf{w}, \mathcal{A}) - \mathbf{t}^m\Big)^2 \tag{1}$$

The task of 'learning' is to find a set of connections $\mathbf{w}$ which gives a mapping which fits the training set well, i.e. has small error E_D; it is also hoped that the learned connections will 'generalise' well to new examples. Plain backpropagation learns by performing gradient descent on E_D in $\mathbf{w}$–space. Modifications include the addition of a 'momentum' term, and the inclusion of noise in the descent process. More efficient optimisation techniques may also be used, such as conjugate gradients or variable metric methods. This paper will not discuss computational modifications concerned only with speeding the optimisation. It will address however those modifications to the plain backprop algorithm which implicitly or explicitly modify the objective function, with decay terms or regularisers.

It is moderately common for extra regularising terms $E_W(\mathbf{w})$ to be added to E_D; for example terms which penalise large weights may be introduced, in the hope of achieving a smoother or simpler mapping (Hinton and Sejnowski, 1986, Ji et. al., 1990, Nowlan, 1991, Rumelhart, 1987, Weigend et. al., 1991). Some of the 'hints' in (Abu–Mostafa, 1990b) also fall into the category of additive weight–dependent energies. A sample weight energy term is:

$$E_W(\mathbf{w}|\mathcal{A}) = \sum_i \frac{1}{2}w_i^2 \tag{2}$$

The weight energy may be implicit, for example, 'weight decay' (subtraction of a multiple of $\mathbf{w}$ in the weight change rule) corresponds to the energy in (2). Gradient–based optimisation is then used to minimise the combined function:

$$M = \alpha E_W(\mathbf{w}|\mathcal{A}) + \beta E_D(D\,|\mathbf{w}, \mathcal{A}) \tag{3}$$

where α and β are 'black box' parameters.

The constant α should not be confused with the 'momentum' parameter sometimes introduced into backprop; in the present context α is a decay rate or regularising constant. Also note that α should not be viewed as causing 'forgetting'; E_D is defined as the error on the entire data set, so gradient descent on M treats all data points equally irrespective of the order in which they were acquired.

---

[1] The framework developed in this paper will apply not only to networks composed of 'neurons,' but to any regression model for which we can compute the derivatives of the outputs with respect to the parameters, $\partial\mathbf{y}(\mathbf{x}; \mathbf{w}, \mathcal{A})/\partial\mathbf{w}$.

What is lacking

The above procedures include a host of free parameters such as the choice of neural network architecture, and of the regularising constant α. There are not yet established ways of objectively setting these parameters, though there are many rules of thumb (see Ji *et. al.*(1990), Weigend *et. al.*(1991) for examples).

One popular way of comparing networks trained with different parameter values is to assess their performance by measuring the error on an unseen test set or by similar cross-validation techniques. The data are divided into two sets, a training set which is used to optimise the parameters **w** of the network, and a test set, which is used to optimise control parameters such as α and the architecture $\mathcal{A}$. However the utility of these techniques in determining values for the parameters α and β or for comparing alternative network solutions, etc., is limited because a large test set may be needed to reduce the signal to noise ratio in the test error, and cross-validation is computationally demanding. Furthermore if there are several parameters like α and β, it is out of the question to optimise such parameters by repeating the learning with all possible values of these parameters and using a test set. Such parameters must be optimised on line.

It is therefore interesting to study objective criteria for setting free parameters and comparing alternative solutions which depend only on the data set used for the training. Such criteria will prove especially important in applications where the total amount of data is limited, so that one doesn't want to sacrifice good data for use as a test set. Rather, we wish to find a way to use *all* our data in the process of optimising the parameters **w** *and* in the process of optimising control parameters like α and $\mathcal{A}$.

This paper will describe practical Bayesian methods for filling the following holes in the neural network framework just described:

1. **Objective criteria for comparing alternative neural network solutions, in particular with different architectures $\mathcal{A}$.** Given a single architecture $\mathcal{A}$, there may be more than one minimum of the objective function M. If there is a large disparity in M between the minima then it is plausible to choose the solution with smallest M. But where the difference is not so great it is desirable to be able to assign an objective preference to the alternatives.

   It is also desirable to be able to assign preferences to neural network solutions using different numbers of hidden units, and different activation functions. Here there is an 'Occam's razor' problem: the more free parameters a model has, the smaller the data error E_D it can achieve. So we cannot simply choose the architecture with smallest data error. That would lead us to an over–complex network which generalises poorly. The use of weight decay does not fully alleviate this problem; networks with too many hidden units still generalise worse, even if weight decay is used (see section 4).

2. **Objective criteria for setting the decay rate α.** As in the choice of $\mathcal{A}$ above, there is an 'Occam's razor' problem: a small value of α in equation (3) allows the weights to become large and overfit the noise in the data. This leads to a small value of the data error E_D (and a small value of M), so we cannot base our choice of α only on E_D or M. The Bayesian solution presented here can be implemented on–line, *i.e.* it is not necessary to do multiple learning runs with different values of α in order to find the best.

3. **Objective choice of regularising function E_W.**

4. **Objective criteria for choosing between a neural network solution and a solution using a different learning or interpolation model**, for example, splines or radial basis functions.

THE PROBABILITY CONNECTION

Tishby *et. al.* (1989) introduced a probabilistic view of learning which is an important step towards solving the problems listed above. The idea is to force a probabilistic interpretation onto the neural network technique so as to be able to make objective statements. This interpretation does not involve the addition of any new arbitrary functions or parameters, but it involves assigning a meaning to the functions and parameters that are already used.

My work is based on the same probabilistic framework, and extends it using concepts and techniques adapted from Gull and Skilling's (1989) Bayesian image reconstruction methods. This paper also adopts a shift in emphasis from Tishby *et. al.*'s paper: their work concentrated on predicting the average generalisation ability of a network trained on a task drawn from a known prior ensemble of tasks. This is called *forward probability*. In this paper the emphasis will be on quantifying the plausibility of alternative solutions to an interpolation or classification task; that task is defined by a single data set produced by the real world, and we do not know the prior ensemble from which the task comes. This is called *inverse probability*.

Let us now review the probabilistic interpretation of network learning.

- **Likelihood.** A network with specified architecture $\mathcal{A}$ and connections $\mathbf{w}$ is viewed as making predictions about the target outputs as a function of input $\mathbf{x}$ in accordance with the probability distribution:

$$P(\mathbf{t}^m | \mathbf{x}^m, \mathbf{w}, \beta, \mathcal{A}, \mathcal{N}) = \frac{\exp(-\beta E(\mathbf{t}^m | \mathbf{x}^m, \mathbf{w}, \mathcal{A}))}{Z_m(\beta)}, \tag{4}$$

where $Z_m(\beta) = \int d\mathbf{t} \exp(-\beta E)$. E is the error for a single datum, and β is a measure of the presumed noise included in $\mathbf{t}$. If E is the quadratic error function then this corresponds to the assumption that $\mathbf{t}$ includes additive gaussian noise with variance $\sigma_\nu^2 = 1/\beta$. The symbol $\mathcal{N}$ denotes the implicit noise model.

- **Prior.** A prior probability is assigned to alternative network connection strengths $\mathbf{w}$, written in the form:

$$P(\mathbf{w} | \alpha, \mathcal{A}, \mathcal{R}) = \frac{\exp(-\alpha E_W(\mathbf{w} | \mathcal{A}))}{Z_W(\alpha)} \tag{5}$$

where $Z_W = \int d^k \mathbf{w} \exp(-\alpha E_W)$. Here α is a measure of the characteristic expected connection magnitude. If E_W is quadratic as specified in equation (2) then weights are expected to come from a gaussian with zero mean and variance $\sigma_W^2 = 1/\alpha$. Alternative 'regularisers' $\mathcal{R}$ (each using a different energy function E_W) implicitly correspond to alternative hypotheses about the statistics of the environment.

- The **posterior probability** of the network connections $\mathbf{w}$ is then:

$$P(\mathbf{w} | D, \alpha, \beta, \mathcal{A}, \mathcal{N}, \mathcal{R}) = \frac{\exp(-\alpha E_W - \beta E_D)}{Z_M(\alpha, \beta)} \tag{6}$$

where $Z_M(\alpha, \beta) = \int d^k \mathbf{w} \exp(-\alpha E_W - \beta E_D)$. Notice that the exponent in this expression is the same as (minus) the objective function M defined in (3).

So under this framework, minimisation of $M = \alpha E_W + \beta E_D$ is identical to finding the (locally) most probable parameters $\mathbf{w}_{MP}$; minimisation of E_D alone is identical to finding the maximum likelihood parameters $\mathbf{w}_{ML}$. Thus an interpretation has been given to backpropagation's energy functions E_D and E_W, and to the parameters α and β. It should be emphasised that 'the probability of the connections $\mathbf{w}$' is a measure of *plausibility* that the model's parameters should have a specified value $\mathbf{w}$; this has nothing to do with the probability that a particular algorithm might converge to $\mathbf{w}$.

This framework offers some partial enhancements for backprop methods: The work of Levin et. al. (1989) makes it possible to predict the average generalisation ability of neural networks trained on one of a defined class of problems. However, it is not clear whether this will lead to a practical technique for choosing between alternative network architectures for real data sets.

Le Cun et. al. (1990) have demonstrated how to estimate the 'saliency' of a weight, which is the change in M when the weight is deleted. They have used this measure successfully to simplify large neural networks. However no stopping rule for weight deletion was offered other than measuring performance on a test set.

Also Denker and Le Cun (1991) demonstrated how the Hessian of M can be used to assign error bars to the parameters of a network and to its outputs. However, these error bars can only be quantified once β is quantified, and how to do this without prior knowledge or extra data has not been demonstrated. In fact β can be estimated from the training data alone.

2. Review of Bayesian regularisation and model comparison

In the companion paper (MacKay, 1992c) it was demonstrated how the control parameters α and β are assigned by Bayes, and how alternative interpolation models $\mathcal{H} = \{\mathcal{A}, \mathcal{N}, \mathcal{R}\}$ can be compared. It was noted there that it is not satisfactory to optimise α and β by finding the joint maximum likelihood value of $\mathbf{w}, \alpha, \beta$; the likelihood has a skew peak whose maximum is not located at the most probable values of the control parameters. The companion paper also reviewed how the Bayesian choice of α and β is neatly expressed in terms of a measure of the number of well–determined parameters in a model, γ. However that paper assumed that $M(\mathbf{w})$ only has one significant minimum which was well approximated as quadratic. (All the interpolation models discussed in (MacKay, 1992c) can be interpreted as two–layer networks with a fixed non–linear first layer.) In this section I briefly review the Bayesian framework, retaining that assumption. The following section will then discuss how the framework can be modified to handle neural networks, where the landscape of $M(\mathbf{w})$ is certainly not quadratic.

DETERMINATION OF α AND β

By Bayes' rule, the posterior probability for these parameters is:

$$P(\alpha, \beta | D, \mathcal{H}) = \frac{P(D|\alpha, \beta, \mathcal{H})P(\alpha, \beta|\mathcal{H})}{P(D|\mathcal{H})} \tag{7}$$

Now if we assign a uniform prior to (α, β), the quantity of interest for assigning preferences to (α, β) is the first term on the right hand side, the **evidence** for α, β, which can be written as[2]

$$P(D|\alpha,\beta,\mathcal{H}) = \frac{Z_M(\alpha,\beta)}{Z_W(\alpha)Z_D(\beta)} \tag{8}$$

where Z_M and Z_W were defined earlier and $Z_D = \int d^N\!D\, e^{-\beta E_D}$.

Let us use the simple quadratic energy functions defined in equations (1,2). This makes the analysis easier, but more complex cases can still in principle be handled by the same approach. Let the number of degrees of freedom in the data set, *i.e.* the number of output units times the number of data pairs, be N, and let the number of free parameters, *i.e.* the dimension of $\mathbf{w}$, be k. Then we can immediately evaluate the gaussian integrals Z_D and Z_W: $Z_D = (2\pi/\beta)^{N/2}$, and $Z_W = (2\pi/\alpha)^{k/2}$. Now we want to find $Z_M(\alpha,\beta) = \int d^k\mathbf{w}\, \exp(-M(\mathbf{w},\alpha,\beta))$. Supposing for now that M has a single minimum as a function of $\mathbf{w}$, at $\mathbf{w}_{\text{MP}}$, and assuming we can locally approximate M as quadratic there, the integral Z_M is approximated by:

$$Z_M \simeq e^{-M(\mathbf{w}_{\text{MP}})}(2\pi)^{k/2}\det^{-\frac{1}{2}}\mathbf{A} \tag{9}$$

where $\mathbf{A} = \nabla\nabla M$ is the Hessian of M evaluated at $\mathbf{w}_{\text{MP}}$.

The maximum of $P(D|\alpha,\beta,\mathcal{H})$ has the following useful properties:

$$\chi_W^2 \equiv 2\alpha E_W = \gamma \tag{10}$$

$$\chi_D^2 \equiv 2\beta E_D = N - \gamma \tag{11}$$

where γ is the effective number of parameters determined by the data,

$$\gamma = \sum_{a=1}^{k} \frac{\lambda_a}{\lambda_a + \alpha} \tag{12}$$

where λ_a are the eigenvalues of the quadratic form βE_D in the natural basis of E_W.

COMPARISON OF DIFFERENT MODELS

To rank alternative architectures, noise models, and penalty functions E_W in the light of the data, we simply evaluate the evidence for $\mathcal{H} = \{\mathcal{A},\mathcal{N},\mathcal{R}\}$, $P(D|\mathcal{H})$, which appeared as the normalising constant in (7). Integrating the evidence for (α,β), we have:

$$P(D|\mathcal{H}) = \int P(D|\alpha,\beta,\mathcal{H})P(\alpha,\beta|\mathcal{H})\,d\alpha\,d\beta \tag{13}$$

The evidence is the Bayesian's transportable quantity for comparing models in the light of the data.

3. Adapting the framework

For neural networks, $M(\mathbf{w})$ is not quadratic. Indeed it is well known that M typically has many local minima. And if the network has a symmetry under permutation of its parameters, then we know that $M(\mathbf{w})$ must share that symmetry, so that every single minimum belongs to a family of symmetric minima of M. For example if there are H

---

[2]The same notation, and the same abuses thereof, will be used as in (MacKay, 1992c).

hidden units in a single layer then each non–degenerate minimum is in a family of size $g = H! 2^H$. Now it may be the case that the significant minima of M are locally quadratic, so we might be able to evaluate Z_M by evaluating (9) at each significant minimum and adding up the Z_Ms; but the number of those minima is unknown, and this approach to evaluating Z_M would seem dubious.

Luckily however, we do not actually want to evaluate Z_M. We would need to evaluate Z_M in order to assign a posterior probability over α, β for an entire model, and to evaluate the evidence for alternative entire models. This is not quite what we wish to do: when we use a neural network to perform a mapping, we typically only implement one neural network at a time, and this network will have its parameters set to a *particular solution* of the learning problem. Therefore the alternatives we wish to rank are the different solutions of the learning problem, *i.e.* the different minima of M. We would only want the evidence as a function of the number of hidden units if we were somehow able to simultaneously implement the entire posterior ensemble of networks for one number of hidden units. Similarly, we do not want the posterior over α, β for the entire posterior ensemble; rather, it is reasonable to allow each solution (each minimum of M) to choose its own optimal value for these parameters. The same method of chopping up a complex model space is used in the unsupervised classification system, AutoClass (Hanson *et. al.*, 1991).

Having adopted this slight shift in objective, it turns out that to set α and β and to compare alternative solutions to a learning problem, the integral we now need to evaluate is a local version of Z_M. Assume that the posterior probability consists of well separated islands in parameter space each centred on a minimum of M. We wish to evaluate how much posterior probability mass is in each of these islands. Consider a minimum located at $\mathbf{w}^*$, and define a solution $S_{\mathbf{w}^*}$ as the ensemble of networks in the neighbourhood of $\mathbf{w}^*$, and all symmetric permutations of that ensemble. Let us evaluate the posterior probability for alternative solutions $S_{\mathbf{w}^*}$, and the parameters α and β:

$$P(S_{\mathbf{w}^*}, \alpha, \beta, \mathcal{H} | D) \propto g \frac{Z_M^*(\mathbf{w}^*, \alpha, \beta)}{Z_W(\alpha) Z_D(\beta)} P(\alpha, \beta) P(\mathcal{H}) \tag{14}$$

where g is the permutation factor, and $Z_M^*(\mathbf{w}^*, \alpha, \beta) = \int_{S_{\mathbf{w}^*}} d^k \mathbf{w} \, \exp(-M(\mathbf{w}, \alpha, \beta))$, where the integral is performed only over the neighbourhood of the minimum at $\mathbf{w}^*$. I will refer to the quantity $g \frac{Z_M^*(\mathbf{w}^*, \alpha, \beta)}{Z_W(\alpha) Z_D(\beta)}$ as the evidence for $\alpha, \beta, S_{\mathbf{w}^*}$. The parameters α and β will be chosen to maximise this evidence. Then the quantity we want to evaluate to compare alternative solutions is the evidence[3] for $S_{\mathbf{w}^*}$,

$$P(D, S_{\mathbf{w}^*} | \mathcal{H}) = \int g \frac{Z_M^*(\mathbf{w}^*, \alpha, \beta)}{Z_W(\alpha) Z_D(\beta)} P(\alpha, \beta) \, d\alpha \, d\beta. \tag{15}$$

This paper uses the gaussian approximation for Z_M^*:

$$Z_M^* \simeq e^{-M(\mathbf{w}^*)} (2\pi)^{k/2} \det^{-\frac{1}{2}} \mathbf{A} \tag{16}$$

---

[3] Bayesian model comparison is performed by evaluating and comparing the evidence for alternative models. Gull and Skilling defined the evidence for a model $\mathcal{H}$ to be $P(D|\mathcal{H})$. The existence of multiple minima in neural network parameter space complicates model comparison. The quantity in (15) is not $P(D|S_{\mathbf{w}^*}, \mathcal{H})$ (it includes the prior for $S_{\mathbf{w}^*} | \mathcal{H}$), but I have called it the evidence because it is the quantity we should evaluate to compare alternative solutions with each other and with other models.

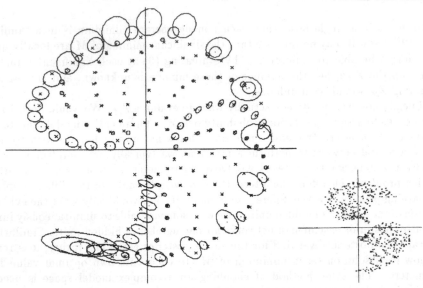

Fig. 1. Typical neural network output. (Inset – training set)

This is the output space (y_a, y_b) of the network. The target outputs are displayed as small x's, and the output of the network with 1σ error bars is shown as a a dot surrounded by an ellipse. The network was trained on samples in two regions in the lower and upper half planes (inset). The outputs illustrated here are for inputs extending a short distance outside the training regions, and bridging the gap between them. Notice that the error bars get much larger around the perimeter. They also increase slightly in the gap between the training regions. These pleasing properties would not have been obtained had the diagonal Hessian approximation of Denker and Le Cun (1991) been used. The above solution was created by a three layer network with 19 hidden units. From (MacKay, 1992b).

where $\mathbf{A} = \nabla\nabla M$ is the Hessian of M evaluated at $\mathbf{w}^*$. For general α and β this approximation is probably unacceptable; however we only need it to be accurate for the small range of α and β close to their most probable value. The regime in which this approximation will definitely break down is when the number of constraints, N, is small relative to the number of free parameters, k. For large N/k the central limit theorem encourages us to use the gaussian approximation (Walker, 1967). It is a matter for further research to establish how large N/k must be for this approximation to be reliable.

What obstacles remain to prevent us from evaluating the local Z_M^*? We need to evaluate or approximate the inverse Hessian of M, and we need to evaluate or approximate its determinant and/or trace (MacKay, 1992c).

Denker and Le Cun (1991) *et. al.* (1990) have already discussed how to approximate the Hessian of E_D for the purpose of evaluating weight saliency and for assigning error bars to weights and network outputs. The Hessian can be evaluated in the same way that backpropagation evaluates ∇E_D (see Bishop (1991) for a complete algorithm and the appendix of this paper for a useful approximation). Alternatively $\mathbf{A}$ can be evaluated by numerical methods, for example second differences. A third option: if variable metric methods are used to minimise M instead of gradient descent, then the inverse Hessian is automatically generated during the search for the minimum. It is important, for the success of this Bayesian method, that the off–diagonal terms of the Hessian should be evaluated. Denker *et. al.*'s method can do this without any additional complexity. The diagonal approximation is no good because of the strong posterior correlations in the parameters.

4. A Demonstration

This demonstration examines the evidence for various neural net solutions to a small inter-polation problem, the mapping for a two joint robot arm,

$$(\theta_1, \theta_2) \rightarrow (y_a, y_b) = (r_1 \cos \theta_1 + r_2 \cos(\theta_1 + \theta_2), r_1 \sin \theta_1 + r_2 \sin(\theta_1 + \theta_2)).$$

For the training set I used $r_1 = 2.0$ and $r_2 = 1.3$, random samples from a restricted range of (θ_1, θ_2) were made, and gaussian noise of magnitude 0.05 was added to the outputs. The neural nets used had one hidden layer of sigmoid units and linear output units. During optimisation, the regulariser (2) was used initially, and an alternative regulariser was in-troduced later; β was fixed to its true value (to enable demonstration of the properties of the quantity γ), and α was allowed to adapt to its locally most probable value.

Figure 1 illustrates the performance of a typical neural network trained in this way. Each output is accompanied by error bars evaluated using Denker *et. al.*'s method, *including off–diagonal Hessian terms.* If β had not been known in advance. it could have been inferred from the data using equation (11). For the solution displayed, the model's estimate of β in fact differed negligibly from the true value, so the displayed error bars are the same as if β had been inferred from the data.

Figure 2 shows the data misfit versus the number of hidden units. Notice that, as expected, the data error tends to decrease monotonically with increasing number of param-eters. Figure 3 shows the error of these same solutions on an unseen test set, which does not show the same trend as the data error. The data misfit cannot serve as a criterion for choosing between solutions.

Figure 4 shows the evidence for about 100 different solutions using different numbers of hidden units. Notice how the evidence maximum has the characteristic shape of an 'Occam hill' — steep on the side with too few parameters, and shallow on the side with too many parameters. The quadratic approximations break down when the number of parameters becomes too big compared with the number of data points.

The next figures introduce the quantity γ, discussed in (MacKay, 1992c), the number of well–measured parameters. In cases where the evaluation of the evidence proves difficult, it may be that γ will serve as a useful tool. For example, sampling theory predicts that the addition of redundant parameters to a model should reduce χ_D^2 by one unit per well–measured parameter; a stopping criterion could detect the point at which, as parameters are deleted, χ_D^2 started to increase faster than with gradient 1 with decreasing γ (figure 6).[4] This use of γ requires prior knowledge of the noise level β; that is why β was fixed to its known value for these demonstrations.

Now the question is how good a predictor of network quality the evidence is. The fact that the evidence has a maximum at a reasonable number of hidden units is promising. A comparison with figure 3 shows that the performance of the solutions on an unseen test set

---

[4]This suggestion is closely related to Moody's (1991) 'generalised prediction error'.

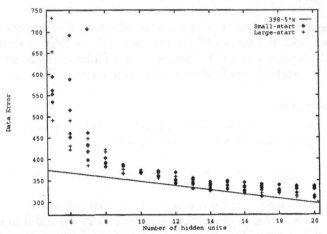

Fig. 2. Data error versus number of hidden units.

Each point represents one converged neural network, trained on a 200 i/o pair training set. Each neural net was initialised with different random weights and with a different initial value of $\sigma_W^2 = 1/\alpha$. The two point–styles correspond to small and large initial values for σ_W. The error is shown in dimensionless χ^2 units such that the expectation of error relative to the truth is 400 ± 20. The solid line is $400 - k$, where k is the number of free parameters.

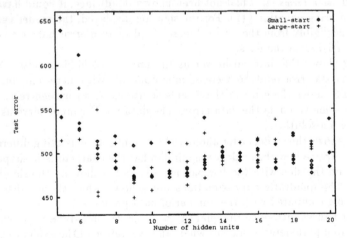

Fig. 3. Test error versus number of hidden units.

The training set and test set both had 200 data points. The test error for solutions found using the first regulariser is shown in dimensionless χ^2 units such that the expectation of error relative to the truth is 400 ± 20.

has similar overall structure to the evidence. However, figure 7 shows the evidence against the performance on a test set, and it can be seen that a significant number of solutions with poor evidence actually perform well on the test set. Something is wrong! It is time for a discussion of the relationship between the evidence and generalisation ability. We will return later to the failure in figure 7 and see that it is rectified by the development of new, more probable regularisers.

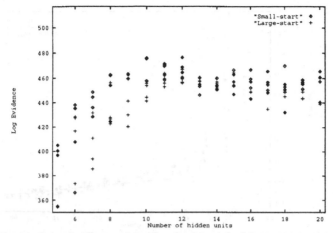

Fig. 4. Log Evidence for solutions using the first regulariser.

For each solution, the evidence was evaluated. Notice that an evidence maximum is achieved by neural network solutions using 10, 11 and 12 hidden units. For more than $\sim$ 19 hidden units, the quadratic approximations used to evaluate the evidence are believed to break down. The number of data points N is 400 (*i.e.* 200 i/o pairs); *c.f.* number of parameters in a net with 20 hidden units $= 102$.

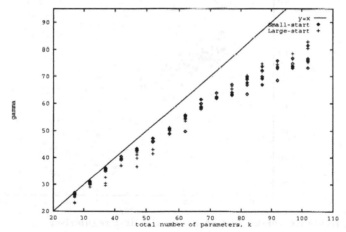

Fig. 5. The number of well–determined parameters.

This figure displays γ as a function of k, for the same network solutions as in figure 4.

RELATION TO 'GENERALISATION ERROR'

What is the relationship between the evidence and the generalisation error (or its close relative, cross–validation)? A correlation between the two is certainly expected. But the evidence is not necessarily a good predictor of generalisation error (see discussion in MacKay, 1992c). First, as illustrated in figure 8, the error on a test set is a noisy quantity, and a lot of data has to be devoted to the test set to get an acceptable signal to noise ratio. Furthermore, imagine that two models have generated solutions to an interpolation problem, and that their two most probable interpolants are completely identical. In this case, the

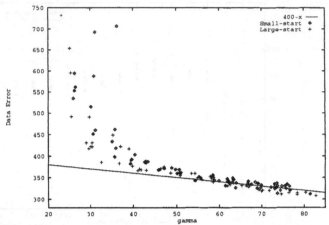

Fig. 6. Data misfit versus γ.

This figure shows χ_D^2 against γ, and a line of gradient -1. Towards the right, the data's misfit χ_D^2 is reduced by 1 for every well–measured parameter. When the model has too few parameters however (towards the left), the misfit gets worse at a greater rate.

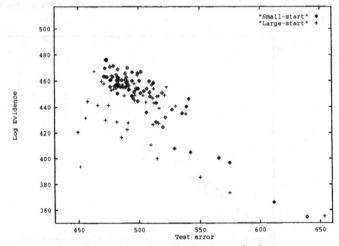

Fig. 7. Log Evidence versus Test error for the first regulariser.

The desired correlation between the evidence and the test error has negative slope. A significant number of points on the lower left violate this desired trend, so we have a failure of Bayesian prediction. The points which violate the trend are networks in which there is a significant difference in typical weight magnitude between the two layers. They are all networks whose learning was initialised with a large value of σ_W. The first regulariser is ill–matched to such networks, and the low evidence is a reflection of this poor prior hypothesis.

generalisation error for the two solutions must be the same, but the evidence will not in general be the same: typically, the model that was a *priori* more complex will suffer a larger Occam factor and will have smaller evidence. Also, the evidence is a measure of plausibility of the whole ensemble of networks about the optimum, not just the optimal network. Thus

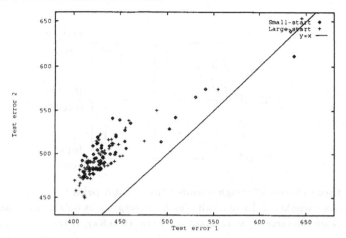

Fig. 8. Comparison of two test errors.

This figure illustrates how noisy a performance measure the test error is. Each point compares the error of a trained network on two different test sets. Both test sets consist of 200 data points from the same distribution as the training set.

there is more to the evidence than there is to the generalisation error.

WHAT IF THE BAYESIAN METHOD FAILS?

I do not want to dismiss the utility of the generalisation error: it can be important for detecting failures of the model being used. For example, if we obtain a poor correlation between the evidence and the generalisation error, such that Bayes fails to assign a strong preference to solutions which actually perform well on test data, then we are able to detect and attempt to correct such failures.

A failure indicates one of two things, and in either case we are able to learn and improve: either numerical inaccuracies in the evaluation of the probabilities caused the failure; or else the alternative models which were offered to Bayes were a poor selection, ill-matched to the real world (for example, using inappropriate regularisers). When such a failure is detected, it prompts us to examine our models and try to discover the implicit assumptions in the model which the data didn't agree with; alternative models can be tried until one is found that makes the data more probable.

We have just met exactly such a failure. Let us now establish what assumption in our model caused this failure and *learn* from it. Note that this mechanism for human learning is not available to those who just use the test error as their performance criterion. Going by the test error alone, there would have been no indication that there was a serious mismatch between the model and the data.

BACK TO THE DEMONSTRATION: COMPARING DIFFERENT REGULARISERS

The demonstrations thus far used the regulariser (2). This is equivalent to a prior that expects all the weights to have the same characteristic size. This is actually an inconsistent prior: the input and output variables and hidden unit activities could all be arbitrarily rescaled; if the same mapping is to be performed (a simple consistency requirement), such transformations of the variables would imply *independent* rescaling of the weights to the

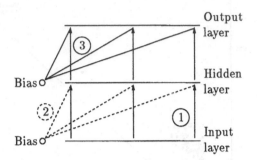

Fig. 9. The three classes of weights under the second prior
1: Hidden unit weights. 2: Hidden unit biases. 3: Output unit weights and biases. The weights in one class c share the same decay constant α_c. From (MacKay, 1992b).

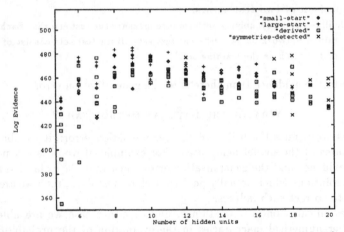

Fig. 10. Log Evidence versus number of hidden units for the second prior
The different point styles correspond to networks with learning initialised with small and large values of σ_W; networks previously trained using the first regulariser and subsequently trained on the second regulariser; and networks in which a weight symmetry was detected (in such cases the evidence evaluation is possibly less reliable).

hidden layer and to the output layer. Thus the scales of the two layers of weights are unrelated, and it is inconsistent to force the characteristic decay rates of these different classes of weights to be the same. This inconsistency is the major cause of the failure illustrated in figure 7. *All the networks deviating substantially from the desired trend have weights to the output layer far larger than the weights to the input layer;* this poor match to the model implicit in the regulariser causes the evidence for those solutions to be small.

This failure enables us to progress with insight to new regularisers. The alternative that I now present is a prior which is not inconsistent in the way explained above, so there are theoretical reasons to expect it to be 'better.' However, we will allow the data to choose, by evaluating the evidence for solutions using the new prior; we will find that the new prior is indeed *more probable*.

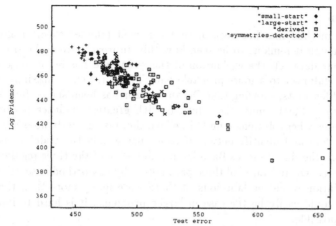

Fig. 11. Log Evidence for the second prior versus test error.
The correlation between the evidence and the test error for the second prior is very good. Note that the largest value of evidence has increased relative to figure 7, and the smallest test error has also decreased.

The second prior has three independent regularising constants, corresponding to the characteristic magnitudes of the weights in three different classes c, namely hidden unit weights, hidden unit biases, and output weights and biases (see figure 9). The term αE_W is replaced by $\sum_c \alpha_c E_W^c$, where $E_W^c = \sum_{i \in c} w_i^2/2$. Hinton and Nowlan (1991) have used a similar prior modelling weights as coming from a gaussian mixture, and using Bayesian re-estimation techniques to update the mixture parameters; they found such a model was good at discovering elegant solutions to problems with translation invariances.

Using the second prior, each regularising constant is independently adapted to its most probable value by evaluating the number of well-measured parameters γ_c associated with each regularising function, and finding the optimum where $2\alpha_c E_W^c = \gamma_c$. The increased complexity of this prior model is penalised by an Occam factor for each new parameter α_c (see MacKay, 1992c). Let me preempt questions along the lines of 'why didn't you use four weight classes, or non-zero means?' — any other way of assigning weight decays is just another model, and you can try as many as you like; by evaluating the evidence you can then find out what preference the data have for the alternative decay schemes.

New solutions have been found using this second prior, and the evidence evaluated. The evidence for these new solutions with the new prior is shown in figure 10. Notice that the evidence has increased compared to the evidence for the first prior. For some solutions the new prior is more probable by a factor of 10^{30}.

Now the crunch: does this more probable model make good predictions? The evidence for the second prior is shown against the test error in figure 11. The correlation between the two is greatly improved. Notice furthermore that not only is the second prior more probable, the best test error achieved by solutions found using the second prior is slightly better than any achieved using the first prior, and the number of good solutions has increased substantially. Thus the Bayesian evidence is a good predictor of generalisation ability, and the Bayesian choice of regularisers has enabled the best solutions to be found.

5. Discussion

The Bayesian method that has been presented is well–founded theoretically, and it works practically, though it remains to be seen how this approach will scale to larger problems. For a particular data set, the evaluation of the **evidence** has led us objectively from an inconsistent regulariser to a more probable one. The evidence is maximised for a sensible number of hidden units, showing that Occam's razor has been successfully embodied with no ad hoc terms. Furthermore the solutions with greatest evidence perform better on a test set than any other solutions found. I believe there is currently no other technique that could reliably find and identify better solutions using only the training set. Essential to this success was the simultaneous Bayesian optimisation of the three regularising constants (decay terms) α_c. Optimisation of these parameters by any orthodox search technique such as cross–validation would be laborious; if there were many more than three regularising constants, as could easily be the case in larger problems, it is hard to imagine any such search being possible.

This brings up the question of how these Bayesian calculations scale with problem size. In terms of the number of parameters k, calculation of the determinant and inverse of the Hessian scales as k^3. Note that this is a computation that needs to be carried out only a very small number of times compared with the immense number of derivative calculations involved in a typical learning session. However, for large problems it may be too demanding to evaluate the determinant of the Hessian. If this is the case, numerical methods are available to approximate the determinant or trace of a matrix in k^2 time (Skilling, 1989b).

APPLICATION TO CLASSIFICATION PROBLEMS

This paper has thus far discussed the evaluation of the evidence for backprop networks trained on interpolation problems. Neural networks can also be trained to perform classification tasks. A future publication (MacKay, 1992f) will demonstrate that the Bayesian framework for model comparison can be applied to these problems too.

RELATION TO V–C DIMENSION

Some papers advocate the use of V–C dimension (Abu–Mostafa, 1990a) as a criterion for penalising over–complex models (Abu–Mostafa, 1990b, Lee and Tenorio, 1991). V–C dimension is most often applied to classification problems; the evidence, on the other hand, can be evaluated equally easily for interpolation and classification problems. V–C dimension is a worst case measure, so it yields different results from Bayesian analysis (Haussler, 1991). For example, V–C dimension is indifferent to the use of regularisers like (2), and to the value of α, because the use of such regularisers does not rule out absolutely any particular network parameters. Thus V–C dimension assigns the same complexity to a model whether or not it is regularised.[5] So it cannot be used to set regularising constants α or to compare alternative regularisers. In contrast, the preceding demonstrations show that careful objective choice of regulariser and α is essential for the best solutions to be obtained.

---

[5] However, E. Levin and I. Guyon *et. al.*(1992) have developed a measure of 'effective V–C dimension' of a regularised model. This measure is identical to γ, equation (12), and their predicted generalisation error based on Vapnik's structural risk theory has exactly the same scaling behaviour as the evidence!

Worst case analysis has a complementary role alongside Bayesian methods. Neither can substitute for the other.

FUTURE TASKS

Further work is needed to formalise the relationship of this framework to the pragmatic model comparison technique of cross–validation. Moody's (1991) work on 'generalised prediction error' (GPE) is an interesting contribution in this direction. However, I have evaluated the GPE= $\frac{1}{N}(\chi_D^2 + 2\gamma)$ for the interpolation models in this paper's demonstration, and found the correlation between GPE and the test error was very poor. More work is needed to understand this.

The gaussian approximation used to evaluate the evidence breaks down when the number of data points is small compared to the number of parameters. For the model problems I have studied so far, the gaussian approximation seemed to break down significantly for $N/k < 3 \pm 1$. It is a matter for further research to characterise this failure and investigate techniques for improving the evaluation of the integral Z_M^*, for example the use of random walks on M in the neighbourhood of a solution.

It is expected that evaluation of the evidence should provide an objective rule for deciding whether a network pruning or growing procedure should be stopped, but a careful study of this idea has yet to be performed.

It will be interesting to see the results of evaluating the evidence for networks applied to larger real–world problems.

6. Appendix: Numerical methods

QUICK AND DIRTY VERSION

The three numerical tasks are automatic optimisation of α_c and β, calculation of error bars, and evaluation of the evidence. I will describe a cheap approximation for solving the first of these tasks without evaluating the Hessian. If we neglect the distinction between well–determined and poorly–determined parameters, we obtain the following update rules for α and β:

$$\alpha_c := k_c/2E_W^c$$
$$\beta := N/2E_D$$

If you want an easy–to–program taste of what a Bayesian framework can offer, try using this procedure to update your decay terms.

HESSIAN EVALUATION

The Hessian of M, $\mathbf{A}$, is needed to evaluate γ (which relates to Trace $\mathbf{A}^{-1}$), to evaluate the evidence (which relates to det $\mathbf{A}$), and to assign error bars to network outputs (using $\mathbf{A}^{-1}$).

I used two methods for evaluating $\mathbf{A}$: a) an approximate analytic method and b) second differences. The approximate analytic method was, following Denker *et. al.*, to use backprop to obtain the second derivatives, neglecting terms in f'', where f is the activation function of a neuron. The Hessian is built up as a sum of outer products of gradient vectors:

$$\nabla\nabla E_D \simeq \sum_{i,m} \mathbf{g}_i^m \mathbf{g}_i^{m\,\mathrm{T}}$$

where $g_i^m = \frac{dy_i(\mathbf{x}^m)}{d\mathbf{w}}$. Unlike Denker *et. al.*, I did not ignore the off–diagonal terms; the diagonal approximation is not good enough! For the evaluation of γ the two methods gave similar results, and either approach seemed satisfactory. However, for the evaluation of the evidence, the approximate analytic method failed to give satisfactory results. The 'Occam factors' are very weak, scaling only as $\log N$, and the above approximation apparently introduces systematic errors greater than these. The reason that the evidence evaluation is more sensitive to errors than the γ evaluation is because γ is related to the sum of eigenvalues, whereas the evidence is related to the product; errors in small eigenvalues jeopardise the product more than the sum. I expect an exact analytic evaluation of the second derivatives would resolve this. To save programming effort I instead used second differences, which is computationally more demanding ($\sim kN$ backprops) than the analytic approach ($\sim N$ backprops). There were still problems with errors in small eigenvalues, but it was possible to correct these errors, by detecting eigenvalues which were smaller than theoretically permitted.

DEMONSTRATIONS

The demonstrations were performed as follows: Initial weight configuration: weights drawn randomly from a gaussian with $\sigma_W = 0.3$. Optimisation algorithm for $M(\mathbf{w})$: variable metric methods, using code from (Press *et. al.*, 1988), used several times in sequence with values of the fractional tolerance decreasing from 10^{-4} to 10^{-8}. Every other loop, the regularising constants α_c were allowed to adapt in accordance with the re–estimation formula:

$$\alpha_c := \gamma_c/2E_W^c$$

PRECAUTION

When evaluating the evidence, care must be taken to verify that the permutation term g is appropriately set. It may be the case (probably mainly in toy problems) that the regulariser makes two or more hidden units in a network adopt identical connection values; alternatively some hidden units might switch off, with all weights set to zero; in these cases the permutation term should be smaller. Also in these cases, it is likely that the quadratic approximation will perform badly (quartic rather than quadratic minima are likely), so it is preferable to automate the deletion of such redundant units.

ACKNOWLEDGMENTS. I thank Mike Lewicki, Nick Weir, and Haim Sompolinsky for helpful conversations, and Andreas Herz for comments on the manuscript. This work was supported by a Caltech Fellowship and a Studentship from SERC, UK.

REFERENCES

Y.S. Abu-Mostafa (1990). 'The Vapnik–Chervonenkis dimension: information versus complexity in learning', *Neural Computation*, 1 3, 312–317.

Y.S. Abu-Mostafa (1990). 'Learning from hints in neural networks', *J. Complexity*, 6, 192–198.

C.M. Bishop (1991). 'Exact calculation of the Hessian matrix for the multilayer perceptron', submitted to *Neural Computation*.

J.S. Denker and Y. Le Cun (1991). 'Transforming neural-net output levels to probability distributions', in *Advances in neural information processing systems 3*, ed. R.P. Lippmann *et. al.*, 853–859, Morgan Kaufmann.

S.F. Gull (1989). 'Developments in Maximum entropy data analysis', in J. Skilling, ed., 53–71.

I. Guyon, V.N. Vapnik, B.E. Boser, L.Y. Bottou and S.A. Solla (1992). 'Structural risk minimization for character recognition', in *Advances in neural information processing systems 4*, ed. J.E. Moody, S.J. Hanson and R.P. Lippmann, Morgan Kaufmann.

R. Hanson, J. Stutz and P. Cheeseman (1991). 'Bayesian classification theory', NASA Ames TR FIA–90-12-7-01.

D. Haussler, M. Kearns and R. Schapire (1991). 'Bounds on the sample complexity of Bayesian learning using information theory and the VC dimension', Preprint.

G.E. Hinton and T.J. Sejnowski (1986). 'Learning and relearning in Boltzmann machines', in *Parallel Distributed Processing*, Rumelhart et. al., MIT Press.

C. Ji, R.R. Snapp and D. Psaltis (1990). 'Generalizing smoothness constraints from discrete samples', *Neural Computation*, **2** 2, 188-197.

Y. Le Cun, J.S. Denker and S.S. Solla (1990). 'Optimal Brain Damage', in *Advances in neural information processing systems 2*, ed. David S. Touretzky, 598–605, Morgan Kaufmann.

W.T. Lee and M.F. Tenorio (1991). 'On Optimal Adaptive Classifier Design Criterion — How many hidden units are necessary for an optimal neural network classifier?', Purdue University TR-EE-91-5.

E. Levin, N. Tishby and S. Solla (1989). 'A statistical approach to learning and generalization in layered neural networks', in *COLT '89: 2nd workshop on computational learning theory*, 245–260.

D.J.C. MacKay (1992b). 'A practical Bayesian framework for backprop networks', *Neural computation*, **4** 3.

D.J.C. MacKay (1992c) 'Bayesian interpolation', this volume.

D.J.C. MacKay (1992f) 'The evidence framework applied to classification networks', in preparation.

J.E. Moody (1991). 'Note on generalization, regularization and architecture selection in nonlinear learning systems', in *First IEEE-SP Workshop on neural networks for signal processing*, IEEE Computer society press.

S.J. Nowlan (1991). 'Soft competitive adaptation: neural network learning algorithms based on fitting statistical mixtures', Carnegie Mellon University Doctoral thesis CS–91–126.

W.H. Press, B.P. Flannery, S.A. Teukolsky and W.T. Vetterling (1988). *Numerical Recipes in C*, Cambridge.

D.E. Rumelhart, G.E. Hinton and R.J. Williams (1986). 'Learning representations by back propagating errors', *Nature*, **323**, 533–536.

D.E. Rumelhart (1987). Cited in Ji et. al. (1990).

J. Skilling, editor (1989). *Maximum Entropy and Bayesian Methods, Cambridge 1988*, Kluwer.

J. Skilling (1989). 'The eigenvalues of mega–dimensional matrices', in J. Skilling, ed., 455–466.

N. Tishby, E. Levin and S.A. Solla (1989). 'Consistent inference of probabilities in layered networks: predictions and generalization', in *Proc. IJCNN*, Washington.

A.M. Walker (1967). 'On the asymptotic behaviour of posterior distributions', *J. R. Stat. Soc. B*, **31**, 80–88.

A.S. Weigend, D.E. Rumelhart and B.A. Huberman (1991). 'Generalization by weight–elimination with applications to forecasting', in *Advances in neural information processing systems 3.*, ed. R.P. Lippmann et. al., 875–882, Morgan Kaufmann.

UNMIXING MINERAL SPECTRA USING A NEURAL NET WITH MAXIMUM ENTROPY REGULARIZATION

Neil Pendock
Department of Computational and Applied Mathematics
University of the Witwatersrand
Johannesburg, South Africa

ABSTRACT. Neural nets are widely used in pattern recognition and classification problems. We consider the problem of unmixing mineral spectra as linear combinations of a set of reference spectra. To do this, we train a neural net, the distributed associative memory [DAM], to remember the reference spectra. If the reference spectra are linearly independent, and in particular, if the number of features characterizing the spectra is greater than the number of reference spectra, the DAM technique produces good results. In the case of linear dependency between reference spectra, we reformulate the DAM methodology as a linear inverse problem of estimating a spectral abundance vector from a matrix of reference spectra and an unclassified input spectral mixture. Two techniques are discussed for estimating a response from this implicit DAM : singular value decomposition and a maximum entropy method. In mineral exploration, the number of spectral values measured [the pattern features] is substantially less than the number of possible constituent minerals and the unmixing problem is under determined. These two techniques were used to unmix *Geoscan* 24-channel imaging spectrometer data and a Thematic Mapper [TM] 6-channel satellite image of Cuprite, Nevada, with promising results.

1. INTRODUCTION

Imaging spectrometry is a useful tool for geological mapping of the earth's surface. Mounted either in an aircraft or an orbiting satellite, a spectrometer may be used to measure reflected sunlight in various spectral bands. The resulting image may be regarded as a cube of data, with x and y axes the usual spatial coordinate system and z axis the particular spectral channel recorded. Each image pixel will be a combination of the spectral response of the object in the field of view of the spectrometer. For quantitative mapping of the mineralogy of rocks and soils at the surface, some type of unmixing of the data into constituent minerals is needed. Much work has been done on the problem by Kruse [1988, 1990], Boardman [1989], Carrere [1989] and others.

Singular value decomposition [SVD] has been proposed for the inversion of imaging spectrometry data [Boardman, 1989]. The basic model is $Ax = b$ where A is an $m \times n$ endmember library matrix [m is the number of spectral features measured, n the number of spectra], x is the $n \times 1$ unknown abundance vector and b is the $m \times 1$ observed data vector. Singular value decomposition is suggested to invert A and yield the 'solution' $x = A^{-1}b$.

C. R. Smith et al. (eds.), *Maximum Entropy and Bayesian Methods, Seattle, 1991*, 185–196.
© 1992 *Kluwer Academic Publishers*.

x has the interpretation of being a linear combination of reference spectra which generates b. This technique neglects an important piece of *a priori* information we have about x, namely that all the components of x are positive. It would not make sense to allow a spectrum to contribute to d by not being present! To remedy this, we could impose a positivity constraint on the components of x and then SVD would produce a positive x that generates d with the property that $\sum x_i^2$ is minimum. There may, however, be other $x's$ that also generate d and are we guaranteed that the SVD solution is the most reasonable one? If we instead choose x such that x has *maximum entropy* amongst all $x's$ that generate d, then we know that our solution will have the least amount of structure not explained by our data. This is the best solution we can hope for.

Unmixing mineral spectra is an example of a pattern classification problem. Neural nets are often used to solve such problems, and we shall briefly discuss a particular one, the distributed associative memory, and show how it may be used to provide an answer to the spectral unmixing problem.

2. DISTRIBUTED ASSOCIATIVE MEMORY

Distributed associative memory [Wechsler, 1990] has been used extensively for classification problems. If we wish to remember n (pattern, recall) vector pairs $\{p^i\}$ and $\{r^i\}$ we may construct a memory matrix M such that $MP = R$ where P is the pattern matrix with columns $p^1 \cdots p^n$ and R is the recall matrix with columns $r^1 \cdots r^n$. Each pattern $\{p^i\}$ has m features. The memory is built in the following manner : each pattern p^i evokes a recall r^i. If we require that the recall of the memory to p^i is p^i, then we have the relationship

$$M_{m \times m} P_{m \times n} = P_{m \times n}$$

and so

$$M = PP^{-g}$$

where P^{-g} is the generalized inverse of P.

A simpler formulation is to require that the evoked recall to p^i is a column vector of length n with a 1 in the $i'th$ position and 0 everywhere else. We now have

$$M_{n \times m} P_{m \times n} = I_{n \times n}$$

where $I_{n \times n}$ is the $n \times n$ identity matrix and so $M = P^{-g}$.

Once the memory has been constructed, if we present an unknown pattern d to the memory, the recall is simply

$$y_{n \times 1} = M_{n \times m} d_{m \times 1}$$

We shall say that d matches pattern j when the euclidean distance between y and the vector $(0_1, \cdots, 1_j, \cdots, 0_n)$ is minimum over all possible recall vectors. This is equivalent to choosing the largest element of y and noting its position since if ω is the euclidean distance between y and pattern j then

$$\omega^2 = \sum_{i \neq j} y_i^2 + (y_j - 1)^2$$

$$= \sum_i y_i^2 + 1 - 2y_j$$

$$= constant - 2y_j$$

ω is minimum when y_j is the largest element of y. Thus the position of the largest element of y is the closest pattern to input d. Similarly, the position of the second largest element of y is the second closest match. Thus the recall y may be interpreted as the distance between an input and each of the pattern exemplars.

IMPLIED DISTRIBUTED ASSOCIATIVE MEMORY

Since M is the generalized inverse of P, we may rewrite the recall equation $y = Md$ as $Py = d$. The recall is now in the form of the linear inverse problem : estimate y given P and d. Since the memory M does not explicitly appear in this formulation, we call this approach the *implicit distributed associative memory*.

CONSTRUCTING AN IMPLICIT DAM

Our implicit DAM has the form

$$P_{m \times n} y_{n \times 1} = d_{m \times 1}$$

We consider two cases for the estimation of y ; $m \geq n$ and $m < n$.

RECALL IN THE CASE $m \geq n$

If $m \geq n$, the estimation of y is an *over determined* linear inverse problem and y may be estimated by minimizing $||Py - d||^2$. Press *et al* [1988] recommend singular value decomposition as the preferred method for solving this least squares problem. P has the following unique decomposition :

$$P = U \Lambda V^t$$

where U is an $m \times n$ column-orthogonal matrix, Λ is an $n \times n$ diagonal matrix with positive or zero elements [singular values] and V is an $n \times n$ orthogonal matrix. The generalized inverse of P [the memory] is thus

$$M = P^{-g} = V \Lambda^{-1} U^t$$

This formulation also has an important interpretation. If P is regarded as a linear mapping of the vector space y to the vector space d, columns of U corresponding to non-zero elements of Λ form an orthonormal set of basis vectors for the linearly independent patterns of P. i.e. we can explicitly construct a set of transformed patterns $P' = U^t P$ that are linearly independent. *Thus U contains the distinguishing features between the pattern exemplars.* Zero [or very small] singular values imply that the memory cannot distinguish between two or more pattern exemplars at the given resolution of m features. Columns of V corresponding to zero elements of Λ form an orthonormal set of basis vectors for the linearly dependent patterns in P. i.e. we can determine which pattern exemplars are linear combinations of other pattern exemplars.

Thus SVD may be used to determine internal consistency amongst the reference patterns and provide information about the separability of the component patterns for an input. Some applications are addressed by Goetz and Boardman [1989] although we should remember that the above results apply to a general linear mixing model, whereas in spectral mixing, a positive linear [or nonlinear] mixing model applies.

RECALL IN THE CASE $m < n$

If $m < n$, the estimation of y is an *under determined* linear inverse problem and no unique solution exists. The problem becomes one of choosing a recall y from a set of possible solutions. A further consideration is that even if $m \geq n$, in the case that some of the exemplar patterns are linearly dependent, the estimation of y will be an ill-posed problem. Menke [1984] describes how P may be partitioned into an over determined and an under determined part. Singular value decomposition may be used to describe the range and nullspace of P in this *mixed determined* case.

How to choose a recall y? We may decide to choose the smallest y [i.e. $\sum y_i^2$ minimum]. To do this we apply the singular value decomposition approach described above, with P and d augmented with $n - m$ rows of zeros. This choice of y is arbitrary; we could, with as much justification, choose the smoothest y or the flattest y. We shall consider how *a priori* information about the recall of may be used to choose a more reasonable memory.

MAXIMUM ENTROPY RECALL

As described above, the implicit DAM has the form

$$P_{m \times n} y_{n \times 1} = d_{m \times 1}$$

In the case that $m < n$ or $m \geq n$ and some of the component patterns of P are linearly dependent, there is no unique y that generates d. We have to choose a best y subject to the constraint $Py = d$. Components y_i of y measure the contribution of reference pattern i to the input d. We shall insist that $y_i \geq 0 \quad 1 \leq i \leq n$ since we do not want a pattern to affect d by not being physically present. Also, the contribution of patterns i and j, $y_{i \cup j}$ is the sum of their individual contributions. In addition, if we normalize y so that the sum of all contributions of all patterns is 1, then y is a positive additive distribution [PAD] and satisfies the Kolmogorov axioms of probability theory :

$$y_i \geq 0 \ \forall \ i$$
$$y_{i \cup j} = y_i + y_j \ \forall \ i \neq j$$
$$\sum_i y_i = 1$$

To choose PAD y_1 in preference to PAD y_2 we need to construct a monotonically increasing function $S(y)$ such that "y_1 better than y_2" is equivalent to $S(y_1) > S(y_2)$. Skilling [1988] shows that if y satisfies four reasonable axioms then the only form for S is

$$S(y, m) = \sum_i [y_i - m_i - y_i log \left(\frac{y_i}{m_i} \right)]$$

or a monotonic function of S. In the above, m is a Lebesgue measure on the space of all PADs. Since S is maximum when $y = m$, m has the interpretation of being a prior model for y. Here we have a way of incorporating *a priori* knowledge of the pattern mixture into the recognition procedure.

We also have the additional information that $Py = d$. So to choose the best feasible recall, we must maximize $S(y, m)$ subject to the constraint $Py = d$.

3. UNMIXING MINERAL SPECTRA

Cuprite, Nevada, has been used extensively as a test site for the main imaging spectrometer systems including the two NASA research instruments *AIS* and *AVIRIS* as well as the commercial systems *Geris* and *Geoscan*. Kruse *et al* [1990] have reported on using the *Geris* for mineral mapping at Cuprite.

Our first task is to convert the airborne and satellite data to reflectances to make them comparable to laboratory measured spectra. Ideally, the data should be calibrated to absolute reflectance which would require onboard calibration information, which was not available. We came across four techniques in the literature for calibrating airborne spectral data :

- *Ground siting :-* This was the technique that Kruse *et al* used at Cuprite. It consists of measuring field spectra for at least two ground areas and calculating a gain and offset for the airborne data to make them fit the field spectra. The method will remove atmospheric effects but requires measurement of field spectra.
- *Flat field :-* Carrere [1989] selects an area which is spectrally flat and homogenous. The entire scene is then divided by this average spectrum which yields an estimate of ground reflectance, although no compensation is made for path radiance which may result in a wavelength shift for some spectral features.
- *Log residuals :-* Green and Craig [1985] use the theoretical relationship between irradiance, topography, reflectance and radiance to calculate reflectance values. The method removes atmospheric effects and reflectance variations occurring from topographic features.
- *Internal average relative [IAR] :-* Kruse [1988] divides each spectrum by the channel average. This method also removes atmospheric effects although problems may arise if the average spectrum has spectral characteristics related to mineral absorbtion.

We chose the IAR method to calibrate the data, as no ground truth information was available. Our data were a 980×1000 pixel section of the central portion of the *Geoscan* image and a co-registered subsection of the TM image, shown in figure 1. The band average spectra contained no obvious mineral absorbtion features. *Geoscan* bands 1 to 18 corresponding to wavelengths 522nm to 2352nm and TM bands 1,2,3,4,5 and 7 corresponding to wavelengths 450nm to 2350nm were used in the analysis.

Eighteen mineral spectra measured in a laboratory by the CSIRO of Australia, were integrated over the same band passes as the *Geoscan* and TM, and were used as the spectral reference matrices. Minerals selected included calcite, muscovite, chlorite, montmorillonite, kaolinite, dickite, alunite, jarosite and hematite. The TM and *Geoscan* laboratory spectral response curves for five minerals are shown in figures 2 to 6. For the unmixing inversion, each mineral was assumed to be equally likely to occur i.e. our initial spectral abundance distribution was uniform [the maximum entropy distribution if no data were measured]. This distribution could be modified according to geological information about the likely mineral composition of the area.

SVD was applied to the two reference libraries and singular values less than one thousandth of the largest singular value were set to zero. The TM library had five non-zero singular values, while the *Geoscan* library had nine.

Unmixing results are presented in figures 2 to 6 for various minerals. The TM unmixing results are surprisingly good for the clay minerals halloysite, kaolinite-1 and kaolinite-3 and the iron mineral hematite, given the relatively coarse spectral characteristics of the thematic

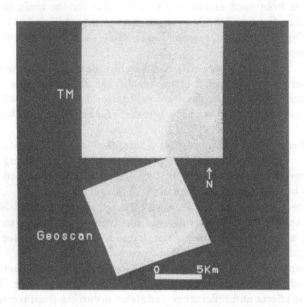

Fig. 1. *Geoscan* [top] and TM [bottom] image of Cuprite, Nevada.

mapper scanner. All the TM mixtures for both SVD and MEM unmixing are too optimistic based on geological expectations and Kruse's 1990 *Geris* results. This is due to the very broad spectral bandwidth of the TM scanner. The TM results show good correlation with the *Geoscan* mxitures, except in the case of alunite, and demonstrate that TM may be used as a 'poor man's imaging spectrometer' for mapping seclected mineral assemblages.

The MEM mixtures show greater discrimination than the SVD results on both the TM and *Geoscan* images. This is due to the exclusion of negative combinations of constituent minerals. Even better results could be expected if a more geologically realistic prior distribution was assigned to the spectral abundance distribution.

4. CONCLUSIONS

Both unmixing procedures gave good results on both data sets. The maximum entropy method [MEM] is the preferred unmixing strategy as it has a convincing reason why its mixtures are the best ones for this ill-posed pattern recognition problem. Another advantage of the MEM is the ability to incorporate prior information about the mixture distribution into the unmixing procedure, in a consistent way.

REFERENCES

Boardman, J.W.: 1989, 'Inversion of imaging spectrometry data using singular value decomposition', Proc. IGARSS '89 : Quantitative Remote Sensing : An Economic Tool for the Nineties, 4.

Carrere, V.: 1989, 'Mapping alteration in the goldfields mining district, Nevada, with airborne visible/infrared imaging spectrometer (AVIRIS)', Proc. VII Thematic Conference on Remote Sensing for Exploration, 1.

Goetz, A.F.H., & J.W. Boardman: 1989, 'Quantitative determination of imaging spectrometer specifications based on spectral mixing models', Proc. IGARSS '89 : Quantitative Remote Sensing : An Economic Tool for the Nineties.

Green, A.A., & M.D. Craig: 1985, 'Analysis of aircraft spectrometer data with logarithmic residuals', Proc. AIS workshop, JPL Publication 85-41.

Kruse, F.A.: 1988, 'Use of airborne imaging spectrometer data to map minerals associated with hydrothermally altered rocks in the northern grapevine mountains, Nevada and California', Remote Sensing of Environment, 24.

Kruse, F.A., Kierein-Young, K.S. & J.W. Boardman: 1990, 'Mineral mapping of Cuprite, Nevada with a 63-channel imaging spectrometer', Photogrammetric Engineering and Remote Sensing, 56.

Menke W.: 1984, Geophysical data analysis : discrete inverse theory, Academic Press.

Pendock N., De Gasparis A.A. & M.A. Brown: 1990, 'Neural net classification of mineral spectra', Proc. Symposium on pattern recognition and neural networks, University of the Witwatersrand.

Press W.H., Flannery B.P., Teukolsky S.A. & W.T. Vetterling: 1988, Numerical recipes in C, Cambridge University Press.

Skilling J.: 1988, 'The axioms of maximum entropy', in Erickson G.J., & Ray Smith, C. (eds.), Maximum Entropy and Bayesian Methods in Science and Engineering, vol. 1 : Foundations, Kluwer, Dordrecht.

Wechsler H.: 1990, Computational Vision, Academic Press.

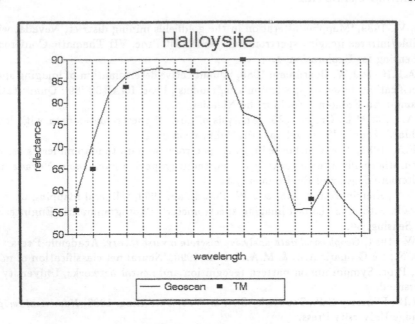

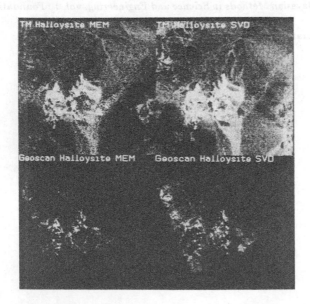

Fig. 2. Halloysite response.

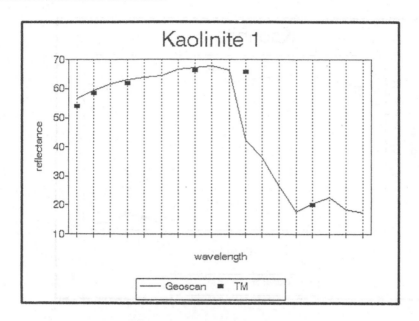

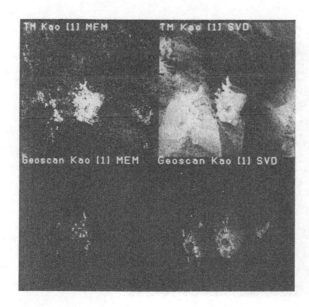

Fig. 3. Kaolinite-1 response.

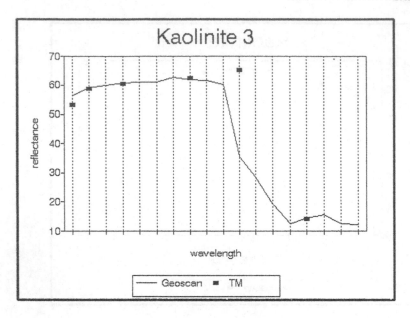

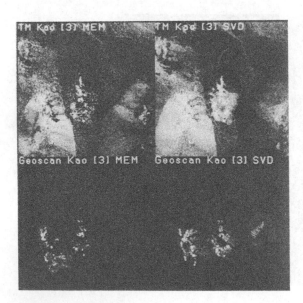

Fig. 4. Kaolinite-3 response.

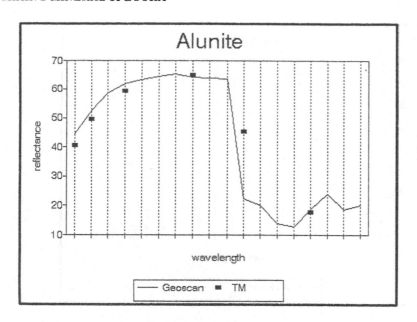

Fig. 5. Alunite response.

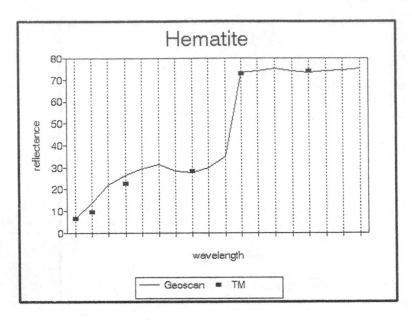

Fig. 6. Hematite response.

BAYESIAN MIXTURE MODELING

Radford M. Neal
Department of Computer Science
University of Toronto
Toronto, Ontario, Canada

ABSTRACT. It is shown that Bayesian inference from data modeled by a mixture distribution can feasibly be performed via Monte Carlo simulation. This method exhibits the true Bayesian predictive distribution, implicitly integrating over the entire underlying parameter space. An infinite number of mixture components can be accommodated without difficulty, using a prior distribution for mixing proportions that selects a reasonable subset of components to explain any finite training set. The need to decide on a "correct" number of components is thereby avoided. The feasibility of the method is shown empirically for a simple classification task.

1. Introduction

Mixture distributions [2,8] are an appropriate tool for modeling processes whose output is thought to be generated by several different underlying mechanisms, or to come from several different populations. One aim of a mixture model analysis may be to identify and characterize these underlying "latent classes", either for some scientific purpose, or as one realization of "unsupervised learning" in artificial intelligence. In other cases, prediction of future observations is the objective — in a "classification" application, for example, we are interested in predicting the category attribute of an item on the basis of various indicator attributes.

The standard Bayesian approach to this problem would be to define a prior distribution over the parameter space of the mixture model and combine this with the observed data to give a posterior distribution over this parameter space. This posterior distribution would then be interpreted for latent class analysis, or used to derive a predictive distribution for use in classification. Unfortunately, since the parameter space is extremely large, this approach is computationally difficult. As discussed later, one current approach is to find the single parameter set with maximum posterior probability and use it as the basis for prediction and interpretation. While this procedure often produces good results, it is not the proper Bayesian solution to the problem, and for some prior distributions gives useless answers.

In this paper, I present a technique for exhibiting the true Bayesian predictive distribution, given a conjugate prior for the parameters, and a set of training items. In this method, the model parameters are first integrated out analytically, reducing the problem to a summation over all possible assignments of mixture components to data items. This

197

C. R. Smith et al. (eds.), Maximum Entropy and Bayesian Methods, Seattle, 1991, 197–211.
© 1992 *Kluwer Academic Publishers.*

still-formidable problem is then solved by the Monte Carlo simulation technique of "Gibbs Sampling".

It turns out that this procedure extends without any computational difficulty to models in which the number of mixture components is countably infinite. Such infinite mixture models are quite natural in many contexts. Consider, for example, an analysis of plant specimens collected at random from some region. We expect *a priori* that the region will harbour many thousands of plant species, and that furthermore there will be many distinct populations within a species, due to varying soil conditions, etc. It therefore seems reasonable to consider the number of latent classes in this example to be effectively infinite. Of course, any finite sample will contain representatives of only a finite number of these classes. Indeed, we will generally wish to explain the sample using many fewer classes than there are data items. As more data is collected, however, more and more of the infinite number of classes will become apparent, as, for example, we obtain significant numbers of specimens of the rarer species.

The empirical results presented in this paper demonstrate that modeling data as an infinite mixture also works well when there are only a small finite number of components in the true mixture. Infinite mixture models are thus an attractive option whenever the true number of components is unknown, since one thereby avoids the problem of selecting between models with different numbers of components.

2. Bayesian mixture modeling

In formalizing the Bayesian mixture modeling task, I will take prediction to be the primary objective. This is natural in a classification application. In a latent class analysis application, we might not have any real interest in predicting values of unknown attributes, but nevertheless it is by their predictive value that the validity of underlying classes is demonstrated.

The Problem and the Model

Assume that some process produces data items that are represented by attribute vectors $\tilde{v}_i = \langle v_{i1}, \ldots, v_{im} \rangle$, which are seen as realizations of corresponding random variables $\tilde{V}_i = \langle V_{i1}, \ldots, V_{im} \rangle$. In this paper, the range of attribute V_{ij} is taken to be the set of integers $\{1, \ldots, N_j\}$, but generalizations to real-valued attributes are possible.

We have some knowledge of, or interest in, n of these vectors: $\tilde{V}_1, \ldots, \tilde{V}_n$. Specifically, we wish to find the predictive distribution for one or more of the unknown attributes of these vectors, given the values of the known attributes.

In a classification application, V_{i1} might be the category of item i and $V_{i2}, \ldots, V_{im}$ be various indicator attributes. Perhaps we know the indicators and category of the first $n-1$ items (the training set) and have seen the indicators for item n (the test item). We wish to find the consequent probability that item n is in category c:

$$P(V_{n1} = c \mid \tilde{V}_i = \tilde{v}_i, \ V_{nj} = v_{nj} : 1 \leq i < n, \ 2 \leq j \leq m)$$

Assume that the parameters of the process generating the data vectors are stable, and that given knowledge of these parameters, the various data items, $\tilde{V}_i$, are independent and distributed identically. In the mixture model approach, we believe, or imagine, that the process generates data items by several different mechanisms — which mechanism being a random variable, G_i — and that given knowledge of which mechanism generated a given item, the various attributes are independent. In other words, the distribution for $\tilde{V}_i$, on the assumption that the process parameters are known, can be expressed as a mixture of component distributions as follows:

$$
\begin{aligned}
P(\tilde{V}_i = \tilde{v}_i) &= \sum_{g=1}^{M} P(G_i = g) \cdot \prod_{j=1}^{m} P(V_{ij} = v_{ij} \mid G_i = g) \\
&= \sum_{g=1}^{M} \phi_g \cdot \prod_{j=1}^{m} \psi_{g,j,v_i,}
\end{aligned}
\tag{1}
$$

Here, M is the number of generating mechanisms, ϕ_g the probability of mechanism g being used, and $\psi_{g,j,v}$ the probability that mechanism g will produce value v for attribute j.

At first, the number of mixture components, M, will be a finite constant, but later it will be set to infinity. In a latent class analysis application, we will be interested in the components of the mixture distribution as possible indicators of real underlying mechanisms. In a classification application we might regard the use of a mixture distribution as just a device for expressing correlations among the attributes.

A Prior for the Mixture Model Parameters

In a Bayesian treatment of this problem, the parameter vectors $\tilde{\phi}$ and $\tilde{\psi}_{g,j}$ are considered to be the values of random variables $\tilde{\Phi}$ and $\tilde{\Psi}_{g,j}$ whose distributions reflect our prior knowledge of the process. It is mathematically convenient to express this prior knowledge via independent Dirichlet probability densities for the parameter vectors, since these are the conjugate priors for this problem. Later, I will briefly discuss the situation where the actual prior knowledge is too unspecific to be captured by densities of this form.

This combined Dirichlet prior density can be written as follows:

$$
\begin{aligned}
p(\tilde{\Phi} = \tilde{\phi}, \ \tilde{\Psi}_{g,j} &= \tilde{\psi}_{g,j} : 1 \leq g \leq M, \ 1 \leq j \leq m) \\
&= p(\tilde{\Phi} = \tilde{\phi}) \cdot \prod_{g,j} p(\tilde{\Psi}_{g,j} = \tilde{\psi}_{g,j}) \\
&= \left(\frac{\Gamma(\alpha)}{\Gamma(\alpha/M)^M} \prod_g \phi_g^{(\alpha/M)-1} \right) \cdot \prod_{g,j} \left(\frac{\Gamma(\beta_j)}{\Gamma(\beta_j/N_j)^{N_j}} \prod_v \psi_{g,j,v}^{(\beta_j/N_j)-1} \right)
\end{aligned}
\tag{2}
$$

where α and the β_j must be greater than zero. Setting $\alpha = M$ and $\beta_j = N_j$ in equation (2) produces a uniform distribution over the parameter space. A smaller value for α produces a prior distribution that favours values for the ϕ_g that are near 0 or 1 — a situation where a few components are much more probable than the others. A larger value for α favours values of ϕ_g near $1/M$ — a situation where the components are all about equally probable. Varying β_j has analogous effects on the priors for the distribution of attribute j in the various mixture components.

Integrating Out the Parameters

The joint distribution of the G_i and the $\tilde{V}_i$ can be found by integrating their probabilities over the region of valid parameter values, as weighted by prior probability density.

This joint distribution will later provide the basis for obtaining the conditional probabilities needed for classification. The $\widetilde{V}_i$ and G_i for different values of i are of course *not* independent — they would be so only if the parameters ϕ_g and $\psi_{g,j,v}$ were known.

The required integration is as follows:

$$P(G_i = g_i,\ \widetilde{V}_i = \widetilde{v}_i\ :\ 1 \leq i \leq n)$$

$$= \int P(G_i = g_i,\ \widetilde{V}_i = \widetilde{v}_i\ :\ 1 \leq i \leq n \mid \widetilde{\Phi} = \widetilde{\phi},\ \widetilde{\Psi}_{g,j} = \widetilde{\psi}_{g,j}\ :\ 1 \leq g \leq M,\ 1 \leq j \leq m)$$

$$\cdot\ p(\widetilde{\Phi} = \widetilde{\phi},\ \widetilde{\Psi}_{g,j} = \widetilde{\psi}_{g,j}\ :\ 1 \leq g \leq M,\ 1 \leq j \leq m)\ d\langle\widetilde{\phi},\ \widetilde{\psi}\rangle$$

$$= C \int \prod_i \phi_{g_i} \cdot \prod_{i,j} \psi_{g_i,j,v_{ij}} \cdot \prod_g \phi_g^{(\alpha/M)-1} \cdot \prod_{g,j} \prod_v \psi_{g,j,v}^{(\beta_j/N_j)-1}\ d\langle\widetilde{\phi},\ \widetilde{\psi}\rangle$$

$$= C \int \prod_g \phi_g^{\Sigma_i \delta(g_i,g)+(\alpha/M)-1} \cdot \prod_{g,j} \prod_v \psi_{g,j,v}^{\Sigma_i \delta(g_i,g)\delta(v_{ij},v)+(\beta_j/N_j)-1}\ d\langle\widetilde{\phi},\ \widetilde{\psi}\rangle$$

$$= C \int_{S_M} \prod_g \phi_g^{\Sigma_i \delta(g_i,g)+(\alpha/M)-1} d\widetilde{\phi} \cdot \prod_{g,j} \int_{S_{N_j}} \prod_v \psi_{g,j,v}^{\Sigma_i \delta(g_i,g)\delta(v_{ij},v)+(\beta_j/N_j)-1}\ d\widetilde{\psi}_{g,j} \quad (3)$$

where $C = (\Gamma(\alpha)/\Gamma(\alpha/M)^M)\cdot\prod_{g,j}(\Gamma(\beta_j)/\Gamma(\beta_j/N_j)^{N_j})$, and $\delta(x,y)$ is one if $x = y$ and zero otherwise. The integrals in the last formula are taken over the simplexes of valid probability distributions, $S_D = \{\langle x_1,\ldots,x_D\rangle\ :\ x_i \geq 0,\ \sum_i x_i = 1\}$.

These integrals are standard for Dirichlet distributions (see [5, section 2.4], for example). Their evaluation gives the following:

$$P(G_i = g_i,\ \widetilde{V}_i = \widetilde{v}_i\ :\ 1 \leq i \leq n)$$

$$= \frac{\Gamma(\alpha)}{\Gamma(n+\alpha)} \prod_g \frac{\Gamma(\Sigma_i \delta(g_i,g) + \alpha/M)}{\Gamma(\alpha/M)}$$

$$\cdot \prod_{g,j} \frac{\Gamma(\beta_j)}{\Gamma(\Sigma_i \delta(g_i,g) + \beta_j)} \prod_v \frac{\Gamma(\Sigma_i \delta(g_i,g)\delta(v_{ij},v) + \beta_j/N_j)}{\Gamma(\beta_j/N_j)}$$

$$= \prod_i \frac{\sum_{k<i} \delta(g_k,g_i) + \alpha/M}{(i-1)+\alpha} \cdot \prod_{i,j} \frac{\sum_{k<i} \delta(g_k,g_i)\delta(v_{kj},v_{ij}) + \beta_j/N_j}{\sum_{k<i} \delta(g_k,g_i) + \beta_j} \quad (4)$$

The above formula can be interpreted as a prescription for generating a sequence of values for the G_i and $\widetilde{V}_i$ according to a distribution picked in accord with the prior (equation (2)) — i.e. for generating a sequence from the prior predictive distribution for sequences of G_i and $\widetilde{V}_i$. This generation procedure uses the following incremental conditional probabilities, identified from equation (4):

$$P(G_i = g_i \mid G_k = g_k,\ \widetilde{V}_k = \widetilde{v}_k\ :\ 1 \leq k < i)\ =\ \frac{\sum_{k<i} \delta(g_k,g_i) + \alpha/M}{(i-1)+\alpha} \quad (5)$$

$$P(V_{ij} = v_{ij} \mid G_i = g_i,\ G_k = g_k,\ \widetilde{V}_k = \widetilde{v}_k\ :\ 1 \leq k < i)\ =\ \frac{\sum_{k<i} \delta(g_k,g_i)\delta(v_{kj},v_{ij}) + \beta_j/N_j}{\sum_{k<i} \delta(g_k,g_i) + \beta_j}$$

$$(6)$$

To generate a sequence of G_i and $\widetilde{V}_i$, we choose values for the G_i in turn, based on the frequencies with which components were previously chosen, biased by the term α/M. Once a G_i is selected, we chose values for the corresponding V_{ij} (independently), based on the frequencies of the various attribute values in vectors previously produced using the same component, biased by the β_j/N_j terms.

Note that by summing over all possible sequences of G_i, we can, via this procedure, define the probability of any sequence of the observable variables, $\widetilde{V}_i$, without any explicit reference to the parameters ϕ_g and $\psi_{g,j,v}$.

MODELS WITH AN INFINITE NUMBER OF COMPONENTS

The above procedure for producing a sequence from the predictive distribution for $\widetilde{V}_i$, remains well-defined even when the number of mixture components, M, is infinite. Values for the underlying G_i are also produced, but are specified only up to identity or non-identity among themselves.

Consider first the choice of a value for G_1. There are an infinite number of candidates for this value, but which of these is chosen has no effect on the values that are then chosen for V_{1j} — equation (6) shows that these are in all cases picked from uniform distributions. Next, consider the choice of a value for G_2. From equation (5), we see that this value should be chosen to be equal to G_1 with probability $(1 + \alpha/M)/(1 + \alpha) = 1/(1 + \alpha)$ (since M is infinite). If this choice is made, values for the V_{2j} will be chosen with probabilities that depend on the corresponding V_{1j}, in accordance with equation (6). Alternatively, a value for G_2 different from G_1 will be picked with probability $\alpha/(1 + \alpha)$. There are an infinite number of such alternatives, but again it makes no difference which is chosen. Proceeding in this fashion, we can pick "values" for the G_i that are defined only in so far as they are specified to be equal or not equal to previously chosen values. This is sufficient to pick values for the corresponding V_{ij}, which are the only observable attributes.

Note that the form of prior distribution chosen here, in which the bias term, α/M, goes to zero as M goes to infinity, ensures that parameter values chosen from the prior distribution will give some of the infinite number of components significant probability. As a result, overfitted solutions in which each data item is attributed to a separate mixture component do not occur.

3. Prediction by Monte Carlo simulation

Although $\widetilde{\phi}$ and the $\widetilde{\psi}_{g,j}$ have been integrated away, calculating $P(\widetilde{V}_i = \tilde{v}_i : 1 \leq i \leq n)$ would still require summing over all M^n possible combinations of values for the G_i. For a classification application, however, all that is needed is the distribution for the category attribute of a test item, conditional on the known attributes of the test item and on the training data. A sample from this conditional distribution can be obtained by a Monte Carlo simulation procedure.

EXHIBITING A DISTRIBUTION VIA MONTE CARLO SIMULATION

Consider the problem of producing a sample from the joint distribution of the random variables $\langle A_1, \ldots, A_n \rangle$ when we can readily compute only the full conditional probabilities, $P(A_i = a_i \mid A_j = a_j : j \neq i)$. This problem arises in many contexts, and can be solved by the Monte Carlo simulation method known as the "Gibbs sampler", which is reviewed in [3].

The simulation starts with arbitrary values $\langle a_1^0, \ldots, a_n^0 \rangle$. In iteration t, new values $\langle a_1^t, \ldots, a_n^t \rangle$ are stochastically chosen in turn, with a_h^t being picked at position h with probability:

$$P(A_h = a_h^t \mid A_i = a_i^t, \ A_j = a_j^{t-1} \ : \ 1 \leq i < h, \ h < j \leq n)$$

In other words, new values are picked at each position from the conditional distribution for that position given the most-recently picked values at all other positions.

It is easy to see that if the distribution of $\langle a_1^T, \ldots, a_n^T \rangle$ for some T is that of $\langle A_1, \ldots, A_n \rangle$, then the same will be true at all later iterations. Furthermore, the method is in fact guaranteed to converge to this equilibrium distribution in the limit as the number of iterations grows, as long as the conditional probabilities used are bounded away from zero. Accordingly, for a large value of T, we can treat the values $\langle a_1^T, \ldots, a_n^T \rangle$ as coming from the desired distribution for $\langle A_1, \ldots, A_n \rangle$. Several independent vectors can be obtained by running several simulations, or by taking vectors from a single simulation at widely spaced times.

The number of iterations required for this method to give a reasonable approximation can be difficult to determine, however, so empirical tests of the practicality of the method in any particular application are necessary. The situation is analogous to the problem of local maxima with deterministic optimization procedures, except that the stochastic aspect ensures that the simulation will escape the local maximum eventually.

SIMULATING THE BAYESIAN PREDICTIVE DISTRIBUTION

This Monte Carlo simulation technique can be used to solve the Bayesian prediction problem for mixture models. Given that we know certain of the V_{ij}, we would like to obtain a sample from the joint distribution for the G_i and the V_{ij} whose values we do not know. This can be done by a simulation in which the values of the V_{ij} that are known are kept fixed, while new values for each G_i and each unknown V_{ij} are chosen repeatedly from their distributions conditional on the current values of all the other variables.

The conditional distributions for the G_i and the unknown V_{ij} are readily obtained from the joint distribution of the G_i and V_{ij} (equation (4)):

$$P(V_{ij} = v \mid G_k = g_k, \ \tilde{V}_k = \tilde{v}_k, \ G_i = g_i, \ V_{il} = v_{il} \ : \ 1 \leq k \leq n, \ k \neq i, \ 1 \leq l \leq m, \ l \neq j)$$

$$= \frac{\sum_{k \neq i} \delta(g_k, g_i) \delta(v_{kj}, v) + \beta_j / N_j}{\sum_{k \neq i} \delta(g_k, g_i) + \beta_j} \tag{7}$$

$$P(G_i = g \mid G_k = g_k, \ \tilde{V}_k = \tilde{v}_k, \ \tilde{V}_i = \tilde{v}_i \ : \ 1 \leq k \leq n, \ k \neq i)$$

$$= \frac{1}{Z} \cdot \left(\sum_{k \neq i} \delta(g_k, g) + \alpha/M \right) \cdot \prod_j \frac{\sum_{k \neq i} \delta(g_k, g) \delta(v_{kj}, v_{ij}) + \beta_j / N_j}{\sum_{k \neq i} \delta(g_k, g) + \beta_j} \tag{8}$$

Here, Z is a factor independent of g that normalizes the distribution to sum to one.

If the number of components is infinite, equation (8) is adapted to allow a new "value" for G_i to be selected that is defined in so far as it is equal to the value for some other G_k, or different from all others, according to the following formulas:

$$P(G_i = g \mid G_k = g_k, \ \tilde{V}_k = \tilde{v}_k, \ \tilde{V}_i = \tilde{v}_i \ : \ 1 \leq k \leq n, \ k \neq i)$$

$$= \frac{1}{Z} \cdot \left(\sum_{k \neq i} \delta(g_k, g) \right) \cdot \prod_j \frac{\sum_{k \neq i} \delta(g_k, g) \delta(v_{kj}, v_{ij}) + \beta_j / N_j}{\sum_{k \neq i} \delta(g_k, g) + \beta_j} \tag{9}$$

$$P(G_i \neq g_k : 1 \leq k \leq n, \ k \neq i \mid G_k = g_k, \ \tilde{V}_k = \tilde{v}_k, \ \tilde{V}_i = \tilde{v}_i : 1 \leq k \leq n, \ k \neq i)$$

$$= \frac{1}{Z} \cdot \alpha \cdot \prod_j \frac{1}{N_j} \tag{10}$$

where again Z is a normalizing factor, the same for both the above formulas. The second formula gives the probability for setting G_i to one of the infinite number of components that are not currently assigned to any of the other G_k. These components do not need to be distinguished, since they all have equivalent effects on the rest of the simulation.

IMPLEMENTATION OF THE METHOD

Figure 1 shows an implementation of the simulation procedure for finite M. This procedure takes as inputs the values of attributes for training items — v_{ij}, for $1 \leq i < n$ and $1 \leq j \leq m$ — along with the values of the indicator attributes for a test item — v_{nj}, for $2 \leq j \leq m$. It outputs an estimate of the distribution for the category attribute of the test item, V_{n1}:

$$q_c \approx P(V_{n1} = c \mid \tilde{V}_i = \tilde{v}_i, \ V_{nj} = v_{nj}, : 1 \leq i < n, \ 2 \leq j \leq m) \tag{11}$$

The number of mixture components, M, the number of simulation iterations, T, and the hyperparameters, α and the β_j, are additional inputs to the procedure.

This implementation incrementally maintains the required occurrence counts for mixture components and for item attributes associated with mixture components. These are kept in the arrays C and A, with

$$C[g] = \sum_{i=1}^{n} \delta(g_i, g), \quad A[g, j, v] = \sum_{i=1}^{n} \delta(g_i, g)\delta(v_{ij}, v) \tag{12}$$

This gives a time complexity of $O((nmM + N_1)T + MN)$, with $N = \sum_j N_j$. On the reasonable assumption that $N_1 < nmM$ and $N < nm$, this reduces to $O(nmMT)$.

If no limit is placed on the number of mixture components, the simulation can be done in time $O(nm\overline{M}T)$, where $\overline{M}$ is the average number of distinct mixture components actually in use (never more than n).

In practice, several modifications to the procedure shown are desirable. Rather than initialize all the g_i to 1, it may be better to set them to sequential or random values. It will also generally be better to let the simulation run for some number of iterations, allowing it to reach equilibrium, before starting to observe which values show up for the category attribute of the test item.

The procedure shown assumes that all the attributes of the training items and all the indicator attributes of the test item are known. If this is not so, values for the unknown V_{ij} could be selected during the simulation, just as is done for V_{n1}. However, if these values are not of interest, the factors that correspond to them in the calculation of the distribution $X[g]$ can instead be simply omitted.

The procedure could easily be modified to observe distributions for the category attributes of several test items as the simulation is run. However, the results would not necessarily be the same as would be obtained by running separate simulations for each test item, since correlations among the indicator attributes of the test items could affect the

Initialize the mixture component associated with each training item, and the unknown v_{n1}.

for $i \leftarrow 1..n$ **do** $g_i \leftarrow 1$ **od**

$v_{n1} \leftarrow 1$

Initialize the occurrence counts for components and for component/attribute combinations.

for $g \leftarrow 1..M$, $j \leftarrow 1..m$, $v \leftarrow 1..N_j$ **do** $C[g] \leftarrow 0$, $A[g, j, v] \leftarrow 0$ **od**

for $i \leftarrow 1..n$ **do**

$\quad\quad$ $C[g_i] \leftarrow C[g_i] + 1$

$\quad\quad$ **for** $j \leftarrow 1..m$ **do** $A[g_i, j, v_{ij}] \leftarrow A[g_i, j, v_{ij}] + 1$ **od**

od

Initialize the frequency counts for the unknown attribute, v_{n1}.

for $c \leftarrow 1..N_1$ **do** $F[c] \leftarrow 0$ **od**

Conduct the simulation for T iterations.

for $t \leftarrow 1..T$ **do**

$\quad$ *Select a new mixture component to go with each item, while updating counts.*

$\quad$ **for** $i \leftarrow 1..n$ **do**

$\quad\quad$ *Remove the old value for g_i from the occurrence counts.*

$\quad\quad$ $C[g_i] \leftarrow C[g_i] - 1$

$\quad\quad$ **for** $j \leftarrow 1..m$ **do** $A[g_i, j, v_{ij}] \leftarrow A[g_i, j, v_{ij}] - 1$ **od**

$\quad\quad$ *Calculate the distribution for the new value of g_i and pick a value from it.*

$\quad\quad$ $Z \leftarrow 0$

$\quad\quad$ **for** $g \leftarrow 1..M$ **do**

$\quad\quad\quad$ $X[g] \leftarrow C[g] + \alpha/M$

$\quad\quad\quad$ **for** $j \leftarrow 1..m$ **do** $X[g] \leftarrow X[g] \cdot (A[g, j, v_{ij}] + \beta_j/N_j) / (C[g] + \beta_j)$ **od**

$\quad\quad\quad$ $Z \leftarrow Z + X[g]$

$\quad\quad$ **od**

$\quad\quad$ **for** $g \leftarrow 1..M$ **do** $X[g] \leftarrow X[g] / Z$ **od**

$\quad\quad$ $g_i \leftarrow$ Value in the range $1..M$ picked according to the distribution X

$\quad\quad$ *Update the occurrence counts to reflect the new value for g_i.*

$\quad\quad$ $C[g_i] \leftarrow C[g_i] + 1$

$\quad\quad$ **for** $j \leftarrow 1..m$ **do** $A[g_i, j, v_{ij}] \leftarrow A[g_i, j, v_{ij}] + 1$ **od**

$\quad$ **od**

$\quad$ *Select a new value for v_{n1}, updating the occurrence and frequency counts accordingly.*

$\quad$ $A[g_n, 1, v_{n1}] \leftarrow A[g_n, 1, v_{n1}] - 1$

$\quad$ **for** $v \leftarrow 1..N_1$ **do** $Y[v] \leftarrow (A[g_n, 1, v] + \beta_1/N_1) / (C[g_n] + \beta_1)$ **od**

$\quad$ $v_{n1} \leftarrow$ Value in the range $1..N_1$ picked according to the distribution Y

$\quad$ $A[g_n, 1, v_{n1}] \leftarrow A[g_n, 1, v_{n1}] + 1$

$\quad$ $F[v_{n1}] \leftarrow F[v_{n1}] + 1$

od

Calculate the observed distribution for the category attribute of the test item, v_{n1}.

for $c \leftarrow 1..N_1$ **do** $q_c \leftarrow F[c] / T$ **od**

Fig. 1. The simulation procedure for finite M.

rest of the simulation. This effect is probably beneficial. It would, however, make comparison with other methods difficult, since the usual criterion for evaluating a classification method is expected performance on a single test item, though naturally many test items are classified in order to obtain significant statistics.

Accordingly, the empirical tests reported below used an implementation in which only the training items, $\tilde{V}_1, \ldots, \tilde{V}_n$, participate in the simulation. The following quantities were calculated with respect to a test item, $\tilde{V}_*$, as the simulation was run:

$$
\begin{aligned}
p_c &= \frac{1}{T} \sum_{t=1}^{T} \sum_{g=1}^{M} \frac{C[g] + \alpha/M}{n + \alpha} \cdot \frac{A[g,1,c] + \beta_1/N_1}{C[g] + \beta_1} \cdot \prod_{j=2}^{m} \frac{A[g,j,v_{*j}] + \beta_j/N_j}{C[g] + \beta_j} \\
&\approx P(V_{*1} = c, \ V_{*j} = v_{*j} : 2 \le j \le m \mid \tilde{V}_i = \tilde{v}_i : 1 \le i \le n)
\end{aligned}
\tag{13}
$$

One can then obtain the desired category probabilities:

$$
P(V_{*1} = c \mid \tilde{V}_i = \tilde{v}_i, \ V_{*j} = v_{*j} : 1 \le i \le n, \ 2 \le j \le m) \approx p_c \, / \, \textstyle\sum_c p_c
\tag{14}
$$

In this way, many test items can be classified simultaneously without any interaction between them. The computation also yields the probability of the complete test item — this is just $p_{v_{*1}}$, where v_{*1} is the true category of the test item.

4. Other mixture modeling methods

In this section, I will briefly review several other methods for modeling data as a mixture, and discuss how they relate to the Bayesian method presented above.

MAXIMUM LIKELIHOOD AND MAP ESTIMATION

Maximum likelihood estimation of the parameters for a mixture model can conveniently be carried out using the EM algorithm [2,8]. In this iterative procedure, expectations for which unobserved mixture component underlies each data item are first computed using the current estimate of parameters (the E step), after which a new parameter estimate is computed using these expectations for the unobserved data (the M step). With repeated application of these steps, the method is guaranteed to converge to a set of parameters that at least locally maximizes the likelihood. By running the algorithm from several starting points, one may be able to find parameter values with likelihood close to the global maximum.

This method can be applied directly only if the number of mixture components, M, is considered fixed. Trying to estimate M itself by maximum likelihood leads to overfitting, with a separate mixture component assigned to each data item. In practice, setting M to a value at the high end of prior expectations sometimes gives acceptable results, with excess components having low probability, or innocuously duplicating other components of the mixture.

Rather than use the parameters that maximize the likelihood, one can instead use the parameters with maximum posterior probability density, the MAP estimate. This method has been used in a latent class analysis context in the early versions of the AutoClass system [1]. In this work, a conjugate prior is used for the model parameters (as in equation (2) here), allowing the MAP estimate to be found using the EM algorithm. If the prior

distribution used has β_j significantly greater than N_j, the problem of overfitting encountered in maximum likelihood estimation is much reduced — there is a pressure for a single
mixture component to underlie many data items, since that allows its posterior attribute
probabilities to overcome the prior bias and accurately fit the observed frequencies.

However, significant deviations from the true Bayesian result would still be expected
if the user has a prior with $\beta_j \leq N_j$ — for example, using $\alpha = M$ and $\beta_j = N_j$ gives
estimates identical to those of maximum likelihood. Furthermore, if several local maxima
are found, selecting between them on the basis of their posterior probability density will
not, in general, give the result which is most likely to be close to the best value.

LOCAL APPROXIMATION OF THE BAYESIAN SOLUTION

In later versions of the AutoClass system [4], the EM algorithm is run from several
starting points, with the number of mixture components set to various values. The local
maxima found in this way are then evaluated by computing an approximation to the total
posterior probability of the region of parameter space in their vicinity. (In this calculation,
the probability is adjusted for the number of permutations of mixture component identities
that give equivalent results, and a prior for the number of components is introduced.) The
local maximum with highest total probability is then considered the best (or they are all
used, with weights given by their probabilities).

The quality of the solutions found with this procedure depends on whether the posterior
parameter probability is in fact concentrated in a set of discrete peaks, whether all the
significant peaks are found, and whether the approximations used to compute the total
probability around each peak are good. In comparison, the quality of the predictions
made by the Bayesian method presented in this paper depends principally on whether
the simulation approaches equilibrium in the time allotted. Which method works best in
practice will have to be discovered by experience.

Both the method of this paper and the later versions of AutoClass can handle the case
where the number of mixture components is unknown, but they do so in rather different
ways. The merits of the two approaches will again have to be evaluated empirically.

5. Performance on a simple classification task

An empirical test of the Bayesian mixture modeling technique was undertaken in order to
determine whether the Monte Carlo simulations involved reach approximate equilibrium in
a reasonable time, and to examine the effect of using a model with an infinite number of
components. Some preliminary comparisons with maximum likelihood and MAP estimation
using the EM algorithm and with nearest-neighbor classification were also done. A more
detailed description of these experiments may be found in [7].

THE EVALUATION METHOD

The Bayesian and other methods were evaluated on the task of modeling data synthetically generated from the mixture distribution shown in Figure 2. This mixture is composed
of four equally-probable components, each of which produces a distribution over nine two-
valued attributes in which each attribute is independent of the others, given knowledge of
the mixture component.

The first of the attributes is considered to be the category of the item. The classification
task is to guess this attribute on the basis of the values for the other eight attributes. The

g	$P(G_i = g)$	$P(V_{ij} = 1 \mid G_i = g)$, $j = 1..9$
1	0.25	1.0 0.8 0.8 0.8 0.8 0.2 0.2 0.2 0.2
2	0.25	1.0 0.2 0.2 0.2 0.2 0.8 0.8 0.8 0.8
3	0.25	0.0 0.8 0.8 0.2 0.2 0.8 0.8 0.2 0.2
4	0.25	0.0 0.2 0.2 0.8 0.8 0.2 0.2 0.8 0.8

Fig. 2.   The true mixture distribution.

best error rate possible for this task with knowledge of the real distribution is 18.6%. Note that none of the indicator attributes provide any information about the category attribute when taken alone, so good performance on this task requires that the procedure uncover the two latent classes underlying each category.

The methods were also evaluated on their ability to predict the entirety of a test item, $\widetilde{V}_*$, whose true value is $\widetilde{v}_*$, with a loss of $-\log_2(\widehat{P}(\widetilde{V}_* = \widetilde{v}_*))$, where $\widehat{P}$ is the estimated probability. This loss function is appropriate for data compression applications. Performance on this task with knowledge of the real distribution is its entropy, which is 7.67 bits.

Each method was applied to three training sets consisting of 12 data items each, and to three training sets consisting of 48 data items each. These sets were randomly generated from the true distribution. For each method and each training set, all $2^9 = 512$ possible test items were used to evaluate performance on classification and whole-item prediction, with each item weighted by its probability under the true distribution. There is thus no sampling error from this source in the evaluations.

A typical run on one of the small training sets with the number of mixture components, M, fixed at 4 is shown in Figure 3. The columns of the figure show which of the mixture components, represented by the symbols ♣, ◇, ♡, and ♠, is associated with each training item at successive iterations of the simulation, starting with an initial state in which all items are associated with ♣.

As can be seen, the simulation soon reaches an apparent equilibrium situation in which the items form relatively stable groupings associated with particular mixture components. These groupings accord with what one would expect from manual examination of the data or knowledge of the true distribution.

CONVERGENCE OF THE SIMULATIONS

Convergence of the simulations to equilibrium was quantitatively evaluated by observing predictive performance as the number of iterations increased. Such tests were performed for $M = 4$, with $\alpha = 1$ and all the $\beta_j = 1$. Simulations using larger values for α or the β_j would presumably reach equilibrium more quickly, since the conditional probabilities that arise are bounded further away from zero. A larger value for M would also be expected to improve convergence to equilibrium, by eliminating some local maxima.

For these tests, the simulation runs began with all items assigned the same mixture component. After each iteration, classification and whole-item prediction for each possible test item were done using equations (13) and (14). For the small training sets, the performance using the exact Bayesian predictive distribution for the 512 possible test items based on the 12 training items in the small data sets was calculated using equation (4) by a summation over all $4^{13} = 67\,108\,864$ possible assignments of mixture components to the training data and a test item (a computation taking sixteen hours for each training set).

#	Data item	Assignments of components to items at successive iterations
1	⟨1 12 11 21 22⟩	♣♣♣♡♡♡♡♡♡♡♡♡♡♡♡♡♡♡♡♡♡♡♡♡♡♡♡♡♡♡♡♡♡♣♣♣♣♣♣♣♣
2	⟨1 22 22 21 11⟩	♣♣♣♣♣♣♣♣♣♣♣♣♣◇◇◇◇♠♠♠♠♠♠♠♠♠♠♠♠♠♠◇◇♣♣◇♠♠♠♣◇◇◇◇♡♡
3	⟨2 11 22 12 22⟩	♣♣♡♠♠♠♠♠♠♠♠♠♠♠♠♠♠♠♠♠♠♠♠♠♠♠♠♠♠♠♠♠♠♠♡♣♣♡♡♡♠♠♠♠
4	⟨1 22 22 12 11⟩	♣♣♣♣♣♣♣♣♣♣♣♣♣◇♣♠♠♠♠♠♠♠♠♠♠♠♠♠♠♠◇♣♣♠♠♦◇◇◇◇◇◇◇♡♡
5	⟨1 12 22 11 11⟩	♣♣♣♣♣♣♣♣♣♣♣♣♠♠♠♠♠♠♠♠♠♠♠♠♠♠♠♠♠♠♠♠♠♠♠♠♠♠♠♠♣◇◇◇♡♡♡
6	⟨2 22 21 21 11⟩	♣♣♣♣◇♣◇◇◇♣◇◇◇◇◇◇◇♠♠◇◇♠♦◇♠◇◇◇◇♠◇♣♦◇♠◇◇◇◇◇◇◇
7	⟨1 11 11 21 12⟩	♣♣♣♡♡♡♡♡♡♡♡♡♡♡♡♡♡♡♡♡♡♡♡♡♡♡♡♡♡♡♡♡♡♡♣♣♣♣♣♣♣♣
8	⟨2 22 11 22 11⟩	♣♣♣♣♣◇◇
9	⟨2 12 22 12 21⟩	♣♣♠♠♠♠♠♠♠♠♠♣♣♣♣♣♣♣♣♣♣♣♣♣♣♣♣♣♣♣♠♠♠♠♣♠♠♠♠♠◇♡♡◇♡♠♠♠♠♠♠
10	⟨1 11 21 22 22⟩	♣♣♡♣♡♠♣♡♡♡♡♡♡♡♡♡♡♣♡♡♣♡♡♡♡♡♡♡♡♡♡♠♡♡♡♣♣♡◇♣♣♣♣♣
11	⟨2 22 12 22 11⟩	♣♣♣♣♣◇◇
12	⟨2 21 11 22 11⟩	♣♣♣♣♣◇◇

Fig. 3. A run of the simulation procedure with $M = 4$, $\alpha = 1$, and $\beta_j = 1$.

Convergence of the simulations on the small training sets to the exact solution was evident within about 200 iterations in each of three independent runs for each of the three training sets, except for one run whose classification performance still differed somewhat from that of the exact solution after 1000 iterations. This one anomaly may be due to the fact that classification performance has a discrete set of possible values. Very small differences in estimated probabilities can therefore have a significant effect on performance.

Simulations for the larger training sets were also run, and appeared to converge at least as quickly as those for the smaller training sets. One might expect that equilibrium would be harder to reach when more data items are involved, but if so, it appears that this was out-weighed by the more rapid accumulation of a significant sample. The exact Bayesian calculation for the larger training sets is infeasible, so only apparent convergence could be tested for.

PERFORMANCE OF THE BAYESIAN METHOD

The Bayesian mixture modeling method was evaluated on the three large and three small training sets, with the number of mixture components (M) set to the true number (4), and to infinity. Various values of α were used, each with a range of values for the β_j (all equal) from $\frac{1}{2}$ to 4 in geometric steps of $\sqrt{2}$.

In the simulations done for these tests, mixture components were assigned sequentially to data items (repeating, for $M < \infty$), and the simulation was run for 100 iterations in order to approach equilibrium. Data was then collected for 400 iterations for use in classification and prediction of the 512 possible test items.

The results for both $M = 4$ and $M = \infty$ were found to be quite insensitive to the value of α, within the range explored (2 to 8 for $M = 4$, $\frac{1}{2}$ to 2 for $M = \infty$). Results for the largest α were slightly better. For $M = 4$ at least, this is as expected, since in the true distribution, the four components are all equally likely, a situation favoured by a high α.

Sensitivity to the value of the β_j was much greater. Since the attribute probabilities in the true distribution are all near 0 or 1, one would expect a values of β_j somewhat less than $N_j = 2$ to be preferred. This was indeed the case for the small training sets. For the large training sets, a value near 2 appeared best, perhaps because these sets have enough data to force most of the probability to the region near the true parameter values without assistance from a prior with $\beta_j < N_j$ — a prior which would also have the undesirable effect of increasing the probability of parameter values more extreme than the true ones.

For both $M = 4$ and $M = \infty$ the best values for α and the β_j were selected based on performance on the true distribution. Results with these values are shown in Figure 4. Of course, this method of selecting α and β_j is not quite legitimate. A proper Bayesian treatment of the situation where α and the β_j are not fixed by prior knowledge would involve treating them as hyperparameters, as is discussed briefly below. The effect of this "cheating" should be small in this case, however, since only two parameters are being set, based on a fair amount of data.

As can be seen, performance with the number of mixture components set to infinity is as good as, even better than, that with the number of components fixed to its true value. This may seem paradoxical, but is explicable if one remembers that the value of α used is not optimal — setting α to 100000 with $M = 4$ in fact gives better results than any others. These results confirm the feasibility of using a model with $M = \infty$ when the number of mixture components is unknown or indefinite.

COMPARISON OF THE BAYESIAN AND OTHER METHODS

A parallel evaluation was done of MAP estimation using the EM algorithm, using the same range of α and the β_j as were used with the Bayesian method, for $M = 4$. In an attempt to mimic the capability of the Bayesian method to give good results with $M = \infty$, runs with M set to the relatively large values of 10 and 100 were also done, with α set to 10 and 50, respectively.

In all cases, the EM algorithm was run for 100 iterations, which was ample to reach convergence, starting from three initial sets of parameter values. The parameters from the run that converged to the point with highest posterior probability density were then selected for use on the test items. When $\alpha < M$ and/or $\beta_j < N_j$, the posterior density function has singularities at points where one or more parameter values are zero. In these cases, the densities at different points were compared as if all such zero parameters were set to the same infinitesimal value.

The results obtained were qualitatively different from those obtained with the Bayesian method. For example, with $\beta_j \leq N_j = 2$ whole-item prediction performance was very poor (in fact, often infinitely poor). This is not surprising, since with $\beta_j < N_j$, the singularities in the prior distribution at zero values of the $\psi_{g,j,v}$ will encourage extreme overfitting. Results for values of α and the β_j for which there were no singularities in the prior also showed differences in both magnitude and trend from the corresponding Bayesian results.

The result of MAP estimation with given α and β_j clearly cannot be regarded as an approximation to the true Bayesian solution. A better correspondence appears if one regards the results of MAP estimation with $\beta_j > N_j$ as an approximation to the Bayesian result with $\beta'_j = \beta_j - N_j$. Such a correspondence might be expected by analogy with the situation with Dirichlet priors for a simple discrete distribution, though in the mixture case there appears to be no exact equivalence. Alternatively, one might simply regard the EM algorithm as performing "maximum penalized likelihood" estimation, abandoning any attempt to interpret it in a Bayesian framework.

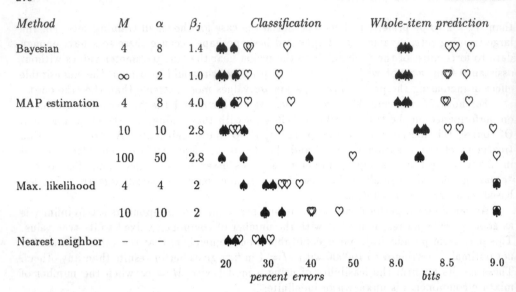

Fig. 4. Comparative performance of Bayesian prediction, MAP estimation, maximum likelihood estimation, and nearest neighbor classification. Results on the small training sets are plotted with ♡; those on the large training sets with ♠.

Nevertheless, for $M = 4$, MAP estimation performs quite well, as seen in Figure 4. Again, results using the best overall values of α and β_j are shown. The attempt at mimicing the success of the Bayesian method when the number of components is regarded as unknown was somewhat less successful, and for $M = 100$ the results shown are also fragile — apparently minor changes to M and/or α can give much worse results.

Figure 4 also shows the performance of maximum likelihood estimation of the mixture distribution parameters, which is equivalent to MAP estimation with $\alpha = M$ and $\beta_j = N_j$. For additional perspective, the performance of a nearest neighbor classifier is shown as well. This classifier attributes a test item to the category of the training item that is closest to it in Hamming distance (with voting to break ties).

Maximum likelihood estimation is not competitive with either the full Bayesian or the MAP methods in any context. Nearest neighbor classification does surprising well, however, especially for the small training sets. This is perhaps an indication that the problem is not tremendously difficult (though in some respects, such as the equal importance of all attributes, the problem is well-suited to this classifier).

Clearly, results such as these on synthetic data can be suggestive only. A true picture of the worth of the Bayesian method can be obtained only by applying it to significant real data sets. Comparison of the Monte Carlo technique presented here with the local approximation approach of [4] would also be of great interest.

6. Conclusion

I have presented a method for performing Bayesian prediction from data modeled by a mixture distribution. The practical benefits of this method include simplicity of implementation, avoidance of overfitting, and some protection from local maxima in the computation.

Of both practical and theoretical significance is the new approach to handling problems where the number of mixture components is either unknown or indefinite. This approach is made possible by the ability of the method to cope easily with countably infinite mixtures.

In this paper, I have assumed that the user's prior knowledge can be adequately captured by a choice of values for α and the β_j in the prior distribution (equation (2)). Often, the real prior will not be so specific. In this case, α and the β_j can be treated as hyperparameters with their own prior distributions. At moderate computational cost, these hyperparameters can be included in the Monte Carlo simulation, allowing Bayesian inference for this hierarchical model to be performed. Modifications to allow Dirichlet distributions that are not necessarily symmetrical with respect to the discrete attribute values are also possible.

Preliminary experiments with an algorithm encorporating these extensions have been performed. They show that one benefit of a hierarchical model is an increased ability to identify "noise" attributes that do not discriminate between any of the mixture components. The distributions for such attributes can be modeled at the hyperparameter level more economically than at the level of parameters for individual components, thereby eliminating the detrimental effects that arise from spurious correlations with such noise attributes.

Although the development in this paper is confined to discrete data, the technique should also be applicable when items have real-valued attributes modeled by independent Gaussian distributions if an appropriate conjugate prior is used. The method can also be extended to the Hidden Markov Models used in speech recognition [6].

ACKNOWLEDGMENTS. I thank David MacKay, Geoff Hinton, and members of the Connectionist Research Group at the University of Toronto for helpful discussions. This work was supported by the Natural Sciences and Engineering Research Council of Canada, and by the Ontario Information Technology Research Centre.

REFERENCES

[1] Cheeseman, P., Kelly, J., Self, M., Stutz, J., Taylor, W., and Freeman, D. (1988) 'AutoClass: A Bayesian classification system', *Proceedings of the Fifth International Conference on Machine Learning.*

[2] Everitt, B. S. and Hand, D. J. (1981) *Finite Mixture Distributions*, London: Chapman and Hall.

[3] Gelfand, A. E. and Smith, A. F. M. (1990) 'Sampling-based approaches to calculating marginal densities', *Journal of the American Statistical Association*, **85**, 398-409.

[4] Hanson, R., Stutz, J., and Cheeseman, P. (1991) 'Bayesian classification with correlation and inheritance', to be presented at the 12th International Joint Conference on Artificial Intelligence, Sydney, Australia, August 1991.

[5] Kai-Tai, F. and Yao-Ting, Z. (1990) *Generalized Multivariate Analysis*, Berlin: Springer-Verlag.

[6] Levinson, S. E., Rabiner, L. R., and Sondhi, M. M. (1983) 'An introduction to the application of the theory of probabilistic functions of a Markov process to automatic speech recognition', *Bell System Technical Journal*, **62**, 1035-1074.

[7] Neal, R. M. (1991) 'Bayesian mixture modeling by Monte Carlo simulation', Technical Report CRG-TR-91-2, Department of Computer Science, University of Toronto.

[8] Titterington, D. M., Smith, A. F. M., and Makov, U. E. (1985) *Statistical Analysis of Finite Mixture Distributions*, Chichester, New York: Wiley.

POINT-PROCESS THEORY AND THE SURVEILLANCE OF MANY OBJECTS

John M. Richardson and Kenneth A. Marsh
Rockwell International Science Center
Thousand Oaks
CA 91360

ABSTRACT. The development of the methodologies of surveillance (i.e., detection and track-ing) must confront emerging situations in which large random numbers of moving objects may be present under conditions of high noise and low resolution. Conventional approaches based upon extended Kalman filter theory require executions of detection and tracking in separate time in-tervals and require relatively favorable noise and resolution conditions. In this paper we discuss a new approach based upon point-process theory that to a significant degree obviates the above requirements. Furthermore, it entails a computational burden that does not increase significantly with the number of objects present. A central feature of our new methodology is an exact infinite hierarchy of coupled equations giving the time evolution of all orders of mean densities of repre-sentative points in single-object state space. A simple closure approximation yields a truncated hierarchy composed of a single equation involving only the first-order density. The corresponding algorithm, called STRIDE (Superresolution in TRacking via Integro-Differential Equations), is also capable of superresolution in both static and dynamic situations.

1. The Problem Of Many Moving Objects

In surveillance (i.e., detection and tracking) problems involving many moving objects (e.g., missiles, decoys, aircraft, submarines, surface ships, torpedoes, ground vehicles, or even spectral lines), the more conventional approaches (Blackman, 1986) require detection before tracking can begin. This is made difficult when there are large numbers of possible targets, the targets are moving, signal-to-noise ratios (SNRs) are poor, there is a mixture of sensor types, and the resolution of the total sensor system is poor.

Our approach involves a combination of conditional Fokker-Planck methodology and point-process theory. This approach overcomes many of the above difficulties and, in par-ticular, achieves the unification of detection and tracking, while maintaining a near-optimal degree of superresolution. It also avoids saturation of the signal processing module with increasingly large numbers of objects under surveillance (although eventually a significant number of neighboring objects cannot be resolved) and avoids the difficulties associated with object labeling. It represents the statistical nature of the many-object system by a set of local ensemble mean densities of all orders in single-object state space. We then derive an infinite hierarchy of equations giving the time evolution of the densities, including both the dynamics of objects and the conditioning on sensor data. The crucial and final step is to reduce this hierarchy to a tractable form by using a suitable closure approximation.

213

C. R. Smith et al. (eds.), Maximum Entropy and Bayesian Methods, Seattle, 1991, 213–220.

2. Derivation And Truncation Of Density Hierarchy

In the multi-object case, one is obliged to label the objects in the initial formulation, although the labelling distinction eventually disappears from the later analytical development. The multi-object state in the initial formulation includes the number of objects (N) and the N state vectors $z_n(t), 1 \leq n \leq N$. The number N is random in this representation. We define the local ensemble-mean density of the first order by the expression

$$\rho(z,t) = E\big[\sum_{n=1}^{N} \delta(z - z_n(t))|y^t\big],$$

where E is the expectation operator, y^t represents all the measurements made up to time t, and $\delta(\cdot)$ is the Dirac delta function. To simplify writing we omit the explicit reference to conditioning in the notation $\rho(z,t)$. We will henceforth refer to $\rho(z,t)$ and its higher order analogs as simply "densities." The higher-order densities are defined by obvious extensions, using multiple summations of products of delta functions. The variable z (or $z, z', z'', \ldots$ for the higher-order densities) is a time-independent reference point in single-object state space. The object density ρ is not a probability density. Its integral over all space should equal the average total number of objects present. It is to be emphasized that the representation of the probabilistic situation in terms of a hierarchy of densities is label-free. This means that different objects are distinguishable only by the positions of their representative points in single-object state space (i.e., z-space). Objects of different types can be handled by including a parameter defining type as a component of the state vector. If this parameter is discrete-valued, suitable changes in the definition of the δ-function must be made. A summary of the derivation is provided in the next section.

The density hierarchy is equivalent to the conventional probabilistic representation, if the latter is invariant to the permutation of object labels. A density of arbitrary order M has the valuable property that it pertains to every possible value of N equal to or greater than M. Because of the above equivalence, it should be possible to derive an equivalent version of the separation theorem. Loosely speaking, this version would imply that the information contained in the set of densities of all orders at time t is so comprehensive that any additional information about past measurements is redundant.

The density hierarchy is beyond any conceivable computational implementation because it is infinite. However, it is possible to obtain approximate solutions by truncating the hierarchy to a finite set of members by applying a suitable closure approximation. In Sec. 4 below, we present an outline of a procedure that leads to a one-member hierarchy involving an equation for the time evolution of $\rho(z,t)$ that depends only upon $\rho(z,t)$ at the same time. This equation is realized in an algorithm called STRIDE (Superresolution in TRacking via Integro-Differential Equations). This algorithm is completely defined (aside from initial conditions) by the models of the dynamics and the measurement process, combined with suitable statistical assumptions.

There is also a static version of STRIDE in which the system of interest is time-independent, but when the inference process evolves in time.

3. Mathematical Derivation Of The Density Hierarchy

- Dynamical model:

$$\dot{z}_n(t) = f[z_n(t), t] + \mu_n(t), \quad n = 1, \ldots, N. \tag{1}$$

- Measurement model:

$$y(t) = \sum_{n=1}^{N} g[z_n(t), t] + \nu(t), \tag{2}$$

where $z_n(t)$ is the state vector of the nth object, $y(t)$ is the measurement vector, the functions $f(.,.)$ and $g(.,.)$ represent the dynamics and measurement of a single object in the absence of noise, and N is the total number of objects, assumed to be random.

- Statistical assumptions (*a priori*):

The vector functions $\mu_n(t)$ and $\nu(t)$ are independent Gaussian random processes with zero means and covariances

$$E\mu_n(t)\mu_{n'}(t')^T = R_\mu \delta_{nn'} \delta(t - t'), \quad n, n' = 1, \ldots, N, \tag{3}$$

$$E\nu(t)\nu(t)^T = R_\nu \delta(t - t'). \tag{4}$$

- Density hierarchy:

$$\frac{\partial}{\partial t}\rho(z, t) + L\rho(z, t) + \phi_1 \rho(z, t) = 0, \tag{5}$$

$$\frac{\partial}{\partial t}\rho(z, z', t) + (L + L')\rho(z, z', t) + \phi_2 \rho(z, z', t) = 0, \tag{6}$$

$$\vdots$$

where it is to be understood that $\rho(z, t)$, $\rho(z, z', t), \ldots$ are conditioned on past values of $y(t)$. The entity L is the Fokker-Planck (FP) operator defined by

$$L(\cdot) = \frac{\partial^T}{\partial z} f(z, t)(\cdot) - \frac{1}{2} Tr \frac{\partial}{\partial z} \frac{\partial^T}{\partial z} R_\mu(\cdot). \tag{7}$$

The definition of L' is the same as (7), except the z is replaced by z' on the right side. We will not discuss in detail the conditioning factors $\phi_1, \phi_2, \ldots$, except to state that, aside from dependence upon $y(t), \phi_1$ is a functional of the 1st, 2nd, and 3rd order densities (all evaluated at time t), ϕ_2 is a functional of the 2nd, 3rd, and 4th order densities, and so forth.

4. Truncation Of The Hierarchy At The First Member (STRIDE)

- Closure approximation:

$$\rho(z, z', t) \rightarrow \rho(z, t)\rho(z', t) \tag{8}$$

$$\rho(z, z', z'', t) \rightarrow \rho(z, t)\rho(z', t)\rho(z'', t) \tag{9}$$

$$\vdots$$

• Truncation at first member:

$$\frac{\partial \rho}{\partial t} + \mathrm{L}\rho + \phi_1 \rho = 0 \tag{10}$$

$$\vdots$$

where $\rho = \rho(z, t)$ and

$$\phi_1 = -\left[y(t) - \int dz' g(z', t)\rho(z', t)\right]^T R_\nu^{-1} g(z, t) + \frac{1}{2} g(z, t)^T R_\nu^{-1} g(z, t) \tag{11}$$

in the case of a bandwidth-limited measured signal $y(t)$.

Several applications are discussed in the following sections.

5. Application Of Static STRIDE To Superresolution

In this application, we consider a diffraction-limited image corresponding to a true image of point sources whose number and positions are random. For the sake of simplicity, we assume all sources have a common known luminosity. The measurement model is assumed to have the form

$$y(\underline{r}, t) = \sum_{n=1}^{N} g(\underline{r} - \underline{s}_n) + \nu(\underline{r}, t), \tag{12}$$

where $y(\underline{r}, t)$ is a possible observed image, $g(\underline{r} - \underline{s}_n)$ is the point-spread function centered at the nth point source whose position is $\underline{s}_n$, and where $\nu(\underline{r}, t)$ is a Gaussian random process with zero mean and a covariance function given by

$$E\nu(\underline{r}, t)\nu(\underline{r}', t') = R_\nu \delta(\underline{r} - \underline{r}')\delta(t - t'), \tag{13}$$

where R_ν is a positive scalar constant. In the above equations $\underline{r}$ and the $\underline{s}_n$ are 2D vectors. Clearly, here the single object state is $\underline{s}$, which is nondynamical, i.e., $\underline{\dot{s}} = 0$. In this problem, the density $\rho(\underline{s}, t)$ is updated by the equation

$$\frac{\partial \rho}{\partial t} + \phi_1 \rho = 0, \tag{14}$$

from which the FP operator has disappeared because of the nondynamical nature of this problem. The factor ϕ_1 is given by the approximate expression

$$\phi_1 = -\int d\underline{r} \left[y(\underline{r}, t) - \int d\underline{s}' g(\underline{r} - \underline{s}')\rho(\underline{s}', t)\right] R_\nu^{-1} g(\underline{r} - \underline{s}) + \frac{1}{2} \int d\underline{r} \, R_\nu^{-1} g(\underline{r} - \underline{s})^2. \tag{15}$$

The Eq. (14) is to be integrated on t from 0 to the terminal time t_f. Here, time can be regarded as proportional to the signal-to-noise ratio (SNR).

The behavior of this algorithm was investigated with noiseless synthetic data for $y(\underline{r}, t)$ (now time-independent) corresponding to a true image consisting of two fixed point sources of equal luminosity. Figures 1a to 1c show the construction of the synthetic data. The

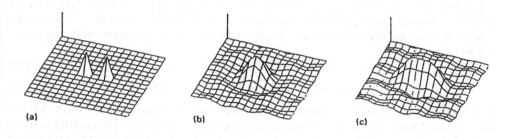

Fig. 1. The "true" image (a) convolved with the point-spread function (b) produces the measured image (c). The measured image is the input to STRIDE.

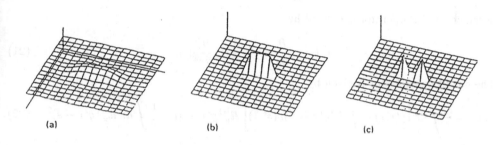

Fig. 2. The output image from STRIDE after one "weak" conditioning* (a), 37 weak conditionings (b), and 128 weak conditionings (c). Note how close this "superresolution" result is to the true image.

evolution of superresolution in $\rho(\underline{s}, t)$ is illustrated in Figs. 2a to 2c. The achievement of a significant degree of superresolution is clearly demonstrated.

6. Application Of Dynamic STRIDE To The Surveillance Of Many Moving Objects

For the sake of simplicity, we limit this treatment to a single scalar position variable s and the associated scalar velocity v in the single-object state space. The dynamical model is represented by the set of equations

$$\dot{s}_n(t) = v_n(t), \quad \dot{v}_n(t) = \mu_n(t), \quad n = 1, \ldots, N, \tag{16}$$

where N, the number of objects, is random and $\mu_n(t)$ is to be defined later. The measurement model is given by the single equation

$$y(r, t) = \sum_{n=1}^{N} g[r - s_n(t)] + \nu(r, t), \tag{17}$$

---

* A "weak" conditioning is one in which the assumed SNR is so small that the conditioning process makes only a small change in its operand.

where $g(r - s_n)$ is the point-spread function centered at the source associated with the nth object at the position $s_n(t)$ at time t. The quantity $y(r)$ is a possible 1-D observed diffraction-limited image given as a function of the 1-D position variable r. The entities $\mu_n(t)$ and $\nu(r, t)$ are statistically independent Gaussian random processes with zero means and covariances:

$$E\mu_n(t)\mu_{n'}(t') = R_\mu \delta_{nn'}\delta(t - t'), \tag{18}$$

$$E\nu(r, t)\nu(r', t') = R_\nu \delta(r - r')\delta(t - t'). \tag{19}$$

This measurement model resembles that described in the previous section, except that the present model has one spatial dimension and has a dynamical (rather than static) position $s_n(t)$ of the nth source. The single object state is now (s, v) and the density now assumes the form $\rho(s, v, t)$, where the conditioning on the past values of $y(r, t)$ is again implicit. The temporal evolution of this density is given by

$$\frac{\partial \rho}{\partial t} + L\rho + \phi_1 \rho = 0, \tag{20}$$

in which the FP operator is defined by

$$L\,(\cdot) = \frac{\partial}{\partial s}v(\cdot) - \frac{1}{2}R_\mu \frac{\partial^2}{\partial v^2}(\cdot). \tag{21}$$

The conditioning factor is given by

$$\phi_1 = -\int dr \left[y(r, t) - \int ds'g(r - s')\rho(s', t) \right] R_\nu^{-1}g(r - s) + \frac{1}{2}\int dr\, R_\nu^{-1}g(r - s)^2, \tag{22}$$

in which $\rho(s, t) = \int dv\rho(s, v, t)$.

We have carried out a number of computational investigations of Eq. 20, using a time sequence of synthetic test images $y(r, t)$. Here, we will show the results obtained when the synthetic images are based upon the assumption that two objects are present with crossing * trajectories as shown in Fig. 3a. The synthesized sequence of diffraction-limited images is shown in Fig. 3b as a function of r and t. With this as an input to Eq. 22, we obtain [with appropriate assumptions concerning R_μ, R_ν, and $\rho(s, v, o)$] the solution $\rho(s, v, t)$, various aspects of which are shown in subsequent figures. The density $\rho(s, t)$ with the velocity v integrated out is shown in Fig. 3c. This result clearly shows the parallel time-distributed processes of detection and tracking which survive the crossing of the assumed trajectories while regaining superresolution. In Fig. 4a, the density $\rho(s, v, t_c)$ is displayed in the full state space (s, v) at the crossing time t_c. At this critical time, one can readily see that superresolution in position is lost but superresolution in velocity is maintained. At the final time t_f, the density $\rho(s, v, t_f)$ is displayed in Fig. 4b with a manifestation of superresolution in both position and velocity.

7. Discussion

We have presented a, perhaps overly brief, derivation of a new algorithm called STRIDE, which accomplishes the detection and location of many stationary objects in

---

* If the position s is a 1-D projection of a higher-dimensional position space, actual physical collisions are unlikely.

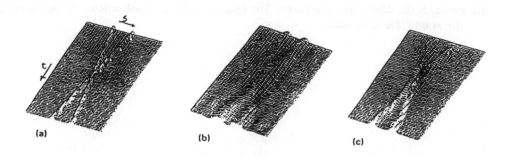

Fig. 3. Position versus time of true state (a),the diffraction-limited sequence of "measurements" (b), and the object density (in position space) history estimated by STRIDE (c).

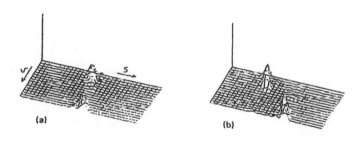

Fig. 4. Object densities in position and velocity space, as estimated by STRIDE at approximate time of crossing (a) and at terminal time (b). Note that crossing objects are distinguishable from their velocities, even though positions are very close.

the static case and the detection and tracking of many moving objects in the dynamic case. These operations are accomplished in a unified way, which means that, in the dynamic case for example, detection and tracking are conducted in parallel in a time-distributed manner. The formalism is free of arbitrary labels for objects, i.e., different objects are distinguished only by the different positions of their representative points in single-object state space. A consequence of this situation is that the data association problem does not exist. A final benefit is that a near-optimal degree of superresolution is attained in both static and dynamic cases.

ACKNOWLEDGMENTS. This investigation was supported by the Independent Research and Development funds of Rockwell International. John M. Richardson is a consultant for Rockwell International Science Center, Thousand Oaks, California 91360. Kenneth A. Marsh is now with the Jet Propulsion Laboratory, Pasadena, California 91103.

REFERENCES

Blackman, S. S.: 1986, *Multiple-Target Tracking with Radar Applications*, Artech House, Norwood (Massachusetts).

A Matlab Program to Calculate the Maximum Entropy Distributions

Ali Mohammad–Djafari
Laboratoire des Signaux et Systèmes (CNRS–ESE–UPS)
École Supérieure d'Électricité
Plateau de Moulon, 91192 Gif–sur–Yvette Cédex, France

ABSTRACT. The classical maximum entropy (ME) problem consists of determining a probability distribution function (pdf) from a finite set of expectations $\mu_n = \mathrm{E}\{\phi_n(x)\}$ of known functions $\phi_n(x), n = 0, \ldots, N$. The solution depends on $N + 1$ Lagrange multipliers which are determined by solving the set of nonlinear equations formed by the N data constraints and the normalization constraint. In this short communication we give three Matlab programs to calculate these Lagrange multipliers. The first considers the case where $\phi_n(x)$ can be any functions. The second considers the special case where $\phi_n(x) = x^n, n = 0, \ldots, N$. In this case the μ_n are the geometrical moments of $p(x)$. The third considers the special case where $\phi_n(x) = \exp(-jn\omega x), n = 0, \ldots, N$. In this case the μ_n are the trigonometrical moments (Fourier components) of $p(x)$. We give also some examples to illustrate the usefullness of these programs.

1. Introduction

Shannon (1948) indicated how maximum entropy (ME) distributions can be derived by a straigtforward application of the calculus of variations technique. He defined the entropy of a probability density function $p(x)$ as

$$H = - \int p(x) \ln p(x) \, dx \qquad (1)$$

Maximizing H subject to various side conditions is well–known in the literature as a method for deriving the forms of minimal information prior distributions; e.g. Jaynes (1968) and Zellner (1977). Jaynes (1982) has extensively analyzed examples in the discrete case, while in Lisman and Van Znylen (1972), Rao (1973) and Gokhale (1975), Kagan, Linjik continuous cases are considered. In the last case, the problem, in its general form, is the following

$$\text{maximize} \quad H = - \int p(x) \ln p(x) \, dx$$
$$\text{subject to} \quad \mathrm{E}\{\phi_n(x)\} = \int \phi_n(x) p(x) \, dx = \mu_n, \quad n = 0, \ldots, N \qquad (2)$$

where $\mu_0 = 1$, $\phi_0(x) = 1$ and $\phi_n(x), n = 0, \ldots, N$ are N known functions, and $\mu_n, n = 0, \ldots, N$ are the given expectation data. The classical solution of this problem is given by

$$p(x) = \exp\left[- \sum_{n=0}^{N} \lambda_n \, \phi_n(x) \right] \qquad (3)$$

221

C. R. Smith et al. (eds.), Maximum Entropy and Bayesian Methods, Seattle, 1991, 221–233.
© 1992 Kluwer Academic Publishers.

The $(N + 1)$ Lagrangien parameters $\boldsymbol{\lambda} = [\lambda_0, \ldots, \lambda_n]$ are obtained by solving the following set of $(N + 1)$ nonlinear equations

$$G_n(\boldsymbol{\lambda}) = \int \phi_n(x) \exp\left[-\sum_{n=0}^{N} \lambda_n \phi_n(x)\right] \mathrm{d}x = \mu_n, \quad n = 0, \ldots, N \tag{4}$$

The distributions defined by (3) form a great number of known distributions which are obtained by choosing the appropriate N and $\phi_n(x), n = 0, \ldots, N$. In general $\phi_n(x)$ are either the powers of x or the logarithm of x. See Mukhrejee and Hurst (1984), Zellner (1988), Mohammad–Djafari (1990) for many other examples and discussions. Special cases have been extensively analyzed and used by many authors. When $\phi_n(x) = x^n, n = 0, \ldots, N$ then $\mu_n, n = 0, \ldots, N$ are the given N moments of the distribution. See, for example, Zellner (1988) for a numerical implementation in the case $N = 4$.

In this communication we propose three programs written in MATLAB to solve the system of equations (4). The first is a general program where $\phi_n(x)$ can be any functions. The second is a special case where $\phi_n(x) = x^n, n = 0, \ldots, N$. In this case the μ_n are the geometrical moments of $p(x)$. The third is a special case where $\phi_n(x) = \exp(-jn\omega x), n = 0, \ldots, N$. In this case the μ_n are the trigonometrical moments (Fourier components) of $p(x)$. We give also some examples to illustrate the usefullness of these programs.

2. Principle of the method

We have seen that the solution of the standard ME problem is given by (3) in which the Lagrange multipliers $\boldsymbol{\lambda}$ are obtained by solving the nonlinear equations (4). In general, these equations are solved by the standard Newton method which consists of expanding $G_n(\boldsymbol{\lambda})$ in Taylor's series around trial values of the *lambda*'s, drop the quadratic and higher order terms, and solve the resulting linear system iteratively. We give here the details of the numerical method that we implemented. When developing the $G_n(\boldsymbol{\lambda})$ in equations (4) in first order Taylor's series around the trial $\boldsymbol{\lambda}^0$, the resulting linear equations are given by

$$G_n(\boldsymbol{\lambda}) \cong G_n(\boldsymbol{\lambda}^0) + (\boldsymbol{\lambda} - \boldsymbol{\lambda}^0)^t \left[\mathbf{grad}\, G_n(\boldsymbol{\lambda})\right]_{(\boldsymbol{\lambda}=\boldsymbol{\lambda}^0)} = \mu_n, \quad n = 0, \ldots, N \tag{5}$$

Noting the vectors $\boldsymbol{\delta}$ and $\boldsymbol{v}$ by

$$\boldsymbol{\delta} = \boldsymbol{\lambda} - \boldsymbol{\lambda}^0$$

$$\boldsymbol{v} = \left[\mu_0 - G_0(\boldsymbol{\lambda}^0), \ldots, \mu_N - G_N(\boldsymbol{\lambda}^0)\right]^t$$

and the matrix $\boldsymbol{G}$ by

$$\boldsymbol{G} = \left(g_{nk}\right) = \left(\frac{\partial G_n(\boldsymbol{\lambda})}{\partial \lambda_k}\right)_{(\boldsymbol{\lambda}=\boldsymbol{\lambda}^0)} \quad n, k = 0, \ldots, N \tag{6}$$

then equations (5) become

$$\boldsymbol{G}\boldsymbol{\delta} = \boldsymbol{v} \tag{7}$$

This system is solved for $\boldsymbol{\delta}$ from which we drive $\boldsymbol{\lambda} = \boldsymbol{\lambda}^0 + \boldsymbol{\delta}$, which becomes our new initial vector $\boldsymbol{\lambda}^0$ and the iterations continue until $\boldsymbol{\delta}$ becomes appropriately small. Note that the matrix $\boldsymbol{G}$ is a symmetric one and we have

$$g_{nk} = g_{kn} = -\int \phi_n(x)\,\phi_k(x) \exp\left[-\sum_{n=0}^{N} \lambda_n \phi_n(x)\right] \mathrm{d}x \quad n, k = 0, \ldots, N \tag{8}$$

So in each iteration we have to calculate the $N(N-1)/2$ integrals in the equation (8). The algorithm of the general Maximum Entropy problem is then as follows:

1. Define the range and the discretization step of x (xmin, xmax,dx).
2. Write a function to calculate $\phi_n(x), n = 0,\ldots,N$ (fin_x).
3. Start the iterative procedure with an initial estimate λ^0 (lambda0).
4. Calculate the $(N+1)$ integrals in equations (4) and the $N(N-1)/2$ distinct elements g_{nk} of the matrix G by calculating the integrals in the equations(8)   (Gn, gnk).
5. Solve the equation (7) to find δ (delta).
6. Calculate $\lambda = \lambda^0 + \delta$ and go back to step 3 until δ becomes negligible.

The calculus of the integrals in equations (4) and (8) can be made by a univariate Simpson's method. We have used a very simplified version of this method.

Case of geometrical moments

Now consider the special case of moments problem where $\phi_n(x) = x^n$, $\quad n = 0,\ldots,N$. In this case equations (3), (4) and (8) become

$$p(x) = \exp\left[-\sum_{m=0}^{N} \lambda_m x^m\right] \tag{9}$$

$$G_n(\lambda) = \int x^n \exp\left[-\sum_{m=0}^{N} \lambda_m x^m\right] dx = \mu_n, \quad n = 0,\ldots,N \tag{10}$$

$$g_{nk} = g_{kn} = -\int x^n x^k \exp\left[-\sum_{m=0}^{N} \lambda_m x^m\right] dx = -G_{n+k}(\lambda) \quad n,k = 0,\ldots,N \tag{11}$$

This means that the $[(N+1)\times(N+1)]$ matrix G in equation (7) becomes a symmetric Hankel matrix which is entirely defined by $2N + 1$ values $G_n(\lambda), n = 0,\ldots,2N$. So the algorithm in this case is the same as in the precedent one with two simplifications

1. In step 2 we do not need to write a seperate function to calculate the functions $\phi_n(x) = x^n, n = 0,\ldots,N$.
2. In step 4 the number of integral evaluations is reduced, because the elements g_{nk} of the matrix G are related to the integrals $G_n(\lambda)$ in equations (10). This matrix is defined entirely by only $2N + 1$ components.

Case of trigonometrical moments

Another interesting special case is the case where the data are the Fourier components of $p(x)$

$$E\left\{\exp\left(-jn\omega_0 x\right)\right\} = \int \exp\left(-jn\omega_0 x\right) p(x)\, dx = \mu_n, \quad n = 0,\ldots,N, \tag{12}$$

where μ_n may be complex–valued and has the property $\mu_{-n} = \mu_n$. This means that we have the following relations

$$\phi_n(x) = \exp(-jn\omega_0 x), \quad n = -N, \ldots, 0, \ldots N,$$

$$p(x) = \exp\left[-\text{Real} \sum_{n=0}^{N} \lambda_n \exp(-jn\omega_0 x)\right],$$

$$G_n(\lambda) = \int \exp(-jn\omega_0 x) \, p(x) \, dx, \quad n = 0, \ldots, N,$$

(13)

$$g_{nk} = \begin{cases} -G_{n-k}(\lambda) & \text{for } n \geq k, \\ -G_{n+k}(\lambda) & \text{for } n < k \end{cases} \quad n, k = 0, \ldots, N,$$

so that all the elements of the matrix G are related to the discrete Fourier transforms of $p(x)$. Note that G is a Hermitian Toeplitz matrix.

3. Examples and Numerical Experiments

To illustrate the usefullness of the proposed programs we consider first the case of the Gamma distribution

$$p(x; \alpha, \beta) = \frac{\beta^{(1-\alpha)}}{\Gamma(1-\alpha)} x^\alpha \exp(-\beta x), \quad x > 0, \alpha < 1, \beta > 0. \tag{14}$$

This distribution can be considered as a ME distribution when the constraints are

$$\int p(x; \alpha, \beta) \, dx = 1 \qquad \text{normalization} \quad \phi_0(x) = 1,$$

$$\int x \, p(x; \alpha, \beta) \, dx = \mu_1 \qquad\qquad\qquad \phi_1(x) = x, \tag{15}$$

$$\int \ln(x) \, p(x; \alpha, \beta) \, dx = \mu_2 \qquad\qquad \phi_2(x) = \ln(x).$$

This is easy to verify because the equation (12) can be written as

$$p(x; \alpha, \beta) = \exp\left[-\lambda_0 - \lambda_1 x - \lambda_2 \ln(x)\right]$$

with $\quad \lambda_0 = -\ln \dfrac{\beta^{(1-\alpha)}}{\Gamma(1-\alpha)}, \quad \lambda_1 = \beta \quad$ and $\quad \lambda_2 = -\alpha.$

Now consider the following problem : Given μ_1 and μ_2 determine λ_0, λ_1 and λ_2.
This can be done by the standard ME method. To do this, first we must define the range of x, (xmin, xmax, dx), and write a function fin_x to calculate the functions $\phi_0(x) = 1$, $\phi_1(x) = x$ and $\phi_2(x) = \ln x$ (See the function fin1_x in Annex). Then we must define an initial estimate λ^0 for λ and, finally, let the program works.

The case of the *Gamma* distribution is interesting because there is an analytic relation between (α, β) and the mean $m = \text{E}\{x\}$ and variance $\sigma^2 = \text{E}\{(x-m)^2\}$ which is

$$\begin{cases} m = (1-\alpha)/\beta, \\ \sigma^2 = (1-\alpha)/\beta^2, \end{cases} \tag{16}$$

or inversely

$$\begin{cases} \alpha = (\sigma^2 - m^2)/\sigma^2 \\ \beta = m/\sigma^2 \end{cases}, \tag{17}$$

so that we can use these relations to determine m and σ^2. Note also that the corresponding entropy of the final result is a byproduct of the function. Table (1) gives some numerical results obtained by ME_DENS1 program (See Annex).

Table 1.

μ_1	μ_2	α	β	m	σ^2
0.2000	-3.0000	0.2156	-3.0962	0.2533	0.0818
0.2000	-2.0000	-0.4124	-6.9968	0.2019	0.0289
0.3000	-1.5000	-0.6969	-5.3493	0.3172	0.0593

The next example is the case of a quartic distribution

$$p(x) = \exp\left[-\sum_{n=0}^{4} \lambda_n x^n\right]. \tag{18}$$

This distribution can be considered as a ME distribution when the constraints are

$$E\{x^n\} = \int x^n p(x)\, dx = \mu_n, \quad n = 0, \ldots, 4 \quad \text{with} \quad \mu_0 = 1. \tag{19}$$

Now consider the following problem: Given $\mu_n, n = 1, \ldots, 4$ calculate $\lambda_n, n = 0, \ldots, 4$. This can be done by the ME_DENS2 program. Table (2) gives some numerical results obtained by this program:

Table 2.

μ_1	μ_2	μ_3	μ_4	λ_0	λ_1	λ_2	λ_3	λ_4
0	0.2	0.05	0.10	0.1992	1.7599	2.2229	-3.9375	0.4201
0	0.3	0.00	0.15	0.9392	0.000	-3.3414	0.0000	4.6875
0	0.3	0.00	0.15	0.9392	0.000	-3.3414	0.0000	4.6875

These examples show how to use the proposed programs. A third example is also given in Annex which shows how to use the ME_DENS3 program which considers the case of trigonometric moments.

4. Conclusions

In this paper we addressed first the class of ME distributions when the available data are a finite set of expectations $\mu_n = E\{\phi_n(x)\}$ of some known functions $\phi_n(x)$, $n = 0, \ldots, N$. We proposed then three Matlab programs to solve this problem by a Newton–Raphson method in general case, in case of geometrical moments data where $\phi_n(x) = x^n$ and in case of trigonometrical moments where $\phi_n(x) = \exp(-jn\omega_0 x)$. Finally, we gave some numerical results for some special examples who show how to use the proposed programs.

REFERENCES

[1] A. Zellnerr and R. Highfiled, "Calculation of Maximum Entropy Distributions and Approximation of Marginal Posterior Distributions," *Journal of Econometrics* **37**, 1988, 195–209, North Holland.

[2] D. Mukherjee and D.C. Hurst, "Maximum Entropy Revisited," *Statistica Neerlandica* **38**, 1984, n 1, 1–12.

[3] Verdugo Lazo and P.N. Rathie, "On the Entropy of Continuous Probability Distributions," *IEEE Trans.*, **vol. IT–24**, n 1, 1978.

[4] Gokhale, "Maximum Entropy Characterizations of some distributions," *Statistical distributions in Scientific work*, **vol. 3**, 299–304 (G.P. Patil et al., Eds., Reidel, Dordrecht, Holland, 1975).

[5] Jaynes, "Papers on probability, statistics and statistical physics," *Reidel Publishing Company, Dordrecht*, Holland, 1983.

[6] Matz, "Maximum Likelihood parameter estimation for the quartic exponential distributions," *Technometrics*, **20**, 475–484, 1978.

[7] Mohammad–Djafari A. et Demoment G., "Estimating Priors in Maximum Entropy Image Processing," *Proc. of ICASSP 1990*, pp: 2069–2072

[8] Mohammad–Djafari A. et Idier J., "Maximum entropy prior laws of images and estimation of their parameters," *Proc. of The 10th Int. MaxEnt Workshop, Laramie, Wyoming*, published in Maximum Entropy and Bayesian Methods, T.W. Grandy ed., 1990.

5. Annex

```
1 function [lambda,p,entr]=me_dens1(mu,x,lambda0)
2 %ME_DENS1
3 %   [LAMBDA,P,ENTR]=ME_DENS1(MU,X,LAMBDA0)
4 %   This program calculates the Lagrange Multipliers of the ME
5 %   probability density functions p(x) from the knowledge of the
6 %   N contstraints in the form:
7 %   E{fin(x)}=MU(n)    n=0:N   with fi0(x)=1, MU(0)=1.
8 %
9 %   MU      is a table containing the constraints MU(n),n=1:N.
10 %   X       is a table defining the range of the variation of x.
11 %   LAMBDA0 is a table containing the first estimate of the LAMBDAs.
12 %           (This argument is optional.)
13 %   LAMBDA  is a table containing the resulting Lagrange parameters.
14 %   P       is a table containing the resulting pdf p(x).
15 %   ENTR    is a table containing the entropy values at each
16 %           iteration.
17 %
18 %   Author: A. Mohammad-Djafari
19 %   Date  : 10-01-1991
20 %
21 mu=mu(:); mu=[1;mu];              % add mu(0)=1
22 x=x(:); lx=length(x);            % x axis
23 xmin=x(1); xmax=x(lx); dx=x(2)-x(1);
24 %
25 if(nargin == 2)                  % initialize LAMBDA
26  lambda=zeros(mu);               % This produces a uniform
27  lambda(1)=log(xmax-xmin);       % distribution.
28 else
29  lambda=lambda0(:);
30 end
31 N=length(lambda);
32 %
33 fin=fin1_x(x);                   % fin1_x(x) is an external
34 %                                % function which provides fin(x).
35 iter=0;
36 while 1                          % start iterations
37   iter=iter+1;
38   disp('--------------------'); disp(['iter=',num2str(iter)]);
39 %
40   p=exp(-(fin*lambda));          % Calculate p(x)
41   plot(x,p);                     % plot it
42 %
43   G=zeros(N,1);                  % Calculate Gn
44   for n=1:N
45    G(n)=dx*sum(fin(:,n).*p);
```

```
46    end
47 %
48    entr(iter)=lambda'*G(1:N);      % Calculate the entropy value
49    disp(['Entropy=',num2str(entr(iter))])
50 %
51    gnk=zeros(N,N);                 % Calculate gnk
52    gnk(1,:)=-G'; gnk(:,1)=-G;      % first line and first column
53    for i=2:N                       % lower triangle part of the
54     for j=2:i                      % matrix G
55      gnk(i,j)=-dx*sum(fin(:,j).*fin(:,i).*p);
56     end
57    end
58    for i=2:N                       % uper triangle part of the
59     for j=i+1:N                    % matrix G
60      gnk(i,j)=gnk(j,i);
61     end
62    end
63 %
64    v=mu-G;                         % Calculate v
65    delta=gnk\v;                    % Calculate delta
66    lambda=lambda+delta;            % Calculate lambda
67    eps=1e-6;                       % Stopping rules
68    if(abs(delta./lambda)<eps),                          break, end
69    if(iter>2)
70     if(abs((entr(iter)-entr(iter-1))/entr(iter))<eps),break, end
71    end
72 end
73 %
74 p=exp(-(fin*lambda));             % Calculate the final p(x)
75 plot(x,p);                        % plot it
76 entr=entr(:);
77 disp('--------  END  ----------')
78 end
```

```
1 %ME1
2 %  This script shows how to use the function ME_DENS1
3 %  in the case of the Gamma distribution. (see Example 1.)
4 xmin=0.0001; xmax=1; dx=0.01;    % define the x axis
5 x=[xmin:dx:xmax]';
6 mu=[0.3,-1.5]';                   % define the mu values
7 [lambda,p,entr]=me_dens1(mu,x);
8 alpha=-lambda(3);   beta=lambda(2);
9 m=(1+alpha)/beta; sigma=m/beta;
10 disp([mu' alpha beta m sigma entr(length(entr))])
```

```
1 function fin=fin1_x(x);
2 % This is the external function which calculates
```

```
3  % the fin(x) in the special case of the Gamma distribution.
4  % This is to be used with ME_dens1.
5    M=3;
6    fin=zeros(length(x),M);
7    fin(:,1)=ones(x);
8    fin(:,2)=x;
9    fin(:,3)=log(x);
10 end
```

```
1  function [lambda,p,entr]=me_dens2(mu,x,lambda0)
2  %ME_DENS2
3  % [LAMBDA,P,ENTR]=ME_DENS2(MU,X,LAMBDA0)
4  % This program calculates the Lagrange Multipliers of the ME
5  % probability density functions p(x) from the knowledge of the
6  % N moment contstraints in the form:
7  % E{x^n}=mu(n)    n=0:N   with  mu(0)=1.
8  %
9  % MU      is a table containing the constraints MU(n),n=1:N.
10 % X       is a table defining the range of the variation of x.
11 % LAMBDA0 is a table containing the first estimate of the LAMBDAs.
12 %         (This argument is optional.)
13 % LAMBDA  is a table containing the resulting Lagrange parameters.
14 % P       is a table containing the resulting pdf p(x).
15 % ENTR    is a table containing the entropy values at each
16 %         iteration.
17 %
18 % Author: A. Mohammad-Djafari
19 % Date  : 10-01-1991
20 %
21 mu=mu(:); mu=[1;mu];              % add mu(0)=1
22 x=x(:); lx=length(x);            % x axis
23 xmin=x(1); xmax=x(lx); dx=x(2)-x(1);
24 %
25 if(nargin == 2)                  % initialize LAMBDA
26  lambda=zeros(mu);               % This produces a uniform
27  lambda(1)=log(xmax-xmin);       % distribution.
28 else
29  lambda=lambda0(:);
30 end
31 N=length(lambda);
32 %
33 M=2*N-1;                         % Calcul de fin(x)=x.^n
34 fin=zeros(length(x),M);          %
35 fin(:,1)=ones(x);                % fi0(x)=1
36 for n=2:M
37  fin(:,n)=x.*fin(:,n-1);
```

```
38 end
39 %
40 iter=0;
41 while 1                          % start iterations
42   iter=iter+1;
43   disp('----------------------'); disp(['iter=',num2str(iter)]);
44 %
45   p=exp(-(fin(:,1:N)*lambda));  % Calculate p(x)
46   plot(x,p);                    % plot it
47 %
48   G=zeros(M,1);                 % Calculate Gn
49   for n=1:M
50    G(n)=dx*sum(fin(:,n).*p);
51   end
52 %
53   entr(iter)=lambda'*G(1:N);    % Calculate the entropy value
54   disp(['Entropy=',num2str(entr(iter))])
55 %
56   gnk=zeros(N,N);               % Calculate gnk
57   for i=1:N                     % Matrix G is a Hankel matrix
58    gnk(:,i)=-G(i:N+i-1);
59   end
60 %
61   v=mu-G(1:N);                  % Calculate v
62   delta=gnk\v;                  % Calculate delta
63   lambda=lambda+delta;          % Calculate lambda
64   eps=1e-6;                     % Stopping rules
65   if(abs(delta./lambda)<eps),                           break, end
66   if(iter>2)
67    if(abs((entr(iter)-entr(iter-1))/entr(iter))<eps),break, end
68   end
69 end
70 %
71 p=exp(-(fin(:,1:N)*lambda));   % Calculate the final p(x)
72 plot(x,p);                     % plot it
73 entr=entr(:);
74 disp('--------  END  ----------')
75 end

1 %ME2
2 %  This script shows how to use the function ME_DENS2
3 %  in the case of the quartic distribution. (see Example 2.)
4 xmin=-1; xmax=1; dx=0.01;            % define the x axis
5 x=[xmin:dx:xmax]';
6 mu=[0.1,.3,0.1,.15]';                % define the mu values
7 [lambda,p,entr]=me_dens2(mu,x);
8 disp([mu;lambda;entr(length(entr))]')
```

```
1  function [lambda,p,entr]=me_dens3(mu,x,lambda0)
2  %ME_DENS3
3  %    [LAMBDA,P,ENTR]=ME_DENS3(MU,X,LAMBDA0)
4  %  This program calculates the Lagrange Multipliers of the ME
5  %  probability density functions p(x) from the knowledge of the
6  %  Fourier moments values :
7  %  E{exp[-j n w0 x]}=mu(n)    n=0:N   with  mu(0)=1.
8  %
9  %  MU     is a table containing the constraints MU(n),n=1:N.
10 %  X      is a table defining the range of the variation of x.
11 %  LAMBDA0 is a table containing the first estimate of the LAMBDAs.
12 %          (This argument is optional.)
13 %  LAMBDA is a table containing the resulting Lagrange parameters.
14 %  P      is a table containing the resulting pdf p(x).
15 %  ENTR   is a table containing the entropy values at each
16 %          iteration.
17 %
18 % Author: A. Mohammad-Djafari
19 % Date   : 10-01-1991
20 %
21 mu=mu(:);mu=[1;mu];                  % add mu(0)=1
22 x=x(:); lx=length(x);                % x axis
23 xmin=x(1); xmax=x(lx); dx=x(2)-x(1);
24 if(nargin == 2)                      % initialize LAMBDA
25   lambda=zeros(mu);                  % This produces a uniform
26   lambda(1)=log(xmax-xmin);          % distribution.
27 else
28   lambda=lambda0(:);
29 end
30 N=length(lambda);
31 %
32 M=2*N-1;                             % Calculate fin(x)=exp[-jnw0x]
33 fin=fin3_x(x,M);                     % fin3_x(x) is an external
34 %                                    % function which provides fin(x).
35 iter=0;
36 while 1                              % start iterations
37   iter=iter+1;
38   disp('---------------------'); disp(['iter=',num2str(iter)]);
39 %
40                                      % Calculate p(x)
41   p=exp(-real(fin(:,1:N))*real(lambda)+imag(fin(:,1:N))*imag(lambda));
42   plot(x,p);                         % plot it
43 %
44   G=zeros(M,1);                      % Calculate Gn
45   for n=1:M
46     G(n)=dx*sum(fin(:,n).*p);
```

```
47  end
48  %plot([real(G(1:N)),real(mu),imag(G(1:N)),imag(mu)])
49  %
50  entr(iter)=lambda'*G(1:N);     % Calculate the entropy
51  disp(['Entropy=',num2str(entr(iter))])
52  %
53  gnk=zeros(N,N);                % Calculate gnk
54  for n=1:N                      % Matrix gnk is a Hermitian
55   for k=1:n                     % Toeplitz matrix.
56    gnk(n,k)=-G(n-k+1);          % Lower triangle part
57   end
58  end
59  for n=1:N
60   for k=n+1:N
61    gnk(n,k)=-conj(G(k-n+1));    % Upper triangle part
62   end
63  end
64  %
65  v=mu-G(1:N);                   % Calculate v
66  delta=gnk\v;                   % Calculate delta
67  lambda=lambda+delta;           % Calculate lambda
68  eps=1e-3;                      % Stopping rules
69  if(abs(delta)./abs(lambda)<eps),                    break, end
70  if(iter>2)
71   if(abs((entr(iter)-entr(iter-1))/entr(iter))<eps),break, end
72  end
73  end
74                                 % Calculate p(x)
75  p=exp(-real(fin(:,1:N))*real(lambda)+imag(fin(:,1:N))*imag(lambda));
76  plot(x,p);                     % plot it
77  entr=entr(:);
78  disp('-------- END ----------')
79  end

1  %ME3
2  %   This scripts shows how to use the function ME_DENS3
3  %   in the case of the trigonometric moments.
4  clear;clg
5  xmin=-5; xmax=5; dx=0.5;        % define the x axis
6  x=[xmin:dx:xmax]';lx=length(x);
7  p=(1/sqrt(2*pi))*exp(-.5*(x.*x));% Gaussian distribution
8  subplot(221);plot(x,p);title('p(x)')
9  %
10 M=3;fin=fin3_x(x,M);           % Calculate fin(x)
11 %
12 mu=zeros(M,1);                 % Calculate mun
13 for n=1:M
```

```
14    mu(n)=dx*sum(fin(:,n).*p);
15 end
16 %
17 w0=2*pi/(xmax-xmin);w=w0*[0:M-1]'; % Define the w axis
18 subplot(222);plot(w,[real(mu),imag(mu)]);title('mu(w)')
19 %
20 mu=mu(2:M);                      % Attention : mu(0) is added
21                                  % in ME_DENS3
22 [lambda,p,entr]=me_dens3(mu,x);
23 disp([mu;lambda;entr(length(entr))]')

 1 function fin=fin3_x(x,M);
 2 % This is the external function which calculates
 3 % the fin(x) in the special case of the Fourier moments.
 4 % This is to be used with ME_DENS3.
 5 %
 6 x=x(:); lx=length(x);           % x axis
 7 xmin=x(1); xmax=x(lx); dx=x(2)-x(1);
 8 %
 9 fin=zeros(lx,M);                %
10 fin(:,1)=ones(x);              % fi0(x)=1
11 w0=2*pi/(xmax-xmin);jw0x=(sqrt(-1)*w0)*x;
12 for n=2:M
13   fin(:,n)=exp(-(n-1)*jw0x);
14 end
15 end
```

MEMSYS AS DEBUGGER

Tj. Romke Bontekoe
European Space Agency, ESTEC-SA
Postbus 299
2200 AG Noordwijk
The Netherlands.

Do Kester
Space Research Laboratory
Postbus 800
9700 AV Groningen
The Netherlands.

ABSTRACT. This paper describes an unusual mode of application of MemSys5, viz. its usefulness in finding inconsistencies in data or noise estimates. Some recommendations for users of MemSys5 is given at the end.

1. Introduction

We found a bug in a complex data-software environment through the application of Mem-Sys5. In simulated data sets such bugs are relatively easy to find, since by construction the data are internally consistent. For real data sets it is the responsibility of the researcher not only to check the internal consistency, but also whether his or her results make sense. Sometimes software or hardware failures are not only difficult to diagnose, but often they occur only under 'special' circumstances. Ultimately, the key ingredient is professional skill.

2. The InfraRed Astronomical Satellite

The InfraRed Astronomical Satellite (IRAS) was the first (non-classified) satellite using liquid helium as cryogen. It was launched in 1983 and operated for ten months, successfully. For the first time the sky was surveyed in four wavelength bands, centered around 12, 25, 60, and 100 μm. Stars, in 12 and 25 μm, and cold dust ($\simeq 40°$ K), in 60 and 100 μm, are the predominant radiators in these wavelengths. The principal goal was to make an all-sky survey of (strong) point sources. At that time it was not expected to detect anything else. As we know now many discoveries were made, e.g. cold dust in the Galaxy resembling atmospheric cirrus clouds, young stars embedded in dark dust clouds not visible at visual wavelengths, comet trails much longer than previously known.

IRAS did not make images like a camera, but was scanning with an always moving telescope. Multiple coverages were made in order to have sufficient redundancy. The sampling of the sky therefore is very non-uniform.

235

C. R. Smith et al. (eds.), Maximum Entropy and Bayesian Methods, Seattle, 1991, 235–240.
© 1992 *Kluwer Academic Publishers.*

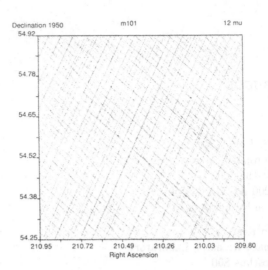

Fig. 1. The dots indicate the positions of the detector centers when they were read out. There are 16 detectors in this 12 μm band, and about 34000 samples in this area in total. The area is 40×40 arc minute. There are three sets of coverages, as can be seen from the three different scan angles.

The total amount of data is of the order of 20 Gbyte. Calibration of the IRAS data turned out to be a non-trivial task; in the US as well as in The Netherlands several tens of man years went into it. As must be obvious, in such a large software environment there can always be bugs remaining. Although the data – image transform is complicated, the construction of images from scan data is quite well possible (Bontekoe *et al.* 1991). In an adapted version, this was the problem of last years image reconstruction contest (Bontekoe, 1991).

3. A problem

The object of astronomical interest is the spiral galaxy M101 at 12μm wavelength. Most galaxies, and this one too, are faint at this wavelength; usually only the brighter, central parts can be reconstructed. In Fig. 1 the dots are the positions on the sky where the detectors have been read out. One can discern three different angles in the scan pattern, as the result of the redundancy in the observing strategy. The area covers 40 x 40 arc minutes, a small area on the sky.

In Fig. 2 the sample positions of only one detector are plotted. For every 50st sample point also an outline of the detector aperture is shown. The thick lines also represent detector scans, but they are 'special', as explained below. Note that the aperture width is three sampling distances in the scan direction.

The re-constructed image of M101 in visible space (not shown), shows a few bright spots coinciding with the central parts of the galaxy, and amplified noise. This noise seems to be correlated with the characteristic length of the Intrinsic Correlation Function (ICF) (Charter, 1990; Gull, 1989). In addition there are some 'funny stripes'. In hidden space

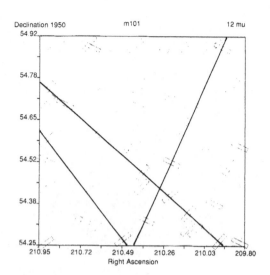

Fig. 2. Same as Fig. 1 but now only 1 detector. The rectangles indicate the size the detector aperture. At every sample point such a detector has been centered. The thick lines represent detector scans related to the problem of this paper, but are otherwise normal.

these stripes are much more prominent (Fig. 3). Obviously, one would have a hard time to convince a colleague astronomer that the infrared sky looks like this!

But what is the cause of this problem?

The first thing which comes to mind is interference. The stripes have about the same width as a detector and run parallel to the scan directions. The 'wavelength' of the pattern is about 4.5 – 5 sampling distances. In sofar we know, this does not correspond with any frequency used on board of IRAS; some cross-talk between various instruments and detectors is known.

Undersampling of the map. One does not need to apply the sampling theorem. If one compares the sampling density (Fig. 1) with the detector aperture (Fig. 2) one can simply *see* that undersampling is not the cause.

Aliasing. Two detector scans, taken at different times, could fall on top of each other. If the detectors had a slight DC offset, e.g. an artifact from the calibration, aliasing can occur. The aliasing frequency, however, would be the same as the sampling frequency. Therefore, no aliasing.

Interference between two, or more, detectors. This is not 'physical' interference but occurs during the data processing, since different scans are made at different times. For instance, a small sinusoidal modulation could be underlying the true signal. Since there are 16*15/2= 120 different combinations of two detectors, checking them all could be quite a substantial job. There is a shortcut, however. Since cross talk requires two detectors, it should disappear when either one of the two is not present. Therefore it should suffice to make 16 different maps in which for each map the data of one detector is omitted. The two maps in which the stripes disappear will identify the two detectors. This, however, did not happen. One 'set' of stripes disappeared when one detector was omitted; the

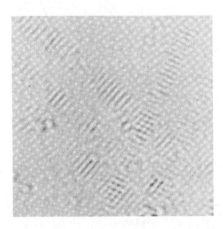

Fig. 3. Hidden space solution of MemSys5. The white dots represent bright parts of the galaxy and other real sources. The 'railroad ties' represent the processing artifacts. The area is the same as in Figs. 1 and 2.

remaining stripes disappeared with another detector was 'switched off'. Clearly, this is not interference but an effect from a single detector, which happened to occur twice. One of those two detectors is the one of Fig. 2, and the thick lines correspond to the scans identified with one set of stripes.

One detector interfering with itself from two different scans. If the two scans make a small angle with each other, the frequency of one scan with respect to the other is slightly different and could give rise to Moire patterns. These patterns have a continuous range of frequencies, depending on the angle. This is not the case, however, since there exists an isolated detector scan which is associated with one of the stripes (see Fig. 2, scan from upper right to lower center).

Amplified glitches in the detectors. Some detectors do have this defect. But we could not find this in the data of the detectors causing the stripes.

4. A solution?

A closer look at the isolated detector scan mentioned above is given in Fig. 4. As is obvious, there is not much signal present. The level is constant at about 15 MJy / Sr, with some digitization noise. There are only five different data values present; the step size represents a single data number. However, we supplied MemSys5 with an erroneous error estimate with a $\sigma = 0.01$ for this detector scan, instead of $\sigma = 0.3$!! Therefore, we are telling MemSys5 that some pieces of data are spectacularly accurate. The consequence of this is an amplification as indicated by the error bar in Fig. 4, as is roughly the amplitude of the artifacts. The total amount of corrupted data, however, is less that one percent.

Why does such a regular pattern of artifacts appear from such irregular data? Our explanation is only qualitative; we cannot exactly account for the amplitude or frequency.

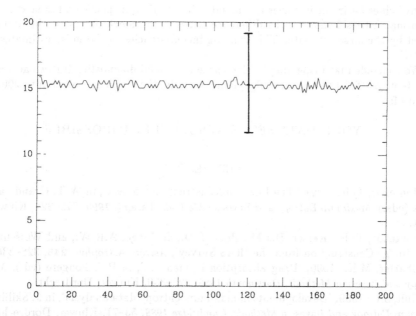

Fig. 4. The thin line represents the input detector data of one of the detector scans of Fig. 2. The error supplied with it is of the order of the thickness of this line. The error bar represents the amplitude to which this got amplified in Fig. 3.

MemSys5 tries to fit the one percent of 'very good' data to the image at almost all cost. However, it is actually fitting noise into the map. As is well known, fitting noisy data is an awkward predicament. But why is the pattern so regular? Usually, amplified noise yields very spiky artifacts but not coherent structure. This is caused by a side effect of the ICF. The ICF is an extra blur on top of the actual instrumental blur. Here the ICF is a gaussian of about two sampling distances. The 'very good' data contain high frequencies, as is common with noise. MemSys5 tries to preserve these frequencies, as it should. But the ICF can not follow these frequencies and acts as a constraint on the first derivative in the image, like a low pass filter. The resulting frequency, of 4.5 – 5 sampling intervals, seems consistent with the gaussian full width of 2 sample intervals.

5. Epilogue

All of this was caused by a bug in the noise estimator, which luckily did not occur in last years image reconstruction contest. As a consequence, we got the answer to our wrong question. After removing this bug the stripes were gone and the image looks better; and the Evidence went up from -200000 to -100000 dB. However, we are not yet satisfied with that result and continue to improve on it.

From discussions with John Skilling and Mark Charter, the following recommendations are made for users of MemSys5. After convergence of MemSys5 one should perform one extra CALL OPUS(ST,11,25). This yields a set of 'mock' data in area < 25 > as if the reconstructed image were the truth. The differences with the data supplied, in area < 21 >,

may give clues for inconsistencies in the data. In addition, inspection of area < 24 > gives
the normalized residuals, i.e. the difference between the mock data and the actual data,
divided by the noise estimate. This will flag inconsistencies in the noise estimates in area
< 22 >.

We conclude that inspecting hidden space is a useful diagnostic. But we also conclude
that MemSys5 finishes ungracefully giving us only the Evidence of about -200000 dB,
instead of:

YOUR DATA SET IS ABSOLUTELY IMPOSSIBLE

REFERENCES

Bontekoe, Tj.R.: 1991, 'The Image Reconstruction Contest', in W.T. Grandy and L.H.
Schick (eds.), *Maximum Entropy and Bayesian Methods, Laramie 1990*, 319–324, Kluwer, Dordrecht.

Bontekoe, Tj.R., Kester, D.J.M., Price, S.D., de Jonge, A.R.W., and Wesselius, P.R.:
1991, 'Image Construction from the IRAS Survey', *Astron. Astrophys.* **248**, 328–336.

Charter, M.K.: 1990, 'Drug absorption in man, ...', in P.F. Fougere (ed.), *Maximum
Entropy and Bayesian Methods, Dartmouth 1989*, 325–339, Kluwer, Dordrecht.

Gull, S.F.: 1989, 'Developments in maximum entropy data analysis', in J. Skilling (ed.),
Maximum Entropy and Bayesian Methods, Cambridge 1988, 53–71, Kluwer, Dordrecht.

ENTROPY AND SUNSPOTS:
THEIR BEARING ON TIME-SERIES

Brian Buck & Vincent A. Macaulay
Theoretical Physics
University of Oxford
1 Keble Road
Oxford, OX1 3NP, England

ABSTRACT. We derive a prior suitable for the reconstruction of a function which is not necessarily positive, such as the Fourier amplitude. It is a close relative of the MaxEnt prior for a positive, additive image. We discuss its *pros* and *cons* in the course of an analysis of the large database of Wolf sunspot numbers. An important ingredient in this analysis is the hypothesis that sunspot number represents a rectified measurement of the underlying solar magnetic cycle.

1. The Bayesian approach to inverse problems

Let us start as we mean to go on by writing down Bayes' theorem for the general inverse problem. It states that

$$\text{prob}(\text{image} \mid \text{data and } \mathcal{K}) = \text{prob}(\text{image} \mid \mathcal{K}) \times \frac{\text{prob}(\text{data} \mid \text{image and } \mathcal{K})}{\text{prob}(\text{data} \mid \mathcal{K})}.$$

Here, the *image* is the object we would like to infer, usually from *data* which are noisy measurements of some transform of that image. The term $\mathcal{K}$ is the all-important background knowledge which makes the probabilities well-defined. The task is to assign the terms on the RHS of this equation. Our major concern will be with the first of these, the *prior* on the image. This term encodes our pre-data knowledge about the image. In inverse problems where the space of possible images is huge, it cannot be swept under the carpet as it often can in problems involving the estimation of a small number of parameters. The remaining two terms are more familiar: the *likelihood* and the *evidence*. The likelihood is the term in which is incorporated the details of the experimental setup. It contains the image-to-data transform and what is known about the errors on the measurements. The presence of this noise implies that, even if we were given the image, we could not deduce the data from it: inference is king even in the apparently 'direct' problem. The evidence, just a normalization factor in our treatment here, takes on a more important rôle when there are different models to compare.

Priors based upon the configurational entropy of the image have been popular since the pioneering work of Gull and Daniell (1978). They have been derived in two main ways. The first has been to construct models of possible image-making processes and from them

241

C. R. Smith et al. (eds.), Maximum Entropy and Bayesian Methods, Seattle, 1991, 241–252.

to produce the prior distribution directly. The models are in no sense meant to be realistic ways of obtaining images, but the philosophy of using them is that a general purpose prior should apply in special cases. The most familiar such argument concerns a team of monkeys throwing image elements into pixels, a whimsical way of identifying the set of images which are, *a priori*, taken to be equally probable. We shall have more to say about these arguments later and indeed shall derive our prior in this way.

The second way, less intuitive, but more useful for seeing what assumptions are being employed, is to consider a small number of special cases where it is clear how the prior should behave. These provide constraints on the possible form of that prior. Sufficient constraints to pin down the MaxEnt prior have been given by Skilling (1989). (See also Skilling (1991b) for another axiomatic derivation.)

This entropic prior, basically $\exp(-\alpha \sum_i f_i \ln f_i)$, can be used in the reconstruction of positive distributions. In many problems, however, there is no positivity constraint; but, as card-carrying Bayesians, we still want to be able to treat such problems. Thus we derive in this paper a different prior applicable to certain inverse problems with not necessarily positive 'images'. Rather amusingly, the basic difference in the result is that 'ln' is replaced by '$\sinh^{-1}$'!

2. Monkeys and entropic priors

To warm up for the derivation of our 'sinh' prior, we shall rehearse the Poisson derivation of the entropic prior given by Skilling (1989). Suppose our image to be discretized into M pixels; then the image becomes an M-dimensional vector with elements f_i. Now, in addition, the image values themselves are made discrete by introducing a small element of image ϵ. Then a particular scene can be represented as an M-dimensional vector of non-negative integers, the numbers of elements n_i in each pixel. At this point, our unpredictable image-producing mechanism makes its appearance—the troupe of monkeys—the knowledge of whose lack of skill at making images, we shall describe with a set of Poisson probability distributions independent for each pixel. We allow that the monkeys might have preferences for some pixels over others and hence each Poisson parameter ν_i can in principle be assigned to be different from any other.

The probability of a particular image is

$$\text{prob}(\{n_i\}) = \prod_{i=1}^{M} \frac{e^{-\nu_i} \nu_i^{n_i}}{n_i!} = \exp\left(\sum_i (-\nu_i + n_i \ln \nu_i - \ln n_i!) \right).$$

For large n_i, the last term in the exponent is approximately $\frac{1}{2} \ln 2\pi + (n_i + \frac{1}{2}) \ln n_i - n_i$, by Stirling's approximation. Inserting this gives

$$\text{prob}(\{n_i\}) = \frac{1}{\prod_i \sqrt{2\pi n_i}} \exp\left(\sum_i \left(n_i - \nu_i - n_i \ln \frac{n_i}{\nu_i} \right) \right).$$

To get the probability (density) of the image proper, we must identify the above with an integral over the equivalent volume in image space:

$$\text{prob}(\{n_i\}) = \int_V \text{prob}(f)\mu(f)\,df,$$

where V is a hypercube of side ϵ centred on the image point and μ is an image space measure which it proves convenient to include.

Attention to all the factors then gives

$$\text{prob}(\boldsymbol{f}) = \left(\frac{\alpha}{2\pi}\right)^{M/2} \exp\left(\alpha S(\boldsymbol{f}, \boldsymbol{m})\right),$$

where

$$S = \sum_i \left(f_i - m_i - f_i \ln \frac{f_i}{m_i}\right)$$

and $f_i = n_i\epsilon$, $m_i = \nu_i\epsilon$, $\alpha = \epsilon^{-1}$. The image space measure is

$$\mu(\boldsymbol{f}) = \prod_i f_i^{-1/2}.$$

Its scaling properties with ϵ allow it to be distinguished (up to a constant) from the terms contributing to the entropy.

The constant α is of some interest but also the cause of some irritation. Its rôle is to govern how strongly the entropic features of the prior are allowed to express themselves. Once this prior has been digested by Bayes' theorem, it becomes clear that α is what is known in the orthodox literature as a 'regularizing' parameter. At the moment, the assignment of this parameter has to be done *a posteriori*: nobody has a good way of assigning it *a priori*.

All this is perfectly fine for positive images. (Remember we are using this term rather generally.) But it would be very useful to have a similar sort of prior for functions which are not required to be positive. The Fourier amplitude is an obvious example of such a function—there are many cases where one might want to infer it from noisy and incomplete measurements of the direct function. The suggestion has been made that the prior we have just derived could be used by offsetting the zero: but the resulting lack of positive/negative symmetry is not at all what we want.

Here is another Poisson monkey argument which produces a prior† which we advocate for many inverse problems. The argument closely mirrors the one above. In this case, let the monkeys be provided with elements of both positive and negative image, both 'quanta' of the same magnitude. Again we describe our knowledge of the possible configurations of elements, after the scattering process, by a set of independent Poisson distributions:

$$\text{prob}(\{n_i^+, n_i^-\}) = \prod_{i=1}^{M} \frac{e^{-\nu_i^+} \nu_i^{+n_i^+}}{n_i^+!} \frac{e^{-\nu_i^-} \nu_i^{-n_i^-}}{n_i^-!},$$

with n_i^+ (n_i^-) the number of positive (negative) elements in the ith pixel and ν_i^+, ν_i^- the Poisson parameters for that pixel. The image itself is to be identified as the difference between the number of positive and negative elements (scaled by the element size ϵ).

---

† Steve Gull and John Skilling have pointed out that a similar prior exists as an option called POSNEG in their MEMSYS code.

For convenience, we define the sum and difference variables with a factor of $\frac{1}{2}$:

$$n_i = \tfrac{1}{2}\left(n_i^+ + n_i^-\right),$$
$$q_i = \tfrac{1}{2}\left(n_i^+ - n_i^-\right).$$

The probability to be calculated is that of the $\{q_i\}$ and, by the addition rule, satifies

$$\mathrm{prob}\left(\{q_i\}\right) = \sum_{\{n_i^+,\,n_i^-\ \text{for fixed } q_i\}} \mathrm{prob}\left(\{n_i^+,n_i^-\}\right).$$

Since each pixel is manifestly independent of the others, we shall just look at the probability of the image in a single pixel. Gathering terms together, we find that

$$\mathrm{prob}\left(q_i\right) = e^{-(\nu_i^+ + \nu_i^-)}\gamma_i^{q_i} \sum_{n_i} \frac{\beta_i^{2n_i}}{(n_i + q_i)!\,(n_i - q_i)!},$$

where $\beta_i^2 = \nu_i^+ \nu_i^-$ and $\gamma_i = \nu_i^+/\nu_i^-$. The sum is over integer or half integer n_i depending on q_i. Now, rather pleasingly, this sum is reexpressible in closed form (Lebedev, 1972, page 108) and yields

$$\mathrm{prob}\left(q_i\right) = e^{-(\nu_i^+ + \nu_i^-)}\gamma_i^{q_i}\, I_{2q_i}\left(2\beta_i\right).$$

Here $I_n(x)$ is the modified Bessel function of order n and argument x. Using a result in the approximation theory of Bessel functions of large order given in Watson's monumental tome (1944), namely Meissel's first extension of Carlini's formula, we can express this probability as

$$\mathrm{prob}\left(q_i\right) \to \frac{1}{\sqrt{2\pi}}\,\frac{1}{\left((2q_i)^2 + (2\beta_i)^2\right)^{1/4}}$$
$$\times \exp\left(\sqrt{(2q_i)^2 + (2\beta_i)^2} - \beta_i\left(\gamma_i^{\frac{1}{2}} + \gamma_i^{-\frac{1}{2}}\right) + q_i \ln \gamma_i - 2q_i \sinh^{-1}\frac{q_i}{\beta_i}\right).$$

To pass back to a continuous amount of stuff in each pixel, we reintroduce the element size ϵ and make the following identifications

$$2q_i \to \frac{f_i}{\epsilon}, \qquad 2\beta_i \to \frac{m_i}{\epsilon}, \qquad \sum_{q_i} \to \int\frac{\mathrm{d}f_i}{\epsilon} \qquad \text{and} \qquad \alpha = \frac{1}{\epsilon},$$

which leads to our central result that

$$\mathrm{prob}\left(f_i\right) = \left(\frac{\alpha}{2\pi}\right)^{1/2}\exp\left(\alpha S_i(f_i, m_i, \gamma_i)\right),$$

where the 'entropy' S_i is given by

$$S_i = \sqrt{f_i^2 + m_i^2} - \tfrac{1}{2}m_i\left(\gamma_i^{\frac{1}{2}} + \gamma_i^{-\frac{1}{2}}\right) + \tfrac{1}{2}f_i \ln \gamma_i - f_i \sinh^{-1}\frac{f_i}{m_i}.$$

The measure on the ith image pixel is

$$\mu_i(f_i) = \frac{1}{\left(f_i^2 + m_i^2\right)^{1/4}}.$$

The probability of the full image and the measure are gotten by multiplying the corresponding quantities for the separate pixels.

It is possible to work out, in Gaussian approximation, the *a priori* expectation of the image in terms of the parameters γ and m and then to eliminate one of the sets of parameters in favour of $\bar{f}$. If we eliminate the γs, the entropy becomes:

$$S(\boldsymbol{f}) = \sum_i \left(\sqrt{f_i^2 + m_i^2} - \sqrt{\bar{f}_i^2 + m_i^2} + f_i \left(\sinh^{-1} \frac{\bar{f}_i}{m_i} - \sinh^{-1} \frac{f_i}{m_i} \right) \right).$$

The $\bar{f}$s are a model which allows for the incorporation of specific information on the magnitude and sign of the fs. The ms set the scale of f on which the entropy is measured; if they are too large, the entropic regularizer becomes effectively quadratic.

A few limiting cases of the entropy are interesting. We shall be concerned solely with the case where we assign each $\bar{f}_i$ to be zero, so that the entropy reduces to

$$S(\boldsymbol{f}) = \sum_i \left(\sqrt{f_i^2 + m_i^2} - m_i - f_i \sinh^{-1} \frac{f_i}{m_i} \right).$$

In this case, around the origin the entropy is effectively quadratic: $S(\boldsymbol{f}) = - \sum_i (f_i^2/(2m_i) + O(f_i^4))$. For large $|f_i|$, an amusing resemblance to the familiar MaxEnt prior emerges: $S \to \sum_i (|f_i| - m_i - |f_i| \ln(2|f_i|/m_i))$.

3. A digression into the history of sunspot investigations

Now we make an abrupt about-face to prepare the way for an application of the above to the time-series of sunspot numbers. From the splendid book of Foukal (1990), we paraphrase a few relevant moments in the history of sunspot observation. The first telescopic observations of the dark patches on the surface on the sun were made around 1610 by Fabricius and Galileo. Twenty years later it was recognized that the motion of the spots across the surface of the solar disk was revealing the rotation of the sun (at a period of 27 days) and the tilt of the rotation axis to the ecliptic. Between 1645 and 1715, there was an solar activity minimum, much argued about and named after Maunder who rediscovered it in solar records in the late 19th century.

It was Schwabe in 1843 who discovered the 11 year cycle in the sunspot record and Wolf who recognized a correlation between peaks in the cycle and geomagnetic storms. The latter's tabulation of sunspot numbers began in 1849. Early this century, Hale identified a 22 year magnetic solar cycle, manifest in the alternation of the polarity of sunspot pairs between cycles. This observation has a direct bearing on our own analysis.

4. Application to an inverse Fourier problem

Many analyses of the large amount of sunspot data exist of which we mention only one, that of Bretthorst (1988). For a variety of models—multiple sinusoids, chirp, etc.—he applies the techniques of Bayesian parameter estimation and model-selection, and shows how, even with quite uninformative priors, it is possible to make sharp estimates of the parameters. His sinusoidal models were of the form:

$$s(t) = \sum_{n=1}^{N} \left(A_n \cos(2\pi\nu_n t) + B_n \sin(2\pi\nu_n t) \right),$$

where the parameters are the As, Bs and the νs. In addition, the number of model terms N must be estimated. This is where Bayesian model comparison is relevant. The Bayesian formalism contains within itself a precise realization of Ockham's razor (Garrett, 1991): more complicated models are less probable unless they can fit the data considerably more closely. Bretthorst manages to get a good description of the mean annual sunspot record with a model containing nine frequencies.

The procedure for the testing of one sinusoid, then two, then three and so on is quite laborious. In addition the sunspot cycle is rather complicated and there are no powerful reasons for expecting a purely sinusoidal model to be a particularly good description of the real variations. Thus a *free-form* model might seem to be a good, if not better, model to try. The point about such a model is that it allows inference of a whole distribution of parameter values (Skilling, 1991a).

With this mind, we replace the above description of the sunspot time-series with

$$s(t) = \int \left(A(\nu) \cos(2\pi\nu t) + B(\nu) \sin(2\pi\nu t) \right) d\nu,$$

a Fourier integral representation, possible for nearly all s of physical interest. It is now the *functions* $A(\nu)$ and $B(\nu)$ which we want to infer. Because of the limitation that emerges from the Nyquist theorem for data points sampled at equally-spaced times, there is no point in extending the limits of integration outside $(0, \nu_c)$, where $\nu_c = 1/(2\tau)$ with τ the time between successive measurements. In effect we are assuming the signal to be band-limited by the Nyquist critical frequency.

The next step is to assign priors to A and B. In fact we will not do that directly. It is more convenient to seek to represent $s(t)$ with one real function in the frequency domain, a function defined on twice the range. The Hartley integral representation (Bracewell, 1986) fits the bill. So we write instead

$$s(t) = T \int_{-\nu_c}^{\nu_c} H(\nu) \operatorname{cas}(2\pi\nu t) \, d\nu,$$

where $\operatorname{cas}(x) = \cos(x) + \sin(x)$ and T is the total time spanned by the data points.

Of course what we do now is to assign as a prior on H the 'sinh' form we derived above. We are not dogmatically asserting that this is the only possible form. Indeed, later on, we shall point out a shortcoming of it and suggest a different prior which better encodes our knowledge about the signal.

5. Some results from a sunspot analysis

The observation of Hale noted above strongly suggests that the sunspot cycle might represent a rectified measurement of the underlying magnetic cycle. If this were the case, many

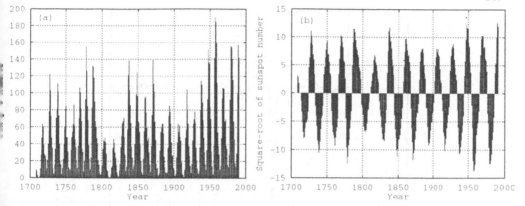

Fig. 1. The mean yearly sunspot numbers (a) and their derectified square-root (b), between 1708 and 1990.

of the large number of lines that appear in Bretthorst's analysis would arise as beating effects between frequencies of the unrectified time-series. Here we will present some results of our entropic analysis of the derectified sunspot record.

The choice of how the derectification is to be performed remains. We have assumed that the measured sunspot numbers are related to the underlying solar variability as the square. Thus we took the square-root of the measured sunspot numbers and restored the missing sign. Experiments with other methods of derectification revealed a remarkable lack of sensitivity to just how it was done. Fig. 1 shows both the original and the derectified data, between 1708 and the present.

We proceed in the Bayesian analysis by multiplying the prior by a likelihood function, which here we take to be Gaussian with a uniform expected error across the whole data record. (This is probably not a terribly realistic assignment, since the old data is considered poor by many authorities.) The most probable image is that which maximizes this product. We used the 'historic' discrepancy method ('$\chi^2 = N$') to determine the regularization constant α: although not optimal it is not too bad. Fig. 2 shows the most probable Hartley spectrum for the derectified sunspot data. To separate out the information on the amplitudes of the 'cas' functions in Fig. 2, we display in Fig. 3 the amplitude spectrum, defined as $\sqrt{H^2(\nu) + H^2(-\nu)}$. Plotted with it is a traditional estimate of amplitudes, the square root of the (Hartley) periodogram.

It is amusing to close the circle of analysis that began with Bretthorst's work and continued through our use of a free-form model, by trying to reconstruct a plausible parametric model of the time-series from the free-form reconstruction. If we identify the peaks in the Hartley spectrum as indications of single 'cas' functions, blurred out by noise and truncation effects, we can make the following approximation:

$$s(t) = T \int H(\nu) \cos(2\pi\nu t) \, d\nu$$

$$\approx T \sum_k \cos(2\pi\nu_k t) \int_{k\text{th line}} H(\nu) \, d\nu.$$

Choosing the four most significant peaks in the reconstructed spectrum leads to the model shown in Fig. 4, where it is shown alongside the data. The lines that emerge when these

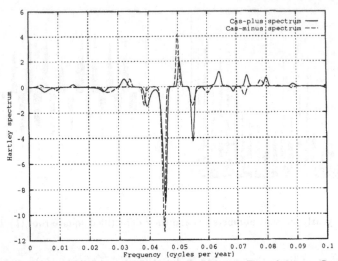

Fig. 2. The most probable Hartley spectrum of the signal of Fig. 1b from a Bayesian calculation with the 'sinh' prior, Gaussian likelihood and 'historic' determination of the regularizing constant. The negative frequency part of the spectrum is shown, reflected through $\nu = 0$, as a dashed line. Only the range of frequencies containing lots of structure is shown.

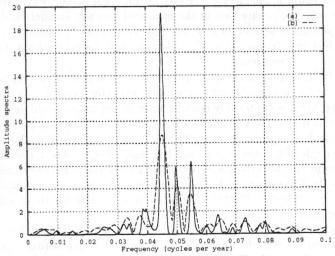

Fig. 3. The amplitude spectrum (a) corresponding to the Hartley spectrum of Fig. 2 and the conventional Hartley periodogram (b).

four are squared-up are in close correspondence with the lines inferred by Bretthorst in his analysis. Indeed we anticipate that a derectified four frequency model would be considerably more probable than this larger model, on Ockham grounds. These model comparisons will be investigated in the future.

6. A shortcoming and a remedy

The form of prior used above has one major shortcoming although, as it turns out in our

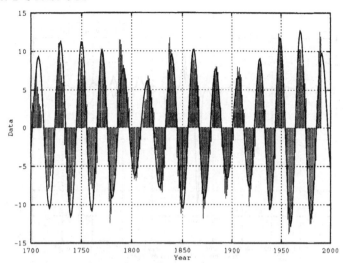

Fig. 4. The four frequency model derived from the most probable Hartley spectrum (solid line), compared with the original data (impulses).

application, not a fatal one. Two real numbers are required to encode information about the amplitude and phase at any frequency. With the Hartley representation, these are $H(\nu)$ and $H(-\nu)$. Now in the prior used above, these two quantities are inferentially independent: knowledge about one does not change our knowledge about the other. This cannot be a good assignment. For we would be rather surprised if signal at $H(\nu)$ did not, in general, coexist with signal at $H(-\nu)$. That is: we expect there to be a positive correlation between $H^2(\nu)$ and $H^2(-\nu)$. Let us see how this can be achieved using an invariance argument to restrict the class of feasible priors.

Stated baldly, the required invariance of the prior is that of changing the origin from which time is measured. Let the time label attached to the signal at a particular instant t be shifted to $t' = t + \tau$ so that $s'(t') = s(t)$. This induces a corresponding transformation in the Hartley spectrum, a mixing of $H(\nu)$ and $H(-\nu)$:

$$H'(\nu) = H(\nu)\cos(2\pi\nu\tau) - H(-\nu)\sin(2\pi\nu\tau).$$

From a candidate prior on H, it is straightforward to work out the corresponding probability of H'. If $\text{prob}\big[H(\nu)\big] = g[H]$ and $\text{prob}\big[H'(\nu)\big] = h[H']$, then g and h must satisfy

$$h[H']\,\mathrm{D}[H'] = g[H]\,\mathrm{D}[H],$$

where $\mathrm{D}[X]$ is the volume element in X-space. Since the transformation on H is a product of 2d rotations, its Jacobian determinant is unity. Thus the volume elements cancel left and right. We have no more nor less knowledge about $H(\nu)$ than about $H'(\nu)$ so, by Jaynes' consistency principle (Jaynes, in preparation), we should assign them the same prior probability functional. That is,

$$h[H'] = g[H'].$$

The upshot of this is that g must satisfy

$$g[H] = g[H'],$$

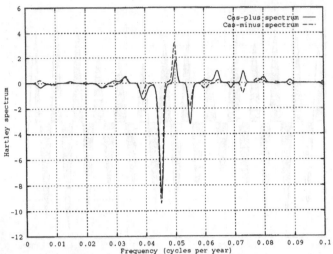

Fig. 5. The most probable Hartley spectrum, as in Fig. 2 but with a prior independent of phase and entropic in the amplitude.

for any τ.

At this point, we make the simplification that our knowledge of the signal at different frequencies is independent, so that the only correlations that appear in the prior occur between $H(\nu)$ and $H(-\nu)$, for each particular ν.† Specializing to a particular $\nu = \nu_i$, the above relation becomes

$$g_i\Big(H(\nu_i), H(-\nu_i)\Big) = g_i\Big(H(\nu_i)\cos(2\pi\nu_i\tau) - H(-\nu_i)\sin(2\pi\nu_i\tau),$$

$$H(\nu_i)\sin(2\pi\nu_i\tau) + H(-\nu_i)\cos(2\pi\nu_i\tau)\Big).$$

This is a functional equation for g_i, the part of the probability associated with frequency ν_i, which has the solution

$$g_i = F\Big(H^2(\nu_i) + H^2(-\nu_i)\Big),$$

where F is an arbitrary function. That this result is in line with common sense is apparent when it recognized that the argument of F is the squared amplitude at frequency ν_i: the prior is *independent* of phase. From this, it is clear that the preceding argument would be simplified if rephrased in terms of amplitude and phase variables throughout: the functional equation that results is much more easily solved.

The function F can be assigned to be a traditional entropic prior $\exp(\alpha S)$ where $S = \sum_i\big(f_i - m_i - f_i\ln(f_i/m_i)\big)$. Fig. 5 shows a reconstruction made with an entropic prior on the amplitude, that is, on the square root of the argument of F. This regularizer is no longer concave in image space so that the uniqueness of the maximum is uncertain, but this seems not to have caused ambiguities in the cases so far tried. This is almost certainly because when the default m is small, as it is for the displayed reconstructions, the region which

---

† The assumption of no interfrequency correlation can in principle be dropped and a fully correlated prior used. The 'hidden image plus ICF' model is an example.

causes concavity to fail is very small. A concave regularizer is the 'sinh' form used with positive argument only. Its smoother behaviour near the origin makes it rather attractive, but in this application it turned out to produce results identical with those produced with the 'ln' form, because of the small default used.

It is interesting to compare Figures 2 and 5. The tendency for peaks to appear at slightly differing frequencies in the two halves of the Hartley spectrum, which is manifest in the reconstruction of Fig. 2, is considerably reduced in Fig. 5 especially for the smaller peaks. This is the result of the prior tying together these two parts of the spectrum. The amplitude spectrum for this reconstruction (not shown) does not contain unfortunate features such as the double hump near $\nu = 0.04$ in Fig. 3a.

7. Comments

Nowhere here have we quantified the relative probability of the different models, using the evidence: this is clearly something which should be done with some urgency. The full, objective quantification that the Bayesian methodology provides, both within the hypothesis space consisting of a single model and its parameters, or in the larger space of competing models is a most powerful tool.

This Hartley analysis does not allow good predictions and retrodictions of the sunspot record that would enable us to answer questions about the Maunder minimum†, for example, or the time of future anomalies in solar activity. The finite width of the reconstructed lines means that the most probable signal decays to zero quickly outside the region where there is data, with, we expect, a corresponding increase in the expected uncertainty. A better model could help us here.

To sum up: we have proposed a new 'entropic' prior for use in inverse problems with images which are just as likely to be positive as negative. We have demonstrated, however, that it cannot be used directly when further prior information is available, such as the phase invariance in the spectrum reconstruction problem which we encountered in our sunspot analysis. However for a problem such as the reconstruction of a charge distribution from some of its Fourier components, it would seem to be ideal.

ACKNOWLEDGMENTS. One of us (V.A.M.) thanks the Science and Engineering Research Council of Great Britain and Merton College, Oxford for financial support.

REFERENCES

Bracewell, R. N. (1986). *The Hartley transform*. Oxford University Press, New York.

Bretthorst, G. L. (1988). *Bayesian spectrum analysis and parameter estimation*, Lecture notes in statistics, Vol. 48. Springer-Verlag, New York.

Foukal, P. V. (1990). *Solar astrophysics*. Wiley, New York.

Garrett, A. J. M. (1991). 'Ockham's razor.' In *Maximum entropy and Bayesian methods, Laramie, Wyoming, 1990* (ed. W. T. Grandy, Jr and L. Schick), pp. 357–64. Kluwer, Dordrecht.

Gull, S. F. and Daniell, G. J. (1978). 'Image reconstruction from incomplete and noisy data.' *Nature*, **272**, 686–90.

---

† What retrodictions are possible in this method are tantalizingly suggestive of a period of unusually low solar activity in the early seventeenth century.

Jaynes, E. T. (in preparation). *Probability theory: the logic of science*. Fragments available
    from the author: Wayman Crow Professor of Physics, Washington University, St. Louis,
    MO 63130, USA.
Lebedev, N. N. (1972). *Special functions and their applications*. Dover, New York.
Skilling, J. (1989). 'Classic maximum entropy.' In *Maximum entropy and Bayesian methods,
    Cambridge, England, 1988* (ed. J. Skilling), pp. 45–52. Kluwer, Dordrecht.
Skilling, J. (1991a). 'On parameter estimation and quantified MaxEnt.' In *Maximum en-
    tropy and Bayesian methods, Laramie, Wyoming, 1990* (ed. W. T. Grandy, Jr and
    L. Schick), pp. 267–73. Kluwer, Dordrecht.
Skilling, J. (1991b). 'Fundamentals of MaxEnt in data analysis.' In *Maximum entropy in
    action* (ed. B. Buck and V. A. Macaulay), pp. 19–40. Clarendon Press, Oxford.
Watson, G. N. (1944). *A treatise on the theory of Bessel functions*, 2nd edn. Cambridge
    University Press.

BASIC CONCEPTS IN MULTISENSOR DATA FUSION*

John M. Richardson†and Kenneth A. Marsh‡
Rockwell International Science Center
Thousand Oaks
CA 91360

ABSTRACT. This paper treats some conceptual aspects of the methodology of fusion of multisensor data in the static (or single-time) case. The paper is divided into two parts: one establishing the general methodology and the other presenting an illustrative application of it. In the general part, we treat several basic questions including (1) requirements on the state vector of the observed system that will allow the formulation of independent observation models for separate sensor systems, (2) extensions of decision theory to handle fusion, and (3) the fusionability of various kinds of reduced signals derived from the different sets of raw data associated with separate sensor systems. We also derive some general results pertaining to the desirability of augmenting the set of measurements. In the applied part, the general methodology established above is employed in the determination of the geometry (i.e., the elevation function) of an object resting upon a flat surface. In the models, two types of measurements are assumed: (1) a single acoustical pulse-echo scattering measurement in which the incident wave propagates downward; and (2) a set of optical measurements involving a downward-looking TV camera imaging the object under sequentially pulsed light sources. In the second type, the raw image data are assumed to be preprocessed according to the principles of photometric stereo. The two types of measured data are combined in accordance with the conceptual methodology discussed in the general part. A computational example using synthetic data is presented and discussed.

I. Introduction

In the last several years, there has been a significant increase in the development of data fusion methodologies with widespread military and commercial applications. For accounts of recent progress in this field, the reader is advised to consult papers by Foster and Hall (1981), LaJeunesse (1986), and Richardson and Marsh (1988).

Data fusion is the process by which data from a multitude of sensors is used to yield an optimal estimate of a specified state vector pertaining to the observed system. There are several kinds of problems involved in this process: 1) the conceptual design of the

---

* This investigation was supported by the Independent Research and Development funds of Rockwell International.

† Consultant.

‡ Now at the Jet Propulsion Laboratory, Pasadena, CA.

C. R. Smith et al. (eds.), Maximum Entropy and Bayesian Methods, Seattle, 1991, 253–271.
© 1992 Kluwer Academic Publishers.

data fusion process; and 2) the implementation of this design in terms of digital hardware without excessive reduction of performance below the optimal level.

In this paper, we will deal exclusively with some of the problems of conceptual design. There are myriads of challenging problems in the implementation area, but to keep our scope within practical bounds we will not deal with them here. We will, within the conceptual domain, further restrict our scope by limiting our discussion to single-time estimation problems, more specifically, to problems in which the state changes negligibly during the time interval in which the various sensor data are taken.

Within this category, we will discuss in Part II the general methodology of fusion, and in Part III an application to the fusion of acoustical and optical data in the imaging of a solid object.

II. Conceptual Problems

1. Introduction

We cover the following topics: in Section 2, the formulation of the problem and a brief review of Bayesian decision theory; in Section 3, the desirability of using more sensors; and in Section 4, the problem of decomposing the total decision system into modules, each of which has stand-alone significance (i.e., each module can perform a useful function by itself for the ultimate user).

In Part III, these results are applied to the fusion of acoustical and optical data in the determination of the elevation function of an object resting on a flat surface.

2. Formulation of the Problem

Let us assume the state of the system of interest to be represented by the m-dimensional vector x and the total set of measured quantities to be represented by the n-dimensional vector y. Unless otherwise specified, we will assume that both x and y are real and continuous-valued in the infinite Euclidean spaces of appropriate dimensionalities. Our problem is to estimate the value of x from a knowledge of y in accordance with some specified optimality criterion. Before proceeding with this task, it is necessary to specify (1) a model of both the system of interest and the sensor system in the absence of experimental errors, and (2) the *a priori* statistical nature of x and y (the random experimental errors are determined by the statistical nature of y when x is given).

The most common procedure in this type of problem is to define a so-called measurement model, for example,

$$y = f(x) + \nu, \tag{1}$$

where the n-dimensional function $f(x)$ represents the ideal operation of the sensor system, and ν is a random n-dimensional vector representing experimental error, model error, background signals, etc. The assumption that the experimental error is simply an additive term is clearly an idealization. It could be multiplicative (or convolutive), or it could be both additive and multiplicative. To achieve complete generality, we should replace $f(x) + \nu$ by $f(x,\nu)$; however, most problems involving this level of generality have not been solved in a practical sense.

The measurement model must include a specification of the *a priori* statistical properties of x and ν from which we could deduce the *a priori* statistical properties of x and y. A common assumption is that x and ν are statistically independent and that the probability densities (p.d.'s) $P(x)$ and $P(\nu)$ are given. We will always assume that $P(\nu)$ has a Gaussian form with zero mean, i.e., ν is a Gaussian random vector with the properties

$$E\nu = 0, \tag{2a}$$

$$E\nu\nu^T = C_\nu, \tag{2b}$$

where E is the *a priori* averaging operator, the superscript T denotes the transpose, and C_ν is the covariance matrix for ν. We will adhere to this assumption in cases where an explicit consideration of ν is involved. The *a priori* p.d. of x given by $P(x)$ is sometimes assumed to be Gaussian, i.e., x is also a Gaussian random vector, but with the different properties

$$Ex = \bar{x}, \tag{3a}$$

$$E\Delta x \Delta x^T = C_x, \tag{3b}$$

where Δ is the operator giving the deviation from the average, $\bar{x}$ is the *a priori* average value of x, and C_x is the *a priori* covariance matrix for x.

A measurement model of particular simplicity is obtained when $f(x)$ is a linear function of x, e.g.,

$$f(x) = Ax, \tag{4}$$

where A is an $n \times m$ matrix representing the characteristics of the sensor system in the linear regime. Here the measurement model is both linear and Gaussian, properties which enable one to obtain a closed-form solution to the estimation problem that is valid for several optimality criteria.

The p.d. $P(y|x)$ can be expressed in terms of the above measurement model in accordance with the expression

$$\log P(y|x) = -\frac{1}{2}\left[y - f(x)\right]^T C_\nu^{-1}\left[y - f(x)\right] + \text{constant}. \tag{5}$$

The joint p.d. $P(x,y)$ can of course be written in the form

$$P(x,y) = P(y\,|\,x)\,P(x) \tag{6}$$

and the *a posteriori* p.d. $P(x|y)$ can be written

$$P(x|y) = P(x,y)P(y)^{-1}, \tag{7}$$

where the second factor on the right-hand side is frequently ignorable because it depends only on y, which in the decision process can be regarded as fixed.

We turn now to the problem of estimating the value of x from the measured value of y. We consider an estimate $\hat{x}(y)$ whose optimal form is determined by an optimality criterion (i.e., a measure of the goodness of the estimate) involving a theoretical ensemble of test data. The first step in defining an optimality criterion is to define a loss function $L(\hat{x}(y), x)$ which is the loss (penalty or cost) incurred when we estimate the state vector to be $\hat{x} = \hat{x}(y)$ when the actual value of the state vector is x. Then, the optimality criterion is given by the negative of the so-called risk R, given by

$$R = E\, L(\hat{x}(y), x), \tag{8}$$

In the above expression, it should be noted that the argument of $\hat{x}$ is the random vector y given by the measurement model [Eq. (2.1)]. When the optimal form of $\hat{x}$ is found, we use the actual measured value of y denoted by $\tilde{y}$ as the argument of $\hat{x}$ to obtain the best estimate based on a particular set of measurements. It is to be emphasized that R is a functional of $\hat{x}(y)$, i.e., it depends on the functional form of $\hat{x}(y)$ or equivalently on the procedure (the policy or rule) for estimating x from y and not on the actual value of $\hat{x}$ in a particular case. The procedure for determining the optimal form of $x(y)$ is straightforward. We note that Eq. (8) can be rewritten in the form

$$E\left\{E\left[L\left(\hat{x}\left(y\right),x\right)|y\right]\right\}. \tag{9}$$

The optimal estimator is determined by minimizing $R(\ |y)$ with respect to $\hat{x}(y)$ with y fixed. $R(\ |y)$ is given by

$$E\left[L\left(\hat{x}(y),x\right)|y\right] \equiv R\left(\ |y\right), \tag{10}$$

where the last quantity is called the *a posteriori* (conditional) risk, which depends on the value of $\hat{x}(y)$ at a specified value of y in contrast with R, which depends on the functional form of $\hat{x}(y)$.

There is, of course, no end of conceivable loss functions that one can consider. However, we will present three such functions:

$$L(\hat{x}, x) = (\hat{x} - x)^T W(\hat{x} - x), \tag{11a}$$

$$= \left[(\hat{x} - x)^T W(\hat{x} - x)\right]^{\frac{1}{2}}, \tag{11b}$$

$$= -\delta(\hat{x} - x), \tag{11c}$$

in which W is a symmetric positive-definite weighting matrix and $\delta(\cdot)$ is an m-dimensional Dirac δ-function. The optimal estimator corresponding to the first loss function is the posterior mean, i.e.,

$$\hat{x}_{opt}(y) = E(x|y), \tag{12}$$

which, surprisingly, is independent of W. In the case of the second loss function, the optimal estimator is the conventional *a posteriori* median when x is one-dimensional and is a generalization thereof when x is multidimensional. Finally, in the case of the third loss function, the optimal estimator is the value of x with the maximum *a posteriori* probability, i.e.,

$$P(\hat{x}_{opt}(y)|y) = \max_x P(x|y). \tag{13}$$

This optimal extimator is sometimes called the *a posteriori* mode.

Our general problem is the determination of $\hat{x}(y)$ (e.g., the *a posteriori* mean, median, or mode of x), when the measurement vector represents the outputs of more than one type of sensor, e.g.

$$y = (\frac{y_1}{y_2}), \tag{14}$$

where y_i is an n_i-dimensional vector representing the outputs of sensors of type i ($i = 1, 2$) and where $n_1 + n_2 = n$. At the present level of generality, the problem of estimating x based on measured data from several types of sensors is no different from the problem involving one type of sensor, or even a single sensor. However, as we will see in later sections, the attempt to modularize the decision process will bring in some nontrivial problems.

3. DESIRABILITY OF USING MORE SENSORS

Before discussing the problem of modularization, we will first consider the basic question of the desirability of using more sensors. It seems self-evident that the use of additional sensors (of the same type or of different types) should give better estimates of the state vector x, or at least estimates that are no worse. However, some investigators have argued otherwise, e.g., that there is a finite optimal number of sensors of a given type in a specified estimation problem (Lee et al, 1984). This could be true if the sensor data were processed suboptimally to yield an estimate of the state. It is to be stressed that here we are not considering the cost of additional sensors in the optimality criterion.

4. THE MODULARIZATION PROBLEM

The primary problem is to estimate the state of the system of interest from two sets of data produced by two separate sensor systems involving, in the general case, different types of sensors. From an abstract point of view, this problem is the same as that involving one set of data. However, significant secondary problems emerge when one investigates the modularization of the estimation process, i.e., how to decompose the total process into parts that have stand-alone significance.

The nature of the modularization problem is illustrated in Fig. 1. It is assumed that the information flows associated with y_1 and y_2 can proceed to a point where certain quantities z_1 and z_2 are produced by processors No. 1 and No. 2. It is further assumed that these quantities provide sufficient information (other than a priori information) to enable the fusion processor to produce the optimal estimate $\hat{x}(y_1, y_2)$ based on the total set of measured data (y_1, y_2). Two questions arise: 1) What requirement is necessary for any separate processing to be done? and 2) What forms of the variables z_1 and z_2 are sufficient?

The first step in the modularization of the estimation process is the imposition of a requirement on the state vector x. This requirement can be expressed in the form

$$P(y_1, y_2 | x) = P(y_1 | x) P(y_2 | x), \tag{15}$$

which means that y_1 and y_2 are statistically independent when x is given*. An alternative form of the above is the relation

$$P(y_1 | x, y_2) = P(y_1 | x), \tag{16}$$

which means that when the probability density of y_1 is conditioned on x (i.e., it is a posteriori relative to the specification of x), then the additional conditioning on y_2 is redundant. Of course, the same relation with y_1 and y_2 interchanged also holds.

---

* This requirement is reminiscent of the definition of state in systems theory, namely, that for a vector quantity to be a state it must make past and future quantities independent when such a quantity is given in the present.

In terms of measurement models of the form [Eq. (1) translated into the present multisensor formalism], i.e.,

$$y_i = f_i(x) + \nu_i, \qquad i = 1, 2, \tag{17}$$

the above requirement means simply that ν_1 and ν_2 are statistically independent. Although this might seem to be almost always true (perhaps because it is almost always assumed), it is not a sterile requirement. For example, there might exist a hidden (or neglected) "common cause" represented by random parameters that induce a correlation between ν_1 and ν_2. A rather commonplace "common cause" is the unknown ambient temperature of the environment in which the two sensor systems operate. The correlation between ν_1 and ν_2 can be eliminated by incorporating the parameters, defining the "common cause," into the state vector.

The investigation of the modularization of the estimation process is conveniently based on the relation

$$\begin{aligned} P(x|y_1, y_2) &= P(y_1|x)\, P(y_2|x) P(x)\, P(y_1, y_2)^{-1} \\ &= \frac{P(x|y_1)\, P(x|y_2)}{P(x)} \cdot \frac{P(y_1)\, P(y_2)}{P(y_1, y_2)} \end{aligned} \tag{18}$$

which can be derived from Eq. (15) with the use of general definitions discussed in Section 2.

In the following sections, we will investigate in more detail the nature of the information flows depicted in Fig. 1. In particular, we will determine what kinds of outputs from the separate processors are required for the fusion module to produce the correct estimate of the state based upon the total set of multisensor data.

4.1 GENERAL CASE

Here, we consider a general situation in which, aside from the factorability requirement [Eq. (15)], no special assumptions are made about the structure of $P(y_1|x), P(y_2|x)$ and $P(x)$. In this case, it is easily seen from Eq. (18) that the p.d.'s $P(x|y_1)$ and $P(x|y_2)$ must be the outputs of processors No. 1 and No. 2, respectively. The entity $P(y_1)P(y_2)P(y_1, y_2)^{-1}$ can be ignored, since it plays only the role of a normalization factor; if it is omitted, this factor can be recovered later (if necessary) by normalizing $P(x|y_1)P(x|y_2)P(x)^{-1}$ with respect to x. The p.d. $P(x)$ is, of course, not ignorable, but it is known a priori.

One might ask why we cannot make separate estimates of the state x, first based on y_1 and then on y_2, and then send these separate estimates to the fusion module which should compute the desired multisensor estimate $\hat{x}(y_1, y_2)$. Although this can be done in the linear-Gaussian case (see the next section), it is clear that this cannot be done in the general case. Let us consider as an example the multisensor a posteriori mode given by the value of x that maximizes Eq. (18) or, alternatively, $P(x|y)$.

It is obvious that in the general case, the maximum on x of $P(x|y_1)P(x|y_2)P(x)^{-1}$ has no relation to the maxima of $P(x|y_1)$ and $P(x|y_2)$ individually. One might ask the related question: "If the values of x that maximize $P(x|y_1)$ and $P(x|y_2)$, respectively, are sufficiently close together, can we average them to get a good approximation to the value of x that maximizes $P(x|y_1, y_2)$?" The answer is no for the general case, although this question has an affirmative answer in a certain limiting form of the Gaussian case.

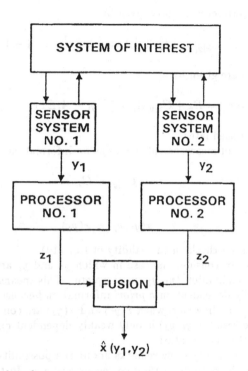

Fig. 1. Information flows in data fusion.

4.2 GAUSSIAN CASE

For the more restricted case in which $P(x|y_1)$ and $P(x|y_2)$ are both Gaussian in x, knowledge of the *a posteriori* means and covariances of x with y_1 and y_2 separately given suffices to determine $P(x|y_1)$ and $P(x|y_2)$, and hence $\hat{x}(y_1, y_2)$. However, even less information related to y_1 and y_2 is necessary. If we assume that both measurement models are linear and Gaussian in x, it is easy to show that the *a posteriori* covariances of x, i.e., $C_{x|y_1}$ and $C_{x|y_2}$, are independent of y_1 and y_2, although they depend on the fact that these quantities are measured. Thus, the quantities $z_1 = E(x|y_1) = \hat{x}(y_1)$ and $z_2 = E(x|y_2) = \hat{x}(y_2)$ provide sufficient information for the fusion module to produce the desired estimate $E(x|y_1, y_2) = \hat{x}(y_1, y_2)$. The explicit formula for combining the separate estimates is

$$\hat{x}(y_1, y_2) = \left[C_{x|y_1}^{-1} + C_{x|y_2}^{-1} - C_x^{-1} \right]^{-1} \cdot \left[C_{x|y_1}^{-1} \hat{x}(y_1) + C_{x|y_2}^{-1} \hat{x}(y_2) - C_x^{-1} \hat{x} \right], \qquad (19)$$

where $C_{x|y_1}$ and $C_{x|y_2}$ are the *a posteriori* covariances of x given y_1 and y_2, C_x is the *a priori* covariance matrix of x and $\hat{x} = Ex$ is the *a priori* average of x. To obtain the explicit forms of the *a posteriori* covariances, we consider the measurement models

$$y_i = A_i x + \nu_i, \qquad i = 1, 2, \qquad (20)$$

where A_i is a constant matrix, and ν_i is a Gaussian random vector with zero mean and covariance C_{ν_i}. We also assume that ν_1 and ν_2 are statistically independent of each other.

The *a posteriori* covariances are now given by

$$C_{x|y_i} = [A_i^T C_{\nu_i}^{-1} A_i + C_x^{-1}]^{-1}, \qquad i = 1, 2. \tag{21}$$

The estimates $\hat{x}(y_i)$ are given by

$$\hat{x}(y_i) = Ex + C_{x|y_i} A_i^T C_{\nu_i}^{-1}(y_i - A_i Ex), \qquad i = 1, 2, \tag{22}$$

a result that will be used later.

If the second set of measurements contains no information, i.e., $A_2 = 0$, we obtain

$$C_{x|y_2} = C_x \tag{23}$$

and

$$x(y_1, y_2) = x(y_1), \tag{24}$$

a result that provides a check on the validity of Eq. (19).

It is interesting to consider the case in which y_1 and y_2 are individually incomplete, but in which the combination (y_1, y_2) is complete. This means that the estimates $\hat{x}(y_1)$ and $\hat{x}(y_2)$ are heavily dependent on *a priori* information because of the rank deficiencies of the matrices $A_i^T C_{\nu_i}^{-1} A_i$. However, when $\hat{x}(y_1)$ and $\hat{x}(y_2)$ are combined in a correct manner using Eq. (19), the result $\hat{x}(y_1, y_2)$ is only weakly dependent on *a priori* information, at least for large signal-to-noise ratios.

In Section 4.1, we made a few remarks about the possibility of averaging the optimal estimates based on the outputs of the two sensor systems. In the present Gaussian case, the result [Eq. (19)] approaches an averaging operation in the limit of very large *a priori* uncertainty. In more explicit mathematical terms, we obtain

$$x\,(y_1, y_2)_\infty \to \left(C_{x|y_1,\infty}^{-1} + C_{x|y_2,\infty}^{-1}\right)^{-1} \cdot \left(C_{x|y_1,\infty}^{-1}\,\hat{x}\,(y_1)_\infty + C_{x|y_2,\infty}^{-1}\,\hat{x}(y_2)_\infty\right) \tag{25}$$

as $C_x \to \infty$ in some suitable sense. The reciprocal matrix weights $C_{x|y_1,\infty}$ and $C_{x|y_2,\infty}$ are the corresponding limiting forms of the *a posteriori* covariance matrices $C_{x|y_1}$ and $C_{x|y_2}$ defined by Eq. (21). The estimators $\hat{x}(y_1)_\infty$ and $\hat{x}(y_2)_\infty$ are also the corresponding limiting forms of the estimates $\hat{x}(y_1)$ and $\hat{x}(y_2)$ given by Eq. (22). The above result [Eq. (25)] is reasonable in the sense that the reciprocal matrix weights are measures of the degrees of uncertainty in the estimates based on y_1 and y_2, respectively. However, we hasten to emphasize that Eq. (25) collapses if the matrix inverses of $C_{x|y_1,\infty}$ and/or $C_{x|y_2,\infty}$ do not exist. The failure of a matrix inverse to exist is directly related to the incompleteness of the corresponding data set.

5. Discussion

These results represent a small fraction of the conceptual problems worthy of further investigation. Even in the relatively limited domain of single-time estimation problems, we can list the following possibilities:

1. The state vector x has components that are all discrete-valued or are partly discrete-valued.

2. The state vector has the structure $x = (x_1, x_{12}, x_2)^T$, where x_1 is unrelated to y_2, x_2 is unrelated to y_1, and x_{12} is related to both.

3. A mixed situation can occur in which the different information channels can accommodate different amounts of processing before fusion [e.g., the entities $P(x|y_1)$ and $\hat{x}(y_2)$ represent the maximal amount of processing in each channel].

4. Experimental error cannot be represented entirely by an additive term in the measurement model. Examples are sensor systems producing two images where one is out of registration with the other by an unknown displacement. Many more examples of this kind can be cited (e.g., inconsistent gray scales).

Clearly, when one enters the much larger domain of time-sequential estimation problems, the kinds of problems, listed above and discussed in earlier sections, become much more complex and ramified.

III. Application to Fusion of Acoustical and Optical Data

1. INTRODUCTION

To illustrate certain aspects of the general methodology of fusion discussed above, we treat the problem of combining acoustical and optical data in the estimation of the elevation function of a solid object resting on a solid surface. In the measurement model, the acoustical data are obtained from a single pulse-echo measurement in which the incident wave propagates downwards through air. The optical data are obtained from a TV camera looking down at the object which is illuminated by sequentially pulsed light sources. In the last case, the raw image data are preprocessed according to the principles of photometric stereo (Horn, 1986) under the assumption that the surface of the object is Lambertian. Using the concepts discussed in Part II, we combine the two sets of data using the elevation function as the state vector. It is possible to formulate two separate measurement models for the two sets of data, with each depending only on the elevation function (aside from fixed experimental parameters and additive noise representing measurement error, extraneous signals, etc.), but on different aspects of this function as discussed in Section 4 below.

2. MATHEMATICAL PRELIMINARIES AND NOTATION

The writing of equations will be simplified significantly by the introduction of a special notation that differentiates between two- and three-dimensional vectors. The desirability of this arises from the fact that the horizontal plane passing through the origin (i.e., xy-space) is a preferred geometrical entity, as we shall see in the later analysis.

The three-dimensional vector $\mathbf{r}$ in xyz-space is defined by the expression

$$\mathbf{r} = \mathbf{e}_x x + \mathbf{e}_y y + \mathbf{e}_z z, \tag{26}$$

where x, y and z are the usual Cartesian coordinates and $\mathbf{e}_x, \mathbf{e}_y,$ and $\mathbf{e}_z$ are unit vectors pointing in the Cartesian coordinate directions. On the other hand, the two-dimensional position vector $\underline{r}$ in xy-space (i.e., in the xy-plane) is defined by

$$\underline{r} = \mathbf{e}_x x + \mathbf{e}_y y. \tag{27}$$

In general, we will denote any three-dimensional vector in xyz-space by a **boldface** symbol and any two-dimensional vector in xy-space by an underline. Clearly, $\mathbf{e}_x$ and $\mathbf{e}_y$ are equal to $\underline{e}_x$ and $\underline{e}_y$, but we will adhere to the boldface notation in this case for the sake of clarity. The vectors $\mathbf{r}$ and $\underline{r}$ are obviously related by the expression

$$\mathbf{r} = \underline{r} + \mathbf{e}_z z. \tag{28}$$

In the case of a general three-dimensional vector $\mathbf{u}$ we can write the relations

$$\mathbf{u} = \mathbf{e}_x u_x + \mathbf{e}_y u_y + \mathbf{e}_z u_z, \tag{29}$$

$$\underline{u} = \mathbf{e}_x u_x + \mathbf{e}_y u_y, \tag{30}$$

$$\mathbf{u} = \underline{u} + \mathbf{e}_z u_z. \tag{31}$$

We will use the vector magnitude operation $||$ for both two- and three-dimensional vectors, e.g.,

$$|\underline{u}| = (u_x^2 + u_y^2)^{\frac{1}{2}}, \tag{32}$$

$$|\mathbf{u}| = (u_x^2 + u_y^2 + u_z^2)^{\frac{1}{2}}. \tag{33}$$

It is also useful to consider the two- and three-dimensional unit tensors defined by

$$\mathbf{I} = \mathbf{e}_x \mathbf{e}_x + \mathbf{e}_y \mathbf{e}_y, \tag{34}$$

$$1 = \mathbf{e}_x \mathbf{e}_x + \mathbf{e}_y \mathbf{e}_y + \mathbf{e}_z \mathbf{e}_z, \tag{35}$$

in which the terms $\mathbf{e}_x \mathbf{e}_x$, etc., are dyadics. The two-dimensional unit tensor $\mathbf{I}$, along with $\mathbf{e}_z$, is useful in the decomposition of a three-dimensional vector into a two-dimensional vector and the z-component, namely

$$\underline{u} = \mathbf{I} \cdot \mathbf{u}, \tag{36}$$

$$u_z = \mathbf{e}_z \cdot \mathbf{u}. \tag{37}$$

These results give the terms on the left-hand side of Eq. (31) in terms of the general three-dimensional vector on the right-hand side.

We close this section with a brief consideration of two- and three-dimensional gradient operators defined by

$$\underline{\nabla} = \mathbf{e}_x \frac{\partial}{\partial x} + \mathbf{e}_y \frac{\partial}{\partial y}, \tag{38}$$

$$\nabla = \mathbf{e}_x \frac{\partial}{\partial x} + \mathbf{e}_y \frac{\partial}{\partial y} + \mathbf{e}_z \frac{\partial}{\partial z}. \tag{39}$$

Other operators, e.g., the divergence and the curl, are related in an obvious way to the above definitions.

3. Representation of the Object

We assume that the object of interest (combined with part of its surroundings within the localization domain) is represented by a single-valued elevation function

$$z = Z(\underline{r}) \tag{40}$$

where z is the vertical coordinate and $\underline{r}$ is a two-dimensional position vector in the xy-plane. The points $\underline{r}$ are limited to the localization domain, D_L, defined by the inequalities

$$-\frac{1}{2}L_x \leq x < \frac{1}{2}L_x, \qquad -\frac{1}{2}L_y \leq y < \frac{1}{2}L_y. \tag{41}$$

Outside of the localization domain, we assume that the elevation is, in effect, given by $z = -\infty$, or at least a negative number sufficiently large that its contribution to the acoustical measurement can be windowed out. In the optical model, the surface outside of D_L is, of course, at least partially visible, no matter how large and negative the elevation function is. We can handle this by requiring this part of the surface to be totally nonreflective. This assumption makes the shadow of the pedestal (associated with D_L), thrown onto the low-lying surface outside of D_L, totally invisible in most cases. It is understood that the object of interest need not fill all of D_L; the elevation function between the object and the boundaries of D_L could, for example, be part of the presentation surface. A typical setup is illustrated in Figs. 2a and 2b.

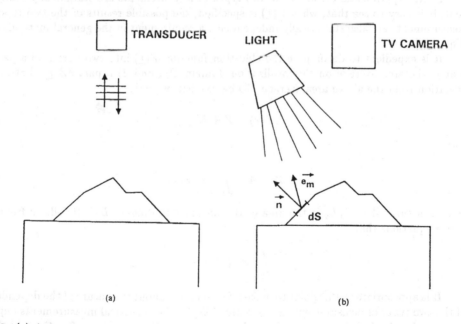

Fig. 2. (a) Acoustic measurement; (b) vision measurement.

The single-valuedness of the elevation function requires further comment. Clearly, many objects are not described faithfully by single-valued elevation functions. We can take two alternative points of view. The first is that we will, for the sake of simplicity, limit our investigation to objects that are actually described by single-valued elevation functions. The second is that, in the case of many objects described by multiple-valued elevation functions, it is possible in an approximate manner to confine attention to the so-called top elevation function (i.e., an elevation function that is equal to the maximum elevation wherever multiple elevations exist) if the lower branches contribute weakly to the measured data.

In the Kirchhoff approximation, the waveform from a pulse-echo scattering measurement with the incident wave propagating in the $-\mathbf{e}_z$ direction depends only on the single-valued elevation function, or on the top elevation function in the case of certain kinds of objects involving multiple-valued elevation functions. It is then clear that the elevation function constitutes an adequate state vector for the interpretation of acoustical measurements.

In general, the way in which the object scatters incident light should be regarded as part of the state of the object, along with the elevation function. However, to simplify the present problem as much as possible, we make the assumption that the surface scatters light diffusely in accordance with Lambert's law. With a sufficient number of illumination directions, one can, under the above assumption, derive a combination of directly measurable quantities that are independent of the local reflectivity. Thus, this measured combination is, in the absence of additive noise, directly related only to the elevation function $Z(\underline{r})$. Hence, $Z(\underline{r})$ constitutes an adequate state vector for interpreting optical measurements.

If the experimental errors in the two types of measurements are statistically independent, it is easy to see that, when $Z(\underline{r})$ is specified, the possible results of the two types of measurements are also statistically independent, a requirement in the general methodology [Eq. (15)].

It is expedient to decompose the elevation function $Z(\underline{r})$ into two parts: 1) a part $\bar{Z}$ that is the area average on the localization domain D_L, and 2) a part $\delta Z(\underline{r})$ that is the deviation from the above area average. To be explicit, we write

$$Z(\underline{r}) = \bar{Z} + \delta Z(\underline{r}), \tag{42}$$

where

$$\bar{Z} = A^{-1} \int_{D_L} d\underline{r} \, Z(\underline{r}), \tag{43}$$

where, in turn, $A = L_x L_y$ is the area of the localization domain D_L. It follows from the above equations that

$$\int_{D_L} d\underline{r} \, \delta Z(\underline{r}) = 0. \tag{44}$$

It is appropriate at this point to make a few remarks about the nature of the dependence of the two types of measurements upon $\bar{Z}$ and $\delta Z(\underline{r})$. The acoustical measurements depend on both $\bar{Z}$ and $\delta Z(\underline{r})$, but only on the horizontal cross-sectional area as a function of height, i.e., $\int d\underline{r} \, \theta(\bar{Z} + \delta Z(\underline{r}) - z^0)$ in which $\theta(\ )$ is the unit step function and z^0 is the height of the cross section. The optical measurements depend only on $\delta Z(\underline{r})$, i.e., they are independent of $\bar{Z}$.

4. ACOUSTICAL MEASUREMENT MODEL

We consider a single pulse-echo measurement of the scattering of acoustical waves from the object of interest. We, of course, assume that the host medium is ordinary air. We further assume that the transducer and the localization domain are in the far field of each other. The experimental setup is depicted schematically in Fig. 2a. The object is assumed to have a very high acoustical impedance and a very high density compared with the corresponding properties of air and thus the surface of the object may be regarded as a rigid immovable boundary.

The appropriate measurement model is given (Richardson et al., 1984) by the expression

$$f(t) = \alpha \int_{D_L} d\underline{r}\, p'(t + 2c^{-1} Z(\underline{r})) + \mu(t), \tag{45}$$

where $f(t)$ = a possible measured waveform, $\mu(t)$ = experimental error and external noise, $p'(t)$ = time-derivative of $p(t)$, the measurement system response function [The latter is defined as the waveform produced by the measurement system if a fictitious scatterer with an impulse response function given by $R(t) = \delta(t)$ is positioned at the origin.], $Z(\underline{r})$ = elevation function discussed in the last section, α = constant dependent upon the acoustical properties of air, and c = propagation velocity of acoustical waves in air.

It is appropriate to consider a discrete version of the above measurement model, namely

$$f(t) = \alpha \sum_{\underline{r}} \delta\underline{r}\, p'[t + 2c^{-1} Z(\underline{r})] + \mu(t), \tag{46}$$

where $\underline{r}$ takes vector values on a suitable grid of points in the xy-plane spanning the localization domain D_L, $\delta\underline{r}$ is the area of one cell (or pixel) in the grid, and the time t is now assumed to take a discrete set of values.

We turn now to a discussion of the *a priori* statistical aspects of the model. The random quantities $\mu(t)$ and $Z(\underline{r})$ are assumed to be statistically independent of each other. The noise $\mu(t)$ is assumed *a priori* to be Gaussian with zero mean and with a covariance matrix given by

$$E\mu(t)\,\mu(t') = \delta_{tt'}\sigma_\mu^2, \tag{47}$$

where $\delta_{tt'}$ is a Kronecker delta generalized for the case of noninteger subscripts ($\delta_{tt'} = 1$ if $t = t'$ and $= 0$ if $t \neq t'$) and σ_μ is the standard deviation of $\mu(t)$.

We now assume that the set of quantities $\{\delta Z(\underline{r})\}$ is a Gaussian random vector with the properties

$$E\delta Z(\underline{r}) = 0, \tag{48}$$

$$E\delta Z(\underline{r})\delta Z(\underline{r}') = C_{\delta Z}(\underline{r} - \underline{r}'). \tag{49}$$

At this point, we introduce the assumption of periodic (or cyclic) boundary conditions, an artifice that gives meaning to nonlocal quantities that depend, at least formally, on points beyond the boundary of D_L. This assumption necessitates the further assumption that $C_{\delta Z}(\underline{r})$ is doubly periodic with the period D_L. Because of the relation (44), we must impose the requirement

$$\sum_{\underline{r}} C_{\delta Z}(\underline{r} - \underline{r}') = \sum_{\underline{r}} C_{\delta Z}(\underline{r}) = 0. \tag{50}$$

The area-average $\bar{Z}$ is assumed to be statistically independent of the $\delta Z(\underline{r})$. One can assume that $\bar{Z}$ is, *a priori*, a Gaussian random variable with zero mean and variance $C_{\bar{Z}}$.

5. OPTICAL MEASUREMENT MODEL

In this case, we assume that a TV camera (or its equivalent) is situated at a high altitude (compared with the *a priori* range of the elevation function of the object) and is pointed straight down. As already noted, we assume that the surface is a diffuse Lambertian scatterer. We make the following additional assumptions: (1) each light source is assumed to be equivalent to a point source at infinity, (2) each element of surface is illuminated by each point source of light, and (3) multiple scattering of light between different elements of surface can be neglected. This model is essentially identical to that treated by Ray, Birk and Kelley (1983). The geometry of the experimental setup is depicted in Fig. 2b. We will first consider a noiseless model of the directly measurable quantities (i.e., the set of image intensities for each illumination direction) from which we ultimately obtain, using the concepts of photometric stereo (Horn, 1986), a set of derived quantities that depend only upon the local slopes. A noisy model is then formulated for these derived quantities. For the sake of completeness and in order to define notation, we present a short derivation below.

The noiseless direct optical measurement model can be expressed in the form

$$\rho(\underline{r}, \mathbf{e}_m) = \beta(\underline{r})\mathbf{e}_m \cdot \mathbf{n}(\underline{r}), \tag{51}$$

where $\rho(\underline{r}, \mathbf{e}_m)$ = possible image intensity at the position $\underline{r}$ in the xy-plane* with the illumination direction $\mathbf{e}_m$, $\mathbf{e}_m$ = unit vector in xyz-space pointing from an element of surface on the object toward the m-th illumination source (since the distance from any part of the object to each light source is assumed to be very large compared with the size of the object and the localization domain, $\mathbf{e}_m$ is independent of $\underline{r}$), $\mathbf{n}(\underline{r})$ = outward pointing normal vector in xyz-space for an element of surface above the position $\underline{r}$ in the xy-plane, and $\beta(\underline{r})$ = factor associated with the surface reflectivity, characteristics of the imaging system, and illumination intensity. The physical basis of Eq. (51) is discussed in a book edited by Kingslake (1965).

The local normal is easily shown to be

$$\mathbf{n}(\underline{r}) = \frac{\nabla[z - Z(\underline{r})]}{|\nabla[z - Z(\underline{r})]|} = (\mathbf{e}_z - \underline{\nabla}Z)(1 + |\underline{\nabla}Z|^2)^{-\frac{1}{2}}, \tag{52}$$

In this model, we assume that the illumination direction, $\mathbf{e}_m$, can assume several vector values.

The above model is suitable for continuous $\underline{r}$. For the case of discrete $\underline{r}$, defined on the xy-grid defined in the last section, certain modifications must be made. The above measurement model can stand as written, except that $\underline{r}$ is discrete. The most satisfactory approximation is obtained from an elevation function for continuous $\underline{r}$ obtained by Fourier interpolation applied to the discrete set of elevations defined on the above xy-grid. This procedure leads to the result

$$\underline{\nabla}Z(\underline{r}) = -\sum_{\underline{r}'} \underline{A}(\underline{r} - \underline{r}')Z(\underline{r}'), \tag{53}$$

---

* Here, we assume that the two-dimensional coordinate system on the image plane and the two-dimensional coordinate system used in the definition of the elevation function are in one-to-one correspondence.

where

$$-\underline{A}(\underline{r}) = Q^{-1} Re \sum_{\underline{k}} i\underline{k} \, \exp(i\underline{k} \cdot \underline{r}). \tag{54}$$

The vector $\underline{k}$ takes the discrete set of values defined by

$$\underline{k} = 2\pi \Big(\frac{p_x}{L_x} \mathbf{e}_x + \frac{p_y}{L_y} \mathbf{e}_y \Big), \tag{55}$$

where p_x and p_y take the integral values corresponding to a suitably defined finite lattice of points in $\underline{k}$-space. Q is the total number of such points. It should be stressed that the above results contain the implicit assumption that $\underline{A}(\underline{r})$ is periodic with a two-dimensional period D_L. This is consistent with the assumption of periodic boundary conditions or toroidal topology.

We turn next to the determination of $\nabla Z(\underline{r})$ in the noiseless case. We assume that the illumination directions $\mathbf{e}_1$, $\mathbf{e}_2$ and $\mathbf{e}_3$ are noncoplanar and we then define a set of reciprocal vectors $\mathbf{e}_1^R$, $\mathbf{e}_2^R$ and $\mathbf{e}_3^R$ by the relations

$$\mathbf{e}_m^R \cdot \mathbf{e}_{m'} = \delta_{mm'}, \qquad m, m' = 1, 2, 3. \tag{56}$$

It is easily shown that

$$\sum_{m=1}^{3} \mathbf{e}_m^R \mathbf{e}_m = \mathbf{1}, \tag{57}$$

where $\mathbf{1}$ is the three-dimensional unit tensor defined by Eq. (35). We now define a new observable quantity

$$\tau(\underline{r}) = \sum_{m=1}^{3} \mathbf{e}_m^R \rho(\underline{r}, \mathbf{e}_m). \tag{58}$$

Using Eq. (5.1), we obtain

$$\tau(\underline{r}) = \beta(\underline{r}) \sum_{m=1}^{3} \mathbf{e}_m^R \mathbf{e}_m \cdot \mathbf{n}(r) = \beta(\underline{r}) \mathbf{1} \cdot \mathbf{n}(\underline{r}) = \beta \mathbf{n}(\underline{r}). \tag{59}$$

We finally consider the ratio of the horizontal and vertical parts of τ to obtain

$$\underline{\sigma}(\underline{r}) \equiv \underline{\tau}(\underline{r})/\tau_z(\underline{r}) = \underline{n}(\underline{r})/n_z(\underline{r})$$
$$= -\underline{\nabla} \delta Z(\underline{r}) = \sum_{\underline{r}} \underline{A}(\underline{r} - \underline{r}') \delta Z(\underline{r}'), \tag{60}$$

where Z has been replaced by δZ since $\nabla \bar{Z}$ vanishes. We have thus found a combination of observable quantities that is independent of the local reflectivity and is dependent only upon the local slope. It should be re-emphasized that, as stated earlier, the object and the illumination directions must be such that no part of the exposed body is in shadow.

We make the approximation that the experimental error [i.e., the combined effects of errors in the $\rho(\underline{r}, \mathbf{e}_m)$] can be represented by additive Gaussian noise, and thus the complete stochastic measurement model for the optical part takes the form

$$\underline{\sigma}(\underline{r}) = \sum_{\underline{r}'} \underline{A}(\underline{r} - \underline{r}') \delta Z(\underline{r}') + \underline{\nu}(\underline{r}). \tag{61}$$

The *a priori* statistical behavior of $\delta Z(\underline{r})$ has been described in the last subsection. The measurement error (or noise) $\underline{\nu}(\underline{r})$ is assumed *a priori* to be statistically independent of $\delta Z(\underline{r})$ and $\bar{Z}$. Also, it is assumed to be Gaussian with the properties

$$E\underline{\nu}(\underline{r}) = 0, \tag{62a}$$

$$E\underline{\nu}(\underline{r})\underline{\nu}(\underline{r}') = \mathbf{I}\delta_{\underline{r}\,\underline{r}'}\sigma_{\underline{\nu}}^2. \tag{62b}$$

The appearance of the two-dimensional unit tensor $\mathbf{I}$ [see Eq. (34)] on the right-hand side of the last expression reflects an assumption of *a priori* statistical isotropy in the xy-plane.

6. Procedure for Estimating the Elevation Function

Here, we treat the problem of combining the two types of measurements, acoustical and optical, in the estimation of the elevation function. Using the best-score optimality criterion, the best estimate of $Z(\underline{r})$ is the most probable form of $Z(\underline{r})$ given the results of the two types of measurements. In more explicit mathematical terms, this means that the best estimate is given by the maximization of

$$P(Z|f,\sigma) = P(f,\sigma|Z)\frac{P(Z)}{P(f,\sigma)}, \tag{63}$$

where Z, f and σ are abbreviated symbols for the sets of values of $Z(\underline{r})$, $f(t)$ and $\underline{\sigma}(\underline{r})$. Using the relation

$$P(f,\sigma|Z) = P(f|Z)P(\sigma|Z), \tag{64}$$

we finally obtain the result that the best estimate is given by the maximization of

$$\phi(Z|f,\sigma) = \log\, P(f|Z) + \log\, P(\sigma|Z) + \log\, P(Z) \tag{65}$$

with respect to Z. In the above expression, an additive constant related to normalization has been ignored. We will use the same practice in all of the subsequent equations. The conditional probability densities are given by

$$\log P(f|Z) = -\frac{1}{2\sigma_\mu^2}\,\sum_t \left[f(t) - \alpha \sum_{\underline{r}} \delta_{\underline{r}}\, p'(t + 2c^{-1}Z(\underline{r}))\right]^2, \tag{66}$$

$$\log P(\sigma|Z) = -\frac{1}{2\sigma_\nu^2}\,\sum_{\underline{r}} \left[\underline{\sigma}(\underline{r}) - \sum_{\underline{r}'} A(\underline{r} - \underline{r}')\delta Z(\underline{r}')\right]^2. \tag{67}$$

The *a priori* p.d. of $\bar{Z}$ and $\delta Z(\underline{r})$ is given by the expression

$$\log P(Z) = -\frac{1}{2}\sum_{\underline{r}\,\underline{r}'} \delta Z(\underline{r})\, C_{\delta z}(\underline{r} - \underline{r}')^+\delta Z(\underline{r}') - \frac{1}{2}C_{\bar{Z}}^{-1}\,\bar{Z}^2, \tag{68}$$

where $C_{\delta z}(\underline{r} - \underline{r}')^+$ is the pseudoinverse of $C_{\delta z}(\underline{r} - \underline{r}')$. The latter quantity is a singular matrix because of the constraint (44).

Our general task of finding the most probable elevation function $Z(\underline{r})$, given the measured values $\{f(t)\}$ and $\{\underline{\sigma}(\underline{r})\}$, is tantamount to the maximization of $\phi(Z|f,\sigma)$ with respect to the set of elevations $Z(\underline{r})$ at the grid points $\underline{r}$.

7. COMPUTATIONAL EXAMPLE

As an illustrative example of the above procedure for estimating the elevation function of an object using a combination of acoustical and optical data, synthetic test data were generated based on an assumed tetrahedron resting on a flat table. A hidden-line representation of the assumed surface is shown in Fig. 3. The surface was defined on a square grid of 16 × 16 points, with a length of 20 mm on each side. The length of one side of the tetrahedron was 15 mm. The coordinate system was chosen such that the table was in the xy-plane at $z = 0$. Optical and acoustical data were generated in accordance with the physical models outlined in the previous section. In the case of the optical data, the two-dimensional intensity distribution, as seen by a camera looking straight down, was calculated for three illumination directions, each having a polar angle of 15°, but with azimuthal angles of 0°, 90° and 270°, respectively. Figure 4 shows gray-scale representations of the three synthetic optical images. In the case of the acoustic data, the assumed measurement system response function $p(t)$ was in the shape of a Hanning window in the frequency domain with a bandpass width between the zero points of 0 – 20 kHz.

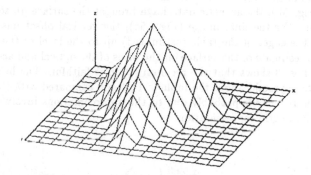

Fig. 3. Hidden-line representation of the assumed surface, corresponding to a tetrahedron whose sides are of length 15mm placed on a flat table of width 20mm.

In the inversion procedure, the standard deviations of the measurement noise were assumed to be 0.01 and 0.1 for the acoustical and optical data, respectively, where the above quantities are relative to the peak absolute value of the synthetic measured data in both cases. The area-average elevation $\bar{Z}$ was assumed to have an ensemble average of zero, and a covariance $C_{\bar{Z}}$ of the square of half the width of the localization domain, i.e., 10 mm, thus providing a slight bias toward a spatial average height of zero in the reconstruction. The covariance of δZ was assumed here to be infinite.

The inversion was performed in three stages:

1. Determine $\delta Z(\underline{r})$ using the optical data alone.
2. Assuming $\delta Z(\underline{r})$ to be fixed at the above values, determine $\bar{Z}$ from the acoustic data alone.
3. Perform the final optimization using both acoustic and optical data by maximizing $P(Z|\sigma, f)$ with respect to δZ and $\bar{Z}$.

The results for the three stages are shown in Fig. 5. In the case of stage 1 (Fig. 5a), the reconstructed surface has a spatial mean elevation of zero, in accordance with our

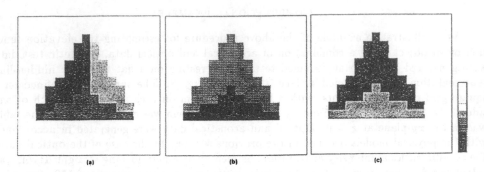

Fig. 4. Synthetically generated optical data representing the intensity distribution which would be seen by a camera looking straight down. The illumination directions correspond to a polar angle of 15° in each case, and azimuth angles of (a) 0°, (b) 90° and (c) 270°.

definition of δZ, and hence contains no information on the absolute vertical position. In the case of stage 2 (Fig. 5b), the acoustic data have brought the surface up to approximately the correct height. For the final image (Fig. 5c), the vertical offset has been improved slightly, bringing the edges of the table (estimated) up to the level of the xy-plane. This represents the best estimate of the surface on the basis of the optical and acoustic data, and comparison with Fig. 3 shows that the reconstruction is faithful. The height of the apex of the reconstructed pyramid above the table is 11.4 mm compared with the true height of 12.2 mm. This discrepancy is probably due to the approximations involved in calculating gradients from a finite-grid representation.

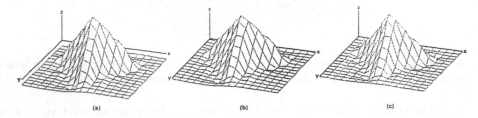

Fig. 5. The results of the three stages of the inversion technique discussed in the text: (a) the relative profile of $\delta Z(x, y)$ deduced from optical data only, (b) an estimate of $Z(x, y)$ in which the acoustic data have been used only to adjust the vertical position offset, and (c) the optimal image used from the combined optical and acoustical data.

8. DISCUSSION

The methodology for the probabilistic estimation of the state of a system from multisensor data has been applied to the estimation of a single-valued elevation function of a rigid object in air based upon a combination of acoustical and optical measurements, the nature of which we discussed in detail in earlier sections. The optical measurement system described here has the property that it provides complete information about the shape of the elevation function, but no information regarding the absolute elevation (or vertical offset) for the object and its immediate surroundings. On the other hand, the acoustical measurement system provides complete information about the vertical offset, but only incomplete information about the shape, i.e., the horizontal cross section as a function of altitude.

REFERENCES

Foster, J. L., and Hall, D. K.: 1981, 'Multisensor Analysis of Hydrologic Features with Emphasis on the SEASAT SAR,' *Eng. and Remote Sensing* **47**, No. 5, 655–664.

Horn, B. K. P.: 1986, *Robot Vision*, MIT Press.

Kingslake, R. (ed.): 1965, *Applied Optics and Optical Engineering*, Vols. I and II, Academic Press, NY.

La Jeunesse, T. J.: 1986, 'Sensor Fusion' in *Def. Sci. and Electr.*, Sept., 21–31.

Lee, D. A., Crane, R. L., and Moran, T. J.: 1984, 'A Practical Method for Viewing Resolution-Noise-Bandwidth Tradeoffs in NDE Data Reductions,' *Rev. of Prog. in QNDE* **3**, D. O. Thompson and D. E. Chimenti, eds., New York, Plenum, 907–915.

Ray, R., Birk, J., and Kelley, R. B.: 1983, 'Error Analysis of Surface Normals Determined by Radiometry,' *IEEE Trans. on Patt. Anal. and Mach. Intel.*, Vol. PAMI-5, No. 6.

Richardson, J. M., Marsh, K. A., Schoenwald, J. S. and Martin, J. F.: 1984, *Proc. of IEEE 1984 Ultrasonics Symp.*, 831–836.

Richardson, J. M., and Marsh, K. A.: 1988, *Int'l. J. of Robotics Research* **7**, No. 6, 78–96.

COMBINING DATA FROM DIFFERENT EXPERIMENTS:
BAYESIAN ANALYSIS AND META-ANALYSIS

A.J.M. Garrett
Department of Physics and Astronomy
University of Glasgow
Glasgow G12 8QQ
Scotland, U.K.

D.J. Fisher
Associate Editor, *Solid State Phenomena*
27, Elderberry Road
Cardiff CF5 3RG
Wales, U.K.

ABSTRACT. The problem of testing a hypothesis with data from a number of distinct experiments
has led, notably in the social sciences, to the development of a new branch of statistical practice
called meta-analysis. However, Bayesian probability theory already provides a uniquely consistent
way of doing this; and is identical in principle to combining repeated measurements from a single noisy experiment. Here, Bayesian and meta-analytical techniques are compared in a specific
problem, so as to highlight the inconsistency of meta-analysis and the advantages of Bayesianism.

Thou shalt not answer questionnaires
Or quizzes upon world affairs,
Nor with compliance
Take any test.
Thou shalt not sit with statisticians nor commit
A social science.
W.H. Auden

1. Introduction

In all statistical problems — that is, problems in which certainty is absent through lack
of information — there arises the question of how much information to include. The answer
to this question is, in principle, easy: all the information you have. If you are estimating
a parameter, or testing a hypothesis, you have no right to complain that the answer looks
crazy if the formal calculation incorporates less data than your intuitive one. (Assuming,
of course, that both the formal calculation and your intuition are both processing the data
'correctly', in a sense to be understood.) If you transport a reasoning automaton from a
rain forest to a desert but don't inform it, you have no right to complain when it assigns a
high probability to rain.

C. R. Smith et al. (eds.), *Maximum Entropy and Bayesian Methods, Seattle, 1991*, 273–286.

In a formal calculation, two facts are crucial: the reasoning must be done correctly; and all of the available data must be

included. The answer is only as good as these are. This paper is concerned with ways of incorporating diverse data, from different experiments, so as to answer a single question. Orthodox sampling theory treats only the problem of incorporating repeated samples of data when the *same* noise process is operating for each; in addition it ignores 'prior' information in forms other than repeated samples, and this is also crucial: you would hardly trust a doctor who advised you, on the grounds that 80% of patients with your symptoms needed surgery, to have an operation, but who refused to examine your personal medical record.

The problem, then, is how to incorporate, into a probability calculation, data collected under differing statistics of noise. Orthodox sampling theory, despite Fisher's proposals [1], stands helpless. In recent years many techniques have been developed for tackling the problem, and are collectively known as *meta-analysis*. This is also known as the 'overview' or 'pooling' method; the name meta-analysis, coined in 1976 [2], refers to the 'analysis of analyses'. Its originator, Gene Glass, has stated: "The findings of multiple studies should be regarded as a complex data set, no more comprehensible without statistical analysis than would hundreds of data points in one study" [ref.3, p12]. This is entirely sound, at least in principle: the combined power of all the data residing in the literature is enormous. The popularity of meta-analysis — over 91 studies had been carried out in the clinical field alone by 1987 [4] — derives from the promise of gaining important new results from existing data. What is needed is a way of automatically placing more weight on those observations which are collected under less noise, since these are more accurate.

Our purpose is to point out that a method already exists for combining data in the necessary way, and that this method is, uniquely, underpinned by elementary requirements of consistency. Anything inequivalent is, therefore, inconsistent, in a precise sense which will be explained. In particular, meta-analytical techniques fail this crucial consistency test; we shall exhibit a particular example in which this failure is made obvious. The correct, consistent technique is Bayesian analysis, and we hope to make its usefulness clear to colleagues working in the traditional applications of meta-analysis.

The areas in which meta-analysis has taken off are precisely those where it is more liable to have an immediate effect on the population, and where sound statistical practice is all the more vital and unsound practice more difficult to spot: medicine [4,5], psychology [6,7], sociology [8,9] and education [10,11]. (Books and reviews on meta-analysis include [3,12-16].) Workers in fields wherein statistics are applied might rightfully feel they should safely be able to leave the foundations to professional statisticians; unfortunately this has not been the case, at least for social scientists. The problem has been exacerbated by the availability of 'off-the-shelf' statistical packages on computers, to which researchers too readily abdicate their thought processes.

It is clear that there is only one answer to any well-defined question, so that the very variety of meta-analytical techniques, which give differing answers to a probability assignment, cannot be correct. (The same criticism applies to the various methods of orthodox sampling theory.) Glass himself has stated that meta-analysis is "not a technique; rather it is a perspective that uses many techniques of ... statistical analysis" [3]. We see this as a weakness rather than a strength. One critic described a meta-analysis as a mere literature review [17]. Perhaps its popularity derives also from its capability of providing methods which give results favouring any researcher's prejudices; personal bias is notoriously relevant in social studies and other fields in which it is difficult to get hard and fast results [18]. Sackett has identified no less than 56 possible sources of bias in 'analytic research' [19].

Meta-analysis almost invariably proceeds by combining, in simple ways, the statistics of the scrutinised reports — chi-square, Student-t value, or whatever — rather than by working with the original data. In this way, information which could help solve the problem is discarded right at the start. Of course, in many problems, including the original application, these methods will give sensible answers —

meaning, with hindsight, close to the Bayesian — but it is always possible to find a dataset which makes the method look arbitrarily illogical. Only the Bayesian technique is immune to this, because it is constructed from criteria of consistency. It is a matter of importance to highlight this discrepancy: there is a danger that, while meta-analysts are active in influencing public policy, Bayesians remain concerned with the finer details of their priors. The proper role of statistics is as a service industry.

Instead of merely bemoaning the shortcomings of meta-analysis, Bayesians are in a position to provide workers in such fields as sociology and medicine with a solidly based rationale for performing future studies of meta-analytical type. We now proceed to this.

2. Bayesian Analysis

Bayesian analysis is based on the assignment and manipulation of probabilities. Assignment of probability takes place via the principle of maximum entropy, and is discussed briefly in the next Section. Probability manipulation is central to the present issue, and proceeds by the product and sum rules

$$p(AB|C) = p(A|BC)p(B|C), \tag{1}$$

$$p(A|C) + p(\bar{A}|C) = 1. \tag{2}$$

These are derived from consistency criteria in the next Section; here, we shall employ them to solve the data incorporation problem. These rules are familiar to probabilists; what may be different is the (Bayesian) interpretation we give them.

The probability $p(A|C)$ that proposition A is true, assuming that proposition C is true, is the *degree of belief* it is consistent to hold in A, given C. This is a number between 0 (certainly false) and 1 (certainly true) and is unique: precise algorithms exist for constructing it, and the result is the same when these are correctly implemented, whether the hardware be one person's brain, another person's brain or an electronic chip. Misunderstanding arises because different persons' brains often contain different information, and thereby assign different probabilities to the same proposition: $p(A|C)$ depends on A *and* C, and in general $p(A|C) \neq p(A|C')$. From this stems much criticism that Bayesianism is flawed through 'subjectivity', motivating the 'frequentist' view that probability is an observable, measurable quality, identical with *proportion*: the proportion of times a particular outcome is found in many 'random' trials (which may be imagined). Bayesians, by contrast, deny that probability is measurable: since different people assign different probabilities to the same event (according to their differing information), its probability clearly can not be measured. In repeated coin tossing, for example, one tests not the hypothesis "$p = 1/2$" but rather "the coin is evenly weighted". Moreover there is no such thing as a 'random' process; only an *unknown* one, and if we had more information on the tossing process, for example, we would assign different probabilities to heads and tails.

Probability, then, is not the same as proportion, either numerically (though it may coincide) or conceptually, for it is equally applicable to one-off events — the weather, the authenticity of an disputed painting — as to coin-tossing. (These examples also show it is as applicable to the past as to the future.) The theory is, quite generally, a mode of reasoning wherever there is insufficient information to allow deductive certainty. It is the theory of inductive logic, and it includes deductive logic as a limiting case. When it is applied to repeated events, like coin tossing, it comprises the results of statistical *sampling theory*, corrected to allow for any prior information we may have. Applied to one-off problems, it is known as *statistical inference*. Conceptually, statistics and statistical inference are one.

Return now to the data incorporation problem. We begin by constructing, from (1) and (2), the *marginalising rule*

$$p(B|C) = [p(A|BC) + p(\bar{A}|BC)]p(B|C) \tag{3}$$
$$= p(AB|C) + p(\bar{A}B|C) \tag{4}$$
$$= p(A|C)p(B|AC) + p(\bar{A}|C)p(B|\bar{A}C). \tag{5}$$

Also, since the logical product is commutative, $AB = BA$ and the product rule allows us to decompose $p(AB|C)$ in two ways, which when equated gives

$$p(A|BC) = Kp(A|C)p(B|AC); \tag{6}$$
$$K^{-1} = p(B|C) \tag{7}$$
$$= p(A|C)p(B|AC) + p(\bar{A}|C)p(B|\bar{A}C) \tag{8}$$

on using (3)-(5). This tells us how to incorporate the information that proposition B is true into the probability that proposition A is true, all the while given C; it is known as Bayes' theorem. The recipe is: multiply the prior probability of truth, $p(A|C)$, by the probability that B is true assuming A is true, $p(B|AC)$ — the *likelihood* — and then renormalise it by the factor

K, so as to ensure the posterior probability of truth, $p(A|BC)$, and the posterior probability of falsehood, $p(\bar{A}|BC)$, sum to one. This relation makes intuitive sense: the more likely we believe A to be true to start with, the more likely we will believe it to be true afterwards, and this is reflected in the dependence of $p(A|BC)$ upon $p(A|C)$. Also, the more likely is B to be true, assuming A is true, then the more likely is A to be true, assuming B is true, and this is reflected in the dependence of $p(A|BC)$ on $p(B|AC)$; this is called the principle of *inverse reasoning*. For example, if you find growing a plant which you know to prefer a damp climate, you suspect that the local climate is damp.

If, now, we eliminate $p(\bar{A}|C)$ from (8) using the sum rule (2), we can rewrite (6), after a little manipulation, as

$$p(A|BC) = \left[1 + \{\frac{1}{p(A|C)} - 1\}\frac{p(B|\bar{A}C)}{p(B|AC)}\right]^{-1}. \tag{9}$$

This is equally well derived by dividing (6) by its conjugate with $A \rightarrow \bar{A}$, so as to eliminate K:

$$\frac{p(A|BC)}{p(\bar{A}|BC)} = \frac{p(A|C)}{p(\bar{A}|C)}\frac{p(B|AC)}{p(B|\bar{A}C)} \tag{10}$$

and then using the sum rule to write, on both the LHS and RHS,

$$\frac{p(A|\bullet)}{p(\bar{A}|\bullet)} = \frac{p(A|\bullet)}{1 - p(A|\bullet)}, \tag{11}$$

and finally extracting $p(A|BC)$. That the result is the same follows from the consistency principles underlying the sum and product rules.

Expression (9) is our key formula. Suppose that A is a hypothesis, which we relabel H; B is the proposition that the data take values D — this can include a single datum or many — and C is the proposition that the prior information is as it is. We assume, at this stage, that data are two-valued, so that we have just D and $\bar{D}$, such as heads and tails in coin tossing; generalisation will be made. Then

$$p(H|DI) = \left[1 + \{\frac{1}{p(H|I)} - 1\}\frac{p(D|\bar{H}I)}{p(D|HI)}\right]^{-1}. \tag{12}$$

The special cases of this all make good sense: if $p(D|\bar{H}I) = 0$, meaning that it is impossible to get the observed data if H is false, then, since we *did* get the observed data, H must be true — and indeed (12) gives $p(H|DI) = 1$. Likewise, if $p(D|HI) = 0$, meaning you can't get the observed data if H is true, then H must be false: and (12) reduces to $p(H|DI) = 0$. If $p(D|HI) = p(D|\bar{H}I)$, so that the data are equally likely whether H is true or false, we expect to learn nothing about H from the data; and indeed (12) tells us that $p(H|DI) = p(H|I)$. Finally, if we begin by being certain that H is true or false so that $p(H|I) = 1$ or 0, no amount of data can change our minds; and (12) confirms that $p(H|DI) = 1$ or 0 as appropriate.

Now move closer to the meta-analysis problem, in which there are many data: $D = D_1 D_2 D_3 \dots$. We specialise by taking these to be independent, given H (or $\bar{H}$): that is,

$$p(D_i|D_j HI) = p(D_i|HI). \tag{13}$$

In words, knowledge of the jth datum, assuming H is true, does not alter the probability we assign to the ith datum D_i. This is what is customarily meant by statistical independence; in this context the assumed truth of H is critical. For example, conditioned on the hypothesis that a coin is not weighted, the probability of getting heads in a toss, whose dynamics are unknown, is not altered by knowing the result of the previous toss: the probability remains $1/2$. (Conditioned on a different hypothesis, that the coin is weighted to an *a priori* unknown extent, the results of previous tosses may be used to infer something about the weighting, and this *will* alter the probability we assign to heads on the next throw.) Quite generally we have, from the product rule,

$$P(D|HI) = p(D_1 D_2 D_3 \dots|HI) = p(D_1|D_2 D_3 \dots HI)p(D_2|D_3 \dots HI)p(D_3|\dots HI)\dots \tag{14}$$

which reduces under statistical independence to

$$p(D|HI) = p(D_1|HI)p(D_2|HI)p(D_3|HI)\dots. \tag{15}$$

It is a further exercise in the sum and product rules to show that, if the data are independent conditioned on H, then they are also independent conditioned on $\bar{H}$. The likelihood of all the data is simply the product of the likelihoods of each individual datum, and (12) reduces to

$$p(H|DI) = \left[1 + \{\frac{1}{p(H|I)} - 1\}\frac{\prod_i p(D_i|\bar{H}I)}{\prod_i p(D_i|HI)}\right]^{-1}. \tag{16}$$

We see from this that if just a single observed datum is impossible, given H, then H is false; and the converse.

We have derived (16) by decomposing the likelihoods under the assumption of statistical independence. However, we could equally well have used (12) iteratively, so as to incorporate each datum in sequence. It is instructive to prove this directly: a new datum D_k merely incorporates an extra product $p(D_k|HI)/(p(D_k|\bar{H}I)$ on the RHS of (10). Once more, the consistency conditions underlying the sum and product rules guarantee that the results will be the same.

Formula (16) solves the meta-analysis problem. It tells us how to incorporate any number of statistically independent data D_k, be these repeated samples from a single experiment in which the noise statistics are fixed (as in orthodox sampling theory); or single results of different experiments on different variables; or several samples from each of

a number of different experiments. The formula is exactly the same, and represents the complete solution.

Having stated our solution and demonstrated its plausibility, we shall now go backwards and justify the product and sum rules on which it is based. Then, in Section 4, we shall apply it to a model problem.

3. Why There Is No Alternative

We have explained what probability is. Here, we demonstrate that it obeys the product and sum rules. Suppose we do not know the product rule in advance, and write the general relation as

$$p(XY|Z) = F\big(p(X|YZ), p(Y|Z)\big) \tag{17}$$

where F is a function to be determined. We find it by decomposing the probability of the logical product of three propositions in different ways and demanding that, since the logical product is associative, the results shall coincide. We have, substituting $X = AB$, $Y = C$, $Z = D$ in (17), the result

$$p(ABC|D) = F\big(p(AB|CD), p(C|D)\big). \tag{18}$$

A further decomposition of $p(AB|CD)$ on the RHS, with $X = A$, $Y = B$, $Z = CD$, gives

$$p(ABC|D) = F\big(F\big(p(A|BCD), p(B|CD)\big), p(C|D)\big). \tag{19}$$

A different decomposition is

$$p(ABC|D) = F\big(p(A|BCD), p(BC|D)\big) \tag{20}$$
$$= F\big(p(A|BCD), F\big(p(B|CD), p(C|D)\big)\big). \tag{21}$$

Equating (19) and (21) gives the functional equation

$$F\big(F(u,v), w\big) = F\big(u, F(v,w)\big). \tag{22}$$

The solution of this equation is a matter of mathematics, and we quote the result from the original derivation [20]:

$$F(\alpha, \beta) = \phi^{-1}(\phi(\alpha)\phi(\beta)) \tag{23}$$

where ϕ is an arbitrary single-valued continuous function. Both sides of (22) are equal to $\phi^{-1}(\phi(u)\phi(v)\phi(w))$. The product rule (17) is, therefore,

$$\phi((pXY|Z)) = \phi(p(X|YZ))\phi(p(Y|Z)) \tag{24}$$

and the $\phi(\bullet)$ can be absorbed into the $p(\bullet)$ without loss of generality, since we would work in applications with $\phi(p(\bullet))$. This gives the product rule. It follows that certainty is represented by $p = 1$: suppose in (1) that A is certain, given C. Then $AB|C = B|C$, and its probability cancels, leaving $p(A|BC) = 1$.

The sum rule is derived by first taking

$$p(\overline{X}|Z) = G(p(X|Z)). \tag{25}$$

One relation of the calculus of propositions — Boolean algebra — is that the logical sum is related to the logical product by $\overline{A + B} = \bar{A}\bar{B}$. We use the product rule and (25) to express $p(A + B|C)$ in terms of $p(A|C)$, $p(B|C)$ and $p(AB|C)$; on interchanging A and B, and equating the results, a functional equation is derived for G [20]:

$$uG\left(\frac{G(v)}{u}\right) = vG\left(\frac{G(u)}{v}\right). \tag{26}$$

This has solutions [20]

$$G(\alpha) = (1 - \alpha^n)^{1/n} \tag{27}$$

which also satisfies $G(G(\alpha)) = \alpha$, since the negation of the negation of a proposition leaves it unchanged. Expression (25) now becomes

$$p(X|Z)^n + p(\overline{X}|Z)^n = 1. \tag{28}$$

Just as we ignored $\phi(\bullet)$ above, we now take $n = 1$ without loss of generality; the sum rule emerges. The decomposition of $p(A + B|C)$ is

$$p(A + B|C) = p(A|C) + p(B|C) - p(AB|C). \tag{29}$$

The sum and product rules, therefore, arise from consistency conditions on the logical sum of propositions and the logical product. Applied to proportions, these rules express trivial algebraic identities; their relevance to probabilities shows them to have far greater significance. If they are not adhered to, actual numerical inconsistencies can arise: by doing the steps of a calculation in a different order, different numerical probabilities result for the same proposition with the same conditioning information.

Before proceeding, we generalise our formalism to allow for multiple exclusive propositions ('outcomes') which together are exhaustive. The generalised sum rule (29) reduces, when A and B are exclusive, so that $AB = FALSE$, to

$$p(A + B|C) = p(A|C) + p(B|C) \tag{30}$$

and, with a greater number of exclusive propositions, to

$$p(\sum_i A_i|C) = \sum_i p(A_i|C). \tag{31}$$

If these are exhaustive, so that $\sum_i A_i = TRUE$, the sum rule tells us that this probability is equal to one. Generalisation to the continuum is immediate.

The sum and product rules are used for manipulating probabilities; we shall see them in action in a moment. They are consequences simply of associating real numbers with propositions, and they are silent as to the interpretation of these numbers as probability, or on how to assign probabilities in the first place. A principle is needed for assigning probability from whatever information we may already have about a distribution (such as: "the standard deviation is 2.1"). This is the principle of *maximum entropy*, an old idea whose relevance to probability was first recognised by Shannon [21] and was placed in the Bayesian format by Jaynes [22]. For example, given a six-sided die and only the information (from whatever source) that the mean is 3.5, nothing prevents us from assigning probability 1/2 to the faces ONE and SIX, and probability 0 to the rest; but clearly the appropriate assignment is probability 1/6 to every face. Maximum entropy achieves this. We do not discuss it further, except to refute the anti-Bayesian criticism that 'prior' probabilities are not well defined. Strictly, there is no such thing as a 'prior', for the very term implies the imminent reception of experimental data, so as to convert it in into a 'posterior'. Yet this, too, could be the 'prior' for still further data. What lies beneath the criticism is the idea that probabilities based on anything other than repeated samples are ill defined: that probability is only meaningful as proportion. We have already refuted this in the previous Section.

The principle of maximum entropy leads to a superior definition of correlation. Two propositions $A, \bar{A}$ and $B, \bar{B}$ are correlated, given C, if

$$S_A + S_B - S_{AB} > 0 \tag{32}$$

where

$$\begin{aligned} S_{AB} \equiv &- p(AB|C)\log p(AB|C) - p(A\bar{B}|C)\log p(A\bar{B}|C) \\ &- p(\bar{A}B|C)\log p(\bar{A}B|C) - p(\bar{A}\bar{B}|C)\log p(\bar{A}\bar{B}|C), \end{aligned} \tag{33}$$

$$S_A \equiv -p(A|C)\log p(A|C) - p(\bar{A}|C)\log p(\bar{A}|C) \tag{34}$$

and S_B is defined in like manner to S_A. The LHS of (32) is non-negative, and is zero only if $p(A|BC) = p(A|C)$ (in which case it equals $p(A|\bar{B}C)$ also), so that the truth value of B does not affect the truth value of A: this is what is meant by *uncorrelated*.

The maximum entropy principle has been reviewed by Garrett [23], and we quote a particular result: the distribution it assigns for a continuous variable, given uniform measure on $(-\infty, \infty)$, and given the mean and the standard deviation, is the Gaussian, or Normal density. This distribution is employed in the following model problem.

4. A Model Problem: The Normal Distribution

Suppose that we have a parameter to estimate from noisy measurements with Gaussian noise statistics; but, unlike the usual problem, each measurement has a different standard deviation in the noise. If the parameter is denoted μ, the probability density for measuring value x_i, in the ith measurement, is proportional to

$$e^{-(x_i-\mu)^2/2\sigma_i^2}. \tag{35}$$

Bayes' theorem gives, for the posterior density of μ, $p_+(\mu)$, in terms of the prior $p_-(\mu)$ and the likelihood $p(\mathbf{x}|\mu)$,

$$p_+(\mu) = K p_-(\mu)p(\mathbf{x}|\mu) \tag{36}$$

where K is determined *a posteriori* by demanding that the posterior be normalised. We have seen that the likelihood of the full dataset is the product of the individual likelihoods:

$$p(\mathbf{x}|\mu) \propto \prod_i e^{-(x_i-\mu)^2/2\sigma_i^2} \tag{37}$$

$$\propto \exp[-\sum_i \frac{(\mu-x_i)^2}{2\sigma_i^2}]. \tag{38}$$

The constant of normalisation is reinserted automatically when we normalise the posterior. Suppose, now, for simplicity's sake, that the prior density of μ is itself Gaussian ('Normal') about μ_-, with standard deviation s_-:

$$p_-(\mu)d\mu = \frac{1}{\sqrt{2\pi s_-^2}} e^{-(\mu-\mu_-)^2/2s_-^2} d\mu. \tag{39}$$

It is readily shown from (36) that in this case the posterior density $p_+(\mu)$ is also Gaussian, with mean μ_+, and standard deviation s_+:

$$p_+(\mu)d\mu = \frac{1}{\sqrt{2\pi s_+^2}} e^{-(\mu-\mu_+)^2/2s_+^2} d\mu, \tag{40}$$

where these parameters are given by

$$\frac{1}{s_+^2} = \frac{1}{s_-^2} + \sum_i \frac{1}{\sigma_i^2}, \tag{41}$$

$$\mu_+ = \frac{\frac{\mu_-}{s_-^2} + \sum_i \frac{x_i}{\sigma_i^2}}{\frac{1}{s_-^2} + \sum_i \frac{1}{\sigma_i^2}}. \tag{42}$$

This posterior density meshes exactly with intuition: if any single σ_i, for $i = j$, say, is far smaller than the others and also than the prior uncertainty s_-, then that measurement is far more reliable and overwhelms all prior knowledge: we expect $s_+ \approx \sigma_j$ and $\mu_+ \approx x_j$, and indeed this is what (41) and (42) tell us. (If there is a large number of measurements, μ_+ is almost certainly close to this value anyway; not so s_+.) Also, if $s_- \ll \sigma_i \forall i$, so that the prior density is like a δ-function relative to the likelihood and we are reasonably confident of the value of μ to begin with, then we are more confident in ascribing the deviations from μ_- in the measurements to noise: $\mu_+ \approx \mu_-$.

In the usual sampling-theoretical case in which all the N measurements are taken under a single noise process, then $\sigma_1 = \sigma_2 = ... = \sigma_N \equiv \sigma$, and

$$\frac{1}{s_+{}^2} = \frac{1}{s_-{}^2} + \frac{N}{\sigma^2}, \tag{43}$$

$$\mu_+ = \frac{\frac{\mu_-}{s_-{}^2} + \frac{N}{\sigma^2}\overline{x}}{\frac{1}{s_-{}^2} + \frac{N}{\sigma^2}} \tag{44}$$

where $\overline{x}$ is the sample mean $\sum_i x_i/N$. If in addition $s_- = \infty$ we have $s_+{}^2 = \sigma^2/N$, $\mu_+ = \overline{x}$. These are well known

formulae in sampling theory, which does not recognise prior information and therefore puts $s_- = \infty$.

How would meta-analysis treat this problem? Since it is not a precisely defined technique, we cannot give a definitive answer. The one we do give, which leads to absurdities, will perhaps be attacked by meta-analysts as a straw man which we have set up deliberately in order to knock down. Yet it is certainly one of the techniques of meta-analysis. Meta-analysts are in effect developing a toolbox of different methods, selecting in each case that method which gives an intuitively plausible (for the less scrupulous, a more desirable) answer. Given a large enough toolbox, this is always possible, but it is equivalent to using intuition in the first place and finding the method to justify it afterwards. A general formula should be exactly what it says; it should not depend on the

numbers which are to be substituted into it.

We suggest that a typical meta-analysis would simply pool the data by taking μ_+ as the sample mean $\overline{x}$ without making any allowance for the differing accuracies of the measurements (or the prior information). This procedure is obviously fallible for the reasons already explained: if any one measurement is much more accurate than the rest, it *should* dominate. Our Bayesian method, by contrast, is underpinned by principles of consistent reasoning; the sum and product rules, applied to this prior and likelihood, automatically assign the appropriate weights to the data in (42), meshing with intuition in every case. This enables us to see when the meta-analytical technique is a good approximation: in this case, when $\sigma_1 \approx \sigma_2 \approx .... \approx \sigma_N \ll s_-$. We are confident that the originators of meta-analytical techniques always applied their methods in such 'good' problems. But approximations they are, always; and embarrassing datasets are

always liable to arise. We see, in highlight, the dangers of *ad hoc* intuition; a really good meta-analyst might have foreseen (41) and (42), but still could not have justified them; and might be at a loss in still more complicated problems. Bayesians, by contrast, did their hard work first, in finding their general formulae, after which the rest is just mathematical manipulation.

A second illustration can be extracted from this calculation. Suppose that the first $N_{(1)}$ samples are collected under the same noise $\sigma_{(1)}$, the next $N_{(2)}$ under $\sigma_{(2)}$, and so on until the last $N_{(M)}$ are collected under noise $\sigma_{(M)}$: we have the total number of samples $N = N_{(1)} + ... + N_{(M)}$. The brackets around the subscript in $\sigma_{(j)}$ distinguishes this quantity from σ_j above. Then, on defining the mean of the samples collected under noise $\sigma_{(j)}$ as

$\bar{x}_{(j)}$, we have from (42)

$$\mu_+ = \frac{\frac{\mu_-}{s_-^2} + \sum_{j=1}^{M} \frac{N_{(j)}\bar{x}_{(j)}}{\sigma_{(j)}^2}}{\frac{1}{s_-^2} + \sum_{j=1}^{M} \frac{N_{(j)}}{\sigma_{(j)}^2}}, \tag{45}$$

and with $s_- = \infty$, this is

$$\mu_+ = \sum_{j=1}^{M} \frac{N_{(j)}\bar{x}_{(j)}}{\sigma_{(j)}^2} \Big/ \sum_{j=1}^{M} \frac{N_{(j)}}{\sigma_{(j)}^2}. \tag{46}$$

Meta-analysis would very likely ignore the weightings $N_{(j)}$ in (46); it would simply pool the sample means without weighting them by the number of measurements in each sample. Even if it did give (46) correctly, it could offer no formal justification.

It is not as easy as it may seem to make further comparisons between Bayesian analysis and meta-analysis. This is, historically,

because the analytical techniques of meta-analysis grew out of orthodox sampling theory, which is dominated by the Normal distribution; so that undoubtedly the previous example is the definitive one. We may, of course, compare Bayesian analysis with orthodox sampling theory. A fine paper, demonstrating with examples the superiority of Bayesianism, is available [24].

5. Limitations

We were careful to state, earlier, that the idea of taking into account all of the data in the literature was sound *in principle*. In practice, this procedure may not permit as definitive a resolution of particular questions as desired. This is because, in the areas of research in which meta-analysis is popular, there are a great many correlated variables of interest. Suppose we are interested in a parameter α, when parameters $\beta, \gamma, \ldots$ also have a bearing on the data. We can find the likelihood $p(D|\alpha, \beta, \gamma.., I)$ readily enough (inserting commas for convenience), and invert this to give

$$p_+(\alpha, \beta, \gamma...|D, I) = K p_-(\alpha, \beta, \gamma...|I) p(D|\alpha, \beta, \gamma..., I) \tag{47}$$

where, as ever, K is fixed by normalisation. On marginalising, we have

$$p_+(\alpha|D, I) = \int \int \ldots \, d\beta d\gamma \ldots \, p_+(\alpha, \beta, \gamma...|D, I). \tag{48}$$

It follows from (47) and (48) that marginalising is equivalent to taking an average over $\beta, \gamma \ldots$ of $p(D|\alpha, \beta, \gamma..|I)$, weighted by the prior $p_-(\alpha, \beta, \gamma...|I)$. All such averaging discards information and smooths out the density for α; the more superfluous parameters to marginalise over, the greater is the smoothing. This is reflected in the mathematical identity

$$S_\alpha + S_\beta - S_{\alpha\beta} \geq 0 \tag{49}$$

where

$$S_{\alpha\beta} \equiv - \int \int d\alpha d\beta \, p(\alpha, \beta) \log \, p(\alpha, \beta) \tag{50}$$

$$S_\alpha \equiv - \int d\alpha p(\alpha) \log \, p(\alpha) \quad ; \quad p(\alpha) \equiv \int d\beta p(\alpha, \beta) \tag{51}$$

and a similar relation defines $p(\beta)$ and S_β; $p(\alpha, \beta)$ here is an arbitrary

probability density. This is the continuum analogue of the correlation relation (32); equality in (49) is attained only if α and β are uncorrelated, so that the joint density $p(\alpha, \beta)$ is separable.

Next, we rewrite (49) as

$$S_\alpha \geq \int d\beta p(\beta). - \int d\alpha p(\alpha|\beta) \log \, p(\alpha|\beta) \tag{52}$$

or, in an obvious notation,

$$S[p(\alpha)] \geq \langle S[p(\alpha|\beta)] \rangle_\beta. \tag{53}$$

This result tells us that marginalising over a parameter, β,

increases the S-value of the resulting density for α over its expectation value conditioned on β. And S —

the *information entropy* — is the most useful measure of the width of a distribution [22,23]. To see its improvement over standard deviation, consider a distribution having two sharp, well-separated peaks.

Proceeding iteratively, it follows that every time we marginalise over a parameter, we lose information (in both the technical and colloquial senses) about the remaining parameters. The greater the number of correlated parameters which are marginalised over, the worse off we are. We therefore doubt that the idea behind meta-analysis, even when transcribed correctly into consistent, Bayesian reasoning, will bear a great deal of fruit: there are many parameters in those problems for which meta-analysis is popular.

An example, in which data of different types could usefully be combined, is the question of whether there is a fifth fundamental force. Diverse experiments have been performed to test this, including the effect of depth underground; of height; of cliffs; and of moveable water masses in hydro-electric storage facilities [25]. Expert input is necessary to assess the noise statistics and the significant experimental parameters. Further details of the Bayesian analysis used for testing this hypothesis are given in [26].

6. Conclusions

Meta-analysis grew out of the *ad hoc* techniques of orthodox sampling theory, and developed in areas removed from the influence of Bayesianism. Like all other non-Bayesian methods it suffers

from internal inconsistencies. The inconsistency of meta-analysis has been masked by the difficulties of reaching clear cut conclusions in its areas of popularity, such as social science. We hope that this paper will, first, encourage meta-analysts to use the original data, rather than summarised 'statistics', from the studies they examine; and second, to analyse this data in a Bayesian fashion. Of course, even the Bayesian result is no better than the data; it should not be blamed for the shortcomings warned against in Section 5. We invite sceptics to seek Bayesian solutions to their problems in addition to meta-analytical results; compare the two; and appraise their preferences.

ACKNOWLEDGMENTS. Anthony Garrett acknowledges the support of a Royal Society of Edinburgh Personal Research Fellowship.

REFERENCES

1. Fisher, R.A.: 1932, *Statistical Methods for Research Workers* (fourth edition), Oliver and Boyd, Edinburgh, U.K.

2. Glass, G.V.: 1976, 'Primary, Secondary and Meta-Analysis of Research', *Educational Researcher* 5, 3-8.

3. Glass, G.V., McGaw, B. and Smith, M.L.: 1981, *Meta-Analysis in Social Research*, Sage Publications, Beverly Hills, California, U.S.A.

4. Chalmers, T.C., Berrier, J., Sacks, H.S., Levin, H., Reitman, D. and Nagalingam, R.: 1987, ' Meta-Analysis of Clinical Trials as a Scientific Discipline II — Replicate Variability and Comparison of Studies that Agree and Disagree', *Statistics in Medicine* 6, 733-744.

5. Lambert, M.J., Hatch, D.R., Kingston, M.D., and Edwards, B.C.: 1986, 'Zung, Beck, and Hamilton Rating Scales as Measures of Treatment Outcome — A Meta-Analytic Comparison', *J. Consulting and Clinical Psychology* 54, 54-59.

6. Smith, M.L. and Glass, G.V.: 1977, 'Meta-Analysis of Psychotherapy Outcome Studies', *American Psychologist* 32, 752-760.

7. Whitley, B.E.: 1983, 'Sex Role Orientation and Self-Esteem — A Critical Meta-Analytic Review', *J. Personality and Social Psychology* 44, 765-778.

8. Rosenthal, R.: 1983, 'Assessing the Statistical and Social Importance of the Effects of Psychotherapy', *J. Consulting and Clinical Psychology* 51, 4-13.

9. Garrett, C.J.: 1985, 'Effects of Residential Treatment on Adjudicated Delinquents — A Meta-Analysis', *J. Research in Crime and Delinquency* 22, 287-308.

10. Kulik, J.A.: 1983, 'How Can Chemists Use Educational Technology Effectively', *J. Chemical Education* 60, 957-959.

11. Blosser, P.E.: 1986, 'What Research Says — Meta-Analysis Research on Science Instruction', *School Science and Mathematics* 86, 166-170.

12. Hunter, J.E., Schmidt, F.L. and Jackson, G.B.: 1982, *Meta-Analysis: Cumulating Research Findings Across Studies*, Sage Publications, Beverly Hills, California, U.S.A.

13. Hedges, L.V. and Olkin, I.: 1985, *Statistical Methods for Meta-Analysis*, Academic Press, San Diego, California, U.S.A.

14. Wolf, F.M.: 1986, *Meta-Analysis: Quantitative Methods for Research Synthesis*, Sage Publications, Beverly Hills, California, U.S.A.

15. Rosenthal, R.: 1987, *Judgement Statistics: Design, Analysis and Meta-Analysis*, Cambridge University Press, Cambridge, U.K.

16. Various authors: 1983, *J. Consulting and Clinical Psychology* 51, 1-75. A collection of meta-analytical papers.

17. Baum, M.: 1991, 'Trials of Homeopathy' (letter), *British Medical Journal* 302, 529.

18. Fisher, D.J.: 1990, 'Pig's Ear into Silk Purse?', *The Skeptic* (Manchester, U.K.), 4(5), 13-15.

19. Sackett, D.L.: 1979, 'Bias in Analytic Research', *J. Chronic Diseases* 32, 51-63.

20. Cox, R.T.: 1946, 'Probability, Frequency and Reasonable Expectation', *Am. J. Phys.* 14, 1-13.

21. Shannon, C.E.: 1948, 'A Mathematical Theory of Communication', *Bell Syst. Tech. J.* **27**, 379-423 and 623-659. Reprinted (1949) in: C.E. Shannon and W. Weaver (eds.), *The Mathematical Theory of Communication*, University of Illinois Press, Urbana, Illinois, U.S.A.

22. Jaynes, E.T.: 1983, *E.T. Jaynes: Papers on Probability, Statistics and Statistical Physics*, R.D. Rosenkrantz (ed.), Synthese Series **158**, Reidel, Dordrecht, Netherlands.

23. Garrett, A.J.M.: 1991, 'Macroirreversibility and Microreversibility Reconciled: The Second Law', in B. Buck and V.A. Macaulay (eds.), *Maximum Entropy In Action*, Oxford University Press, Oxford, U.K., pp139-170.

24. Jaynes, E.T.: 1976, 'Confidence Intervals vs Bayesian Intervals', in W.L. Harper and C.A. Hooker (eds.), *Foundations of Probability Theory, Statistical Inference and Statistical Theories of Science*, Reidel, Dordrecht, Netherlands, pp175-257. Largely reprinted as Chapter 9 of Reference [22].

25. Gillies, G.T.: 1990, 'Resource Letter: Measurements of Newtonian Gravitation', *Am. J. Phys.* **58**, 525-534.

26. Garrett, A.J.M.: 1991, 'Ockham's Razor', in W.T. Grandy and L.H. Schick (eds.), *Maximum Entropy and Bayesian Methods, Laramie, Wyoming, 1990*, Kluwer, Dordrecht, Netherlands, pp357-364.

MODELLING DRUG BEHAVIOUR IN THE BODY WITH MAXENT

M.K. Charter
Mullard Radio Astronomy Observatory
Cavendish Laboratory
Madingley Road
Cambridge CB3 0HE, U.K.

ABSTRACT. Description of the passage of drugs into, around and out of the body is conventionally based on simple compartmental models, often with only one or two compartments. These models are clearly gross simplifications of such a complex biological system. Continuous ('free-form') distributions, appropriately used, seem more suitable for summarising the complexities of such systems. The use of distributions for this purpose, and their reconstruction from experimental data, is illustrated. The response (concentration of drug in blood) to a rapid intravenous injection of drug is often taken to vary linearly with the injected dose, which suggests characterising the system by its impulse response function, i.e., the concentration in blood as a function of time after a rapid injection of a unit dose. A distribution closely related to the inverse Laplace transform of this impulse response function forms the starting point for an improved model, but is too empirical to have a physiologically useful interpretation. This model can, however, be re-parameterised in a way which has more physiological meaning, and which immediately suggests an extension to take account of the finite capacity of the enzyme systems which metabolise the drug. The system may become markedly nonlinear when this capacity is approached, and so this type of model has important applications in the design of those dosing regimens where this happens.

1. Introduction

For most drugs there is a range of blood concentrations (the 'therapeutic window') which is associated with the desired therapeutic effect. Below this range, the drug will be ineffective, and above it there may be unwanted or toxic effects. During the drug's development, therefore, a suitable dosing regimen must be designed to maintain the drug concentration as nearly as possible within this range. The design of a dosing regimen is greatly assisted if a model is available which predicts drug concentration in response to a given input or inputs. The purpose of the work described here is to illustrate the development of such a model, using as an example the minor analgesic paracetamol (acetaminophen, APAP).

2. A distribution of decay times

The decline in blood concentration after a rapid intravenous dose is often described by a sum of decaying exponentials, of the form $c(t) = DR(t)$ where D is the dose administered

C. R. Smith et al. (eds.), Maximum Entropy and Bayesian Methods, Seattle, 1991, 287–302.

and

$$R(t) = \sum_{j=1}^{N} A_j e^{\lambda_j t}. \tag{1}$$

The function $R(t)$ is thus the impulse response function, i.e., the concentration of the drug in blood in response to a rapid intravenous administration of a unit dose. Such functions, with suitable choices for N, $\{A_j\}$ and $\{\lambda_j\}$ are used as the impulse response functions for many drugs.

Although the function (1) is often a convenient empirical summary of the 'response' to a rapid intravenous dose, its parameters have no direct physiological interpretation, as noted previously (Charter, 1990). Choosing the number N of exponential terms illustrates this difficulty. Of course, *if* such a model is adopted, then probability theory provides the mechanism to select the most probable number of terms for a given set of data (Skilling, 1991). But this number does not reflect specific anatomical features, and may well vary between similar individuals given the same dose of drug under identical conditions. Rather than using a small number of discrete exponentials, it therefore seems attractive to use a distribution of exponentials, of the form

$$R(t) = \int_{-\infty}^{+\infty} f(\log \lambda) \, e^{-\lambda t} \, d\log \lambda. \tag{2}$$

The distribution $f(\log \lambda)$ is thus closely related to the inverse Laplace transform of the impulse response $R(t)$. This is illustrated by the same data as in the earlier work, obtained when a healthy fasted volunteer was given 1000 mg APAP by a two-minute intravenous infusion. The analysis was performed using the MemSys5 quantified maximum entropy program (Gull and Skilling, 1991). The observed data and predicted concentration are shown in Figure 1, and the corresponding distribution of decay times $f(\log \lambda)$ is shown in Figure 2.

All the usual quantities characterising the drug's kinetic behaviour can be obtained from $f(\log \lambda)$. For example, the clearance CL is the constant of proportionality between the rate at which the drug is eliminated and its measured concentration in blood. For a drug input rate $g(t)$, the blood concentration is $c(t) = R(t) * g(t)$, and the elimination rate is $CL\, c(t)$. If $\tilde{c}(s)$, $\tilde{R}(s)$ and $\tilde{g}(s)$ are the Laplace transforms of $c(t)$, $R(t)$ and $g(t)$ respectively, then by the Convolution Theorem $\tilde{c}(s) = \tilde{R}(s)\, \tilde{g}(s)$ and so

$$CL \int_0^\infty c(t)\, dt = CL\, \tilde{c}(0)$$

$$= CL\, \tilde{R}(0)\, \tilde{g}(0)$$

$$= CL \int_0^\infty R(t)\, dt \int_0^\infty g(t)\, dt.$$

If the total drug input is eventually eliminated from the body, then

$$CL \int_0^\infty R(t)\, dt \int_0^\infty g(t)\, dt = \int_0^\infty g(t)\, dt$$

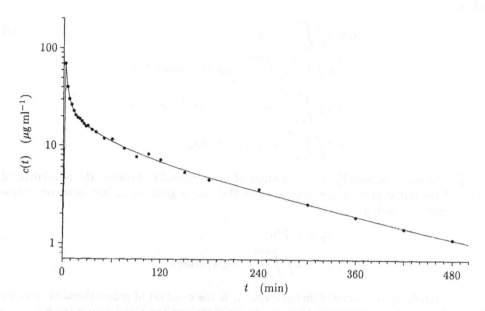

Fig. 1. Concentration of APAP in blood after giving 1000 mg APAP by two-minute infusion.

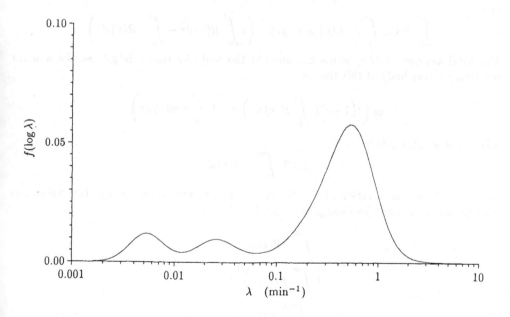

Fig. 2. Distribution $f(\log \lambda)$ of decay times reconstructed from the data shown in Figure 1.

and so

$$\text{CL} = 1 \Big/ \int_0^\infty R(t)\, dt \tag{3}$$

$$= 1 \Big/ \int_0^\infty dt \int_{-\infty}^{+\infty} d(\log \lambda)\, f(\log \lambda)\, e^{-\lambda t}$$

$$= 1 \Big/ \int_{-\infty}^{+\infty} d(\log \lambda)\, f(\log \lambda) \int_0^\infty e^{-\lambda t}\, dt$$

$$= 1 \Big/ \int_{-\infty}^{+\infty} \lambda^{-1} f(\log \lambda)\, d(\log \lambda).$$

The 'plasma volume' V_p is the constant of proportionality between the administered dose and the initial blood concentration when the dose is given as an instantaneous intravenous injection, so that

$$V_p = 1/R(0)$$

$$= 1 \Big/ \int_{-\infty}^{+\infty} f(\log \lambda)\, d(\log \lambda).$$

The steady-state volume of distribution V_{ss} is the constant of proportionality between the total amount of drug in the body at steady state and the blood concentration. At a time t after the start of a constant-rate input g_0 the amount of drug eliminated from the body is

$$\int_0^t d\tau\, \text{CL} \int_0^\tau dv\, R(v)\, g_0 = g_0\, \text{CL} \left(t \int_0^t R(\tau)\, d\tau - \int_0^t \tau\, R(\tau)\, d\tau \right).$$

The total amount of drug which has entered the body by time t is $g_0 t$, so the amount remaining in the body at this time is

$$g_0 \left(t \left(1 - \text{CL} \int_0^t R(\tau)\, d\tau \right) + \text{CL} \int_0^t \tau\, R(\tau)\, d\tau \right)$$

which, using (3), tends to

$$g_0\, \text{CL} \int_0^\infty \tau\, R(\tau)\, d\tau$$

as $t \to \infty$. The concentration at steady state is g_0/CL, and so the steady-state volume of distribution is, as given by (Vaughan, 1982, 1984),

$$V_{ss} = \frac{\displaystyle\int_0^\infty t\, R(t)\, dt}{\left(\displaystyle\int_0^\infty R(t)\, dt \right)^2}$$

$$= \frac{\displaystyle\int_{-\infty}^{+\infty} \lambda^{-2} f(\log \lambda)\, d(\log \lambda)}{\left(\displaystyle\int_{-\infty}^{+\infty} \lambda^{-1} f(\log \lambda)\, d(\log \lambda) \right)^2}.$$

Table 1. Parameters obtained from the data in Figure 1 using the model (2).

Parameter	Units	Estimate	S.D.
CL	$l\,min^{-1}$	0.325	0.009
V_p	l	8.8	1.5
V_{ss}	l	59.6	8.0

These quantities, together with their uncertainties, are given in Table 1.

3. A distribution of peripheral volumes

The use of (2) removes the difficulties associated with interpretation of the number of exponential terms, but the time constants λ themselves, and the distribution $f(\log \lambda)$ still have no obvious physiological interpretation, except as an empirical summary of the decay of drug concentration after a rapid intravenous injection.

To make further progress, it is useful to examine a model which gives an interpretation of the use of functions such as (1), i.e., a sum of a small number of discrete exponentials. Such a model is the two-compartment system shown in Figure 3.

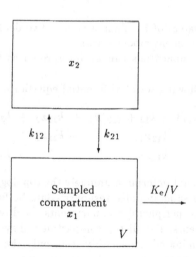

Fig. 3. The two-compartment model.

The differential equations describing this system are as follows:

$$\begin{cases} \dot{x}_1 = -(K_e/V + k_{12})x_1 + k_{21}x_2 \\ \dot{x}_2 = k_{12}x_1 \phantom{+k_{21}} - k_{21}x_2 \end{cases} \tag{4}$$

or

$$\dot{\mathbf{x}} = \mathbf{E}\mathbf{x}$$

where

$$\mathbf{E} = \begin{pmatrix} -(K_e/V + k_{12}) & k_{21} \\ k_{12} & -k_{21} \end{pmatrix}.$$

Each compartment is assumed to be homogeneous and well-stirred, and drug can move between the two compartments with the first-order rate constants shown. In general, $k_{ij} \neq k_{ji}$. The amounts of drug in the two compartments are x_1 and x_2. The measured concentrations of drug in the blood samples are interpreted as observations of the concentration in compartment 1, which is therefore called the 'sampled compartment', and given the attribute of a volume V. This volume is thus equal to the plasma volume V_p. The other compartment is known as the 'peripheral compartment'. Irreversible elimination from the system occurs from the sampled compartment at a rate proportional to the concentration x_1/V in it. The constant of proportionality K_e is therefore equal to the clearance CL. This elimination may take the form of physical removal of drug from the body, for example excretion by the kidneys into the urine, or chemical conversion, for example metabolism by the liver to a different chemical entity. The solution of the system (4) (for the concentration in the sampled compartment after a delta function input of unit dose) is of the form

$$R(t) = x_1(t)/V = \sum_{j=1}^{2} A_j e^{-\lambda_j t}$$

where the $\{-\lambda_j\}$ are the eigenvalues of $\mathbf{E}$. Thus a sum of two decaying exponentials may be interpreted as the two-compartment model shown.

Similarly, a sum of three exponentials can be interpreted as the three-compartment model shown in Figure 4.

This is described by the following set of differential equations:

$$\begin{cases} \dot{x}_1 = -(K_e/V + k_{12} + k_{13})x_1 + k_{21}x_2 + k_{31}x_3 \\ \dot{x}_2 = k_{12}x_1 - k_{21}x_2 \\ \dot{x}_3 = k_{13}x_1 - k_{31}x_3 \end{cases} \tag{5}$$

The response function $R(t)$ does not determine uniquely the topology of these compartmental systems; a model having the same impulse response could be found with, for example, irreversible elimination from the peripheral compartments, or direct transfer of drug between the peripheral compartments. For all the compartmental systems considered in this paper, however, the drug is eliminated irreversibly from only the sampled compartment, and it does not move directly between peripheral compartments.

In the same way that an impulse response function with N exponential terms can be derived from a compartmental model with $N - 1$ peripheral compartments, the distribution $f(\log \lambda)$ can be seen in terms of a model involving a distribution of peripheral compartments. The wavelength or frequency of a spectral line is the property which determines the position of that line in a spectrum, but it is not immediately clear what property of each peripheral compartment should determine the position of that compartment in the proposed distribution. The relationship between the amounts x_1 and x_j (for $j > 1$) in

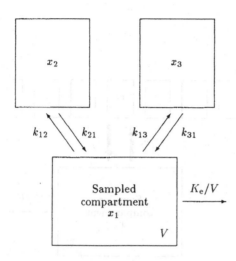

Fig. 4. The three-compartment model.

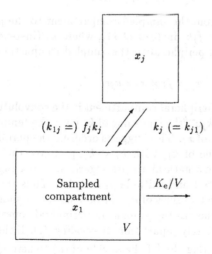

Fig. 5. The relationship between sampled and peripheral compartments.

the sampled and jth peripheral compartments respectively, shown in Figure 5, gives some

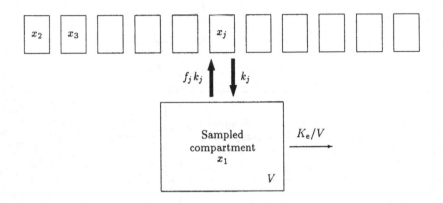

Fig. 6. The linear disposition model.

insight into this:

$$\dot{x}_j = f_j k_j x_1 - k_j x_j. \tag{6}$$

The rate constant for transfer from the sampled compartment to the peripheral compartment is now re-parameterised as $f_j k_j$ instead of k_{1j}, where k_j (instead of k_{j1}) is the rate constant for return from the jth peripheral to the sampled compartment. Then it follows from (6) that

$$x_j = f_j \times x_1 * k_j e^{-k_j t}.$$

Hence the amount x_j in the jth peripheral compartment is the convolution of the amount in the sampled compartment with $k_j e^{-k_j t}$. The value of k_j gives the temporal variation of x_j, while f_j describes the magnitude of x_j. This suggests ordering the peripheral compartments in the distribution by their value of k_j, with $k_j < k_{j+1}$. Since the $\{k_j\}$ are effectively temporal scaling factors, it seems natural to use logarithmic spacing for them, so that $\log k_j = y_0 + (j - 3/2)\Delta y$ for $j > 1$ and suitable y_0 and Δy. Thus the amount of drug in peripheral compartments with small values of k_j will vary slowly compared with the amount in the sampled compartment, whereas the amount in peripheral compartments with large values of k_j will follow x_1 more closely (apart from the scaling f_j). If the amounts at steady state are $x_1^{(ss)}$ and $x_j^{(ss)}$, then setting the left-hand side of (6) to zero gives

$$x_j^{(ss)} = f_j x_1^{(ss)}. \tag{7}$$

This suggests interpreting f_j as the amount of drug in the jth peripheral compartment at steady state, expressed with respect to the amount in the sampled compartment. Put another way, $f(\log k)$ is the distribution of peripheral volume as a function of the compartmental time constant, with the volume expressed as a multiple of the volume of the sampled compartment. This is clearly a positive quantity and is also additive, since the sum of the $\{f_j\}$ over any set of peripheral compartments has a physical interpretation as the total volume of that set of peripheral compartments. The complete disposition model, shown in Figure 6, is thus:

$$
\begin{cases}
\dot{x}_1 = -\sum_{j>1} f_j k_j x_1 + \sum_{j>1} k_j x_j - \dfrac{K_e}{V} x_1 \\
\dot{x}_j = \quad f_j k_j x_1 \quad - \quad k_j x_j
\end{cases}
\tag{8}
$$

Thus the distribution $\mathbf{f}$, or $f(\log k)$ in the continuous case, may be reconstructed using the normal MaxEnt formalism, incorporating expectations of local smoothness in $f(\log k)$ by constructing it by blurring an underlying 'hidden' function $h(x)$ with a suitable intrinsic correlation function (ICF) $C(\log k, x)$

$$
f(\log k) = \int C(\log k, x) \, h(x) \, dx.
$$

The reconstruction of $f(\log k)$ for the same set of data as before is shown in Figures 7 and 8. Since the model in (8) is merely a re-parameterisation of (2), the predicted concentration, and estimates of parameters such as the clearance, the plasma volume and and the steady-state volume of distribution should differ only very slightly, reflecting the different priors (and perhaps the approximations in the estimation procedures). Comparison of Figures 1 and 7 shows that the predicted profiles are very similar. The clearance and plasma volume are estimated directly as K_e and V, and the steady-state volume, using (7), is

$$
V_{ss} = \frac{x_1^{(ss)} + \sum_{j>1} x_j^{(ss)}}{x_1^{(ss)}/V} = V\left(1 + \sum_{j>1} f_j\right).
$$

The parameter estimates from this model, shown in Table 2, are indeed very similar to those from the model (2) in Table 1.

Table 2. Parameters obtained from the data in Figure 7 using the model (8).

Parameter	Units	Estimate	S.D.
CL	$l\,min^{-1}$	0.328	0.007
V_p	l	8.9	1.7
V_{ss}	l	57.0	5.1

4. Saturable elimination kinetics

The set of intravenous data discussed above was in fact taken from an experiment in which six doses of APAP (100, 200, 500 and three replicates of 1000 mg) were given on

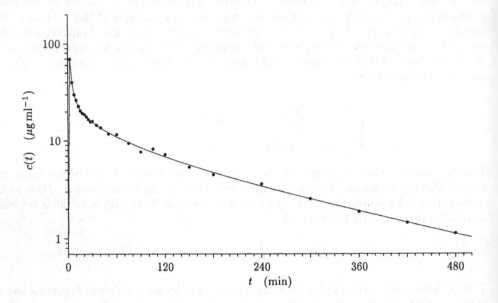

Fig. 7. The data of Figure 1 fitted by the model (8).

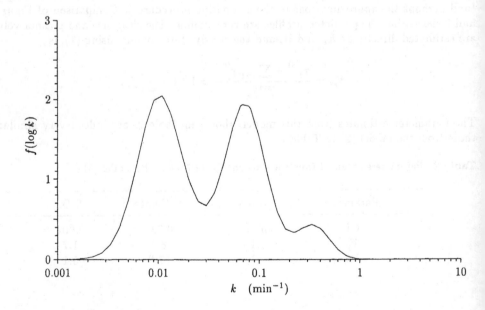

Fig. 8. Distribution $f(\log k)$ of peripheral volumes reconstructed from the data shown in Figure 7.

separate occasions to the same volunteer. All six sets of data are shown in Figure 9, with

Table 3. Parameters obtained from the data in Figure 9 using the linear model (8).

Parameter	Units	Estimate	S.D.
CL	$1\,\text{min}^{-1}$	0.369	0.008
V_{p}	1	14.6	2.1
V_{ss}	1	64.8	4.0

the concentrations predicted using the model (8).

The reconstruction of $f(\log k)$ is shown in Figure 10, and the corresponding parameter values are given in Table 3. Apart from the day-to-day variation which is evident in the differences between the three 1000 mg doses, the model predicts systematically low concentrations for the highest dose, and systematically high concentrations for the lower three. This would be expected if irreversible elimination from the sampled compartment, represented in (2) by the term $K_e x_1/V$, were not in fact directly proportional to the concentration x_1/V in the sampled compartment, but occurred instead by a process with a finite capacity. This is more realistic for a process catalysed by an enzyme, as is most of the elimination of APAP. A simple model based on Michaelis-Menten kinetics can be used to describe this process:

$$\text{Elimination rate} = \frac{K_e}{V}\left(\frac{x_1}{1 + x_1/(c_0 V)}\right)$$

in which one extra parameter, a characteristic concentration c_0, is introduced. When the concentration x_1/V in the sampled compartment is much smaller than c_0, the elimination rate tends to $K_e x_1/V$, i.e., it becomes directly proportional to x_1/V, as in the linear model (2). When x_1/V is much larger than c_0, the elimination rate tends to $K_e c_0$, i.e., it becomes independent of x_1/V, corresponding to the situation where the enzyme is working at full capacity.

The complete disposition model analogous to (8) thus becomes

$$\begin{cases} \dot{x}_1 = -\sum_{j>1} f_j k_j x_1 + \sum_{j>1} f_j k_j x_j - \dfrac{K_e}{V}\left(\dfrac{x_1}{1 + x_1/(c_0 V)}\right) \\ \dot{x}_j = \quad f_j k_j x_1 \quad - \quad k_j x_j \end{cases} \tag{9}$$

The results of using this model are shown in Figure 11. The systematic under-prediction of the concentration at the highest dose and over-prediction of concentration at the lower doses seen in Figure 10 are removed. The reconstruction of $f(\log k)$ is shown in Figure 12, and the derived parameter values are given in Table 4.

The largest difference between the linear and saturable elimination models (8) and (9) is in the estimates of $\text{CL} = K_e$, as might be expected. It is substantially higher for the nonlinear model because it is no longer depressed by the effects of saturation at higher concentrations. The characteristic concentration, c_0, is approximately $18\,\mu\text{g ml}^{-1}$. This can be interpreted as a 'Michaelis constant', i.e., the concentration at which the enzyme system is running at half its maximum rate. This value is higher than the levels usually achieved by oral, therapeutic doses (except perhaps transiently shortly after dosing), indicating that

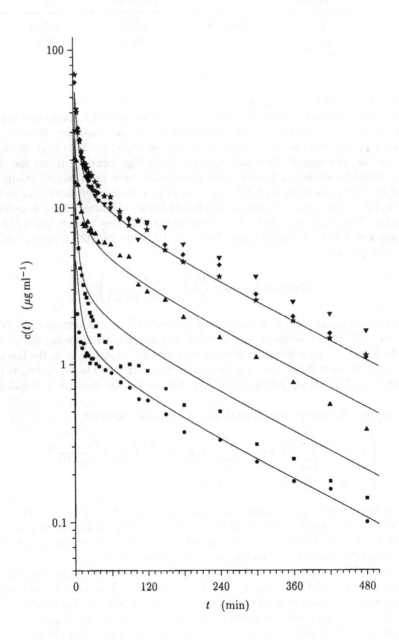

Fig. 9. The six sets of APAP data, with concentrations predicted from the linear model (8).

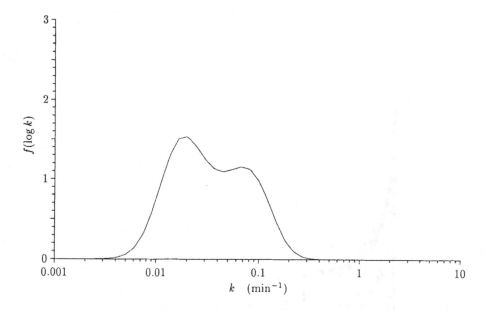

Fig. 10. Distribution $f(\log k)$ of peripheral volumes reconstructed from the data shown in Figure 9.

the nonlinearity is unlikely to be particularly noticeable under these circumstances, and also justifying the use of a linear model for studies of APAP absorption kinetics (Charter and Gull, 1987; Charter, 1991).

The data can be used to help choose quantitatively between the linear model (8), denoted by M_{lin}, and the saturable elimination model (9), denoted by M_{sat}. This can be done independently of any particular parameter values by calculating the ratio of the posterior probabilities of the two models:

$$\frac{\Pr(M_{\text{sat}} \mid \mathbf{D}, H)}{\Pr(M_{\text{lin}} \mid \mathbf{D}, H)} = \frac{\Pr(\mathbf{D} \mid M_{\text{sat}}, H)}{\Pr(\mathbf{D} \mid M_{\text{lin}}, H)} \frac{\Pr(M_{\text{sat}} \mid H)}{\Pr(M_{\text{lin}} \mid H)} \tag{10}$$

where $\mathbf{D}$ are the data and H is the background information. If the prior probabilities $\Pr(M_{\text{sat}} \mid H)$ and $\Pr(M_{\text{lin}} \mid H)$ for the two models are equal, then (10) reduces to the ratio $\Pr(\mathbf{D} \mid M_{\text{sat}}, H)/\Pr(\mathbf{D} \mid M_{\text{lin}}, H)$. The quantity $\Pr(\mathbf{D} \mid M_{\text{sat}}, H)$ is called the 'evidence' (Skilling, 1991) for the model M_{sat}, and is calculated from

$$\Pr(\mathbf{D} \mid M_{\text{sat}}, H) = \int \Pr(\mathbf{D}, \boldsymbol{\theta}_{\text{sat}} \mid M_{\text{sat}}, H) \, d\boldsymbol{\theta}_{\text{sat}}$$

$$= \int \Pr(\mathbf{D} \mid \boldsymbol{\theta}_{\text{sat}}, M_{\text{sat}}, H) \, \Pr(\boldsymbol{\theta}_{\text{sat}} \mid M_{\text{sat}}, H) \, d\boldsymbol{\theta}_{\text{sat}}$$

where $\boldsymbol{\theta}_{\text{sat}}$ is the parameters K_{e}, V, c_0 and $\mathbf{h}$ for the model M_{sat}. The evidence $\Pr(\mathbf{D} \mid M_{\text{lin}}, H)$ is calculated similarly by integrating the joint posterior over the parameters K_{e},

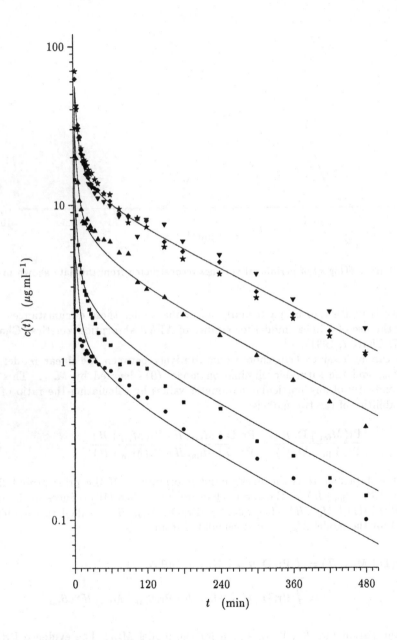

Fig. 11. The same data as in Figure 9, and the predictions from the nonlinear model (9).

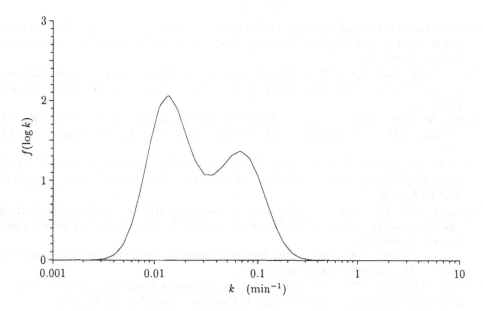

Fig. 12. Distribution $f(\log k)$ of peripheral volumes reconstructed from the data using the nonlinear model (9).

Table 4. Parameters obtained from the data in Figure 11 using the nonlinear model (9).

Parameter	Units	Estimate	S.D.
CL	$l\,min^{-1}$	0.476	0.016
V_p	l	14.1	1.6
V_{ss}	l	75.6	4.1
c_0	$\mu g\,ml^{-1}$	18.0	3.1

V and $\mathbf{h}$ for the model M_{lin}. For these six sets of data the ratio (10) (with equal priors) is approximately e^{40} or 174 dB, indicating that the nonlinear model (9) is a much more plausible description of these data than the linear model (8). The degree of nonlinearity is small, but nonetheless clearly detectable in this analysis.

5. Conclusions

This paper shows the use of MaxEnt to develop a series of models of drug concentration in the blood, using free-form distributions to summarise the complexities of such a system. At each stage the models are fully quantitative, and estimates of parameter values and their uncertainties are provided. The benefit of including an extra parameter in the model to allow for saturable elimination of the drug can also be tested quantitatively, and is shown to be considerable for these data.

REFERENCES

Charter, M.K.: 1990, 'Drug Absorption in Man, and its Measurement by MaxEnt', in *Maximum Entropy and Bayesian Methods, Dartmouth College 1989*, P. Fougère (ed.), Kluwer, Dordrecht.

Charter, M.K.: 1991, 'Quantifying Drug Absorption', in *Maximum Entropy and Bayesian Methods, Laramie, Wyoming, 1990*, W.T. Grandy, Jr., L.H. Schick (eds.), Kluwer, Dordrecht.

Charter, M.K. and Gull, S.F.: 1987, 'Maximum Entropy and its Application to the Calculation of Drug Absorption Rates', *J. Pharmacokinetics and Biopharmaceutics* **15**, 645.

Gull, S.F. and Skilling, J.: 1991, *Quantified Maximum Entropy* MemSys5 *Users' Manual*, Maximum Entropy Data Consultants Ltd., 33 North End, Meldreth, Royston, Herts, SG8 6NR, U.K.

Skilling, J.: 1991, 'On Parameter Estimation and Quantified MaxEnt', in *Maximum Entropy and Bayesian Methods, Laramie, Wyoming, 1990*, W.T. Grandy, Jr., L.H. Schick (eds.), Kluwer, Dordrecht.

Vaughan, D.P.: 1982, 'Theorems on the Apparent Volume of Distribution of a Linear System', *J. Pharmaceutical Sciences* **71**, 793.

Vaughan, D.P.: 1984, 'Theorem on the Apparent Volume of Distribution and the Amount of Drug in the Body at Steady State', *J. Pharmaceutical Sciences* **73**, 273.

INFORMATION ENTROPY AND DOSE-RESPONSE FUNCTIONS FOR RISK ANALYSIS

Jay R. Lund and James D. Englehardt
Department of Civil Engineering
University of California
Davis, California 95616(916) 752-5671, -0586.

ABSTRACT. . Estimating functions to predict dose-response functions below available experimental or epidemiological data is a great cause of controversy and economic impact. Information entropy is used to evaluate different forms of dose-response function for doses below epidemiological and experimental levels. When both the threshold dose and the mean response at some upper-end dose are known, the maximum-entropy dose-response function is found to be a horizontal line at the response related to the upper-end dose. However, when the threshold dose is uncertain as well, linear and various exponential dose-response functions can result.

1. Introduction

The selection of dose-response functions for small doses of potential carcinogens and other toxics is a highly controversial subject in the risk analysis literature. The problem is of great economic importance, since many economic activities involve or potentially involve low exposure levels of people to these undesirable substances. Risk analyses for these substances and economic activities are often very sensitive to the dose-response function assumed. The problem is also difficult to address with conventional statistics, since there is little data for low dosage levels and scant understanding of dose-response mechanisms. Also, experiments to collect data and generate knowledge of toxicity for low doses are both expensive and highly specific to the particular toxin and organism of concern. Thus, the problem is characterized by great importance, little knowledge, and great controversy.

An approach to evaluating the information entropy of proposed dose-response functions is presented here. Information entropy represents a criterion for evaluating different candidate dose-response functions. Dose-response functions for responses and doses below available epidemiologic data are examined.

The functions examined all meet several constraints themselves. 1) Response at a dose of zero is zero. This constrains the lower end point of the dose-response function. 2) Response at a given (non-zero) dose is known, or the probability distribution of the response for this dose is known from epidemiologic studies. Thus, the upper-dose end of the dose-response function is also fixed. 3) Response cannot decrease with increasing dose. (The dose-response function may be only flat or increasing with dose.) These functions are illustrated in Figure 1.

C. R. Smith et al. (eds.), Maximum Entropy and Bayesian Methods, Seattle, 1991, 303–312.

The entropy maximizing dose-response function for deterministic end points is found to be a constant response for all doses, equal to the response found at the upper end point from epidemiologic studies.

A more interesting entropy-maximizing dose-response function results when the response at the upper-end dose is described by a probability distribution, rather than being fixed, and when the threshold dose, below which there is no effect also is described probabilistically.

Some conclusions are made regarding the limitations of this work and possible extensions.

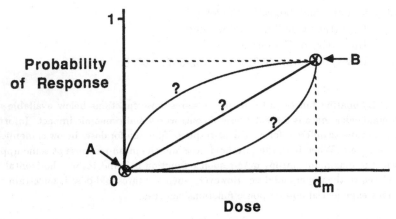

Fig. 1. Realm of possible dose-response functions.

2. Information Entropy of Functions

Like many equations, a dose-response function summarizes a probabilistic relationship between two variables. In this case, the dose-response function describes the response to a particular level of exposure or dose of a potential risk. Common examples would include equations to predict the probability of a person dying of cancer based on the daily exposure of that person to different daily doses of radiation. However, the prediction for a given dose is uncertain, both in an aggregate sense and for particular individuals. A more complete dose-response function would then provide a probability distribution expressing the relative likelihood of different cancer probability for any given radiation dose. This is particularly the case for low-dose, dose-response functions, where there are few or no observations. More typically, and practically, the dose-response function represents the expected value of some response for a given dose. Seeing dose-response functions in this light enables one to describe the entropy of different proposed forms for dose-response functions.

The information entropy of a given probabilistic function is the sum of the entropies of the probability distributions of response at different doses. If the response predicted by a particular dose-respone function is the mean, or expected value, of the response for a given dose, then the maximum entropy distribution of response for a given dose is distributed exponentially (Tribus, 1969). For cases where the response is a probability (of cancer, death, etc.), the range of possible responses to each dose lies between zero and one. The entropy of each dose-specific distribution of response is, up to a sign, (Shannon, 1948)

$$H_i = \sum_{j=1}^{m} p_{ij} \ln p_{ij}, \tag{1}$$

where p_{ij} is the probability of response level j for dosage level i and m is the number of response levels.

Since information entropy is additive (Shannon, 1948), the total information entropy of a dose response function is:

$$H = \sum_{n=1}^{n} H_i = \sum_{i=1}^{n} \sum_{j=1}^{m} p_{ij} \ln(p_{ij}), \tag{2}$$

where n is the number of dose intervals between zero and some maximum dose. This provides a measure of comparing different proposed forms for dose-response functions, particularly in the low-dose range, below dosages examined in epidemiologic or laboratory studies.

The entropy of response probabilities for a given dose, H_i, requires estimation of the response probability distributions. Response probabilities for a given dose are characterized by a mean, given by the dose-response function, and are non-negative. Where responses are probabilities of an outcome, say probabilities of a person exposed to some dose contracting cancer, the responses are also less than or equal to one. For the case where there is no clear upper bound, the maximum entropy distribution that results from the above information is the exponential probability density function (Tribus, 1969):

$$P(p) = ae^{-bd}, \tag{3}$$

where p is the response, d is the dose level, and a and b are parameters. The parameters a and b are functions of the mean of the distribution, given by the proposed dose-response function, $a = b = 1/\mu_p(d)$, where $\mu_p(d)$ is the dose-response function. Tribus finds the information entropy of this distribution to be $H_i = \ln(\mu_p(d_i)) + c$, where c is a constant.

The total entropy over the entire range of dosages $(0, d_m)$ would then be:

$$H = (n + c + \sum_{i=0}^{n} \ln(\mu_p(d_i))) = (n+1)c + \sum_{i=0}^{N} \ln(\mu_p(i\Delta d)), \tag{4}$$

where n is the number of dosage intervals and Δd is the width of the dosage interval. Integrating the entropies over the range of dosages would yield the continuous formulation:

$$H_C = \int_{0}^{\infty} \ln(\mu_p(d))dd, \tag{5}$$

neglecting the constant and treating it as a constant or base potential.

Where responses are in terms of the probability of a particular response, the probability density function for the response should be bounded at the upper end at one. However, for many low-dose risk analysis problems, the mean response probabilities being considered are so low that this bound should be effectively infinite. Therefore, the above probability distributions and entropy measures will be employed directly to evaluate the relative entropy of some commonly proposed dose-response functions.

3. Relative Entropy of Common Dose-Response Functions with Fixed Ends

Several proposed dose-response functions can now be compared in terms of the information entropy inherent in their low-end ranges. Each proposed function has fixed upper and lower end-points. The upper end-point is fixed from epidemiologic data at some dosage level where epidemiologic data exists. Therefore, the mean response at this upper dose, d_m, is assumed to be known, $\mu_p(d_m)$ (point B in Figure 1). The lower end-point is fixed by a threshold or zero response at a dose of d_0, where $d_0 > 0$ (Point A in Figure 1). Five potential forms of dose-response function are examined below. All functions will be examined for threshold response levels of $d_0 = 0$.

Since there is no data within this region and very little other knowledge, the probability distribution of response for each dose within this low range is assumed to follow the exponential form described above. The mean response at each dose is assumed to be the value given by the proposed dose-response function.

Straight-Line

The straight-line form of dose response function interpolates response as a function of dose between the known mean upper and lower responses (the middle curve in Figure 1). If the threshold dose level is d_0 and the dose level of the epidemiological data is d_m with a corresponding mean response $\mu_p(d_m)$, then the linear dose-response function has the form:

$$\mu_p(d) = \frac{\mu_p(d_m)}{d_m - d_0}(d - d_0), \quad \text{for } d_0 \le d \le d,, \tag{6}$$

and

$$\mu_p(d) = 0, \quad \text{for } d \le d_0.$$

The entropy of this dose response function is found by applying Eq. 4 or Eq. 5. Applying Eq. 5, for $d_0 = 0$ gives an entropy for the linear dose-response function of

$$H_s = d_m\left(\ln(\mu_p(d_m))d_m - 1\right). \tag{7}$$

Exponential Convex

The exponential convex function is defined for the same dose range with the same endpoints used for the linear dose-response function defined above. For exponents k, $1 < k$, the following function is exponential convex (Figure 1, lower curve):

$$\mu_p(d) = a(e^{k(d-d_0)} - 1), \tag{8}$$

where a and k are constants set so that $\mu_p(d_m) = a(e^{k(d_m-d_0)} - 1)$ or,

$$a = \mu_p(d_m)d^{k(d_m-d_0)} - 1. \tag{9}$$

Using Eq. 5 to measure the entropy of the distribution between $d_0 = 0 \le d \le d_m$ yields:

$$H_{ex} = \int_0^{d_m} \ln(a(e^{kd} - 1))dd. \tag{10}$$

Note that the entropy of the response for each value of d associated with the convex exponential curve ($k > 1$) is less than the entropy of the response for each value of d along the straight-line function above. Therefore, $H_{ex} < H_s$, overall.

Exponential Concave

The exponential concave has the same function form as the exponential convex, except that the exponent k has values $0 < k < 1$.

$$H_{ec} = \int_0^{d_m} \ln(a(d^{kd} - 1))dd. \tag{11}$$

For the exponential concave dose-response function the entropy of the response for each value of d is greater than the response entropy of the corresponding dose using the linear function. Therefore, $H_{ec} > H_s$, overall.

Maximum Response

The maximum response within the dose-response range $d_0 \leq d \leq d_m$, is $\mu_p(d_m)$. The corresponding dose-response function is $\mu_p(d) = \mu_p(d_m)$, for $d_0 < d \leq d_m$, and $\mu_p(d) = 0$, for $d \leq d_0$. This amounts to a horizontal dose-response function at the level of the mean response observed at the epidemiological dose.

The entropy for this dose response function is again found using Eq. 5,

$$H_M = d_m \ln(\mu_p(d_m)). \tag{12}$$

Minimum Response

The minimum response for any dose is assumed to be zero. Therefore the minimum dose-response function has a response of zero within the range $d_0 \leq d \leq d_m$.

The entropy of this function, using Eq. 5, would be:

$$H_m = -\infty. \tag{13}$$

Maximum-Entropy Dose-Response with Fixed Ends

The ordering of the entropies of these five different dose-response functions is then:

$$H_M > H_{ec} > H_S > H_{ex} > H_m. \tag{14}$$

Of these potential low-dosage dose-response functions, the maximum response function clearly represents the function with the maximum information entropy. This conclusion seems intuitively correct as well: Since the maximum response function has the largest mean response for each dosage level and since mean responses are very low relative to the maximum response of one, the response distribution for each dosage level will have a higher entropy than distributions with lesser means.

Thus, the dose-response function $\mu_p(d) = \mu_p(d_m)$, for $d_0 < d \leq d_m$, and $\mu_p(d) = 0$, for $d \leq d_0$ represents available information most exclusively for situations where only the mean response at the upper end of the dose-response curve, $\mu_p(d_m)$, and the threshold dose d_0 are known with certainty. This conclusion is now applied to several common cases where the ends of the dose-response function are known with less certainty.

4. Maximum Entropy Dose-Response With a Probabilistic Upper End Mean

Quite often the mean of the response associated with the upper-end dose is not known with certainty. The mean response at this dose really represents the expected value of

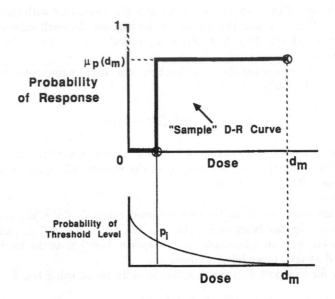

Fig. 2. Possible dose-response functions with probabilistic threshold dose.

the mean response at this dose. When viewed in this way, there is a probability distribution of $\mu_p(d_m)$, each of which implies a different maximum-entropy dose-response function. $P(\mu_p(d_m))$, the probability density function of $\mu_p(d_m)$, can be used to describe the likelihood of each of these possible dose-response curves. In the case where only the mean of the mean response is known, the maximum-entropy distribution for $P(\mu_p(d_m))$ is exponential. If the threshold dose $d_0 = 0$, the mean dose-response curve is then given by:

$$EV[\mu_p(d)] = \int_0^1 \mu_p(d_m)P(\mu_p(d_m))d\mu_p(d_m), \tag{15}$$

for each value of $d > 0$, where $EV[\ ]$ is the expected value operator. The right-hand-side of the above equation is the mean of the mean response at the upper-end dose, which we assume to be known. Therefore,

$$EV[\mu_p(d)] = EV[\mu_p(d_m)]. \tag{16}$$

This conclusion is similar to that found by Howard (1988); there is little reason to introduce frequency concepts regarding the response probabilities, other than the mean response or the mean of the mean response.

5. Maximum Entropy Dose-Response With a Probabilistic Threshold Level

Another common case is where the upper end of the dose-response function is fixed ($\mu_p(d_m)$ is known), but the threshold dose is uncertain. Certainly, we know that at zero dose, there is no response. Also, at the upper end dose, there is a mean response $\mu_p(d_m)$. Yet, this leaves some uncertainty where the threshold dose d_0 lies within this interval, $0 \le d_0 < d_m$ (Figure 2).

If a maximum entropy distribution were to be assigned to d_0 within this range, in the absence of any other information, a uniform distribution results, $P(d_0) = 1/d_m$. From the maximum entropy dose-response function found above, the mean response for a given dose is either zero, if $d \leq d_0$, or $\mu_p(d_m)$, if $d_0 < d \leq d_m$. As illustrated in Figure 2, there is one such function for each possible value of the threshold dose.

The expected value of response for each dose is then the sum of the probability of each response times the probability of each response. The probability that a given dose d has a zero response is the probability that that dose lies at or below the uncertain threshold dose:

$$P(\mu_p(d) = 0) = \int_d^{d_m} P(d_0)dd_0. \tag{17}$$

For a uniform distribution of threshold doses between zero and d_m, this becomes:

$$P(\mu_p(d) = 0) = \frac{d_m - d}{d_m}. \tag{18}$$

The probability that the response has a value of $\mu_p(d_m)$ is the remaining probability,

$$P(\mu_p(d) = \mu_p(d_m)) = 1 - P(\mu_p(d) = 0). \tag{19}$$

The mean response for each dose is then:

$$EV[\mu_p(d)] = 0 * P(\mu_p(d) = 0) + (1 - P(\mu_p(d) = 0)) * \mu_p(d_m) \tag{20}$$

or

$$EV[\mu_p(d)] = \left(1 - \int_d^{d_m} P(d_0)dd_0\right)\mu_p(d_m). \tag{21}$$

With the linear probability density function for d_0 (Eq. 18), this simplifies to:

$$EV[\mu_p(d)] = \left(1 - \frac{d_m - d}{d_m}\right)\mu_p(d_m). \tag{22}$$

This result, which assumes a uniform distribution of threshold doses between zero and d_m, is a linear dose response function.

Assuming different knowledge to describe the distribution of threshold doses results in different mean dose-response functions.

6. Maximum Entropy Dose-Response With Both Ends Probabilistic

The situation where both the threshold dose and the mean of the mean response at the upper-end dose are uncertain is depicted in Figure 3. Following an approach analogous to that used to find the mean response for each dose in the previous case results in:

$$EV[\mu_p(d)] = \int_0^{d_m} \int_0^1 P[\mu_p(d)|d_0, \mu_p(d_m)]P(d_0)P(\mu_p(d_m))d\mu_p(d_m)dd_0. \tag{23}$$

This equation can be decomposed into two cases, the expected value of the response if the dose lies at or below the threshold dose, $\mu_p(d) = 0$, weighted by the probability of this event $P(d \leq d_0)$ and the expected value of the response if the dose lies above the threshold dose multiplied by the remaining probability, $1 - P(d \leq d_0)$. This becomes:

$$EV[\mu_p(d)] = 0*\int_d^{d_m} P(d_0)dd_0 + \left(1 - \int_d^{d_m} P(d_0)dd_0\right)\int_0^1 \mu_p(d_m)P(\mu_p(d_m))d\mu_p(d_m). \quad (24)$$

The first term becomes zero and the second part of the third term becomes $EV[\mu_p(d_m)]$, which is known, so,

$$EV[\mu_p(d)] = \left(1 - \int_d^{d_m} P(d_0)dd_) \right) EV[\mu_p(d_m)] \quad (25)$$

Again, the effect of uncertainty in the mean of the mean response at the upper-end dose is unimportant, so long as the expected value of the mean response is known. Also again, a uniform distribution of d_0 between zero and d_m results in a linear dose-response function within this range.

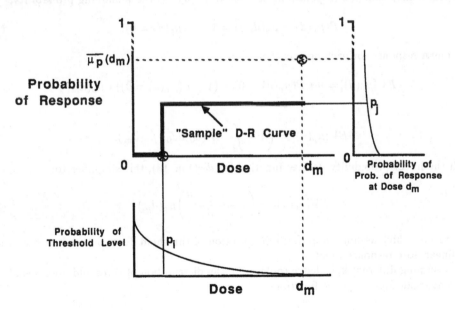

Fig. 3. Possible dose-response functions with probabilistic end points.

7. Introducing Knowledge Regarding the Mean Threshold Value

There may be occasion where the developers of a dose-response curve feel they know more than just the mean response at some epidemiologic dose and feel there is some greater knowledge about the threshold dose than is implied by a uniform distribution. Having an estimate of the mean threshold dose changes the resulting entropy-maximizing dose-response function.

If the mean threshold dose, μ_{d_0}, is known, the entropy-maximizing probability distribution for the threshold dose $P(d_0)$ is no longer necessarily uniform, but becomes more generally exponential, $P(d = d_0) = ae^{bd}$ (Tribus, 1969). This exponential distribution will decrease from $d = 0$ (i.e., $b < 0$) if $\mu_{d_0} < d_m/2$ and will increase until d_m (i.e., $b > 0$) if $\mu_{d_0} > d_m/2$. When the cumulative of this result is entered into either Eq. 25 or Eq. 21, the resulting dose-response curve also becomes exponential,

$$\int_d^{d_m} P(d_0)dd_0 = \frac{a}{b}(e^{bd_m} - e^{bd}). \tag{26}$$

The three resulting cases are depicted in Figure 4. For the dose-response Curve A, the mean threshold dose $\mu_{d_0} < d_m/2$, creating an increasing exponential dose-response curve. Dose-response Curve B in the figure results from the absence of information on the mean threshold dose or having the mean threshold dose $\mu_{d_0} = d_m/2$. Dose-response curves of the general shape of Curve C result when the mean threshold dose $\mu_{d_0} < d_m/2$. The closer the mean threshold dose is to zero or d_m, the steeper the dose-response function.

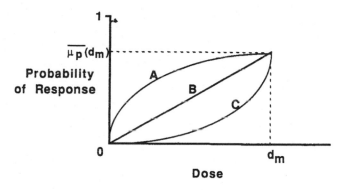

Fig. 4. Dose-response functions with threshold information.

8. Conclusions

This paper demonstrates the use of information entropy maximization to select and derive dose-response functions for doses below prevailing epidemiological or experimental data. Very little knowledge was required to estimate these functions. At the very least, only the mean of the mean response at some upper-end dose is needed. In this case, a linear dose-response function results. Additional knowledge, in the form of estimates of the mean threshold dose can also affect the shape of the low-dose dose-response function, resulting in various exponential forms of dose-response (Figure 4).

Perfect knowledge of the mean response at an upper-end dose and perfect knowledge of the exact threshold dose result in a horizontal maximum-entropy dose-response function at the level of the response to the mean upper-end dose. Some forms of knowledge do not affect the maximum-entropy dose-response function. Since the dose-response function is assumed to represent the mean response for any given dose, the probability distribution of the response at the upper-end dose is irrelevant, as is the distribution of the mean of the mean response at this dose.

REFERENCES

Howard, Ronald A. (1988), 'Uncertainty About Probability: A Decision Analysis Perspective,' *Risk Analysis*, **8**, No. 1, March, pp. 91-98.

Shannon, Claude E. (1948), 'A Mathematical Theory of Communications,' *Bell System Technical Journal*, **27**, pp. 379-423 and pp. 623-656.

Tribus, Myron (1969), *Rational Descriptions, Decisions, and Designs*, Pergamon Press, N.Y., 478 pp.

MAKING BINARY DECISIONS BASED ON THE POSTERIOR PROBABILITY DISTRIBUTION ASSOCIATED WITH TOMOGRAPHIC RECONSTRUCTIONS

Kenneth M. Hanson
Los Alamos National Laboratory, MS P940
Los Alamos, New Mexico 87545 USA
kmh@lanl.gov

ABSTRACT. An optimal solution to the problem of making binary decisions about a local region of a reconstruction is provided by the Bayesian method. Decisions are made on the basis of the ratio of the posterior probabilities for the two alternative hypotheses. The full Bayesian procedure requires an integration of each posterior probability over all possible values of the image outside the local region being analyzed. In the present work, this full treatment is approximated by using the maximum value of the posterior probability obtained when the exterior region is varied with the interior fixed at each hypothesized functional form. A Monte Carlo procedure is employed to evaluate the benefit of this technique in a noisy four-view tomographic reconstruction situation for a detection task in which the signal is assumed to be exactly within a local region.

1. Introduction

When interpreting reconstructed images, it is often desired to make a decision about a small region of interest without regard to the rest of the image. A standard approach to this problem might be to reconstruct the full image from the available data and then make the decision on the basis of how closely the reconstruction resembled the suspected object in the region of interest. Such an approach is not guaranteed to yield an optimal decision.

We desire a computational method that achieves the optimal performance of a binary decision task. Such an 'ideal observer' has been useful in the past to help define the ultimate precision with which one can interpret data of a given type (Hanson, 1980, 1983; Wagner et al., 1989; Burgess et al., 1984a, 1984b; Burgess, 1985). A fully Bayesian approach is proposed in which the decision is based on the posterior probability. Much of this work is based on the Bayesian concepts developed by Gull and Skilling and their colleagues (Gull, 1989a, 1989b; Gull and Skilling, 1989; Skilling, 1989), albeit under the assumption of a Gaussian distribution for the prior probability rather than their preferred entropic form.

Examples of this Bayesian decision procedure are presented for a computed tomographic situation in which a nonnegativity constraint on the image is incorporated. The performance of the comprehensive Bayesian procedure is compared to that of the traditional two-step approach using a Monte Carlo simulation of the entire imaging process, including the decision process (Hanson, 1988a, 1990a). The present work is an extension of the previous study by Hanson (1991).

C. R. Smith et al. (eds.), Maximum Entropy and Bayesian Methods, Seattle, 1991, 313–326.

2. The Bayesian Approach

We briefly present the concepts of the Bayesian approach relevant to our problem. Much more complete discussions of the fundamentals can be found in other contributions to the proceedings of this workshop series. The essence of the Bayesian approach lies in the steadfast use of the posterior probability, which is assumed to summarize the full state of knowledge concerning a given situation.

The posterior probability properly combines the likelihood, which is based on recently acquired measurements, with the prior probablity, which subsumes all information existing before the new data are acquired.

Making a binary decision is the simplest possible type of hypothesis testing, because there are just two alternative models between which to choose. According to basic probability theory, binary decisions should be made on the basis of the ratio of the probabilities for each of the hypotheses (Van Trees, 1968; Whalen, 1971). In the context of Bayesian analysis, then, a binary decision should be based on the ratio of posterior probabilities. This decision strategy is altered when there are asymmetric cost functions, indicating a difference in the relative value of making correct versus incorrect decisions for each state of truth.

When a continuum of possible outcomes exists, as in the estimation of one (or many) continuous parameters, the best possible choice of parameter values depends upon the type of cost function that is appropriate. It may be argued that for general analyses, the most appropriate rule is to find the parameters that maximize the posterior probability, which is called the maximum *a posteriori* (MAP) solution (Van Trees, 1968).

In many problems there exist parameters that may be necessary to fully describe the solution, but whose values are of no interest. These unnecessary parameters can transform a simple hypothesis test into one of testing composite hypotheses. In such cases the proper approach is to integrate the probability density distribution over these unwanted variables. The result of this integration is called the marginal probability. irrelevant parameters

2.1 POSTERIOR PROBABILITY

We assume that there exists a scene that can be adequately represented by an orderly array of N pixels.

We are given M discrete measurements that are linearly related to the amplitudes of the original image. These measurements are assumed to be degraded by additive noise with a known covariance matrix $\mathbf{R_n}$, which describes the correlations between noise fluctuations. The measurements, represented by a vector of length M, can be written as

$$\mathbf{g} = \mathbf{Hf} + \mathbf{n}, \tag{1}$$

where $\mathbf{f}$ is the original image vector of length N, $\mathbf{n}$ is the random noise vector, and $\mathbf{H}$ is the measurement matrix. In computed tomography the jth row of $\mathbf{H}$ describes the weight of the contribution of image pixels to the jth projection measurement.

Because the probability is a function of continuous parameters, namely the N pixel values of the image and the M data values, it is actually a probability density, designated by a small $p()$.

By Bayes' theorem the negative logarithm of the posterior probability is given by

$$-\log\left[p(\mathbf{f}|\mathbf{g})\right] \propto \phi(\mathbf{f},\mathbf{g}) = -\log\left[p(\mathbf{g}|\mathbf{f})\right] - \log\left[p(\mathbf{f})\right], \tag{2}$$

where the first term is the probability of the observed data for any particular image $\mathbf{f}$, called the likelihood, and the second term is the prior probability of the image $\mathbf{f}$. For additive Gaussian noise, the negative log(likelihood) is just half of chi-squared

$$-\log[p(\mathbf{g}|\mathbf{f})] = \frac{1}{2}\chi^2 = \frac{1}{2}(\mathbf{g} - \mathbf{Hf})^{\mathrm{T}}\mathbf{R}_{\mathbf{n}}^{-1}(\mathbf{g} - \mathbf{Hf}) , \tag{3}$$

which is quadratic in the residuals. Instead of a Gaussian distribution assumed here, the Poisson distribution is often a better model for expected measurement fluctuations. The choice should be based on the statistical characteristics of the measurement noise, which we assume are known *a priori*. To simplify matters, we will make the standard assumption that the measurement noise is stationary and uncorrelated, so $\mathbf{R}_{\mathbf{n}} = \mathrm{diag}(\sigma_{\mathbf{n}}^2)$, where $\sigma_{\mathbf{n}}$ is the rms deviation of the noise.

The prior-probability distribution should incorporate as much as possible the known statistical characteristics of the original image. original image so we envision an ensemble of images from which the original image $\mathbf{f}$ is assumed to belong. The prior-probability distribution describes the properties of that ensemble. We use a Gaussian distribution for the prior, whose negative logarithm may be written as

$$-\log[p(\mathbf{f})] = \frac{1}{2}(\mathbf{f} - \bar{\mathbf{f}})^{\mathrm{T}}\mathbf{R}_{\mathbf{f}}^{-1}(\mathbf{f} - \bar{\mathbf{f}}) , \tag{4}$$

where $\bar{\mathbf{f}}$ is the mean and $\mathbf{R}_{\mathbf{f}}$ is the covariance matrix of the prior-probability distribution. As we have done before (Hanson and Myers, 1990c), we invoke the prior knowledge that the image $\mathbf{f}$ cannot possess any negative components and institute nonnegativity as a separate constraint.

The Bayesian approach does not specify any particular choice of prior. Another choice for prior, ubiquitous at this workshop, is that of entropy. The entropic prior has been argued by Skilling (1989) to play a unique role for additive positive distributions. Whatever prior is used, its strength affects the amount the reconstruction is offset from the true image (Hanson, 1990b; Myers and Hanson, 1990). It is important to understand the characteristics of solutions obtained regardless of the prior chosen. It is recognized that the prior provides the regularization essential to solving ill-posed problems (Nashed, 1981; Titterington, 1985), which arise because $\mathbf{H}$ possesses a null-space (Hanson and Wecksung, 1983; Hanson, 1987).

With the posterior probability specified, we have the means for deciding between two possible images $\mathbf{f}_1$ and $\mathbf{f}_2$, given set of measurements $\mathbf{g}$. The decision should be based on the ratio of the posterior probabilities, or equivalently, the difference of their logarithms

$$\psi_{21} = \phi(\mathbf{f}_2,\mathbf{g}) - \phi(\mathbf{f}_1,\mathbf{g}) = \frac{1}{2\sigma_{\mathbf{n}}^2}\left[2\mathbf{g}^{\mathrm{T}}(\mathbf{g}_1 - \mathbf{g}_2) + |\mathbf{g}_2|^2 - |\mathbf{g}_1|^2\right] + \text{constant} , \tag{5}$$

where $\mathbf{g}_k = \mathbf{Hf}_k$ and only the data-dependent terms are explicitly written out. The only part of this expression that depends on the data is the inner product between data vector $\mathbf{g}$ and the difference of the measurements predicted by $\mathbf{f}_1$ and $\mathbf{f}_2$, that is, $(\mathbf{g}_1 - \mathbf{g}_2)$. We note that this inner product represents the familiar cross correlation between the data and the difference between the alternative signals, which is called the matched filter (Van Trees, 1968; Whalen, 1971). The constant in Eq. (5) depends solely on $\mathbf{f}$, $\mathbf{f}_1$, and $\mathbf{f}_2$. It provides an offset to ψ_{21} indicating a prior preference for one of the two choices.

As ψ_{21} is linearly dependent on the data, it too has a Gaussian-shaped probability distribution for any particular image. When the image is one or the other of the two hypothesized images, two distinct probability distributions of ψ_{21} will result. A measure of the degree of distinguishability between the two hypotheses is the difference of their mean values divided by their rms width, which, from (3) and (5) is

$$d'_{SKE} = \frac{|g_1 - g_2|}{\sigma_n} \, , \tag{6}$$

where the subscript SKE indicates the signal is known exactly in the data. See (Van Trees, 1968) for details. This situation provides the best possible discrimination since the full image is specified. Any lack of knowledge about $\mathbf{f}$ can only introduce more uncertainty into the interpretation. Note that as $\sigma_n \to 0$, $d'_{SKE} \to \infty$ implying perfect discrimination. This derivation obviously ignores the potential ambiguities caused by artifacts in reconstructions that can occur because of limited data and only includes uncertainties in the measurements.

2.2 RECONSTRUCTION PROBLEM

In the reconstruction problem, we seek to estimate all pixel values in the original scene. An appropriate Bayesian solution to this problem is the image that maximizes the posterior probability or, equivalently, minimizes the negative logarithm of the posterior probability. For the **unconstrained** MAP solution $\hat{\mathbf{f}}$, it is necessary that

$$\nabla_{\mathbf{f}}\phi = \mathbf{R}_{\mathbf{f}}^{-1}(\mathbf{f} - \bar{\mathbf{f}}) + \mathbf{H}^T\mathbf{R}_n^{-1}(\mathbf{g} - \mathbf{H}\mathbf{f}) = 0 \, . \tag{7}$$

However, under the constraint that the solution should be nonnegative, the derivative with respect to f_i must be zero only when $f_i > 0$; a negative derivative is permissible on the boundary $f_i = 0$. In computed tomography (CT), the matrix operation $\mathbf{H}^T$
is the familiar backprojection process.

A consequence of the prior is to pull the reconstruction away from the actual value in the original image, an effect studied by Hanson (1990b) in unconstrained tomographic reconstructions. The extent of this biasing effect depends on the relative weights of the two terms in Eq. (2). As the prior contribution vanishes, the MAP result approaches the maximum likelihood (or least-square residual) solution.

2.3 ANALYSIS OF A LOCAL REGION

Instead of asking for an estimate of the original image, suppose that we ask a different question: which of two possible objects exists at a specific location in the image? The rest of the image suddenly becomes irrelevant. To address this question, we assume that within the image domain, a local region $\mathcal{D}$ is to be analyzed. Inside $\mathcal{D}$ the image $\mathbf{f}$ is assumed to be given by either $\mathbf{f}_{\mathcal{D}1}$, and $\mathbf{f}_{\mathcal{D}2}$. Now the parameters in the problem are not the full set of image values $\mathbf{f}$, but rather $\mathbf{f}_{\mathcal{E}}$, the image values in the disjoint exterior region $\mathcal{E}$ and the two possible choices for $\mathcal{D}$. With Bayes' law the posterior probability may be written as $p(\mathbf{f}_{\mathcal{E}}, \mathbf{f}_{\mathcal{D}k}|\mathbf{g}) \propto p(\mathbf{g}|\mathbf{f}_{\mathcal{E}}, \mathbf{f}_{\mathcal{D}k})p(\mathbf{f}_{\mathcal{E}}, \mathbf{f}_{\mathcal{D}k}) \propto p(\mathbf{g}|\mathbf{f}_{\mathcal{E}}, \mathbf{f}_{\mathcal{D}k})p(\mathbf{f})$. In the last step we have chosen to avoid explicit specification of a prior on $\mathbf{f}_{\mathcal{D}k}$, allowing it to be implicitly included in the general prior for $\mathbf{f}$.

As the new question regards only the region $\mathcal{D}$, the image values $\mathbf{f}_{\mathcal{E}}$ outside $\mathcal{D}$ are irrelevant. Probability theory specifies that we integrate the posterior probabilities over the unwanted variables in the problem, namely over the image values outside $\mathcal{D}$. If the problem at hand is to decide between two possible subimages, f_{D1} or f_{D2}, the decision variable should be the ratio of the two marginal posterior probabilities (Van Trees, 1968; Whalen, 1971), or equivalently its logarithm

$$\psi = \log \left[\frac{\int_{\mathcal{E}} p(f_{\mathcal{E}}, f_{D1}|g) df}{\int_{\mathcal{E}} p(f_{\mathcal{E}}, f_{D2}|g) df} \right] , \tag{8}$$

where the integrals are to be carried out only over the external region $\mathcal{E}$ and include all possible image values not disallowed by constraints. Within the context of Bayesian analysis, this decision variable logically follows from the statement of the problem. Hence, we assert that it should yield optimal decisions. The ideal observer uses Eq. (8) to make binary decisions regarding a local region.

Under certain circumstances these integrals may be difficult to estimate accurately. However, when dealing with the Gaussian prior- and likelihood-probability density distributions presented in Sec. 2.1, we expect the posterior-probability density $p(f_{\mathcal{E}}, f_{Dk}|g)$ to decrease rapidly from a unique maximum. Using $\hat{f}_{\mathcal{E}k}$ to designate the image in the exterior region that maximizes the posterior probability for the subimage f_{Dk}, we are prompted to rewrite the above ratio as,

$$\psi = \log \left[\frac{p(\hat{f}_{\mathcal{E}1}, f_{D1}|g) K_1}{p(\hat{f}_{\mathcal{E}2}, f_{D2}|g) K_2} \right] , \tag{9}$$

where the phase-space factor is

$$K_k = \frac{1}{p(\hat{f}_{\mathcal{E}k}, f_{Dk}|g)} \int_{\mathcal{E}} p(f_{\mathcal{E}}, f_{Dk}|g) df , \tag{10}$$

which accounts for the extent of the spread in f-space of the posterior-probability density distribution about its constrained peak value $p(\hat{f}_{\mathcal{E}k}, f_{Dk}|g)$.

Generally $\hat{f}_{\mathcal{E}1} \neq \hat{f}_{\mathcal{E}2}$, because a change in the model f_D alters the projections, implying that a different exterior image will minimize the posterior probability. In many situations, however, replacing the local region of the MAP solution with either f_{D2} or f_{D2} may have little effect on the predicted projection values. Then, $p(f_{\mathcal{E}}, f_{Dk}|g)$ is independent of f_{Dk} and, to good approximation, $\hat{f}_{\mathcal{E}1} = \hat{f}_{\mathcal{E}2} = \hat{f}_{\mathcal{E}}$, so both K factors in Eq. (8) are the same and

$$\psi = \log \left[\frac{p(\hat{f}_{\mathcal{E}}, f_{D1}|g)}{p(\hat{f}_{\mathcal{E}}, f_{D2}|g)} \right] . \tag{11}$$

In these situations, the decision variable can be given adequately by the change in the log(posterior probability) induced by replacing the MAP solution $\hat{f}$ in D with the two models, leaving the exterior region unchanged.

For unconstrained solutions of Eq. (7), the K factor is independent of f_{Dk}, because the shape of the Gaussian posterior-probability distribution is governed by the full curvature of ϕ, namely $R_n^{-1} + R_f^{-1}$. Then the K factors in Eq. (9) cancel and

$$\psi = \log \left[\frac{p(\hat{f}_{\mathcal{E}1}, f_{D1}|g)}{p(\hat{f}_{\mathcal{E}2}, f_{D2}|g)} \right] . \tag{12}$$

The argument of the logarithm is called the generalized posterior-probability ratio (Van Trees, 1968). Equation (12) may not be a good approximation to (8) for constrained solutions, as the contribution to the phase-space K factor from the integral over each f_i depends on the relation of the peak in $\hat{f}_{\mathcal{E}i}$ to the constraint boundary. Nonetheless, because of its simplicity, we use Eq. (12) and reserve for the future an investigation of a better approximation.

To evaluate Eq. (12) for subimages $f_{\mathcal{D}1}$ and $f_{\mathcal{D}2}$, it is necessary to find the pair of exterior images, $f_{\mathcal{E}1}$ and $f_{\mathcal{E}2}$, that maximize the posterior-probability density. In other words, one must find the maximum *a posteriori* or MAP reconstruction in the exterior region with the image inside the local region fixed by the parameter values. To extend the binary decision problem to one in which the local region is described by a model whose parameters are to be estimated, it becomes necessary to simultaneously estimate the parameters and reconstruct the exterior region with the aim of minimizing the posterior probability (Hanson, 1991).

We employ the iterative method described by Butler, Reeds, and Dawson (1981) to find the constrained MAP solutions. See (Hanson and Myers, 1991; Hanson, 1991) for more details.

3. Methodology

We demonstrate the use of the Bayesian approach to making decisions
about a local region in a reconstructed image with a very simple example: detection of disks based on a very limited number of noisy projections. This binary discrimination task is employed because it is theoretically tractable, it is easy to perform the required decision-making procedure, and it is possible to summarize the results simply.

3.1 Monte Carlo Method to Evaluate Task Performance

The overall method for evaluating a reconstruction algorithm used here has been described before (Hanson, 1988a, 1990a). In this method a task performance index for a specified imaging situation is numerically evaluated. The technique is based on a Monte Carlo simulation of the entire imaging process including random scene generation, data taking, reconstruction, and performance of the specified task. The accuracy of the task performance is determined by comparison of the results with the known original scene using an appropriate figure of merit. Repetition of this process for many randomly generated scenes provides a statistically significant estimate of the performance index (Hanson, 1990a).

3.2 Specifications of Detection Tests

The imaging situation is chosen in an attempt to maximize the possible effect of re-estimation of the exterior region implied by the full Bayesian treatment. The original scenes contain either one or two disks, all with amplitude 0.1 and diameter 8 pixels. The disks are randomly placed, but not overlapping, within the circle of reconstruction of diameter 64 pixels. The background level is zero. Enough scenes are generated in the testing sequence to provide 100 disks with amplitude 0.1 and 100 null disks placed in the background region.

The measurements consist of four parallel projections, each containing 64 samples, taken at $45°$ increments in view angle. Measurement noise is simulated by adding to each measurement a pseudorandom number taken from a Gaussian distribution with a standard deviation of 2. The peak projection value of each disk is 0.80. From Eq. (6), it is possible the calculate the signal-known-exactly dectectability, $d'_{\text{SKE}} = 1.89$. This result defines the upper limit to the dectectability index that should be achievable for the stated measurement situation. in To reduce aliasing artifacts in the reconstruction, the projection data used for reconstruction are presmoothed using a triangular convolution kernel with a FWHM of 3 sample spacings. As a result, the expected rms noise value in the smoothed data is reduced very nearly to 1.0. Thus for all cases studied we use the noise covariance matrix $R_n = \mathrm{diag}(\sigma_n^2) = (1.0)^2$. With this assumption we are ignoring the correlations in the data caused by presmoothing.

For the Gaussian prior probability distribution we employ the ensemble mean $\bar{f}_i = 0.0031 = $ constant, which is the average value of the scenes containing two disks. We assume the ensemble covariance matrix is diagonal with $R_f = \mathrm{diag}(\sigma_f^2)$ and explore the effect of choosing different values of σ_f.

The stated task is to detect the presence of the disks under the assumption that the signal and background are known exactly (SKE) only in a 2D local region. As the rest of the image is unspecified, the the measurements are not SKE, as assumed in the derivation of Eq. (6). The various strategies for making this binary decision are presented in the next section. A useful measure to summarize the performance of binary decisions is the detection index d_A, which is based on the area under the Receiver Operating Characteristic (ROC) curve. The ROC curve is obtained in the usual way (Hanson, 1990a) from the histograms in the decision variable for the signal-known-present and the signal-known-absent tests. Once the ROC curve is generated and its area A determined, then d_A is found using $d_A = 2\ erfc^{-1}\{2(1 - A)\}$, where $erfc^{-1}$ is the inverse complement of the error function. There are good reasons for not using the detectability index d', which is based on the first and second moments of the histograms of the decision variable (Wagner et al., 1990). For a fixed number of binary tests, the relative statistical error in d_A is smallest when d_A is about 2.2 (Hanson, 1990a). The imaging situation should be arranged to keep d_A roughly between 1 and 3.5 to optimize the statistical value of the testing procedure.

3.3 DECISION STRATEGIES

For the simple binary discrimination tests performed here, only two parameters are needed to describe the model for the local region – the background level and the disk amplitude relative to the background. The background is assumed to be constant. The position and diameter of the disk are assumed to be known. The edge of the disk is linearly ramped over 2 pixels in radius to roughly match the blur caused by the reconstruction process. The local region of analysis is assumed to be circular with a diameter of 14 pixels and centered on the test position. When the disk is assumed present, the amplitude is set to 0.1 and when assumed absent, 0. The background level is 0 for both tests. Because of this choice for the model, it should be understood that all references to the amplitude of the disk implicitly mean relative to the surrounding background, that is, the disk contrast.

In all the decision strategies except method E, a decision variable

is evaluated for each of the two hypotheses, and the difference between the two values is used to make the decision whether a disk is present or not.

The following decision strategies are employed in this study:

Method A) In the simplest possible approach, one uses the projection data directly. The decision is based on the difference in χ^2 for the two hypotheses. Explicitly, Eq. (5) is evaluated under both hypothesized subimages for the local region of analysis $\mathcal{D}$. The image values outside the analysis region are implicitly assumed to be zero. If the background

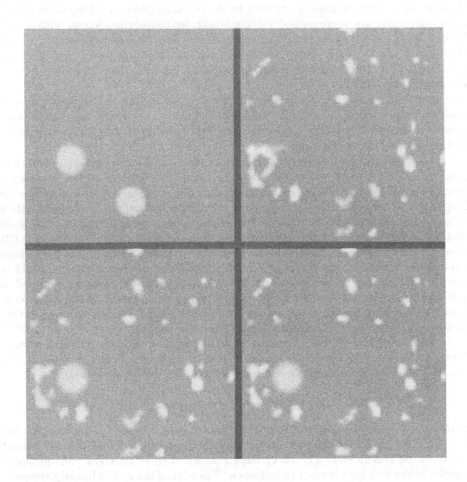

Fig. 1. This composite image shows the process used to make the binary decision regarding the presence of a disk. The original scene (upper left) is reconstructed from four projections using constrained maximum *a posteriori* reconstruction (upper right) with ensemble standard deviation $\sigma_f = 1$. To test the possible presence of a disk, that disk is placed into the reconstruction (lower left). Then the image outside the local region of the disk is 're-reconstructed' to obtain the image (lower right) that maximizes the posterior probability with the disk present. This procedure is repeated with the same region replaced by the background value (zero). The difference in the logarithms of the two resulting posterior probabilities is used as the decision variable.

is truly zero and only one disk is present in the scene, this decision variable operates at the statistical limit attainable in the absence of prior information as defined by Eq. (6). However, it is obviously deficient for complex scenes as it ignores the contributions to the projections arising from features outside the local region.

Method B) By Bayesian reckoning, the best possible decision variable for local analysis is given by Eq. (8). For this method we use the approximation given by the generalized posterior-probability ratio Eq. (12), which implies that for each choice of image for D, the exterior region is reconstructed to maximize $p(f_{\mathcal{E}}, f_{Dk}|g)$. In actual practice, this second reconstruction step follows a preliminary constrained MAP reconstruction of the whole image as pictorially described in Fig. 1.

Method C) This method uses Eq. (11) for the decision variable based on the posterior-probability distribution associated with the MAP reconstruction. Readjustment of the reconstruction external to the analysis region for each test hypothesis is not required. This method was introduced by Gull and Skilling (1989) and studied by Myers and Hanson (1990) for an entropy prior.

Method D) Method D proceeds from the constrained MAP reconstruction $\hat{f}_{MAP}$ from the data. The decision variable is taken as the difference in $|f - \hat{f}_{MAP}|^2$ for the two models hypothesized for the local region. This method was used by Hanson and Myers (1991a) to compare performance of the Rayleigh task using MAP reconstructions based on Gaussian and entropy priors. It corresponds to using a likelihood approach using the reconstruction as the data assuming that the noise fluctuations in the reconstruction are uncorrelated and Gaussian distributed. This method therefore ignores the correlations in the posterior-probability distribution, shown in Fig. 2, that are incorporated to various degrees by methods B and C.

Method E) Method E also proceeds from the constrained MAP reconstruction $\hat{f}_{MAP}$. Unlike the preceding methods, the amplitude and background are varied to find the combination of values that minimizes $|f - \hat{f}_{MAP}|^2$. In this fitting process, both the relative amplitude and the background are constrained to be nonnegative. The amplitude so determined is used as the decision variable. This method was used by Hanson in many earlier studies (1988a, 1988b, 1990a, 1990b, 1990c). It is closely related to the non-prewhitening matched filter, which would be optimal if the fluctuations in the reconstruction were uncorrelated and Gaussian distributed.

4. Example

A constrained MAP reconstruction of the first scene of the testing sequence for two disks is shown in Fig. 1. Because of the noise in the projection data, the presence of the disks in the original scene is obscured in the reconstruction. An interesting aspect of the posterior-probability approach is that one may calculate the probability of a disk being present at any location in the reconstruction. Even though the reconstruction

might be zero (the lower limit decreed by the constraint of nonnegativity) throughout a certain region, the probability of a disk being present in that region is finite and calculable. By contrast, any analysis method based solely on the reconstruction

would not be able to distinguish two different regions that are completely zero. This point is emphasized by the contour plot in Fig. 2, posterior-probability distribution when the values of two nearby pixels are varied. The MAP solution for one of the pixels is zero, although both pixels actually fall within a disk in the original scene and should have the value 0.1. The plot shows how the prior probability shifts the posterior away from the likelihood.

The test sequences generated to demonstrate the use of posterior probability in decision making are analyzed for several different values of the ensemble covariance matrix σ_f. We

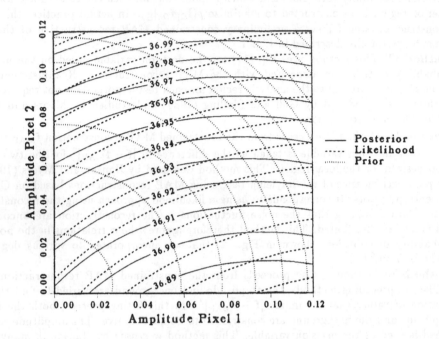

Fig. 2. Contour plot showing the correlation in −log(posterior probability) (solid line) for fluctuations in two pixel values about the MAP solution for an assumed value of the ensemble
standard deviation $\sigma_f = 1.0$. The first pixel is centered on the lower middle disk in the first
scene (Fig. 1) and the other is three pixels down and three pixels to the left of the first. The
dotted contours are for the likelihood and the dashed contours for the prior.

have found before (Hanson, 1989b, 1990b; Hanson and Myers, 1991a; Myers and Hanson,
1990) that the performance of vision-like tasks usually varies with the parameters that
control the rms residual achieved by the reconstruction algorithm. For the present MAP
algorithm, that parameter is the ratio σ_f/σ_n. Recall that σ_n is fixed at its expected value
of 1.0. The strength of the prior is proportional to $1/\sigma_f^2$. As the prior becomes stronger,
the rms residuals of the constrained MAP reconstructions increase. The disk amplitudes,
measured as the average value over each disk relative to the average over its surrounding
annulus (essentially method E), are steadily reduced. These amplitudes never come close
to the actual value of 0.10, probably because there are so few views, giving rise to a gigantic
null space (Hanson, 1987), together with so much noise. When a Gaussian prior with $\bar{f}_i = 0$
is employed, which is nearly the case here, the MAP algorithm amounts to using minimum-
norm regularization. Therefore, control of the noise, which dominates the reconstructed
field, can only be achieved by reducing the sensitivity of the reconstruction (Hanson, 1990b).

Table 1 summarizes the detectability results obtained in the tests described above for

Method	Decision Variable	d_A			
		$\sigma_f = 0.02$	$\sigma_f = 0.1$	$\sigma_f = 0.2$	$\sigma_f = 1$
A	$\Delta\chi^2$ (use data only)	1.75	same	same	same
B	$\Delta\log(\text{posterior probability})$ (exterior re-estimated)	1.80	1.87	1.82	1.74
C	$\Delta\log(\text{posterior probability})$ (exterior fixed at $\hat{f}_{MAP}$)	1.81	1.87	1.81	1.70
D	$\Delta|f - \hat{f}_{MAP}|^2$ (use reconstruction only)	1.80	1.76	1.67	1.47
E	Disk amplitude (constrained fit to $|f - \hat{f}_{MAP}|^2$)	1.01	1.09	1.01	0.96
	RMS residual	0.914	0.823	0.774	0.725
	$<\text{amplitude}>_{\text{disk}}$	0.0005	0.0068	0.0134	0.0247
	$<\text{amplitude}>_{\text{bkg}}$	0.0002	0.0014	0.0024	0.0039

Table 1. Summary of the performance of the detection task for scenes containing two disks each obtained using the decision methods described in the text.

two disks per scene. The absolute statistical accuracy of these d_A values is about 0.25. Much better accuracy should prevail in comparisons between entries in the table, however, because they are obtained by analyzing the exact same data sequence. The d_A value for the two-disk scenes based on using just the measurement data (Method A) is 1.75, in good agreement with the value of 1.89 estimated in Sec. 3.2. As only the likelihood is involved, this value is independent of σ_f. tests is sets. Both methods of using the posterior probability (methods B and C) provide nearly the same detectability over a large range of σ_f values. Perhaps this consistent behavior stems from the ability of the posterior probability to fully retain the available information even though σ_f changes. There seems to be little advantage to re-estimation of the exterior of the local region to minimize the posterior probability implied by Eq. (12) in this imaging situation. There is a slight trend toward better detectability as σ_f gets smaller. The force of regularization imposed by the prior is overwhelming at $\sigma_f = .02$. For example, the reconstruction values lie between 0.0005 and 0.0046; the nonnegativity constraint is not even engaged. We observe very similar trends for the single disk scenes as well.

Method D, which is based only on the reconstruction, yields performance comparable to the methods based on the posterior probability for small σ_f values, but its performance drops off as σ_f increases. Basing the decision on the estimated disk amplitude (method E) significantly ($\approx 45\%$) reduces detectability compared to the other methods.

For **unconstrained** MAP with $\sigma_f = 0.1$, the d_A values are nearly the same as those in Table 1, so the nonnegativity constraint has little effect on detectability in the present situation. In previous work involving a limited number of views, we have seen remarkable improvements in detectability wrought by the nonnegativity constraint (Hanson, 1988a, 1988b, 1990c). Although the less efficient method E was used in those studies, the principal reason for the ineffectiveness of nonnegativity in the present case is that it is more limited by noise than by the limited nature of the data. The large amount of noise is needed to limit d_A within the range of reasonable accuracy as discussed in Sec. 3.2. The effects of artifacts were enhanced in previous studies by adding several disks with large amplitude to the scene. In the present study there is only the presence of a second disk with the same amplitude outside a given analysis region. This extra disk can hardly give rise to significant artifacts.

5. Discussion

We have compared several methods for detecting small disks in tomographic reconstructions. The worst performance is provided by method E in which the amplitude obtained by fitting the MAP reconstruction is used as the decision variable. This choice is the same as the matched filter for uncorrelated, Gaussian-distributed, noise fluctuations, so it is probably more appropriate for unconstrained reconstructions than for constrained reconstructions. A better decision variable is the mean-square difference between the model and the reconstruction $|\mathbf{f} - \hat{\mathbf{f}}_{MAP}|^2$ (Method D), which is equivalent to using a log(likelihood ratio) if the image is taken to be the input measurements and correlations in the reconstruction fluctuations are ignored. This method provides much better results, especially for small σ_f values. However, the performance of method D varies the most any of the methods over the range of σ_f tried dropping significantly as σ_f approaches the maximum likelihood limit. The best detectabilities are achieved by basing decisions on the calculated posterior probability, which takes fully into account the information contained in the measurements as well as in the prior knowledge. In the present tests, however, there is little benefit in re-estimating the exterior region.

We note that as an image containing N pixels, the MAP solution (or a reconstruction of any type) corresponds to a singe point in an N-dimensional space. Any analysis based solely on such a reconstruction must necessarily ignore the complexity of the full posterior-probability distribution, which corresponds to a cloud in the same N-dimensional space. It is the correlations embodied in the posterior-probability distribution that presumably set the ideal observer apart from mortals. A human observer viewing a reconstruction is, in a sense, handicapped by not having access to the full posterior probability distribution and thus may be limited to the use of a decision method similar to D or E.

The full Bayesian treatment codified by Eq. (8) is expected to represent the ideal observer.

ACKNOWLEDGMENTS. I am indebted to Stephen F. Gull and John Skilling for their inciteful work, especially their MEMSYS 3 Manual (1989), which substantially motivated this work. I have had many stimulating discussions and worthwhile suggestions from Kyle J. Myers and Robert F. Wagner. This work was supported by the United States Department of Energy under contract number W-7405-ENG-36.

REFERENCES

Burgess, A. E. and Ghandeharian, H.: 1984a, 'Visual signal detection. I. Ability to use phase information', *J. Opt. Soc. Amer.* **A1**, 900–905.

Burgess, A. E. and Ghandeharian, H.: 1984a, 'Visual signal detection. II. Signal-location identification', *J. Opt. Soc. Amer.* **A1**, 906–910.

Burgess, A. E.: 1985, 'Visual signal detection. III. On Bayesian use of prior information and cross correlation', *J. Opt. Soc. Amer.* **A2**, 1498–1507.

Butler, J. P., Reeds, J. A., and Dawson, S. V.: 1981, 'Estimating solutions for first kind integral equations with nonnegative constraints and optimal smoothing', *SIAM J. Numer. Anal.* **18**, 381–397.

Gull, S. F.: 1989, 'Developments in maximum entropy data analysis', in *Maximum Entropy Bayesian Methods*, J. Skilling (ed.), Kluwer, Dordrecht, 53–71.

Gull, S. F.: 1989, 'Bayesian inductive inference and maximum entropy', in *Maximum Entropy and Bayesian Methods in Science and Engineering (Vol. 1)*, G. J. Erickson and C. R. Smith (ed.), Kluwer, Dordrecht, 53–74.

Gull, S. F. and Skilling, J.: 1989, *Quantified Maximum Entropy - MEMSYS 3 Users' Manual*, Maximum Entropy Data Consultants Ltd., Royston, England.

Hanson, K. M.: 1980, 'On the optimality of the filtered backprojection algorithm', *J. Comput. Assist. Tomogr.* **4**, 361–363.

Hanson, K. M.: 1983, 'Variations in task and the ideal observer', *Proc. SPIE* **419**, 60–67.

Hanson, K. M.: 1987, 'Bayesian and related methods in image reconstruction from incomplete data', in *Image Recovery: Theory and Application*, H. Stark (ed.), Academic, Orlando, 79–125.

Hanson, K. M.: 1988a, 'Method to evaluate image-recovery algorithms based on task performance', *Proc. SPIE* **914**, 336–343.

Hanson, K. M.: 1988b, 'POPART – Performance OPtimized Algebraic Reconstruction Technique', *Proc. SPIE* **1001**, 318–325.

Hanson, K. M.: 1990a, 'Method to evaluate image-recovery algorithms based on task performance', *J. Opt. Soc.* **7A**, 1294–1304.

Hanson, K. M.: 1990b, 'Object detection and amplitude estimation based on maximum a posteriori reconstructions', *Proc. SPIE* **1231**, 164–175.

Hanson, K. M.: 1990c, 'Optimization of the constrained algebraic reconstruction technique for a variety of visual tasks', in *Proc. Information Processing in Medical Imaging*, D. A. Ortendahl and J. Llacer (ed.), Wiley-Liss, New York, 45–57.

Hanson, K. M.: 1991, 'Simultaneous object estimation and image reconstruction in a Bayesian setting', *Proc. SPIE* **1452**, 180–191.

Hanson, K. M. and Myers, K. J.: 1991, 'Rayleigh task performance as a method to evaluate image reconstruction algorithms', in *Maximum Entropy and Bayesian Methods*, W. T. Grandy and L. H. Schick (ed.), Kluwer Academic, Dordrecht, 303–312.

Hanson, K. M. and Wecksung, G. W.: 1983, 'Bayesian approach to limited-angle reconstruction in computed tomography', *J. Opt. Soc. Amer.* **73**, 1501–1509.

Myers, K. J. and Hanson, K. M.: 1990, 'Comparison of the algebraic reconstruction technique with the maximum entropy reconstruction technique for a variety of detection tasks', *Proc. SPIE* **1231**, 176–187.

Myers, K. J. and Hanson, K. M.: 1991, 'Task performance based on the posterior probability of maximum entropy reconstructions obtained with MEMSYS 3', *Proc. SPIE* **1443**, 172–182.

Nashed, M. Z.: 1981, 'Operator-theoretic and computational approaches to ill-posed problems with applications to antenna theory ', *IEEE Trans. Antennas Propagat.* **AP-29**, 220–231.

Skilling, J.: 1989, 'Classic maximum entropy', in *Maximum Entropy Bayesian Methods*, J. Skilling (ed.),Kluwer, Dordrecht, 45–52.

Titterington, D. M.: 1985, 'General structure of regularization procedures in image reconstruction', *Astron. Astrophys.* **144**, 381–387.

Van Trees, H. L.: 1968, *Detection, Estimation, and Modulation Theory - Part I*, John Wiley and Sons, New York.

Wagner, R. F., Myers, K. J., Brown, D. G., Tapiovaara, M. J., and Burgess, A. E.: 1989, 'Higher order tasks: human vs. machine performance', *Proc. SPIE* **1090**, 183–194.

Wagner, R. F., Myers, K. J., Brown, D. G., Tapiovaara, M. J., and Burgess, A. E.: 1990, 'Maximum a posteriori detection and figures of merit for detection under uncertainty', *Proc. SPIE* **1231**, 195–204.

Whalen, A. D.: 1971, *Detection of Signals in Noise,* Academic, New York.

THE APPLICATION OF MAXENT TO ELECTROSPRAY MASS SPECTROMETRY

A.G. Ferrige and M.J. Seddon
The Wellcome Research Laboratories
Langley Court
Beckenham
Kent BR3 3BS, UK

S. Jarvis
VG Biotech
Tudor Road
BroadHeath
Altringham WA14 5RZ, UK

J. Skilling
Department of Applied Mathematics and Theoretical Physics
University of Cambridge
Cambridge CB3 9EW, UK

J. Welch
Mullard Radio Astronomy Observatory
Cavendish Laboratory
Madingley Road
Cambridge CB3 0HE, UK

ABSTRACT. Electrospray Mass Spectrometry is a relatively new ionisation technique which has the advantage of allowing compounds of large molecular weight to be studied. However, it has the serious disadvantage that the spectra are complicated and difficult to analyse. The application of MaxEnt, specifically the Cambridge MEMSYS5 software, has allowed electrospray spectra to be automatically analysed to produce a single, zero charge, spectrum which also exhibits dramatic enhancements in signal to noise ratio and resolution, making interpretation much more straight-forward.

1. Introduction

THE ELECTROSPRAY TECHNIQUE

Only ionised species will pass down the flight tube of a mass spectrometer and arrive at the detector. Ionisation is usually accomplished through electron impact which, at the same time, breaks bonds and fragments the molecule. Softer ionisation techniques are less destructive, but generally less sensitive. With all the conventional ionisation methods, the

327

C. R. Smith et al. (eds.), Maximum Entropy and Bayesian Methods, Seattle, 1991, 327–335.
© 1992 Kluwer Academic Publishers.

molecule or fragment almost invariably carries only a single charge, commonly the proton, so observed masses are that of species present plus one proton. The upper mass limit that can be studied even with powerful and costly magnetic sector instruments is of order 10 kDa and sensitivity, for a number of reasons, also seriously diminishes with increasing mass. Therefore, large macromolecules of biological interest are excluded from investigation.

The electrospray technique overcomes most of there problems since multiple ionisation (protonation) occurs, and large molecules will collect a whole range of charges from just a few, up to, in the case of very large molecules, in excess of 100. The process is relatively efficient and since the observed mass axis is actually mass divided by charge, then multiply charged macromolecules are observed as a whole sequence of peaks (one for each charge value) at apparently relatively low mass. Typically, peaks are observed in the range 500–2500 Da. This means that conventional quadrupole instruments may be used as the mass analyser, their main advantage being that they are much less susceptible to contamination than magnetic sector instruments, Their main disadvantage is that their resolution is relatively low compared with magnetic sector instruments, but this is essentially overcome through the gains achieved by applying the MaxEnt technique.

CONVENTIONAL ANALYSIS OF ELECTROSPRAY DATA

What is required is the full mass of any compound present, for a single compound, in a pure state, the conventional method of analysis is to solve pairs of simultaneous equations fort the number of charges and the mass associated with each peak; the masses being combined by the knowledge that two adjacent peaks are probably separated by one charge. Having assigned the charge number to each peak in the series, the full mass may be obtained from each peak using the equation:

$$M = z(m - p) \tag{1}$$

Where: M = full mass = average molecular weight
z = number of charges carried by the peak in question
m = observed mass/charge ratio of peak
p = mass of the species providing the charge (normally the proton)

However, this rather simplistic state rarely occurs in practice, the technique often being used to study impure biological specimens. Consequently, many components of different molecular weight are normally present, each giving rise to its own sequence of peaks in the spectrum. Additionally, the signal to noise ratio (S/N) may be poor, so that noise spikes and genuine signals become confused. The task of analysing an impure macromolecular mixture of, say, ten major components each of molecular weight around 50 kDa is tedious and time consuming. Indeed, it frequently defeats the spectroscopist, mainly through his having to much rather than too little data to deal with. In such a situation, there could easily be up to 1000 peaks in the observation window with many peaks from one series overlapping with peaks from the distinct series of another compound.

There are obvious situations where analysis would be relatively straightforward but for inadequate resolution. In these circumstances, simple one-dimensional MaxEnt deconvolution has been shown to be a valuable aid to interpretation (Ferrige *et al*, 1991). However, for more complicated spectra, a different approach is necessary, A typical electrospray spectrum of a supposedly reasonably pure sample containing only two components, is shown as figure 1. This figure illustrates the spectral complexity that can be encountered.

The Maxent approach

In other applications of MaxEnt such as image reconstruction and spectroscopic decon-volution, the visible results suffice to predict the data. One blurs a trial image or spectrum with the appropriate point spread function (PSF) and adds noise appropriate to that esti-mated from the data, to produce simulated data. Then MaxEnt is used to discover which images or spectra are the most probable in the light of the data. With electrospray mass spectrometry, the mass spectrum alone does not suffice to predict the data. Each species can and does enter the data several times, according to its charge state. One needs the two-dimensional distribution of mass and charge in order to predict the data. MaxEnt is then used to discover which of these 'hidden' distributions (Gull, 1989) of M vs. z are the most probable.

Moreover there is structure in the hidden distributions due to charge correlation; if mass M appears with charge z, then it should also appear at $z + 1$ and $z - 1$, thence $z + 2$, $z - 2$ etc. These correlations form part of the 'intrinsic correlation function' which takes an underlying hidden distribution of M and z (originally drawn from an entropic prior distribution), and convolves it in z to produce the 'visible' distribution of M and z which might enter the spectrometer flight tube to produce the observed data. It is the intrinsic correlation that allows the procedure to recognise series of peaks in the data as representing the same parent mass.

Although information on the charge states is present in the two-dimensional results, it is simplest to project this out in order to present the zero charge (molecular) mass spectrum alone. After all, this is what the spectroscopist wants to see.

Test Data

In order to test the ability of the MEMSYS5 software to automatically construct a highly resolved zero-charge mass spectrum, data were acquired of known mixtures. To this end, mixtures of normal and mutant globins were used since these would generate at least three distinct series of peaks., By using a conventional quadrupole spectrometer and globins of similar molecular weight, the ability of the algorithm to separate the individual components was tested. Candidate mixtures were normal globins plus either β–Montreal Chori or β–Hafnia. The known masses are shown below.

Component	Mass/Da
α–globin	15126.4
β–globin	15867.2
β–Montreal Chori	15879.3
β–Hafnia	15858.2

Table 1. True Mass of Sample Constituents.

2. Experimental

Electrospray mass spectra of the globin mixtures were obtained from a VG BIO-Q spec-trometer equipped with a VG electrospray interface. The instrument was tuned to give good resolution and lineshape, and data were acquired over a mass range of 980 to 1320 Da. The data were transferred to a SUN Sparcstation 2 for MaxEnt processing using the MEMSYS5 algorithm.

Noise is primarily chemical rather than electronic, arising from random ion arrivals at the detector. Consequently the noise is always positive, which might bias the results. To eliminate this, a smooth baseline was set through the average chemical noise, prior to MaxEnt processing.

The width of the PSF was taken as the average width of known single peaks in the data – those attributed to the α–globin component. The observed width includes the isotope broadening which is always present for large molecules. Since both the peak profiles and the spectrometer lineshape functions are closely Gaussian in shape, a cubic B-spline provides an adequate approximation to the lineshape and has the additional benefit of greater computational efficiency.

The only other requirement is to set out the output mass window for the results. Initially the output range can be set deliberately wide in order to locate the peaks. Also, for computational speed the resolution of the output was set low, around 20 Da per point. For the final run, the window is adjusted to cover just the region of interest and the resolution increased to 2 Da per point.

Among the various diagnostics the programme outputs the is Evidence which informs the operator of how plausible his assumptions about the PSF were, in the light of the data. The PSF can therefore be tuned to provide the greatest Evidence and hence a more probable result.

As a final exercise, the complex and noisy data of a chromatographic fraction from mutase shown in figure 1 were processed using this technique.

3. Results and Discussion

Figure 2 shows the starting data for the normal globins/β–Montreal chori mixture. The inset is a ×20 horizontal expansion of the lowest charge β peaks. Figure 3 is the zero charge mass spectrum and the inset is a ×10 horizontal expansion of the B peaks.

Figures 4 and 5 show the corresponding data and results for the normal globins/β–Hafnia mixture.

The fully quantified MaxEnt results are shown in table 2. Masses have been calibrated to the true mass of the α–globin peak.

Peak	True Mass/Da	MaxEnt Mass/Da	Error/Da
Normal + Montreal-Chori			
α–globin	15126.4	15126.4±0.2	0.0
Normal β	15867.2	15867.6±0.3	−0.4
Montreal–Chori β	15879.3	15880.4±0.4	−0.7
Normal + β-Hafnia			
α–globin	15126.4	15126.4±0.2	0.0
Normal β	15867.2	15867.0±0.3	+0.2
Hafnia β	15858.2	15858.2±0.4	0.0

Table 2. Calibrated MaxEnt Masses.

As expected from experience of other applications, the MaxEnt spectra show dramatic gains in resolution and S/N. Also there is no evidence of artefacts because the weak peaks may be accounted for as arising from alkali meta adducts and small quantities of other components.

In addition to the obvious benefits of fully quantified results, the main advantage of MaxEnt is its ability to provide a reconstruction of enhanced resolution. When large molecules are studied, the resolution observed in the data is not just limited by the instrument resolution. The isotope broadening can be very significant and, depending on the molecular formula, peaks for a compound of molecular weight 50 kDa can be up to 50 Da wide at half-height, At an observed mass/charge ratio of 1000 the peaks will be around 1 Da wide from the isotopic effect alone. The instrument linewidth is typically about the same, so peaks would be approximately 2 Da wide which may seriously limit the spectroscopist's ability to interpret the data.

In the case of the globins, the peaks would be expected to be around 10 Da wide at the full mass. However, the peaks of the MaxEnt spectra are only around 1.9 Da wide at half-height, and the locations can be identified to almost another order of magnitude in accuracy. This improvement is resolving power allows the components of similar molecular weight to be readily resolved and quantified.

The MaxEnt spectrum of the data shown in figure 1 is shown in figure 6 and the complexity is clearly seen to arise from the two major components with masses around 80 kDa and 69 kDa.

This exciting new development is data processing open the door to detailed studies of macromolecular compounds of biological significance and will undoubtedly become a routinely adopted analytical tool.

4. Conclusions

In the work described here the Maxent technique has been shown to be capable of:
- Automatically assigning the appropriate charge to each peak of the various series of peaks present and generating a zero-charge mass spectrum.
- Providing substantial gains in both resolution and S/N with respect to the starting data.

This novel application of MaxEnt therefore has the potential to become the method of choice for disentangling electrospray data and to become the standard by which other methods are judged.

REFERENCES

Ferrige, A.G., Jarvis, S. and Seddon, M.J.: 1991, 'Maximum Entropy Deconvolution in Electrospray Mass Spectroscopy', *Rapid Communications in Mass Spectroscopy* 5, 374—379.

Gull, S.F.: 1989, 'Developments in Maximum Entropy Data Analysis' in J. Skilling (ed.), *Maximum Entropy and Bayesian Methods*, Kluwer, Dordrecht.

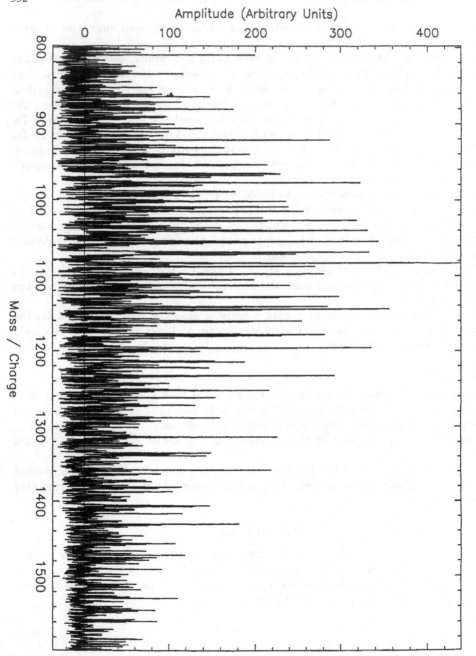

Fig. 1. Electrospray mass spectrum of chromatographic fraction from Mutase.

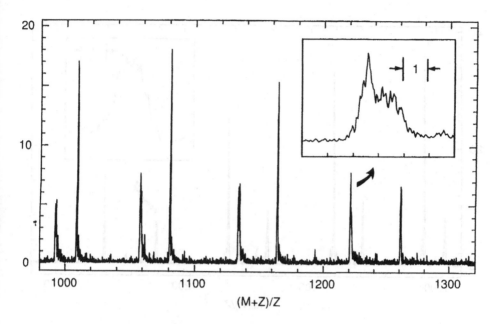

Fig. 2. Electrospray mass spectrum of a mixture of normal and Montreal–Chori globins.

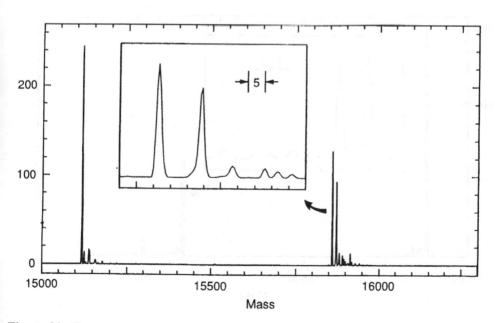

Fig. 3. MaxEnt reconstruction of normal and Montreal–Chori globin mixture.

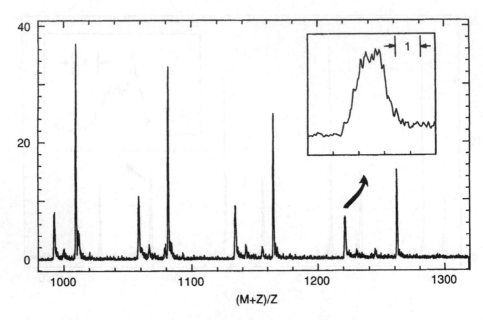

Fig. 4. Electrospray mass spectrum of a mixture of normal and β–Hafnia globins.

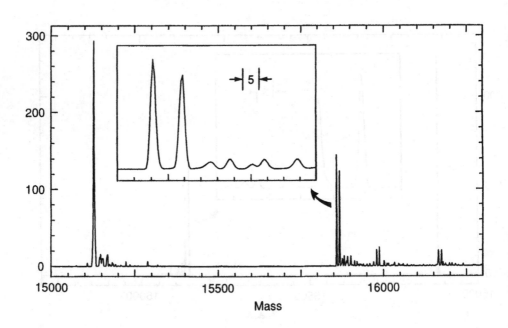

Fig. 5. MaxEnt reconstruction of normal and β–Hafnia globin mixture.

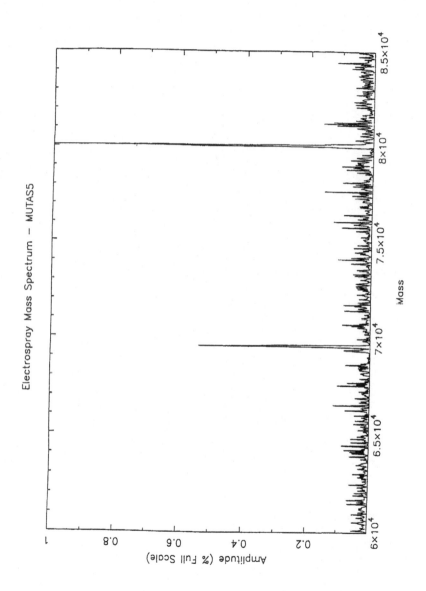

Fig. 6. MaxEnt reconstruction of mass spectrum of chromatographic fraction from Mutase.

THE APPLICATION OF MAXENT TO ELECTRON MICROSCOPY

A.G. Ferrige and M.J. Seddon
The Wellcome Research Laboratories
Langley Court
Beckenham
Kent BR3 3BS, UK

S. Pennycook and M.F. Chisholm
Oak Ridge National Laboratory
P.O. Box 2008
Oak Ridge
Tennessee 37831-6030, USA

D.R.T. Robinson
Mullard Radio Astronomy Observatory
Cavendish Laboratory
Madingley Road
Cambridge CB3 0HE, UK

ABSTRACT. MaxEnt has been used to great effect, for optical deblurring, medical image enhancement and minimising noise and improving the resolution in radio astronomy. More recently we have successfully applied the Cambridge 'MemSys5' MaxEnt algorithm to extending the resolution of high resolution electron microscopes by up to an order of magnitude.

1. Introduction

The work described here is concentrated on data acquired from ultra high resolution scanning transmission electron microscopes capable of providing images which can be described as a convolution between an object function and a suitable point spread function (PSF), close to the intensity profile of the incident probe (Pennycook and Jesson, 1990). The PSF can adequately be described by a Gaussian profile for images of this type. Although this is far from ideal, the signal to noise ratio (S/N) of the measurement is sufficiently poor that the difference between the true blurring function and the assumed Gaussian is well within the noise. Since the data are in photographic form the true image background is unknown, this information being lost between the instrumental measurement and the photographic process.

The parameters that the operator must establish are therefore limited to the PSF radius and the image background. Of course, where the S/N is particularly low it is necessary to introduce blurring in the MaxEnt result through the 'Intrinsic Correlation Function' or 'ICF' (Gull, 1989). Without this, the noise in the MaxEnt result would be unacceptably high, signals of interest being confused with noise and *vice versa*.

337

C. R. Smith et al. (eds.), Maximum Entropy and Bayesian Methods, Seattle, 1991, 337–344.
© 1992 *Kluwer Academic Publishers.*

Experience has shown that the Evidence diagnostic (Skilling, 1991) output by the program is only a reliable guide to parameter optimisation when the noise level is accurately known and can be set. The internal noise calculation facility provided within the MaxEnt interface has unfortunately been shown to have some dependence on the applied PSF radius. This presents a minor difficulty in that the calculated noise level is only approximate. However, a reasonable starting point for the PSF radius is readily obtained by direct examination of the data and the noise may initially be fixed at the level corresponding to this radius. The optimum value for the image background may then be determined from the Evidence.

Optimising the PSF radius is subjective in that this is not performed from the Evidence at present. In the work described here the samples under examination are thin sections of single crystals, the thickness typically corresponding to between 50 and 100 atoms. The specimen is aligned to a major zone axis, so that the image reflects the arrangement of atomic columns, seen in projection. The scattering power of these columns is mapped by the scanning probe, with the resolution determined by the probe PSF. The intrinsic width of the object function is an order of magnitude below currently available probe sizes (Pennycook and Jesson, 1990) so that it is a straightforward matter of optimising the PSF radius to generate the sharpest points in the reconstruction.

Where the S/N is low, the reconstruction is complicated by excessive noise. However, the optimum PSF radius may still be accurately obtained by the process described. However, by incorporating an ICF a compromise between the displayed size of the reconstructed object function and the residual noise may be reached.

2. Experimental

The specimens studied were samples of single crystal silicon viewed along the $\langle 110 \rangle$ direction and $YBa_2Cu_3O_{7-x}$ (YBCO) high T_c superconductors. The images were acquired using a VG Microscopes H8501UX scanning transmission electron microscope using a 75-150 mrad annular detector and were recorded on Polaroid type 52 film direct from the display monitor using 20 second scan times. The theoretical resolution limit for this 100kV instrument is 2.2Å, under Scherier optimum in coherent conditions (Pennycook and Jesson, 1990).

The photographs were converted to digital images using an AI Cambridge video unit equipped with a frame grabber. The images were then transferred to an IBM compatible PC, equipped with the 'MemSys Flyer' accelerator card, and converted into a format suitable for processing by the 'MemSys5' MaxEnt algorithm.

The 'MemSys Flyer' is an Intel I860 based accelerator which allows the MaxEnt algorithm to run at up to 45 MFlops. This increase in speed reduces the time taken for each trial MaxEnt run to less than 10 minutes on a 256 × 256 image. In the case of the silicon data, the final run required approximately 50 minutes for convergence. The final run for the poorer S/N superconductor data, in which a larger ICF was applied, converged in 25 minutes.

3. Results

The starting photographic data for single crystal silicon and the MaxEnt result are shown in figures 1a and 1b. For the reconstruction shown in figure 1b a small ICF was required to minimise noise in the reconstruction. The high noise with no ICF was attributed to the background intensity not being constant throughout the image which arises from sample thickness variations. The significant feature observed in the MaxEnt reconstruction is that the silicon dumb-bell, characteristic of the $\langle 110 \rangle$ projection, has been resolved.

By capturing a much smaller part of the image with a more uniform background the image background may be optimised with much greater reliability without compromising the MaxEnt result. The starting data, again captured as a 256 × 256 pixel image, and the MaxEnt result for the smaller image area are shown in figures 2a and 2b. In this case it was unnecessary to incorporate an ICF and there is no doubt that the crystal projects as closely spaced pairs of atoms.

The MaxEnt reconstruction has used the shapes of image features to extend the high frequency information with no prior knowledge. The reconstruction appears to have rotated the dumb-bells somewhat, though with hindsight this asymmetry can also be seen in the data, and is most probably a result of some sample tilt. Even so, the reconstruction still gives an average separation of 1.33Å ± 0.20Å for the two columns of a dumb-bell, remarkably close to the true value of 1.36Å. Since the reconstruction is noise limited, improved detector efficiencies should result in accuracies such as this to be achievable from individual columns in the image.

Note that the analysis also quantifies the relative strengths of the scattering centres, which is a great advantage in quantifying the compositional information in the image. To illustrate this, figure 3 compares the raw data with the MaxEnt reconstruction of an interface between a YBCO film and a $KTaO_3$ substrate. Again, the maximum entropy result shows substantially enhanced resolution, and can also provide the positions and intensities of the various columns with associated individual error bars. It demonstrates clearly that for one or two unit cells at the interface a compound is formed that appears to project as $BaTaO_3$. This is not stable in bulk form but is stabilised in some way at the interface.

Figure 4 shows the raw image and the MaxEnt reconstruction of a high angle grain boundary in a YBCO film. Viewed along the c-axis, the superconductor projects as mixed columns of Y and Ba, which are seen bright in the raw image, with Cu columns in between. The Cu columns are not visible in the image, even though the distance to the adjacent Y/Ba columns is 2.72Å, well above the 2.2Å probe size. This is because of the presence of an amorphous surface layer on the sample created during ion milling, which broadens the probe before it reaches the crystal.

An advantage of the MaxEnt software is that the actual PSF can be measured using the image from a known structure. In this case it was estimated by scanning along the $\langle 100 \rangle$ directions through the centres of the Y/Ba columns. This scan would pass 1.93Å away from the lighter Cu column and therefore give a good first approximation of the actual PSF. The resulting image reconstruction is shown in Figure 4b. Additional columns have been introduced by MaxEnt to account for the higher than expected background in the vicinity of the Cu columns. The reconstruction is still limited severely by the noise in the original data however, highlighting the importance of improving the efficiency of the annular detector. In fact, it may well be beneficial to operate the microscope at a lower source demagnification, which would degrade the image resolution, but give substantially improved image statistics, so that the MaxEnt software might give a higher final resolution in the reconstruction.

4. Conclusions

Although present results are noise limited, MaxEnt image restoration offers a promising route to higher resolution imaging of crystalline materials, based on incoherent Z-contrast imaging in the STEM. The high angle annular detector effectively replaces the crystal by an array of sharply peaked spikes centred on each projected atom column. Each individual spike is weighted by the high angle scattering power of the column, and the image is given to the first order by simply convolving with an appropriate PSF (Pennycook and Jesson, 1990).

It is because the width of these spikes is an order of magnitude less than available probe sizes that the MaxEnt reconstruction can enhance the resolution to such a remarkable degree. At some point however, the second order effects in the Z-contrast image will presumably become first order again, and it is unclear at present just how far the resolution can be enhanced before the restoration becomes sensitive to sample thickness and defocus.

Present results are very encouraging however, and clearly indicate a factor of two is quite straightforward to achieve. With the advent of 300kV STEMs having probe size in the range of 1.3Å, MaxEnt reconstruction will allow sub-Angstrom microscopy to be achieved in a rather simple and direct manner.

5. Future Work

This work has highlighted a number of deficiencies in the interface to the algorithm and future work will concentrate on rigorous methods which will:

(a) Cope with non-uniform sample thickness.
(b) Reliably estimate the noise level automatically.
(c) Measure experimental PSF from line scans.
(d) Input theoretical PSFs.

REFERENCES

Gull, S.F.: 1989, 'Developments in Maximum Entropy Data Analysis' in J. Skilling (ed.), *Maximum Entropy and Bayesian Methods*, Kluwer, Dordrecht.

Pennycook, S.J and Jesson, D.E.: 1990, 'High-Resolution Incoherent Imaging of Crystals', *Phys. Rev. Lett.* **64**, 938.

Pennycook, S.J and Jesson, D.E.: 1991, *Ultramicroscopy* **37**, 14.

Skilling, J.: 1990, 'Quantified Maximum Entropy' in Paul F. Fougère (ed.), *Maximum Entropy and Bayesian Methods*, Kluwer, Dordrecht.

Skilling, J: 1991, 'On Parameter Estimation and Quantified MaxEnt' in W.T. Grandy, Jr. and L.H. Schick (eds.) *Maximum Entropy and Bayesian Methods*, Kluwer, Dordrecht.

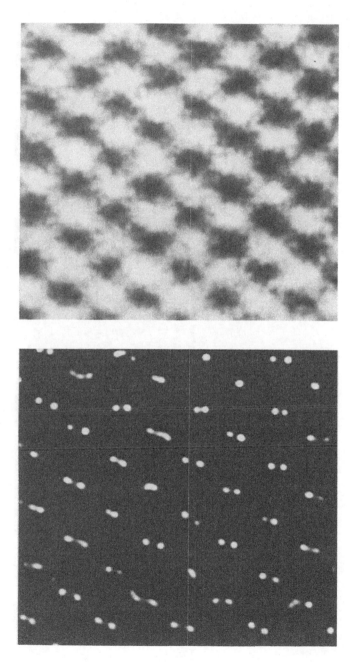

Fig. 1. (a) (above) Silicon dataset and (b) (below) MaxEnt reconstruction.

Fig. 2. (a) (above) Sub-image of silicon dataset and (b) (below) MaxEnt reconstruction.

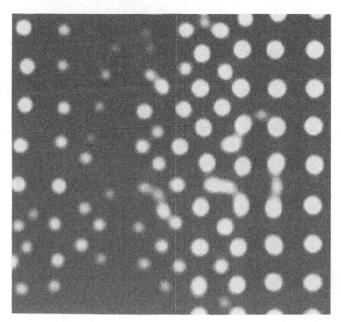

Fig. 3. YBCO film on $KTaO_3$ substrate; (a) (above) Original dataset and (b) (below) MaxEnt reconstruction.

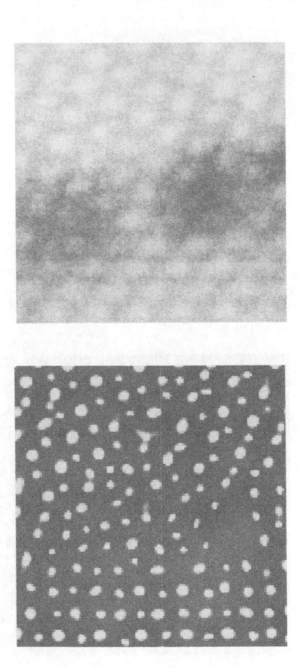

Fig. 4. High angle grain boundary in YBCO; (a) (above) Original dataset and (b) (below) MaxEnt reconstruction.

The Inference of Physical Phenomena in Chemistry:
Abstract Tomography, Gedanken Experiments, and Surprisal Analysis

Charlie E.M. Strauss
Los Alamos National Laboratory
Chemical and Laser Sciences
Photophysics and Photochemistry
J567, Los Alamos NM 87545

Paul L. Houston
Cornell University
Department of Chemistry
Baker Lab, Ithaca NY 14853

ABSTRACT. The maximum entropy method for tomographic reconstruction of a joint probability distribution from its marginals can be applied to problems frequently encountered in molecular dynamics. Once obtained the posterior joint probability can be used to predict the outcomes of untested gedanken experiments. These predictions can often be more readily assimilated than the raw data itself, allowing the scientist to tap his or her intuition further. Surprisal analysis of the gedanken measurements can reveal how many hidden constraints are active in the experimental process. Two problems often face the experimentalist: first, it may be infeasible to directly measure a quantity of interest although various projections of it can be sampled. Second, the sheer complexity of multidimensional projection data prohibits an intuitive grasp of the data's scientific content. We recommend breaking the analysis into three parts: tomographic back-projection to manage the data, gedanken experiment prediction to stimulate discovery of unused prior information, and surprisal analysis for discovery of hidden constraints. Two new methods of applying graphical surprisal analysis to multidimensional data are described. Example applications to five molecules are given.

1. Introduction

When one sets out to discover how a certain process, say the photodissociation of a particular molecule, might occur, quite often one will organize one's thoughts around a gedanken experiment. The thought experiment will not necessarily be feasible in the laboratory, but it is valued for its clear-cut nature and the intuition it promotes. For instance, a physical chemist might imagine a single molecule, AB, falling apart into two molecular fragments, A + B, in specific quantum states (i,j): $AB \longrightarrow A_i + B_j$. In this picture concepts like conservation of energy and angular momentum, or geometrical entities like impact parameters and recoil angles are tangibly related to the omniscient hypothetical 'observations'(i,j). Of particular chemical interest, the degree to which the AB molecule bends as it falls apart

345

C. R. Smith et al. (eds.), Maximum Entropy and Bayesian Methods, Seattle, 1991, 345–357.
© 1992 Kluwer Academic Publishers.

relates directly to the rotational states of the fragment pairs (Shinke, 1986). In practice, instead of a single molecule, perhaps 10^{16} AB molecules will dissociate concurrently and though it may be possible to measure the distribution of the quantum states in each of the A and B fragment populations separately, it is impossible to observe which pairs of fragments came from the same parent AB. Besides not knowing the pair correlations it is even possible that certain coordinates, such as the relative orientation of the reacting molecules, will not be amenable to measurement at all. The dilemma is that the form of the feasible measurements thwarts straightforward analysis, yet the data set may be insufficient to uniquely predict the outcome of theoretically more tractable measurements.

This work will advocate that situations of this nature can be approached as tomographic inverse problems. The measured A and B fragment state distributions, $P_a(A_i)$ and $P_b(B_j)$, are projections of a higher dimensional joint probability describing complete correlation of all the degrees of freedom, $P(A_i, B_j, ...|I)$. Moreover, any imaginable measurement, feasible or not, is just another projection of $P(A_i, B_j, ...|I)$. Information (I) about the initial quantum states of the parents and conservation laws provide prior information. The way to proceed is to back-project from whatever set of measurements is available to the joint posterior probability distribution then to reproject onto the axes appropriate to a cogent gedanken experiment. In particular we will discuss the reprojection of an impact parameter distribution. The impact parameter is the torque-arm length exerted between the fragments that precipitates their rotation. The back-projection is performed using standard maximum entropy techniques, treating the data as constraints, and incorporating whatever prior information is available.

By way of contrast, the customary approach of physical chemists to deciphering indirect measurements is to compare them to the predictions of popular mechanistic models, and then, having obtained a successful fit, to embrace the physical premises of that model as bonafide. However, it is not routinely appreciated by chemists that there exist an infinite number of possible joint probability distributions that agree with the data. The selection of one model, however popular, is implicitly an ad-hoc selection of a specific joint probability distribution. While sufficient prior confidence in a given modeling scheme might justify a bayesian model comparison approach (Bretthorst, 1990), the potentially large number of degrees of freedom and their high degree of coupling normally found in molecular dynamics warrants an entropic selection.

Unfortunately, the physical chemist feels uneasy with the maxent back-projection precisely due to its inherent lack of mechanistic recommendations about the process. It is not a simple matter to recognize the expression of a physical model in $P(A_i, B_j, ...|I)$. While some patterns might be seen directly in the correlations, in practice, they tend to be mercurial and subdued. We recommend reprojections of the joint probability distribution as a way of attaching intuition. What one hopes to find is that when projected into the basis of the gedanken experiment the patterns are simple and therefore recognizable. Put another way, since the gedanken basis was deliberately chosen to give a clear cut picture, if the patterns remain incomprehensible then either the process is too complex to resolve from the available measurements or the thought experiment is mistaken.

An alternative means of extracting the residue of sense from the joint probability distributions is surprisal analysis (Levine, 1981). The principle is to look for the 'simplest' set of constraints for which the joint probability distribution is the maximum entropy distribution: thus all the data could derive from a simple set of physical constraints. We

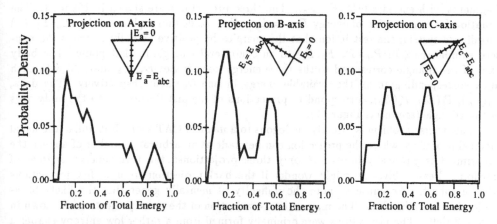

Fig. 1 Three projections of the planar triangular region onto the three body axes: A,B, and C. Insets show orientation of projection axis with respect to the triangle.

will briefly discuss two new approaches in surprisal analysis: synthesized surprisal and decoupled surprisal. Again the emphasis will be to maximally utilize information theory but finish in a basis where the chemist has physical intuition.

There are other reasons, aside from its sheer problablistic weight, to prefer the maxent choice from the feasible set of back-projections. Perhaps the strongest is that of Levine and co-workers who proved that 1) any ensemble of initial molecular conditions that is reproducible in a lab can be written formally as a maxent distribution with respect to some closed set of constraints, and 2) a maxent distribution evolved under the Hamiltonian must remain a maxent distribution with the same set of constraints. Therefore the question is not whether a quantum state distribution under consideration is one of maximum entropy but whether the set of operative constraints is limited to a small enough subset so that the states might be elucidated in a finite set of experiments. However the 'least biased' property of the maxent distribution is also relevant here: given the intent to reproject this inferred joint probability distribution one would do well to choose the least-biased joint probability distribution so that there will be evidence in the data for any striking features in the new projection (Gull, 1984).

In the remainder of this work we demonstrate these ideas in context. We give three distinct examples of the application of this procedure to photodissociating molecules.

2. Fragment Correlations Without Coincidence Measurements

THE INVERSE PROBLEM,
ROBUSTNESS, AND CHEMICAL RELEVANCE.

We begin with a simple illustration of the procedure (see (Strauss, 1990a) for a discussion of methods). A parent molecule, ABC, prepared with a fixed total energy, E_{abc}, falls apart into fragments A, B and C. The experiment is able to measure the distribution of energy into each of the fragment species, $P_a(E_a)$, $P_b(E_b)$, $P_c(E_c)$. The goal is to find the back-projection, $P_{abc}(E_a, E_b, E_c)$, which is the probability one parent molecule produced fragments with energies (E_a, E_b, E_c). The three projections are shown in Figure 1. The equation for conservation of energy $E_{abc} = E_a + E_b + E_c$ defines a plane in the (E_a, E_b, E_c) coordinate system; the restriction that the energies be positive actually restricts the non-zero sample space of $P_{abc}(E_a, E_b, E_c)$ to an equilateral triangle in that plane. The body axes of the triangle correspond to the three energy coordinates E_a, E_b, and E_c, with each apex corresponding to all the available energy in one fragment respectively. The data, $P_a(E_a)$, $P_b(E_b)$, $P_c(E_c)$, correspond to projections of the distribution on to the body axes of the triangle (see insets Figure 1).

The inverse problem is directly analogous to a medical CAT scan. But unlike the usual medical procedure where the projection can be made at an arbitrary number of angles, the experiment is physically limited to only three projections: the three energy degrees of freedom. Reasonably, one might wonder if the back-projection from so few projections will be robust. We pause to note the difficulty of mentally generating the likely back-projection from this data. The maxent back-projection of the three projections is shown in Figure 2 (left). The projections were originally formed from a rather low entropy shape: a question mark. The three projections constrict the set of possible back-projections so that the majority resemble each other.

This higher entropy reconstruction not only superficially resembles the test object but has also resolved chemically relevant fragment correlation features. In particular, bimodality is seen by the dip in the center, and asymmetry by the gap. Chemically, one might expect related dissociative pathways to be connected by a continuum of probability as seen in the contiguous snaking curve, and chemically distinct pathways might be expected to segregate themselves in islands seen in the two components of the question mark. This provides reassurance that the qualitative characteristics seen in the reconstruction from actual molecules (which tend to be naturally high entropy objects) will be chemically meaningful.

TRIPLE WHAMMIES:
DISSOCIATION INTO THREE FRAGMENTS

One of the more aggressively pursued fields in chemistry regards dissociations in which more than a single bond is broken or formed (Woodward, 1971, Strauss, 1990a). In such reactions, the intriguing possibility for a 'concerted' rearrangement exists: the bonds may break or form with the electrons and nuclei moving in an integrated fashion as opposed to rearranging in a stepwise fashion with the steps well separated in time and only grossly influencing each other.

Consider, a reaction breaking two bonds and producing three fragments: ABC $\longrightarrow$ A+B+C. Conceptually, the course of the stepwise reaction can be imagined to begin with a specific bond breaking and ejecting a fragment BC which tumbles away from an A fragment. The BC fragment subsequently dissociates with the B and C fragments recoiling apart along the instantaneous orientation of the B-C bond axis. (We shall employ a colloquial usage of the term 'bond' to denote the force axis between two objects.) In contrast, for a concerted dissociation, the A,B and C fragments recoil apart beginning from their relatively static positions in the excited parent molecule (or 'transition state geometry').

This picture leads to the gedanken experiment distinguishing the two cases: if one examined the angle between the recoil direction of fragment A and the line connecting the corresponding B and C fragments one would expect a peaked distribution in the concerted case. The peak reflects the well defined parental geometry from which all three fragments

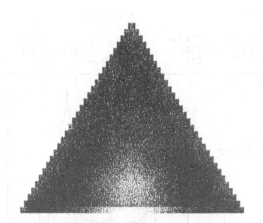

Fig. 2 Density Plot of the back-projections for the Question Mark (left) and Carbon Suboxide (right). On right: bottom apex corresponds to all energy in carbon fragment; top apecies correspond to all energy in one CO fragment.

'push off'. In the stepwise case, tumbling of the BC fragment would effectively scramble any memory of the A fragment's recoil direction from the parental geometry. Since this angular distribution cannot be *deduced* from experimentally feasible measurements, the most plausible distribution must be *inferred* by maximum entropy.

The data is, as before, the total energy distribution in each of the three fragments: $P_a(E_a)$, $P_b(E_b)$, and $P_c(E_c)$. The back-projection to the joint probability energy distribution $P_{abc}(E_a, E_b, E_c|I)$ must be determined before making the projection to the angular distribution. The prior information (I) is the available energy, the momentum of the parent, and the vibrational, rotational, translational, and electronic state densities of the fragments. Lacking further information the default prior distribution treats every allowed fragmentation state as equally likely. Figure 2 displays the maxent back-projection for the photodissociation of Carbon Suboxide (Strauss, 1990b):

$$O = C = C = C = O \longrightarrow CO + CO + C$$

The density displays the difference between the maximum entropy back-projection and the prior distribution. Thus it shows the 'flow' away from the prior distribution: the darker regions are positive, the lighter regions are negative and middle grey is zero. In particular note that there are two fingers of probability flow into the corners corresponding to high energy in the carbon monoxide fragments with a deficit of probability in the middle. This bimodal tendency indicates a greater than prior statistical preference for a hot carbon monoxide fragment to be paired with a cold carbon monoxide fragment. A plausible mechanistic interpretation of this tendency toward non-equivalent carbon monoxide pairs is that they do not break off of the parent molecule symmetrically and simultaneously.

However, these patterns are very subtle presented in this multidimensional energy space. If there had been more than three coordinates the hidden physical mechanisms would be even more elusive. This motivates a more revealing presentation provided by the gedanken recoil angle distribution. Details of the computation of the posterior recoil angle

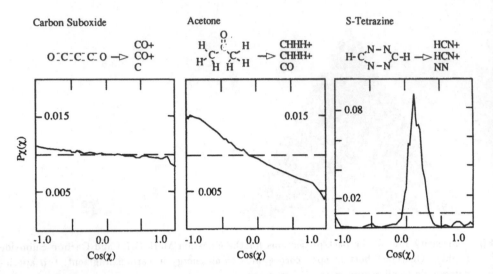

Fig.   3 Posterior predicted recoil angle distributions for three dissociation cases. Dashed line is
prior (spatially uniform) recoil angle distribution.

distribution from $P_{abc}(E_a, E_b, E_c | I)$ are given in the appendix of (Strauss, 1990a).

Figures 3 shows the recoil angle distribution $P_\chi(cos(\chi))dcos(\chi)$ for the triple fragmen-
tations of carbon suboxide and two other molecules: acetone (CH_3COCH_3) and symmetric
tetrazine (Strauss, 1990a). The angle χ is the recoil direction of the 'first' fragment to depart the molecule measure with respect to the line (ie. the bond) connecting the remaining
two fragments. If this fragment pair rotates much before falling apart the distribution of
the angle χ will flatten out towards the uniform distribution represented by the flat, dashed
line in the figures. (Mathematically, the angle calculation can be performed designating
any of the three fragments as 'first', however there is usually prior information available to
make the choice exclusive.)

These results for the three different parent molecules span a wide range of behavior.
The symmetric tetrazine case is sharply peaked indicating a rapid, concerted dissociation
from a rigidly defined parent geometry. The carbon suboxide case is approaching the flat line
indicating a stepwise dissociation. The acetone case lies intermediate, most likely indicating
a short but stepwise delay between the two bonds breaking. These indications are consistent
with the conclusions reached by more extensive studies using more traditional chemical
physics methods (Trentleman, 1989; Hall, 1991; Strauss, 1990b; Zhao, 1989). Qualitatively,
one might even redefine 'concertedness' as the memory one bond shows about the other;
the entropy deficiency of the angular distribution with respect to the non-informative flat
distribution would then serve as a monotonic measure.

IMPACT PARAMETERS
FOR BIMOLECULAR EVENTS

More common than the triple whammy case are molecules which break into just two fragments. Two particularly intriguing classes of these are dissociations which must break two

bonds, such as in a ring structure, and those that simultaneously form and break bonds to produce the fragments. Formaldehyde (H_2CO) photodissociated by 340nm wavelength light produces H_2 and CO fragments, and is an example of the latter. One knows that something peculiar must be taking place because the hydrogen atoms in the parent molecule lie on opposite sides of the carbon atom and are not bonded to each other, yet they cleave from the parent H_2CO as a bonded pair. The chemist imagines that there could exist a transition state where the carbon and two hydrogen atoms become temporarily bonded in a triangle before the hydrogen atoms sever their bonds to the carbon.

Qualitatively, a chemist may expect that a ring form provides a restricted and possibly stiffer parent molecular structure than an open form. This steric restriction will influence the way the fragments push apart. A particular characterization of this restriction is the impact parameter distribution. The impact parameter, b, is the length of the lever arm between the fragments by which they can exert a torque on each other; it is defined by the expression $L = mvb$ where L is the orbital angular momentum of the fragment center of masses, v is the recoil velocity between the fragments, and m is the reduced mass, and it is equivalent to the shortest perpendicular distance between the two fragment trajectories. Typically the impact parameter will range from zero to distances on the scale of the interatomic separation in the parent molecule (a few angstroms). For example, if the fragments pushed apart along the axis of a linear parent molecule, they could not exert a torque on each other thus giving a zero impact parameter. Conversely, a severely bent parent structure could cause a large torque between fragments and correspond to a large impact parameter. In general, there will be a distribution of impact parameters and the physical chemist would like to know the location of the peak and whether the distribution is sharp or broad.

This gedanken impact parameter distribution cannot be measured experimentally nor deduced from other measurables except in simple cases such as a triatomic photodissociation. In the face of the uncertainty in the fragment pair correlations, the back-projection is regularized with maximum entropy. The feasible experiment can simultaneously measure the rotational state, the vibrational state, and possibly the speed (V) of each fragment. So one has $P_{H_2}(E_{H_2}^{rot}, E_{H_2}^{vib}, V_{H_2})$ and $P_{CO}(E_{CO}^{rot}, E_{CO}^{vib}, V_{CO})$ and would like to infer $P_{H_2CO}[H_2(E_{H_2}^{rot}, E_{H_2}^{vib}), CO(E_{CO}^{rot}, E_{CO}^{vib}), v, L]$. As before, the prior information is that all states allowed by the total energy, conservation of linear momentum, and conservation of angular momentum are equally likely.

The six dimensional joint probability distribution is awkward to interpret, so we proceed directly to the reprojection onto the chosen gedanken basis $P_b(b)db$:

$$P_b(b) = \int P_{H_2CO}[H_2(E_{H_2}^{rot}, E_{H_2}^{vib}), CO(E_{CO}^{rot}, E_{CO}^{vib}), v, L] \times \delta[\frac{L}{v*m} - b]dH_2dCOdLdV \quad (1)$$

This posterior projection for the formaldehyde molecule's dissociation may be compared with the same projection of the prior distribution as displayed in figure 4. Clearly, the data have pulled this posterior away from the prior. It is more sharply peaked and that peak is displaced by half an angstrom. The well defined single peak is consistent with the simple picture of the sterically constrained triangular transition state; more extensive theoretical calculations of transition state potential surface back this up as well.(Osamura, 1981)

The distribution also agrees very well with an excellent, extensive non-maxent analysis of the same data by Debarre et al. In particular, the model those authors chose assumed

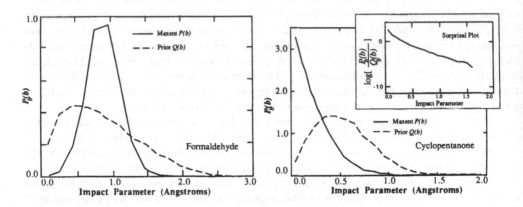

Fig. 4 Posterior predicted impact parameter distributions for formaldehyde (left) and cyclopentenone (right) dissociation cases. Dashed line is prior impact parameter distribution.

that there was no correlation between the fragment distributions. This leads to joint probability distributions that must be truncated arbitrarily to obey the conservation laws; such
distributions have lower entropy and are thus less plausible aprioi than the maxent distribution. The agreement between Debarre's assumed model and the maxent result is a
tribute to the data's constriction of the space of possible back-projections, and is not a
justification of the premises of that model.

For comparison, figure 4 also shows the maxent impact parameter distribution for the
photodissociation of cyclopentenone (Jimenez, 1991). Cyclopentenone is a ring compound
and must break two bonds to form two fragments. The clustering of the impact parameter
near zero is consistent with a picture that the bonds break in a simultaneous, counterbal-
anced manner imparting little net torque to the fragment mass centers.

3. Inference of Constraints

In trying to elucidate the nature of a physical process the scientist is asking *"why is my
data different from my prior expectations?"*. In general, because it is constrained to agree
with the data, the maxent back-projection is different from the prior distribution. The
mathematical reply to the question of *"why the difference"* is a set of additional constraints
for which the maxent back-projection and the prior distribution are identical. Since there
are an infinite number of constraints fulfilling this criteria one should look for the set
that is physically relevant yet simplest in some sense. We present here two approaches
to identifying simple constraints hidden in complex multidimensional data sets: posterior
decoupled surprisal and synthetic surprisal analysis.

Posterior Decoupled Surprisal

Reference (Levine, 1981) contains an excellent discussion of the principles and applications
of surprisal mechanics. Simplistically, the ordinary surprisal is defined as the log ratio of
an observation to its prior expectation. While the mathematics of surprisal analysis are

inherently multidimensional, its most powerful incarnation is in graphical presentation of the surprisal on a single coordinate. Perhaps as a result, surprisal analysis has been applied most often to simple systems commonly with one or two measured degrees of freedom (Engel, 1986; Levine, 1981). When a measured coordinate is directly related to a physically relevant constraint a simple surprisal plot can often be isolated, but multidimensional data are usually complicated by conservation law coupling. While mathematically it may be possible to decompose the data into orthogonal coordinates, the authors feel this leads one away from the coordinates where one has some intuition. Analogously, only viewing the singular value decomposition space of a blurred photograph would inhibit the use of one's crucial though nebulous prior knowledge of 'realistic looking' scenes.

We prefer to decouple the measured coordinates by considering them one at a time while treating the remainder as prior information. The decoupled surprisal is then defined as the log ratio of the observed distribution to the posterior distribution given all the other data. To illustrate, consider a three coordinate system (A, B, C): the measurements are the projections $P_a(a)$, $P_b(b)$ and $P_c(c)$, and additionally there can be conservation laws, such as $A + B + C = constant$, which couple the coordinates. The maximum entropy joint probability distribution will have the form:(Frieden, 1982)

$$P(a,b,c) = Q(a,b,c)e^{\mu_0 + \lambda_a(a) + \lambda_b(b) + \lambda_c(c)} \tag{2}$$

where μ_0 is a normalization constant and λ_a, λ_b, and λ_c are undetermined lagrange functions to be chosen to satisfy the three projections. The conservation laws can be absorbed into the known prior distribution $Q(a,b,c)$.

The prior distribution on coordinate A, $Q_a(a)$, is obtained by marginalizing the joint prior distribution:

$$Q_a(a) = \int db \int dc Q(a,b,c) \tag{3}$$

The posterior distribution on coordinate A given the data for the b and c axes, $Q_a(a|b,c)$ is given by the following construction:

$$Q_a(a|b,c) = \int db \int dc Q(a,b,c)e^{\mu_0 + \lambda_b(b) + \lambda_c(c)} \tag{4}$$

Considering $P_a(a)$, the constraint imposed by this data requires

$$\int db \int dc P(a,b,c) = P_a(a) \tag{5}$$

substituting [2] and then [4] gives,

$$P_a(a) = Q_a(a|b,c)e^{\lambda_a(a)} \tag{6}$$

In these terms the ordinary surprisal is then expressed:

$$\log[\frac{P_a(a)}{Q_a(a)}] = \log[\frac{Q_a(a|b,c)}{Q_a(a)}] + \lambda_a(a) \tag{7}$$

And the decoupled surprisal is expressed:

$$\log[\frac{P_a(a)}{Q_a(a|b,c)}] = \lambda_a(a) \tag{8}$$

In order to appreciate the difference between the ordinary and decoupled surprisal, consider a simple linear constraint $< A >= const$ which would correspond to $\lambda_a(a) = k_1 * a + k_0$ where k_1 and k_0 are constants. In general the ordinary surprisal could be non-linear due to the complicated first term. However, in the decoupled surprisal, the constraints and conservation laws on the other coordinates do not influence the surprisal equation, and so a linear constraint produces a linear plot

$$\log[\frac{P_a(a)}{Q_a(a|b,c)}] = \lambda_a(a) = k_1 * a + k_0. \tag{9}$$

Thus the decoupled surprisal plot graphically displays the 'simple' constraint on that coordinate. The analogous constructions can be formed for the other two coordinates. Note that while $Q_a(a)$ can be found from direct integration of the known prior distribution, determination of $Q_a(a|b,c)$ follows solving for the maximum entropy joint probability since the constraint functions must be computed.

We note that lacking any prior knowledge the maximum entropy form produces uncoupled constraints $\lambda_a(a)$, $\lambda_b(b)$, and so forth. This is undesirable if one has prior knowledge that a constraint will take a coupled form such as $< a * b >= const$, and this knowledge should be incorporated at the beginning of the maximum entropy procedure. In this case Equation [2] is no longer the proper form for the maximum entropy joint posterior distribution.

Synthetic Surprisal

Commonly, the prior knowledge will consist of imprecise notions about the important constraint coordinates but will be insufficient to state a specific functional form. In this soft knowledge regime the surprisal analysis of synthesized distributions on the gedanken coordinates may be simpler than on the measurement coordinates. To illustrate, consider the surprisal of the previously projected impact parameter distribution.

The geometry of the transition state constrains the molecule away from uniformly populating all the final fragment states, and this is reflected in the difference between the back-projected impact parameter distribution and the prior density of states expectation. The anticipation that the parent has a single preferred geometry for its transition state suggests there will be a preferred impact parameter. The detailed shape of the distribution of impact parameters is determined by a combination of this preference and the density of states for the differing geometries. A quantification of this preference is a series of expectation values on the moments of the distribution. For instance a constraint on the first moment–the mean impact parameter– is written:

$$< b >= \int_{b=0}^{b=\infty} db P_b(b) * b = b_0 \tag{10}$$

where $P_b(b)$ is the probability of b, the impact parameter, and b_0 is the mean value. If this were the only operative constraint imposed by the mechanics of the dissociation, the impact parameter distribution would be have the form:

$$P_b(b) = Q_b(b)e^{\mu_0 - \mu_1 b} \tag{11}$$

where Q_b is the density of states and μ_1, μ_0 are constants chosen to satisfy the constraint and implicit normalization. The surprisal plot, $\log[P_b(b)/Q_b(b)]$ versus b, would be a straight line: $\mu_0 - \mu_1 * b$. The surprisal plot of the impact parameter distribution of cyclopentenone is roughly a straight line (see Figure 4 inset). Therefore the data are consistent with a maximum entropy distribution including that constraint. This inference combined with a chemists intuitive prior knowledge raises the plausibility of the hypothesis that cyclopentenone has a single preferred transition state geometry.

There are of course many possible constraints which when projected onto the impact parameter basis would produce a straight line and any could be the true physical constraint. Indeed if one examined the full joint probability distribution one might find that this particular constraint was unnecessarily restrictive, having a lower entropy than needed to fulfill the minimum requirement that the observed and expected data be identical. However, the scientist believes he has the difficult to quantify prior knowledge that if the constraints appear 'simple' and 'physically realistic' in the gedanken basis they are overwhelmingly likely to be correct. In practice if the surprisal plot had been composed of two line segments with different slopes the chemist might suspect that the molecule dissociates from two preferred geometries; on the other hand if the plot had appeared even more complex the inclination would be say there must be a more important dynamical constraint than the geometry and this too-simple gedanken frame was poorly chosen.

4. Discussion

An attractive feature of this approach is its rational consistency. A variety of chemical processes can be examined by the procedure of back-projection from the measurements to the joint probability distribution and then reprojected onto physically tangible axes. While one could, in principle, directly compute the posterior distribution on the gedanken axis this two step procedure adds uniformity and versatility. The construction of the gedanken experiment is most affected by physical intuition while the back-projection is primarily affected by the known experimental conditions. The back-projection procedure is general and adapts to available molecular data; different experimental apparatus may gather different or larger subsets of the possible measurements yet from the back-projection one may make reprojections onto the same gedanken experiment axes. Also, once obtained, the joint probability distribution may be projected on to more than one trial gedanken frame.

Traditionally, physical chemical analysis grapples with intangible measurements by looking for a mechanistic model that fits them. We find the maximum entropy approach more satisfying since the most salient predictions such as the recoil angle and impact parameter distributions can be obtained from the data without first imposing a strong physical model. These cogent predictions can then guide a more sophisticated search for a physical mechanism.

Simply put, the human mind is not adept at recognizing physical mechanisms intertwined in multidimensional data; an easier task is the recognition of mechanisms that produce characteristic constraints. The key feature is that the scientist's intuition consists of important but difficult to quantify prior knowledge. Used strongly, to pick a particular mechanistic model and fit it to the data without considering the prior likelihood of the model is putting the cart before the horse: one is guessing at models before understanding the information in the data. A better way is to use it weakly to guide a choice of coordinate frame in which the physics will be most important and most explicitly "simple" but to make no strong statements about exactly what form it will take in this space.

Finally there are some general features and caveats that should be noted. Projections tend to be smoothing operations. This built in smoothness combined with the natural tendency of maximum entropy back-projections to err on the side of smoothness assures than any sharp features in the projections normally can be considered real and not artifacts of the regularization procedure. Of course one must be careful that the gedanken projection does not lie entirely in the singular value decomposition null-space of the of the measurement directions or else the results will be meaningless. Similarly, if the number of degrees of freedom in the back-projection minus the degrees of freedom restricted by conservation laws is greater than the number of data projections, the back-projection is likely to be poorly constricted. In this regime one is effectively imposing a statistical model for the unmeasured coordinates, and surprisal analysis of reprojections must be considered cautiously. For instance, in the 'Question Mark' example, discarding one projection results in a poorly constricted solution space and the maxent solution is abysmal. Fortunately, molecular events tend to be controlled by very few constraints, so by prior experience it is plausible to assume most unmeasured coordinates to lack hidden constraints. (As has been noted (Levine, 1981), if this were not the case little could ever be discovered).

5. Conclusions

Two problems often face the experimentalist: first, it may be infeasible to measure a quantity of interest and only various projections of it can be sampled. Second, the sheer complexity of multidimensional projection data prohibits an intuitive grasp of the data's scientific content. We recommend breaking the analysis into three parts: tomographic back-projection to infer inter-dimensional correlations amongst the data, gedanken experiment prediction to stimulate discovery of unused prior information, and surprisal analysis for discovery of hidden constraints. We emphasize that this process in not a unique transformation of the data but rather is a prediction of the outcome in other coordinates.

This formalism was applied in a consistent manner inherently adaptable to diverse data formats and chemical systems. In each case highly intangible data was combined with prior experimental information to predict the most salient dynamical characteristics: the impact parameter and the recoil angle distributions. The indications of these distributions were found to be in agreement with previous non-maxent investigations.

Two new ways of applying surprisal analysis to multi-dimensional data were discussed. First, in decoupled surprisal analysis the distribution in each coordinate is separately examined for hidden constraints by comparison to its posterior prediction based on all the other data. Second, it may facilitate the untangling of complicated surprisals in the measured coordinates to synthesize a prediction of them in more physically relevant coordinates.

REFERENCES

Bretthorst, G.L.: 1990, 'Bayesian Analysis II: Model Selection', *J. Magnetic Resonance* **88**, 552.

Debarre D., Lefebvre, M., Paelat, M., Taran, J-P.E., Bamford, D.J., Moore, C.B.: 1985, *J. Chemical Physics* **83**, 4476.

Engel, Y.M. and Levine, R.D.: 1986,'Surprisal Analysis and Synthesis for the F+H2 Reaction and its Isotopic Variants',*Chemical Physics Letters* **123**, 42. note: surprisal synthesis is the logical converse of a synthetic surprisal.

Frieden, B.R.:1982, 'Estimating Occurrence Laws with Maximum Probability and the Transition to Entropic Estimators', in C.R. Smith and W.T. Grandy (eds.), *Maximum-Entropy and Bayesian Methods in Inverse Problems*, Kluwer, Dordecht.

Gull, S.F. and Skilling, J.: 1984, 'Maximum Entropy Method in Image Processing', IEE proceedings 131, 646.

Hall, G.E., Vanden, B.D., Sears, T.J.:1991, 'Photodissociation of Acetone at 193nm: Rotational- and Vibrational-State Distributions of Methyl Fragments by Diode Laser Absorption/Gain Spectroscopy', *J. Chemical Physics* **94**, 4182.

Jiminez, R., Kable, S.H.,Loison, J-C, Simpson, C.J.S.M., Adam, W., Houston, P.L.: (1991), 'The Photodissociation dynamics of 3-Cyclopentenone: using the Impact Parameter Distribution as a Criterion for Concertedness', *J. Physical Chemistry*, Submitted October 1991.

Levine R.D.: 1981, 'The Information Theoretic Approach to Intramolecular Dynamics', *Adv. in Chemical Physics: Photoselective Chemistry* **47**, 239.

Osamura, T., Schaefer, H.F., Dupuis, M., Lester, W.A.: 1981, *J. Chemical Physics* **75**, 5828.

Shinke, R.: 1986, *J. Chemical Physics* **85**, 5049.

Strauss, C.E.M. and Houston, P.L.: 1990a, 'Correlations Without Coincidence Measurements: Deciding Between Stepwise and Concerted Dissociation Mechanisms for ABC $\longrightarrow$ A+B+C', *J. Physical Chemistry* **94**, 8751.

Strauss, C.E.M., Jimenez, R., Houston, P.L.: 1991, 'Correlations Without Coincidence Measurements: Impact Parameters Distributions for Dissociations', *J. Physical Chemistry* , in press (Bernstein Issue).

Strauss, C.E.M., Kable S.H., Chawla G.K., Houston P.L., Burak I.R.: 1990b, 'Dissociation Dynamics of C3O2 excited at 157.6nm', *J. Chemical Physics* **94**, 1837.

Trentleman, K.A., Kable S.H., Moss, D.B., Houston, P.L.: 1989, 'Photodissociation Dynamics of Acetone at 193nm: Photofragment Internal and Translational Energy Distributions.' *J. Chemical Physics* **91**, 7498.

Zhao, X., Miller W.B., Hinsta, R.J., Lee, Y.T.:1989, *J. Chemical Physics* **90**, 5527.

Woodward, R.B. and Hoffmann R.: 1971, *The Conservation of Orbital Symmetry*, Verlag Chemie, Academic Press, Weinheim, Germany.

THE MAXIMUM ENTROPY RECONSTRUCTION OF PATTERSON AND FOURIER DENSITIES IN ORIENTATIONALLY DISORDERED MOLECULAR CRYSTALS: A SYSTEMATIC TEST FOR CRYSTALLOGRAPHIC INTERPOLATION MODELS

R.J. Papoular
Laboratoire Léon Brillouin
CEN-Saclay
91191 Gif-sur-Yvette cedex, France

W. Prandl and P. Schiebel
Institut für Kristallographie, Universität Tübingen
Charlottenstrasse 33
D-7400 Tübingen, Germany

ABSTRACT. Accurate density data are the primary information required for a thermodynamic model of molecular disorder. Neutron or X-ray Bragg scattering yields truncated and noisy data sets of unphased Fourier components. In the specific case of disordered molecular crystals, the phase problem can be bypassed by means of a density interpolation model using Frenkel atoms. The fitting to the data is usually quite good, and the validity of such a parametric model follows from the stability of the recovered phases with respect to the parameters. It is well established by now that MaxEnt is the best tool to retrieve 3-dim or projected 2-dim densities from phased data.

But MaxEnt helps a great deal more. By removing most of the noise and series truncation effects, a direct analysis of 2-dim and 3-dim Patterson functions, which relate to the autocorrelation function of the sought density, yields a direct check of the density interpolation model: one can observe directly the disordered density looked for.

The points mentioned above will be illustrated on a specific example, that of disordered ammine molecules in $Ni(ND_3)_6Br_2$.

Moreover, a possible pitfall leading to spurious line-splitting in the standard case of back transforming phased Fourier data using MaxEnt will be addressed.

1. Introduction

The application of Maximum Entropy to linear inverse problems such as interferometric measurements has been extensively used since the pioneering work by Gull & Daniell (1978) in radioastronomy. It was readily applied to crystallography by Collins (1982) in his seminal work, soon followed by Bricogne, Bryan, Gilmore, Gull, Lehmann, Livesey, Navaza, Sivia, Skilling, Steenstrup, Wilkins and others, who essentially tried to tackle the still open "Phase Problem" (see Skilling, 1989a and references therein). More recent crystallographic

C. R. Smith et al. (eds.), Maximum Entropy and Bayesian Methods, Seattle, 1991, 359–376.

work deals with already phased Fourier data, either merely concerned with the retrieval of the Patterson density autocorrelation function (David, 1990) or with an accurate density analysis (Sakata & Sato, 1990, Sakata *et al.*,1990). By comparison with the Standard Fourier Synthesis, the great success of MaxEnt in these latter studies is due to its huge reduction of termination effects as well as its noise suppressing role.

Our own work, to be presented hereafter, makes use of a preliminary Least-Squares fitting to phase measured Fourier components in a *ad hoc* but effective manner, specific to the physical problem under study: Molecular Disorder.

2. Molecular Disorder: from Thermodynamics to Crystallography

Often, disorder in a molecular crystal may be described by a rigid group or molecule acting as a top in an effective crystal-potential to be determined. Possible motions of this top involve the translation of its center of mass and rotations about the latter. One complexity of this potential results from the non-decoupling between translational and rotational motions. Thermodynamics connects the probability density of a given configuration (position, orientation) of the top to the overall effective potential via a mere Boltzmann factor. If the rigid group or molecule can be assumed to be composed of identical atoms (e.g., deuterons), the above mentioned potential may be further assumed to result from an effective single-atom potential $V(\vec{r})$. The density of such single atoms is now connected to the single-atom potential through an integral equation. Retrieval of this single-atom potential $V(\vec{r})$ is the end game, since it is a key to understand the dynamics of the disordered top. Amazing is the fact that it can be retrieved directly from the averaged disordered density, which is obtained via the use of a static diffraction experiment (Neutrons, X-rays or else). Indeed, thermally induced disordered motions affect the (elastic) Bragg peaks via a generalized Debye-Waller factor. Excellent quality diffraction data and ensuing reconstructed disordered density are mandatory in order to retrieve the single-atom potential from the aforementioned integral equation. This is where MaxEnt enters the scene, since the density is derived from the data through an Inverse Fourier transform.

3. Densities and Patterson densities from Phased Fourier Data sets using MaxEnt: Projections and Sections.

The Experiment

The dedicated instrument used to measure (unphased) Fourier components (also called structure factors) from a given single crystal is the 4-circle diffractometer. A crystal is, by definition, a 3-dimensional periodic arrangement based on an elementary entity of finite size: the unit cell. A horizontal monochromatic neutron or X-ray beam, of wavelength λ, is directed onto the sample under study, and the scattered beam in a given direction is measured. Usually, the detection takes place in the horizontal plane and the scattered beam is defined by only one angle: the scattering angle θ (corresponding to one circle). The modulus of the Fourier scattering vector $\vec{K}$ is equal to $4\pi\sin(\frac{\theta}{2})/\lambda$ and its direction lies along the bisector of the incoming and outgoing beams. The orientation of the crystal is described by its three Euler angles, which correspond to the remaining three circles of the spectrometer. A convenient basis, fixed with respect to the sample, is the so-called reciprocal basis $(\vec{a^*}, \vec{b^*}, \vec{c^*})$. The measurement of one datum consists in *i)* selecting $\vec{K}$ corresponding to a so-called Bragg reflection for *ii)* a given orientation of the reciprocal basis and *iii)* counting for a given amount of time. Let $I(\vec{K})$ and $\sigma(\vec{K})$ be the measured intensity (square of the modulus of the Fourier component) and the related experimental error bar respectively. A Bragg reflection is entirely defined by its 3 integer Laue indices h, k, ℓ such that $\vec{K} = h\vec{a^*} + k\vec{b^*} + \ell\vec{c^*}$. The direct basis $(\vec{a}, \vec{b}, \vec{c})$, which defines the unit cell of the crystal, is associated to the reciprocal basis mentioned above. It is also fixed with respect to the crystal, and any point $\vec{r}$ inside the crystal is described by its 3 dimensionless coordinates X, Y, Z such that $\vec{r} = X\vec{a} + Y\vec{b} + Z\vec{c}$.

MODEL FITTING: SPLIT AND FRENKEL MODELS

The underlying assumptions beyond model fitting are:
- The number of atoms, their species and the related scattering powers (scattering lengths for neutrons) inside the unit cell are known. The parameters of the unit cell itself are known.
- What is not known, and hence looked for, is the position of each atom (described by $\vec{r}$), the thermal motion about the atom equilibrium position assumed to be harmonic (described by the symmetric tensor $\bar{\bar{U}}$) so that the spread of the atom is 3-dim Gaussian, and the total number of unit cells inside the sample under investigation. For the sake of clarity, miscellaneous corrections will be omitted here. They usually are taken care of in the widespread crystallographic *Prometheus* and *ShelX* routines used to carry out the fittings.

The physics of a disordered molecular crystal may involve non-harmonic potentials and motions for part of the atoms or part of the molecule (e.g., a rigid ammine group) within the unit cell. The idea behind ad hoc model fitting consists in replacing a disordered atom or molecule by a fictitious larger number of fractional atoms (for the Split Model) or fractional groups (for the Frenkel Model) undergoing harmonic motion. The sum of these fractional atoms or groups are then made equal to the actual number existing in the unit cell. All positions and thermal motion tensors are then fitted to the data as closely as possible. The procedure just described is known as Density Interpolation.
A typical expression for the structure factor, which is defined for one unit cell, is thus:

$$F(\vec{K}) = \sum_{\alpha} b_{\alpha} \cdot \exp\left\{-\frac{1}{2}\vec{K} \cdot \bar{\bar{U}}_{\alpha} \cdot \vec{K}\right\} \cdot \exp\left\{i\vec{K} \cdot \vec{r_{\alpha}}\right\} \tag{1}$$

where the sum is over the real or fictitious atoms α. An essential feature of the Density Interpolation is the **stability of the phases** of the calculated structure factors (using the fitted parameters) with regard to the choice and number of the arbitrary parameters involved in the description of disorder and/or anharmonic motions. The admixture of the moduli of measured structure factors together with these reconstructed stable phases is thus expected to yield a "model-free" scattering density, provided that the Linear Inverse Fourier Transform can be handled satisfactorily.

NOTATIONS AND BASIC EQUATIONS

Let V and S be the volume of the unit cell and its projected surface along the $\vec{c}$ axis respectively. Let $\rho(\vec{r})$ and $P(\vec{r})$ be the scattering density and the Patterson density respectively. The latter is the autocorrelation function of the former, namely:

$$P(\vec{u}) = \int_{\wp} \rho(\vec{r})\rho(\vec{r} + \vec{u}) \, d\vec{r} \tag{2}$$

The structure factor for a given scattering vector $\vec{K}$ is defined by:

$$F(\vec{K}) = \int\int\int_p \exp\left\{i\vec{K}\cdot\vec{r}\right\}\rho(\vec{r})\,d\vec{r} \tag{3a}$$

$$= V\int_0^1\int_0^1\int_0^1 \exp\left\{2\pi i(hX + kY + \ell Z)\right\}\rho(X,Y,Z)\,dX\,dY\,dZ \tag{3b}$$

Furthermore, when $\vec{K}$ is orthogonal to a given crystallographic direction such as the $\vec{c}$ axis (in which case $\ell = 0$), one can write:

$$F(\vec{K}) = F(h,k,0) = S\int_0^1\int_0^1 \exp\left\{2\pi i(hX + kY)\right\}s(X,Y)\,dX\,dY \tag{3c}$$

where the projected density $s(\vec{r}) = s(X,Y)$ is defined by:

$$s(X,Y) = c\int_0^1 \rho(X,Y,Z)\,dZ \tag{3d}$$

Very similar relations hold for the Patterson function defined by:

$$I(\vec{K}) = \int\int\int_p \exp\left\{i\vec{K}\cdot\vec{r}\right\}P(\vec{r})\,d\vec{r} \tag{4}$$

At this point, it is useful to distinguish between retrieving a projection (say, $S(X,Y)$) of a density $\rho(\vec{r})$, which can be considered as a 2-dim inverse problem, from retrieving a section (say, $\rho(X,Y,Z_0)$ at constant Z_0) of the same density, which is a 3-dim inverse problem. In practice, all the sections are computed at the same time when using MaxEnt.

INCLUDING CRYSTAL SYMMETRY

The scattering density $\rho(\vec{r})$ as well as the Patterson density $P(\vec{u})$ inside the unit cell must obey spatial symmetry requirements: they must remain invariant when transformed through the g symmetry operations $\hat{\mathcal{R}}_s = (\hat{\alpha}_s, \vec{\beta}_s)$ of a suitable space group $\mathcal{G}$, where $\hat{\alpha}_s$ and $\vec{\beta}_s$ relate to the rotational and translational parts of the $s-th$ symmetry operation $\hat{\mathcal{R}}_s$. When a 2-dim projection along a crystallographic axis is sought, only those symmetry operations compatible with the projection should be kept (Papoular, 1991).
There are, at least, two ways of including this hard symmetry constraint into MaxEnt reconstructions:
- First, one may replace the density $f(\vec{r})$ (Patterson or not, 2-dim or 3-dim) by its spatial average $< f(\vec{r}) >$ defined as:

$$< f(\vec{r}) > = \frac{1}{g}\cdot\sum_{s=1}^g f(\hat{\mathcal{R}}_s\vec{r}) = \frac{1}{g}\cdot\sum_{s=1}^g f(\hat{\alpha}_s\vec{r} + \vec{\beta}_s) \tag{5a}$$

and subsequently use a Fast Fourier Transform.
- Alternatively, one may replace the Fourier integrand by its totally symmetric average defined as:

$$< \exp\left\{i\vec{K}\cdot\vec{r}\right\} > = \frac{1}{g}\cdot\sum_{s=1}^g \exp\left\{i\vec{K}\cdot(\hat{\alpha}_s\vec{r} + \vec{\beta}_s)\right\} \tag{5b}$$

thereby generalizing the definition of the Fourier transform (Papoular, 1991). It has been proved that the MaxEnt solution automatically satisfies the symmetry hard constraints (Papoular & Gillon, 1990).

The Fourier integrals must be computed numerically, and hence the Image $\vec{r}$-space must be discretized into M pixels. In cases of high symmetry, we find procedure (b) more advantageous since the $M \cdot \ln M$ typical factors of procedure (a) can then be larger than the $M \cdot N \cdot \phi$ multiplications required for procedure (b). In the latter expression, ϕ is the fraction of the unit cell that needs to be reconstructed and N is the number of independent data points (also called unique reflections). For instance, in the example to be discussed later, $g = 192, M = 64 * 64 * 64, \phi = \frac{1}{32}$ and $N = 232$. The cost in multiplications for the FFT procedure is thus 3270679. By contrast, only 1900544 multiplications are required by the alternative procedure. Moreover, the latter does not demand a binning of the unit cell using powers of 2 for each dimension, and the distinct (averaged) integrand values need only be computed once and then be stored.

Running Historic MaxEnt

The Cambridge algorithm as described by Skilling & Gull (1985), nowadays known as Historic MaxEnt, has been used in the present study. Based upon the maximization of the Shannon-Jaynes entropy over χ^2, it is characterized by the non-Bayesian $\chi^2 = N$ soft constraint. More theoretical (Skilling, 1989b, Gull, 1989) as well as practical ($MEMSYS5$ code) advances have since been made but were disregarded in the present preliminary study. In particular, the proper "Bayesian choice of α", characteristic of the new Quantified MaxEnt, would correct the underfitting of χ^2 and use up all the information contained in the less accurate data.

Back to Historic practice, one classically defines the entropy of the density $f(\vec{r})$ to be reconstructed, given the prior density model $m(\vec{r})$ and a suitable $\vec{r}$-space discretization, as:

$$S(\{f\}) = -\sum_{i=1}^{M} p_i \ln \left(\frac{p_i}{q_i}\right) \quad \text{where} \quad p_i = \frac{f_i}{\sum_{i=1}^{M} f_i} \quad \text{and} \quad q_i = \frac{m_i}{\sum_{i=1}^{M} m_i}. \quad (6a)$$

In the case when the sought physical density $f(\vec{r})$ is expected to go negative, a possible trick is to define it as the the difference of two strictly positive densities $f^+(\vec{r})$ and $f^-(\vec{r})$ and reconstruct $2 \cdot M$ discretized f_i values (or M pairs (f_i^+, f_i^-)). Negative densities may occur in neutron nuclear scattering due to atoms with scattering powers (scattering lengths) of different signs. The ensuing Patterson density $P(\vec{u})$, though being an autocorrelation function, can then also go negative (David, 1990).

In the present paper, unless otherwise specified, only positive physical densities $f(\vec{r})$ will be considered. Our prior knowledge will be taken highly uninformative: $m(\vec{r})$ will be taken to be uniform within the unit cell. As a consequence, all the q_i's are equal in $(6a)$, hence contribute a known additive constant to the entropy and consequently they can be dropped. We shall only assume that some parts of the unit cell are empty, so that the constant value m of the m_i's, toward which the f_j's tend in the absence of information from the data, is expected to be as small as possible, if required.

To any choice of a density $f(\vec{r})$ and a scattering vector $\vec{K}$ corresponds a mock datum $D^c(\vec{K})$ which stands either for a calculated structure factor $F^c(\vec{K})$ or for a calculated intensity $I^c(\vec{K})$. Defining $\chi^2(\{f\})$ by:

$$\chi^2(\{f\}) = \sum_{k=1}^{N} \left| \frac{D_k - D_k^c}{\sigma_k} \right|^2 \tag{6b}$$

one then maximizes the entropy $S(\{f\})$ over $\chi^2(\{f\})$ via extremizing the Lagrangian:

$$\mathcal{L}(\{f\}) = \alpha S(\{f\}) - \frac{\chi^2(\{f\})}{2} \tag{6c}$$

where the regularizing parameter α is chosen so that $\chi^2(\{f\})$ be equal to N, the number of independent data points.

4. A worked-out example: Deuteron disorder in $Ni(ND_3)_6 Br_2$

WHY STUDY THIS SYSTEM ?

The deuterated Nickel Hexammine Bromide system was chosen for the following reasons:
- Extensive and very accurate neutron data exist for this system (Hoser *et al.*, 1990). Hydrogen atoms were replaced by deuterons for a mere practical purpose: enhancing the signal-to-noise ratio of the measured Fourier components. The latter SNR is of the order of 1% for the whole data set.
- The Physics of this system is reasonably well understood (Schiebel *et al.*, 1990). In particular, the phasing of the structure factors using Frenkel models or Split atom models was found to be insentive to the models: 225 out of the 232 measured reflections could be phased unambiguously. The remaining 7 reflections were ignored for the density reconstructions, but not for the Patterson densities. Moreover, the dynamical nature of the Deuteron disorder could be checked directly by inelastic neutron scattering experiments (Janik *et al.*, 1988).
- The scattering lengths b_s for the different species s of atoms are all positive, resulting in positive scattering densities and Patterson densities. Using MaxEnt is consequently even more appropriate.
- Last but not least, the $Ni(ND_3)_6 Br_2$ crystal is centrosymmetric (space group $Fm3m$), resulting in real structure factors: the phase factors to be determined are either 1 or -1.

THE MOLECULAR DISORDER OF $Ni(ND_3)_6 Br_2$

Let us first describe the unit cell of the crystal (Figure 1). It consists of four cubic subunits arranged in a face-centered cubic cell. A Nickel atom sits at the center of each of these subunits (darkened on Figure 1, top) while 8 Bromine atoms sit on the corners of the subunit. The Br atoms are not represented on Figure 1 for the sake of clarity. Six identical $Ni-N$ arms extend from the Nickel atom along 4-fold axes and are ended by an ND_3 ammine group. Frustration arises from the local 3-fold symmetry of the ND_3 group and the 4-fold symmetry of its site inside the subunit cubic cell, resulting in the dynamical disorder of the ammine group, and hence in its delocalization within the unit cell. As will be shown below, the 3-dim Patterson density directly establishes the 4-fold symmetry of the disordered deuteron density, confirming conclusions drawn from the previous analyses mentioned above, based on least-squares fitting (Figure 1, bottom ‡).

---

‡ This figure is reproduced from P.Schiebel *et al.* (1990) (Courtesy Zeitschrift für Physik B) .

RESULTS

Let us first consider the 3-dim reconstruction of the density autocorrelation (Patterson) function directly from the measured reflection intensities and the associated experimental error bars. No pre-phasing operation is required, and the sought Patterson density is known to be positive everywhere. It is known from Buerger (1959) that, in our instance, the symmetry of the Patterson function is also described by the space group $Fm3m$. Only that portion of the unit cell corresponding to $0 \leq Y, Z \leq \frac{1}{4}$ and $0 \leq X \leq \frac{1}{2}$ needs to be reconstructed due to symmetry considerations. $X, Y, Z = 0$ corresponds to a Ni atom. The unit of length is the side $a = 10.382$ Å of the cubic unit cell. The latter was discretized into a $64*64*64$ pixel grid, $\frac{1}{32}$ of which was effectively reconstructed through 16 sections from $Z = 0$ to $Z = \frac{1}{4}$, or more appropriately from $Z \approx 0.008$ to $Z \approx 0.242$ since the Patterson density is sampled at the center of each pixel. In an autocorrelation density map $P(X, Y, Z)$, a peak centered at (X, Y, Z) is due to contributions of the densities both at (X_0, Y_0, Z_0) and at $(X_0 + X, Y_0 + Y, Z_0 + Z)$. Mere inspection from Figure 1 therefore suggests that sections 15 and 16, corresponding to $Z \approx \frac{1}{4}$ should yield 3-dim information about $Ni - D$ distances. Since Ni atoms are known to merely experience harmonic thermal motion about their equilibrium positions, the Patterson density in the mentioned sections close to the 4-fold ($\vec{c}$) axis should reveal the deuteron density, albeit slightly distorted due to the thermal motion of Ni atoms. The results for section 16 are displayed on Figure 3 for two experimental resolutions: $h, k, l \leq 8$ (54 data points) and $h, k, l \leq 18$ (232 data points). The plane squared deuteron density is now directly visible from high resolution data both on Standard Fourier and MaxEnt reconstructions (Figures 3a,b respectively) whereas it is only apparent on the MaxEnt reconstruction from low resolution data (Figure 3d). The peak at the center of the square stems from the $Ni - N$ correlations. It is enhanced with respect to the direct deuteron density due to the thermal motion of Ni atoms. Similar results were obtained for section 15.

Technically, the procedure to be described in section 5 of the text was used: an optimal constant 2-dim default of 500 was obtained from the projection along the $\vec{c}$ axis, yielding a constant 3-dim default of $\frac{500}{10.382} \approx 50$. Lower values resulted in distorted reconstructions.

This direct check of the result obtained from the Frenkel and split-atom ad hoc procedures substantiates their quasi model-free calculated phases of the measured unphased structure factors. The next step is thus to reconstruct the overall scattering density of the crystal using this latter combination. Once again, an optimum 2-dim default of 2.5 yielded a 3-dim default of ≈ 0.25. It was checked that lower default values result in distorted reconstructions. Figure 2b displays such an example: the 2-dim MaxEnt projection was reconstructed with a default of 10^{-4}. Figures 7a,b,c,d show how the optimal value 2.5 was obtained. The larger well-defined peak pertains to Ni and the smallest, on the opposite corner, to Br.

Note that the $F(000)$ value can be estimated from both the Least-Squares fitting approach and the MaxEnt reconstruction: in the case of the 2-dim projected density reconstruction the agreement was found to be of the order of 1%, the two methods yielding ≈ 799 and ≈ 809 respectively. The average scattering density can thus be estimated to be $\approx \frac{809}{10.832*10.832} \approx 7$. This latter value corresponds to our reconstruction displayed on Figure 7b: this choice of the default is clearly not optimal.

The 16 sections were reconstructed at the same time. First, that same part of section 16 as was displayed for the Patterson density autocorrelation function is displayed on Figures

4a,b,c,d, showing both the improvement over Standard Fourier reconstructions as well as high and low resolution results. Only those 225 reliably phased structure factors were kept for this analysis. As expected and seen from Figures 4c,d, the low resolution improvement resulting from using MaxEnt is obvious. By contrast with the previous Patterson results, the high resolution Standard Fourier (Figure 4a) and MaxEnt (Figure 4b) reconstructions now differ substantially: the peak due to Nitrogen atoms is still present but flattened out in the MaxEnt reconstruction. Its presence in the standard Fourier reconstruction is due to series termination effects. A pending problem is to disentangle the Nitrogen from the Deuterium contributions to the scattering density. A seemingly possible way of doing so is to use the Least-Squares fitted parameters (from the split or Frenkel models) describing all the atoms but the deuterons and reconstruct from them a 3-dim scattering density to be used as a non-uniform prior model $m(\vec{r})$.

Finally, Figures 5&6 display the 16 sections as retrieved using the Standard Fourier and MaxEnt procedures respectively. The contour levels are equi-spaced, positive and negative for Standard Fourier, strictly positive for MaxEnt. The results speak for themselves, yielding a much cleaner MaxEnt reconstruction with no spurious features popping up. The inquisitive reader may wonder why the three distinct deuteron "rings" appearing in section 16 are not equally intense. This stems from our discretized grid: again, section 16 does not correspond to $Z = \frac{1}{4}$ since it is offset by half a pixel. A second reason is that the distance of the Ni atom to the D_3 ring is equal to $0.24a$ rather than $0.25a$ as shown on Figure 1b for the sake of simplicity. Consequently, the averaged Z coordinates for the three D_3 rings appearing in section 16 are $0.26, 0.24$ and 0.26 from left to right, respectively.

5. A practical Historic MaxEnt problem: selecting the DEFAULT parameter

GENERAL

As mentioned in the above section, MaxEnt can yield very poor results if misused. The Cambridge Historic algorithm features one free parameter DEF, the DEFAULT (Skilling and Gull, 1985, page 94). This parameter can be understood:
- as the value towards which the local density will tend, if there is no information in the data pertaining to this part of the image to be reconstructed.
- as the weighted logarithmic **average** of the density over the **total** image. More specifically, $\ln(DEF)$ is equal to $\sum p_i \ln f_i$ using the standard notations of the aforementioned reference. Although not contradictory with the previous statement, this weighted average is reminiscent of the $-\sum p_i \ln p_i$ form of the Shannon-Jaynes entropy. It is thus very plausible that, as we find in practice for phased Fourier data sets, a given value of the non local quantity DEF will constrain the total flux $\sum f_j$. Note that the total flux is not constrained by the data, since only non trivial Fourier components are measured.

If DEF is chosen to be too small, not enough total flux will be allowed in the reconstruction in order to accommodate the data and χ^2 may never reach N, the number of independent data points. By contrast, if DEF is too large, a lot of flux $\{f_j\}$ will be injected in the reconstructed density. The modulations of the density, required by the data, will not be hindered by the positivity constrain and χ^2 will easily reach N, yielding a modulated density on top of a huge constant "background". As the latter rises, along with DEF, the entropy becomes less and less sensitive to the modulation and regularization becomes less and less effective.

Even though a fully Bayesian determination of DEF should be desirable, our prior knowledge borne from Physics must also be accommodated: we know that there is no atom in some parts of the unit cell, so that the "background" mentioned just above must be kept as small as possible.

EXAMPLE

We consider the projection of the scattering density along the $\vec{c}$ axis of the unit cell of $Ni(ND_3)_6Br_2$ depicted on Figure 1. The data consist of 42 unique reflections measured within 1%. A $100 * 100$ pixel grid, restricted to $-\frac{1}{4} \leq X, Y \leq \frac{1}{4}$ from symmetry considerations, was selected and the Historic MEMSYS algorithm was run with the much too small DEF value of 10^{-4}, yielding the Image shown on Figure 2b. The choice of this very small default was prompted by the wish to have no density at all where there is no atom.

The ripples from the Standard Inverse Fourier Transform displayed on Figure 2a, due to truncation effects are now replaced on Figure 2b by split peaks at the center and the corners, as well as edge-centered distorted ones. The reconstructed projected density pertains to the bottom left darkened cube shown on Figure 1, projected along the vertical axis. Correspondingly, the center peak is due to Ni atoms and the corners to Br atoms. The highly edge-centered distorted peaks relate to the information actually sought for: the Deuteron density.

It was then checked, through numerical simulations, that the above described syndrom was genuine. In one instance, the mock data was generated using the calculated fitted parameter values regarding the well-behaved atoms Ni, N and Br and the ensuing calculated structure factors were used to reconstruct the projected density, yielding split peaks for too small default values.

The following two-step procedure was used to solve this problem and might be used if analogous difficulties occur. It is illustrated on Figure 7.

- First, we choose an initial very small default value, say 10^{-4}. The default is thus upgraded after each MaxEnt iterate, using $DEF = \exp\{\sum p_i \ln f_i\}$. The default is gradually increased until χ^2 is equal to N (or the normalized χ^2 to 1). The algorithm is then run at constant default until convergence is achieved. As can be seen on Figure 7a, a large constant background shows up and some spurious ripples are still present, due to the ensuing poor entropic regularization.

- Second, the default is suitably decreased (Figures 7b,c) until the background BACK becomes negligible: the entropic prior becomes more and more effective and truncation effects vanish (Figure 7d). Numerically, one finds $DEF = 20, 7, 2.5$ and $BACK = 13.1, 1.73, 0.00218$ for Figures 7b,c,d respectively.

Whilst this procedure is valid for densities as well as Patterson densities, projections or sections, it becomes more and more tedious with the number of involved pixels. If a suitable projection that gathers enough information from the data can be found and the ensuing optimal default A obtained, a very good guess for the corresponding 3-dim optimal default B can be found using:

$$B = A \cdot \frac{volume\ of\ the\ projected\ unit\ cell}{volume\ of\ the\ unit\ cell}$$

That B is optimal can be further checked by inspecting the MaxEnt reconstructed sections and looking for the background level in each one of these. If this is not the case, the general procedure can then be followed.

6. Conclusions

The present paper demonstrates the use of MaxEnt to yet another phased crystallographic problem: the study of disorder in simple molecular crystals. In this latter instance, the 3-dim Patterson (autocorrelation) density function, as retrieved by MaxEnt from measured intensity data sets, provides a direct and model-free estimate of the disordered density itself. In turn, this information corroborates the *ad hoc* but consistent least-squares fitting procedure used to phase the structure factors. These ensuing structure factors with measured moduli and computed phases yield, via a second use of MaxEnt, the disordered density sought for, from which relevant thermodynamic quantities can be estimated.

Technically, the DEFAULT value from the Historic MaxEnt algorithm must be carefully chosen: if too small, the algorithm may never meet the $\chi^2 = N$ criterion or will at least produce a severely spuriously distorted image. If too large, the algorithm will take a long time to converge and will become entropy insensitive, hence inefficient. Albeit interested in the 3-dim density reconstruction (sections), we advocate tuning the default value through the reconstruction of one suitable projection, which reduces to a 2-dimensional problem.

ACKNOWLEDGMENTS. Thanks are due to Steve Gull for useful remarks and suggestions. The use of the Cambridge MEMSYS code is gratefully acknowledged.

REFERENCES

Buerger, M.J.: 1959, 'Vector space', New York: Wiley. (Chapter9, table 1, page 210.)

Collins, D.M.:1982, 'Electron density images from imperfect data by iterative entropy maximization', *Nature* **298**, 49.

David, W.I.F.:1990, 'Extending the power of powder diffraction for structure determination', *Nature* **346**, 731.

Gull, S.F. & Daniell, G.J.: 1978, 'Image reconstruction from incomplete and noisy data', *Nature* **272** 686.

Gull, S.F.: 1989, 'Developments in Maximum Entropy Analysis', in J. Skilling (1989a).

Hoser, A., Prandl, W., Schiebel, P. & Heger, G.: 1990, 'Neutron diffraction study of the orientational disorder in $Ni(ND_3)_6Br_2$', *Z. Phys. B.* **81**, 259.

Janik, J.A., Janik, J.M., Migdal-Mikuli, A., Mikuli, E. & Otnes, K.: 1988, 'Incoherent quasielastic neutron scattering study of NH_3-reorientations in $[Ni(NH_3)_6]Br_2$ in relation to the phase transition', *Acta Phys. Pol.* **A74**,423.

Papoular, R.J. & Gillon, B.: 1990, 'Maximum Entropy reconstruction of spin density maps in crystals from polarized neutron diffraction data', in M.W. Johnson (ed.), *Neutron Scattering Data Analysis 1990*, Inst. Phys. Conf. Ser. **107**, Bristol.

Papoular, R.J.: 1991, 'Structure factors, projections, inverse Fourier transforms and crystal symmetry', *Acta Cryst.* **A47**, 293.

Sakata, M. & Sato, M.: 1990 'Accurate Structure Analysis by the Maximum Entropy Method', *Acta Cryst.* **A46**, 263.

Sakata, M., Mori, R., Kumazawa, S, Takata, M. & Toraya, H.: 1990, 'Electron-Density distribution from X-ray Powder Data by use of Profile fits and the Maximum-Entropy Method', *J. Appl. Cryst.* **23**, 526.

Schiebel, P., Hoser, A., Prandl, W. & Heger, G.: 1990, 'Orientational disorder in $Ni(ND_3)_6Br_2$: a two-dimensional thermodynamic rotation-translation model for the deuterium disorder', *Z. Phys. B.* **81**, 253.

Skilling, J. & Gull, S.F.: 1985, 'Algorithms and applications', in C. Ray Smith & W.T. Grandy, Jr.
    (eds.), *Maximum Entropy and Bayesian Methods in Inverse problems*, D. Reidel, Dordrecht.
Skilling, J., ed.: 1989a, *Maximum Entropy and Bayesian Methods*, Kluwer, Dordrecht.
Skilling, J.: 1989b, 'Classic Maximum Entropy', in J. Skilling (1989a).

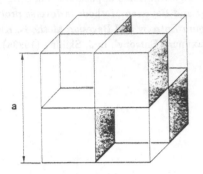

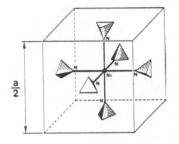

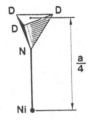

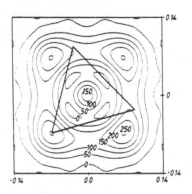

Fig. 1.   The $Ni(ND_3)_6Br_2$ unit cell. a) It consists of 4 cubic subunits. b) Each subunit corresponds to one molecule. The Bromine atoms, sitting at the corners, are not represented. The disordered parts are the rigid ND_3 ammine molecules. c) The D-triangle samples a squared potential (from Schiebel *et al.*, 1990).

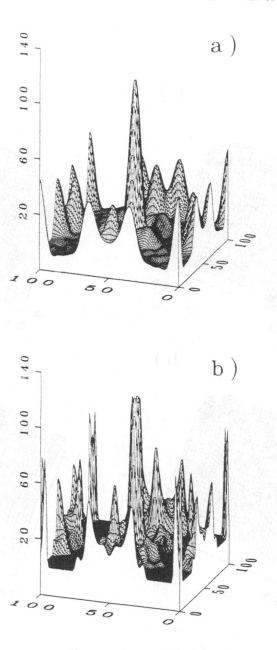

Fig. 2. Reconstruction of the 2-dim projected density along $\vec{c}$. a) Standard Inverse Fourier reconstruction with truncation effects. b) MaxEnt reconstruction corresponding to too small a default: it is split and distorted.

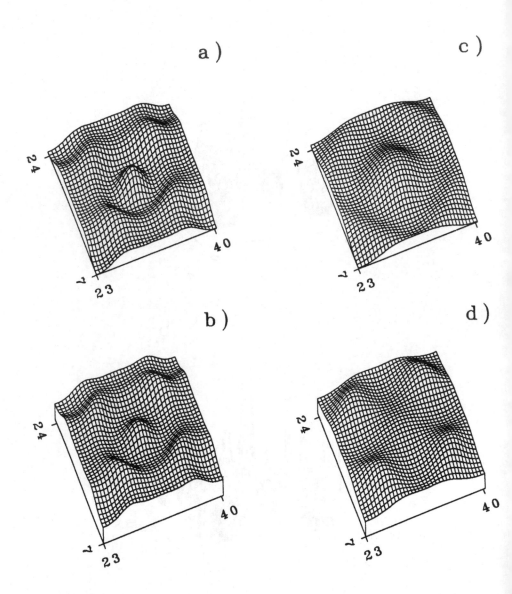

Fig. 3.   3-Dim reconstruction of the Patterson function. Part of section 16 showing directly the squared disordered deuteron density.
a) & b) High Resolution (232 unique reflections): Standard Fourier and MaxEnt respectively.
c) & d) Low Resolution (54 unique reflections): Standard Fourier and MaxEnt respectively.

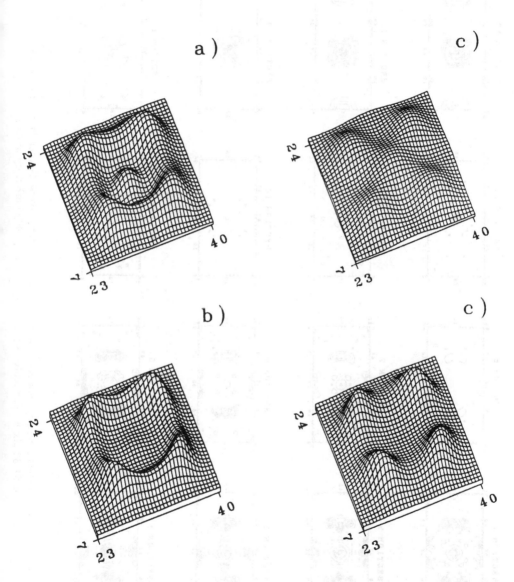

Fig. 4.   3-Dim reconstruction of the scattering density. Part of section 16 showing the squared disordered deuteron density.

   a) & b) High Resolution (225 unique reflections): Standard Fourier and MaxEnt respectively.

   c) & d) Low Resolution (54 unique reflections): Standard Fourier and MaxEnt respectively.

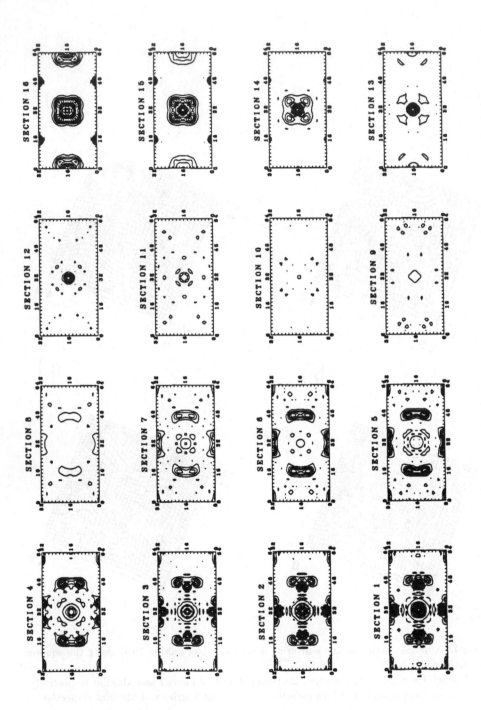

Fig. 5. 3-Dimensional reconstruction of the density using a Standard Inverse Fourier Transform. The size of each section is 64*32 pixels. The contours are equispaced (step: 40 in arbitrary units).

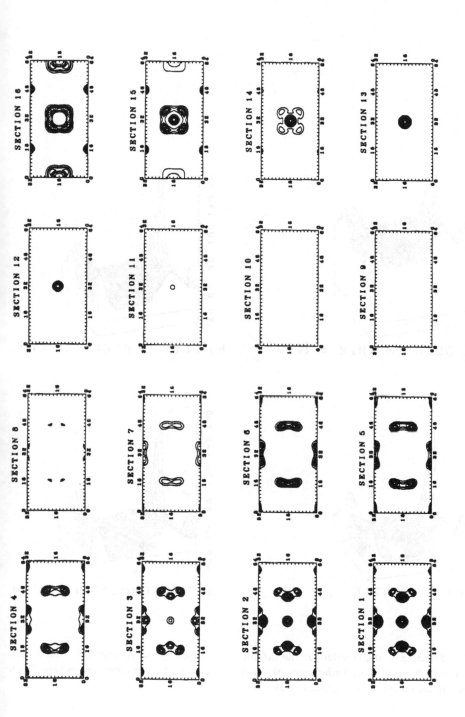

Fig. 6.   3-Dimensional reconstruction of the density using Historical MaxEnt: There is no spurious contour. The size of each section is 64*32 pixels. The contours are equispaced (step: 40 in arbitrary units).

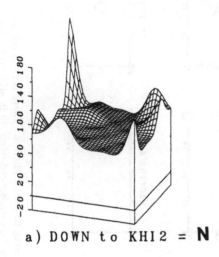

a) DOWN to KHI2 = **N**

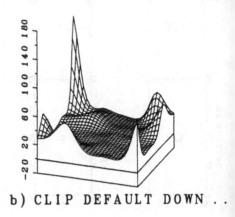

b) CLIP DEFAULT DOWN ..

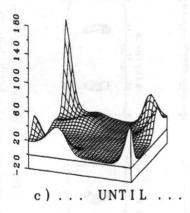

c) ... UNTIL ...

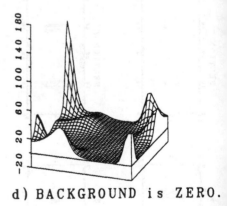

d) BACKGROUND is ZERO.

Fig. 7.   Selecting a proper value for the default: a) meet the $\chi^2 = N$ criterion with a variable default. b),c) & d) Dichotomize the default down until the background is zero. Each reconstruction is run at constant default.

ON A BAYESIAN APPROACH TO COHERENT RADAR IMAGING

D. Styerwalt *
G. Heidbreder †
Department of Electrical and Computer Engineering
University of California, Santa Barbara
Santa Barbara, CA 93106

ABSTRACT. In coherent radar imaging the object of interest is complex but only its amplitude is to be estimated. The object phase yields nuisance variables and in a proper Bayesian approach it is necessary to obtain a phaseless likelihood function. We investigate a two-dimensional case in which the target object is modelled as a collection of point scatterers having independent random phases. The phaseless likelihood function is determined exactly for a configuration of data samples in a uniformly spaced square array in spatial frequency when the target scatterers are confined to lattice positions of a "matched" spatial array. It is determined approximately when the target scatterers are arbitrarily positioned, at most one per conventional resolution cell. The relation between maximum likelihood and conventional Fourier transform imaging is explored and the feasibility of a CLEAN algorithmic technique in coherent imaging is considered.

1. Introduction

Two-dimensional high resolution radar imaging is now widely used in radar scattering diagnostics. The target of interest is mounted on a turntable which is minimally reflective and the radar line-of-sight is normal to the axis of turntable rotation. High resolution along the line-of-sight is effected by time delay sorting of the wideband radar returns. High resolution in a direction normal to both the line of sight and the axis of target rotation is obtained by inverse synthetic aperture (ISAR) processing of data taken as the target is rotated through a small angle. The target is treated as a planar object with a complex reflectivity distribution which is in fact the marginal distribution obtained by projecting the actual three-dimensional distribution on a plane normal to the axis of rotation. With the assumptions of plane wave illumination and no multiple scattering, shadowing or other interactions between scatterers affecting the radar returns, it can be shown (Mensa, 1981) that the radar data are samples of the Fourier transform of the (complex) radar reflectivity as described above. Similarly related data may be obtained from a moving target in an operational environment if it is turning relative to the line-of-sight and if the motion of its center of rotation can be compensated (Wehner, 1987). On appropriate selection of radar

---

* Currently at Hughes Aircraft Co., Canoga Park, CA.
† Currently at David Taylor Research Center, Bethesda, MD.

C. R. Smith et al. (eds.), Maximum Entropy and Bayesian Methods, Seattle, 1991, 377–396.
© 1992 *Kluwer Academic Publishers.*

wavelengths and target aspect angles the data samples are grouped at less than Nyquist spacing within a keystone shaped area in the spatial frequency domain.

The conventional approach to image reconstruction in this application has been to interpolate the data onto a rectangular array of equi-spaced elements, window, and inverse Fourier transform, retaining only the resultant magnitudes. In imaging reflectivity magnitude, interest has centered on providing high resolution and wide dynamic range, i.e., the ability to distinguish closely spaced concentrations of reflectivity and to identify weakly scattering concentrations in the presence of strong concentrations. Relative (but not absolute) location information with accuracy approaching the order of a radar wavelength is desired. It is not clear that the reflectivity phase distributions cannot be effectively used in radar cross section diagnostics, but they are typically discarded.

We are concerned herein with how Bayesian estimation might be employed to improve the radar images of reflectivity magnitude. We draw significantly from earlier work of Jaynes (1983) and Bretthorst (1985) in one-dimensional spectral analysis.

2. The Target Model

The conventional image data processing as described above models only the mapping relationship between target and data. No statistical modelling of the target reflectivity is involved. Indeed this is a significant shortcoming of the procedure since coherent radar images are not ordinarily repeatable with precision because of the phenomenon commonly known as speckle noise. Sometimes images derived from independent data sets are averaged to minimize the speckle effect. Bayesian reconstruction forces this important issue at the outset. It will be necessary to assign a priori distributions to both scatterer amplitude and phase variables in order to obtain Bayesian estimates of scatterer amplitudes.

We assume a two-dimensional (2-D) target object discretely modelled with ith scatterer complex reflectivity

$$f_i = a(\bar{x}_i)e^{j\theta(\bar{x}_i)} = a_i e^{j\theta_i}, \tag{1}$$

where $\bar{x}_i$ is a vector describing the ith scatterer position. The data are noisy samples of the Fourier transform of the target object. The sample at spatial frequency $\bar{f}_k$ is

$$\begin{aligned} D_k &= F_k + E_k \\ &= \sum_{i=0}^{N-1} a(\bar{x}_i) \exp\left[j\theta(\bar{x}_i) - j2\pi\bar{x}_i \cdot \bar{f}_k\right] + E_k, \end{aligned} \tag{2}$$

where E_k is a zero mean complex Gaussian noise variable each component of which has variance σ^2. Data samples result from independent measurements; hence, the various E_k are independent. We treat $\bar{x}_i$ as measured from the target center of rotation. This neglects a phase gradient with spatial frequency which has only the effect of shifting the position of the reconstructed image in spatial coordinates.

No a priori knowledge of scatterer amplitudes and phases is assumed. We assume only that scatterers are independent with phase and amplitude independent of each other, that phases are uniformly distributed in $(0, 2\pi)$, and that amplitudes are uniformly distributed over the range of amplitudes which are feasible, i.e., compatible with the data.

3. The Phaseless Likelihood Function

We seek the set of values of N, the a_i, and the $\bar{x}_i$ having maximum probability density, given the data vector $\bar{D} = [D_0, D_1, \ldots, D_{N-1}]^T$, the measurement noise standard deviation σ, and general background information I (the target model and the relationship mapping it to data). Let $\{w\}$ represent the set of unknown image parameters to be estimated. Bayes' theorem yields the relation of probability densities

$$p(\{w\}|\bar{D}I\sigma) = \frac{p(\{w\}|I\sigma)\,p(\bar{D}|\{w\}I\sigma)}{p(\bar{D}|I\sigma)} \tag{3}$$

$$\propto p(\{w\}|I)\,p(\bar{D}|\{w\}I\sigma). \tag{4}$$

In writing (4) we have noted that our *a priori* knowledge of $\{w\}$ is independent of σ and have omitted $p(\bar{D}|I\sigma)$ which is merely a normalizing constant not affecting maximization over $\{w\}$.

To obtain the likelihood function $p(\bar{D}|\{w\}I\sigma)$, we first express the more easily recognized (Gaussian) likelihood function conditioned on both $\{w\}$ and the vector of scatterer phases $\bar{\theta} = [\theta_0, \theta_1, \ldots, \theta_{N-1}]^T$. Thus, in lieu of (4) we have

$$p(\{w\}\bar{\theta}|\bar{D}I\sigma) \propto p(\{w\}\bar{\theta}|I)\,p(\bar{D}|\{w\}\bar{\theta}I\sigma). \tag{5}$$

The components of $\bar{\theta}$ are treated as nuisance variables to be removed by the marginalization

$$p(\{w\}|\bar{D}I\sigma) \propto \int d\bar{\theta}\,p(\{w\}\bar{\theta}|I)\,p(\bar{D}|\{w\}\bar{\theta}I\sigma). \tag{6}$$

The likelihood function $p(\bar{D}|\{w\}\bar{\theta}I\sigma)$ in the integrand of (6) is the familiar joint Gaussian density for the M^2 independent measurements, namely

$$p(\bar{D}|\{w\}\bar{\theta}I\sigma) = (2\pi\sigma^2)^{-M^2} \exp[-\frac{1}{2\sigma^2}\sum_k |D_k - F_k|^2]. \tag{7}$$

Note that each measured value $D_k = I_k + jQ_k$ is complex and that σ^2 is the variance for each of the I_k and Q_k. The F_k are dependent on the a_i and the θ_i as noted in (2). The a priori density for the latter parameters must be inferred from the target model. We have chosen a model consisting of independent scatterers each with random phase independent of other scatterer parameters and uniformly distributed. Thus

$$p(\{w\}\bar{\theta}|I) = (2\pi)^{-N}p(\{w\}|I). \tag{8}$$

Recalling our earlier assumption that the parameters $\{w\}$ are uniformly distributed over their range of feasibility (as determined by their non-zero likelihood values), we treat (8) as constant in (6). Then on incorporating (7), (6) becomes

$$p(\{w\}|\bar{D}I\sigma) \propto \int d\bar{\theta}\,\exp[-\frac{1}{2\sigma^2}\sum_k |D_k - F_k|^2]. \tag{9}$$

It will be convenient to separate unknown parameter dependent terms from constant ones in the expansion of the sum in (9). Thus

$$\sum_k |D_k - F_k|^2 = M^2(D^2 + Q), \tag{10}$$

where

$$D^2 = M^{-2} \sum_k |D_k|^2$$

and

$$Q = M^{-2} \sum_k \left\{ |F_k|^2 - 2Re[D_k^* F_k] \right\}. \tag{11}$$

Further expansion yields

$$M^{-2} \sum_k |F_k|^2 = M^{-2} \sum_k \sum_{i=0}^{N-1} \sum_{l=0}^{N-1} a_i a_l \exp[j(\theta_i - \theta_l) - j2\pi(\bar{x}_i - \bar{x}_l) \cdot \bar{f}_k]$$

$$= \sum_{i=0}^{N-1} \sum_{l=0}^{N-1} a_i a_l M_{il} \exp[j(\theta_i - \theta_l)] \tag{12}$$

and

$$M^{-2} \sum_k Re[D_k^* F_k] = M^{-2} Re \left\{ \sum_k D_k^* \sum_{i=0}^{N-1} a_i \exp\left[j\theta_i - j2\pi\bar{x}_i \cdot \bar{f}_k\right] \right\}$$

$$= \sum_{i=0}^{N-1} a_i Re \left\{ \exp(j\theta_i) M^{-2} \sum_k D_k^* \exp\left[-j2\pi\bar{x}_i \cdot \bar{f}_k\right] \right\}$$

$$= \sum_{i=0}^{N-1} a_i Re \left\{ \exp(j\theta_i) d_i \exp(-j\phi_i) \right\} \tag{13}$$

$$= \sum_{i=0}^{N-1} a_i d_i \cos(\theta_i - \phi_i),$$

where

$$M_{il} = M^{-2} \sum_k \exp\left[j2\pi(\bar{x}_l - \bar{x}_i) \cdot \bar{f}_k\right] \tag{14}$$

is the response at $\bar{x}_l$ to a unit amplitude zero phase point target at $\bar{x}_i$ for the imaging system which simply does inverse Fourier transformation of the data and

$$d_i \exp(j\phi_i) = M^{-2} \sum_k D_k \exp(j2\pi \bar{x}_i \cdot \bar{f}_k) \tag{15}$$

is the inverse Fourier transform of the data evaluated at $\bar{x}_i$. The Fourier transforms in (14) and (15) are inverse discrete Fourier transforms (IDFT's) but incorporate the frequency offset in the data.

Evaluation of a phaseless likelihood function requires that we perform the integration in (9). This is best done by converting the integral into the product of N one-dimensional integrals and in general requires a transformation of the variables $\{\exp(j\theta)\}$ through diagonalization of the matrix whose elements are the $\{M_{il}\, a_i\, a_l\}$. We evade this daunting task however and note that the desired reduction of the integral is immediate if $M_{il} = \delta_{il}$. We consider the form taken by M_{il} in conventional imaging practice. Normally the $\bar{f}_k$ are arranged on the lattice points of a square grid (sometimes after a necessary interpolation). On describing the vectors $(\bar{x}_i - \bar{x}_l)$ and $\bar{f}_k$ by their rectangular coordinates, we have

$$\bar{x}_i - \bar{x}_l = (x_i - x_l, y_i - y_l)$$
$$\bar{f}_k = (f_0 + k_x\,\Delta f,\ k_y\,\Delta f), \qquad k_x, k_y = -\frac{(M-1)}{2}, \ldots, \frac{(M-1)}{2},$$

where k_x and k_y are integers, Δf is data sample spacing and f_0 is the radial offset in spatial frequency resulting from the use of a band of radar frequencies offset from zero frequency. With these substitutions (14) becomes

$$M_{il} = M^{-2} \sum_{k_x=-(M-1)/2}^{(M-1)/2} \sum_{k_y=-(M-1)/2}^{(M-1)/2} \exp\left[-j2\pi\{(x_i - x_l)(f_0 + k_x\Delta f) + (y_i - y_l)k_y\Delta f\}\right]$$

(16)

from which we obtain

$$M_{il} = s(x_i - x_l)\, s(y_i - y_l)\, \exp\left[\frac{j4\pi(x_l - x_i)}{\lambda_0}\right], \tag{17}$$

where

$$s(x) = \frac{\sin\left[M\pi\Delta f x\right]}{M\,\sin\left[\pi\Delta f x\right]}$$

and $\lambda_0 = 2/f_0$ is the radar wavelength at band center.

We note immediately that $M_{ii} = 1$ and that when $x_i - x_l$ and $y_i - y_l$ are integer multiples (other than 0 or a multiple of M) of $1/M\Delta f$, $M_{il} = 0$. Thus if all the scatterers are located on lattice points of a regular grid with spacing $1/M\Delta f$ (the sampling points of an inverse discrete Fourier transform of the data), $M_{il} = \delta_{il}$. In the more general case of arbitrary scatterer locations separated by more than $1/M\Delta f$, M_{il} will be small when M is large, -13 dB relative to M_{il} in the worst case. From (12) we note that M_{il} expresses a coupling between scatterers which is to be taken into account in computing the likelihood in (9). With M_{il} as in (17) and considering that the sum in (12) is to be taken over random phases, we expect that sum to be dominated by the terms for which $i = 1$.

Consider now the integration in (9) when $M_{il} = \delta_{il}$:

$$\int d\bar{\theta}\, \exp\left\{-\frac{M^2}{2\sigma^2}(D^2 + Q)\right\} \propto \int d\bar{\theta}\, \exp\left\{-\frac{M^2}{2\sigma^2}\sum_{i=0}^{N-1}[a_i^2 - 2a_i d_i\, \cos(\theta_i - \phi_i)]\right\}$$

$$\propto \prod_{i=0}^{N-1}\left\{\exp\left[-\frac{M^2 a_i^2}{2\sigma^2}\right]\int_0^{2\pi} d\theta_i\, \exp\left[\left(\frac{M^2 a_i d_i}{\sigma^2}\right)\cos(\theta_i - \phi_i)\right]\right\}$$

$$\propto \prod_{i=0}^{N-1} \exp\left[-\frac{M^2 a_i^2}{2\sigma^2}\right] I_0\left[\frac{M^2 a_i d_i}{\sigma^2}\right]. \tag{18}$$

In writing (18) we have used (10) through (15) and have neglected to express terms which are merely constant multipliers not affecting the maximization of (9) over the set $\{w\}$. $I_0(\cdot)$ is the modified Bessel function of first kind and order zero. Since (18) is proportional to the desired phaseless likelihood function we refer to it as the quasi-likelihood function

$$L = \prod_{i=0}^{N-1} \exp\left(-\frac{M^2 a_i^2}{2\sigma^2}\right) I_0\left(\frac{M^2 a_i d_i}{\sigma^2}\right). \tag{19}$$

It is exact for the case of $M_{il} = \delta_{il}$, i.e., that of a target model consisting of $N = M^2$ point scatterers on the lattice points of the regular array described above. In that case the unknowns of interest are only the set $\{a_i\}$ of scatterer amplitudes. But we regard (19) as approximately correct in the case of other configurations of scatterers with N, the set $\{a_i\}$, and the set of scatterer positions $\{x_i\}$ as unknowns. Scatterer locations appear in (19) through the d_i, each of which depends on the parameters of all the scatterers.

4. A Maximum Likelihood Image

Consider the case of a target consisting of point scatterers at the regular grid lattice points. It might be argued naively that, since a continuous image may be well represented by samples on this grid, that an arbitrary planar target may be similarly well represented. Such a rationale presupposes erroneously that the field sampled by the measurements is essentially limited to the measurement bandwidth (the spatial frequency window containing data sample points) and that the sample values of the image are actually to be associated with elemental scattering distributions of finite dimensions associated with that bandwidth. In short it relates to a model representable by a set of sample values, not by a set of point scatterers. We choose the lattice point array of scatterers as a model with no justification other than the exactness which it attaches to (19).

It is more convenient to work with the quasi-log-likelihood function

$$\log L = \sum_{i=0}^{N-1} \left\{ \ln I_0\left(\frac{M^2 a_i d_i}{\sigma^2}\right) - \frac{M^2 a_i^2}{2\sigma^2} \right\} \tag{20}$$

than with L; $\log L$ (and hence L) will be maximum when each of the terms of the sum in (20) is maximum. We find that the expression

$$\ln I_0\left(\frac{M^2 a_i d_i}{\sigma^2}\right) - \frac{M^2 a_i^2}{2\sigma^2}$$

is stationary in a_i for values which are solutions of

$$\frac{a_i}{d_i} = \frac{(a_i M/\sigma)}{(d_i M/\sigma)} = \frac{I_1\left[\left(\frac{a_i M}{\sigma}\right)\left(\frac{d_i M}{\sigma}\right)\right]}{I_0\left[\left(\frac{a_i M}{\sigma}\right)\left(\frac{d_i M}{\sigma}\right)\right]}, \tag{21}$$

where $I_1(x) = dI_0(x)/dx$. For $d_i^2 < 2\sigma^2/M^2$ there is a single solution, $a_i = 0$, which yields a maximum. For $d_i^2 > 2\sigma^2/M^2$, $a_i = 0$ remains a solution but yields a minimum and there is an additional solution which approaches $a_i = d_i$ asymptotically with increasing d_i. This additional solution is readily shown to yield a maximum. Figure 1 is a plot, obtained by numerical solution of (21), of the normalized value of the a_i maximizing (20) versus the correspondingly normalized value of d_i. The normalization is explicitly indicated in (21) to emphasize that a single non-linear function mapping d_i to a_i can be applied on a pixel-by-pixel basis regardless of measurement array parameter M or noise level σ^2.

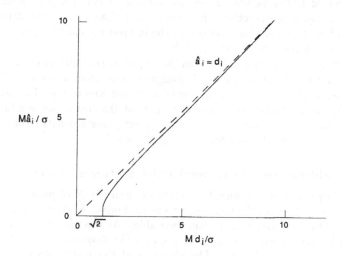

Fig. 1. Scatterer amplitude ML estimate vs. IDFT amplitude.

The cutoff value of d_i in Figure 1, i.e., the value of d_i which is so low that Bayesian analysis rejects it as evidence of a scatterer at the ith lattice position, corresponds to the condition

$$\frac{M^2 d_i^2}{2\sigma^2} = 1, \qquad (22)$$

which may be interpreted as unity ratio of signal-plus-noise to noise in the inverse discrete Fourier transform (IDFT) of the data (The IDFT may be thought of as a coherent processor with processing gain M^2). The cutoff value of d_i also sets the lower end of the dynamic range of the image reconstruction. This dynamic range is

$$10 \log_{10}\left(\frac{M^2 d_i^2 \, max}{2\sigma^2}\right) dB$$

and is to be contrasted with the usual dynamic range limited on the lower end, not by measurement error, but by sidelobe levels of the strong scatterers. We note that the dynamic range is determined entirely by the sampled quantized data if roundoff errors in computation of the IDFT are negligible.

We conclude that the IDFT, without the usual processor windowing, but with a cutoff at noise level, is a maximum likelihood estimator of the amplitudes of the scatterers in our simplistic point scatterer lattice target model. Further, with the IDFT computed on the target lattice points there is no leakage between IDFT coefficients representing pixel amplitudes, i.e., there are no sidelobe effects.

The above results seem rather trivial for it is well known that an N-point DFT exhibits no leakage when the input consists of sample sequences whose frequencies are matched, i.e., having an integer multiple of $2\pi/N$ samples per radian. The sidelobes of the rectangular measurement window transform have not disappeared; we have merely done our sampling at its nulls. The objectionable sidelobe effects appear when scatterers are shifted away from lattice points or even when zero padding of the data is employed to interpolate the image between ideally spaced lattice points. Even the noise level cutoff seems inconsequential since in practical image reconstruction, sidelobes rather than noise level determine low level pixel amplitudes. It remains however one of the interesting and unexpected subtleties revealed by Bayesian analysis (Bretthorst, 1985).

One significant conclusion does follow from the apparent triviality of the above results. It is that simple IDFT image reconstruction is designed to be optimal for a trivially simple target object. This fact is not altered by the use of processor windowing. The latter merely provides some robustness against more general targets at the cost of some additional loss of resolution. Should we not be able to do better by designing reconstruction algorithms which are optimal for somewhat more general target models?

5. Maximum Likelihood and the General Point Scatterer Model

Suppose the target model is changed to allow the number N of point scatterers and their locations and amplitudes as the unknowns to be estimated. Scatterer phases, modeled as random and independent, remain as nuisance variables. As was noted above computation of the phaseless likelihood function in general requires the diagonalization of the $N \times N$ matrix whose ilth element is $M_{il}a_ia_l$. The elements of the matrix and indeed its size depend on the unknown parameters. A search for the maximum likelihood values of the parameters is accordingly difficult. Rather than embarking on this search we ask if a useful approximate result may be found. We have previously cited (17) as justification for the approximation $M_{il} = \delta_{il}$ for the case of a square 2-D array of sampling points in spatial frequency. Then (19) and (20), which depend on all the unknowns of interest, approximate the correct forms for the phaseless quasi-likelihood function and its logarithm. It is easy to show, subject to the approximations now inherent in (20), that the IDFT again produces the desired results. At a stationary point of (20) the first partial derivative with respect to each unknown must be zero and the second partial derivative must be negative. We find that the spatial coordinates of scatterers yielding maxima of $\log L$ are precisely the spatial coordinates yielding maxima of d_i, i.e., the locations of the N highest peaks of IDFT magnitude. [Recall the implicit assumption in (20) of a single scatterer per resolution cell.] In accurately locating IDFT magnitude peaks it will be necessary to compute the IDFT on a fine grid by appending a large field of zero samples to the data array (zero padding). With scatterer locations determined the analysis leading to (21) applies to optimizing over the $\{a_i\}$. The IDFT values for the maximum scatterer amplitudes are essentially optimum. The value of the discrete variable N which maximizes $\log L$ is readily determined by individually testing the terms of (19) or (20), each of which is identified with a scatterer producing an IDFT peak. Having determined the values of d_i and a_i to be associated with a scatterer we compute the factor

$$\exp\left(\frac{M^2a_i^2}{2\sigma^2}\right) - I_0\left(\frac{M^2a_id_i}{\sigma^2}\right).$$

If the factor exceeds unity it tends to increase (19) and (20), i.e., there is evidence in the data for the existence of its associated scatterer. The value of N maximizing (20) is the number of scatterers passing the individual scatterer test. We have already encountered this test in solving (22), having found that

$$\ln I_0 \left(\frac{M^2 a_i d_i}{\sigma^2} \right) - \frac{M^2 a_i^2}{2\sigma^2}$$

has a single maximum which occurs for $a_i = 0$ (and has value 0) unless $d_i^2 > 2\sigma^2/M^2$, in which case the maximum exceeds zero. Thus the test of a factor in the log-likelihood function reduces to the simpler one of testing individual IDFT peak signal-to-noise ratio.

What may we conclude from these approximate results? Certainly we cannot claim that they produce good approximations to maximum likelihood estimates of all image parameters. It would be folly to claim identification and parameter specification of a low level scatterer in the presence of a number of much higher level ones. The matrix M_{il} has potential off diagonal elements only 13 dB below the diagonal element level. We do gain some understanding of how conventional IDFT image reconstruction relates to maximum likelihood estimation. We also suspect that the approximation does provide near maximum likelihood estimates of the parameters of the strongest scatterer and perhaps those of some others which are above the troubling -13 dB relative magnitude level. Consider Q expressed in terms of (12) and (13) as a single sum over the scatterers:

$$Q = \sum_{i=0}^{N-1} a_i \left[\left\{ \sum_{l=0}^{N-1} a_l \, \exp\left[j\left(\theta_i - \theta_l\right)\right] M_{il} \right\} - 2d_i \, \cos\left(\theta_i - \phi_i\right) \right]. \tag{23}$$

In the approximation we neglect terms of the l-summation for $l \neq i$. On comparing terms of the i-summation, we observe that the greater the value of a_i, the less the term is degraded by the approximation. Thus in the representation of Q as a sum of independent terms the term identified with the strongest scatterer is most accurate.

The above approximate results may be viewed as maximum likelihood estimates of the parameters of single scatterer models associated with the respective IDFT peaks. The parameters found, in the case of the highest peak, are those of the single scatterer whose data contribution provides the best least squares fit to the actual data. Suppose we wish to find the location, amplitude, and phase of the single scatterer most likely to have produced our data. Our choices should minimize (11) which is now

$$Q = a^2 - 2ad \, \cos\left(\theta - \phi\right), \tag{24}$$

where d and ϕ, the amplitude and phase of the IDFT are functions of scatterer location $\bar{x}$. We have written the quasi-log-likelihood Q as a function of actual scatterer phase to show that not only do our earlier results apply showing that the IDFT yields the optimum scatterer location and essentially the optimum amplitude, but also to show that the IDFT phase corresponding to its peak amplitude is the optimum phase. Thus to find the single scatterer optimum parameters we simply search the IDFT for its highest peak amplitude value and record position, amplitude, and phase. By restricting the IDFT search to the vicinity of any peak we can simply obtain the parameters of the single scatterer model which best accounts for that peak. We hasten to add, however, that the parameters for individual peak fitted models are not those of the best multiple peak fitted model. Further in comparing individual peak fitted models to more proper multiple peak fitted models we may expect greatest correspondence for the parameters associated with the highest peak.

6. Is a coherent imaging CLEAN algorithm feasible?

The computational complexity associated with finding the maximum likelihood parameters of a multiple point scatterer model results from the large size of our image parameter hypothesis space. One may reasonably ask if the task may be simplified by asking simpler questions (ones with fewer alternative answers) of our Bayesian analysis. We reason that a search of the larger hypothesis space might be simplified using the results of Bayesian analyses on simpler hypothesis spaces. In particular we ask how we might use the simply obtained IDFT results associated with the above approximation to the likelihood function.

When images are reconstructed from high signal-to-noise ratio data using the IDFT, the dynamic range is limited not by the signal-to-noise ratio but by the sidelobe levels of the strong scatterers obscuring the weaker scatterers which might otherwise stand out amidst the noise. If however the parameters of the strong scatterers are accurately known along with their point spread functions, their effects including sidelobes, may be removed from the image, revealing lower amplitude scatterers previously obscured. The strong scatterers may then be reinserted, sans objectionable sidelobes. This principle is embodied in the so-called CLEAN algorithm (Högbom, 1974) used to advantage in radio astronomy.

We now investigate the feasibility of a CLEAN algorithmic technique. For this purpose we assume that single point scatterer modelling is appropriate for the highest IDFT peaks and that the point spread function is strictly aperture dependent and spatially invariant. These assumptions of course are not met in practice. Single resolution cells, even if appropriately modelled with point scatterers, generally contain several scatterers and point spread functions are altered by such effects as shadowing and multiple scattering. Our idealistic determination of sub-sidelobe target visibility will only be indicative of what is possible with appropriate models, which in general may be much more complex than the ones we use herein.

If a CLEAN algorithm is to be feasible, the residual difference between the response to an actual scatterer and the response to one with its estimated parameters (which is to be subtracted from the IDFT image) ought to be quite small, so small in fact that it is exceeded by the response to any other scatterer of interest. Ideally it should be at the level of system noise or lower; otherwise it limits system dynamic range in the same way that sidelobes of the strongest scatterers may do in imaging systems employing windowing of the data. We now examine this residual difference. Let the actual scatterer parameters be x_1, y_1, a_1, and θ_1 with estimates as follows:

$$\hat{x}_1 = x_0, \quad \hat{y}_1 = y_0$$
$$\hat{a}_1 = d(x_0, y_0) = d_0 \tag{25}$$
$$\hat{\theta}_1 = \phi(x_0, y_0) = \phi_0.$$

On applying the point spread function in (17) we have for the residual

$$r(x,y) = \left| a_1 s(x - x_1) s(y - y_1) \exp\left[j\theta_1 + \frac{j4\pi(x - x_1)}{\lambda_0} \right] \right.$$
$$\left. - d_0 s(x - x_0) s(y - y_0) \exp\left[j\phi_0 + \frac{j4\pi(x - x_0)}{\lambda_0} \right] \right| \tag{26}$$

Of interest is the expectation of $r^2(x,y)$ obtained by averaging over the unknown actual parameters x_1, y_1, a_1, and θ_1. $r^2(x,y)$ may also, for the small anticipated errors, be expressed in terms of amplitude and phase errors

$$\delta a = |a_1 s(x - x_1) s(y - y_1)| - |d_0 s(x - x_0) s(y - y_0)|$$
$$= (a_1 - d_0)|s(x - x_0) s(y - y_0)| \tag{27}$$
$$+ a_1 \{|s(x - x_1) s(y - y_1)| - |s(x - x_0) s(y - y_0)|\}$$

and

$$\delta\theta = \theta_1 - \phi_0 + \frac{4\pi(x_0 - x_1)}{\lambda_0}. \tag{28}$$

Then, as Figure 2 implies,

$$r^2(x,y) \approx (\delta a)^2 + a^2(\delta\theta)^2, \tag{29}$$

where $a \approx d_0|s(x - x_0)s(y - y_0)|$ is the response amplitude at the point (x,y).

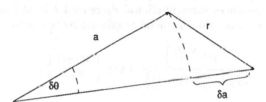

Fig. 2. Residual error.

In (27) the amplitude error is expressed as the sum of two terms, one dependent only on amplitude estimation error, the other dependent on error in the position estimate. Unlike the amplitude error, the phase error (28) is independent of the point (x,y) at which the residual is examined. Further it is jointly dependent on range and phase estimation errors. This joint dependence is of particular concern since (28) indicates that the range error contribution to $d\theta$, taken alone, is excessive unless the range estimation error $(x_0 - x_1)$ is controlled to within a small fraction of a half wavelength.

To evaluate the residual errors we consider the likelihood function (7), under the condition that a single scatterer produced the data. Then we have, using (12) and (13),

$$p(\bar{D}|x_1, y_1, a_1, \theta_1, I, \sigma) \propto \exp\left\{ -\frac{M^2}{2\sigma^2} [a_1^2 - 2a_1 d \cos(\theta_1 - \phi)] \right\}$$
$$= L(x_1, y_1, a_1, \theta_1), \tag{30}$$

where $d = d(x_1, y_1)$ and $\phi = \phi(x_1, y_1)$. Assuming uniform prior distributions for all the unknown parameters, the quasi-likelihood $L(x_1, y_1, a_1, \theta_1)$ is also proportional to the a posteriori distribution for the parameters. Thus

$$p(x_1, y_1, a_1, \theta_1 | \bar{D} I \sigma) = \frac{L(x_1, y_1, a_1, \theta_1)}{\int\int\int\int dx_1\, dy_1\, da_1\, d\theta_1\, L(x_1, y_1, a_1, \theta_1)}. \tag{31}$$

It will be helpful to note the very high degree of concentration of probability in the vicinity of the set of optimal parameters (x_0, y_0, d_0, ϕ_0). The quasi-likelihood (30) can be rewritten

$$\begin{aligned}
L(x_1, y_1, a_1, \theta_1) &= \exp\left\{\frac{M^2 d_0^2}{2\sigma^2}\left[1 - 1 - \frac{a_1^2}{d_0^2} + \frac{2a_1 d}{d_0^2}\cos(\theta_1 - \phi)\right]\right\} \\
&= \exp\left\{\frac{M^2 d_0^2}{2\sigma^2}(1 - \delta)\right\},
\end{aligned} \tag{32}$$

where

$$\begin{aligned}
\delta &= 1 + \frac{a^2}{d_0^2} - \frac{2a_1 d}{d_0^2}\cos(\theta_1 - \phi) \\
&= \left(1 - \frac{a_1}{d_0}\right)^2 + \frac{2a_1}{d_0}\left[1 - \frac{d}{d_0}\cos(\theta_1 - \phi)\right] \\
&\geq 0,
\end{aligned} \tag{33}$$

since d_0 is by definition the largest value of d. Equality in (33) occurs when all unknown parameters are at their optimum values. Any departure of one of the unknown parameters from its optimal value produces some fractional decrement δ in the log-likelihood and a reduction in the likelihood from the likelihood maximum by the factor

$$\frac{\exp\left\{\frac{M^2 d_0^2}{2\sigma^2}\right\}}{\exp\left\{\frac{M^2 d_0^2}{2\sigma^2}(1 - \delta)\right\}} = \exp\left\{\frac{M^2 d_0^2 \delta}{2\sigma^2}\right\}. \tag{34}$$

$M^2 d_0^2 / 2\sigma^2$ is ordinarily very large so that an extremely small fractional decrement δ will produce a very large decrease in likelihood. The quantity $a_1^2/2\sigma^2 \approx d_0^2/2\sigma^2$ is the ratio, in each measured datum, of the scatterer's signal power to expected noise power. $M^2 d_0^2 / 2\sigma^2$ is the signal-to-noise power ratio in the image peak at (x_0, y_0). (The IDFT coherently integrates the M^2 contributions of the scatterer to the data to produce the peak image value.) Putting some numbers in (34) illustrates the sensitivity of the likelihood to the fractional decrement δ. From (34) we note that the likelihood is below its maximum value by 100:1 when $\delta M^2 d_0^2 / 2\sigma^2 = 4.6$. The following table lists values of δ and signal-to-noise ratio yielding this result when $M = 64$:

δ	$\frac{M^2 d_0^2}{2\sigma^2}$	$\frac{d_0^2}{2\sigma^2}$ dB
.001	4600	0.5 dB
.01	460	-9.5 dB
.1	46	-19.5 dB

We see that, even at rather low values of signal-to-noise ratio and with relatively modest sized data arrays, a very small decrement will effect a reduction in likelihood by two orders of magnitude. As a consequence we may use Taylor series approximations with at most quadratic terms in the exponential argument of (30) and in the integrands occurring in the various integrals we require. In addition we may replace finite limits on these integrals by infinite ones.

The likelihood function dependence on position (x_1, y_1) is expressed through the functions $d(x, y)$ and $\phi(x, y)$. These are the amplitude and phase of the response to the scatterer as perturbed by noise and interference from other scatterers. Thus,

$$d(x,y) \exp[j\phi(x,y)] = a_1 s(x - x_1) s(y - y_1) \exp\left[j\theta_1 + \frac{j4\pi(x - x_1)}{\lambda_0}\right]$$
$$+ \eta(x,y) \exp[j\gamma(x,y)],$$

(35)

where $\eta(x, y) \exp[j\gamma(x, y)]$ represents noise plus interference. In the absence of noise and interference, $d(x, y)$ peaks at (x_1, y_1) and has peak amplitude a_1 whereas $\phi(x_1, y_1) = \theta_1$. The primary effect of the noise and interference is to alter the amplitude and phase slightly and to shift the position of the peak response. Since $a_1 >> \eta$ in the vicinity of the peak, the shape of the peak response in that vicinity is not significantly altered. The phase gradient with range is common to the scatterer of interest and all interfering scatterers; hence it remains in the sum response. It follows that reasonable approximations for $d(x, y)$ and $\phi(x, y)$ in the immediate vicinity of (x_0, y_0) are

$$d(x,y) \approx d_0 s(x - x_0) s(y - y_0)$$

and

$$\phi(x,y) \approx \phi_0 + \frac{4\pi(x - x_0)}{\lambda_0}.$$

(36)

We use these expressions to evaluate $d(x_1, y_1)$ and $\phi(x_1, y_1)$ in (30).

We are interested in the joint distribution of x_1 and θ_1 so that we may assess the phase error $\delta\theta$. This distribution is obtained by marginalization of (31), i.e.,

$$p(x_1, \theta_1 | \bar{D}\sigma I) \propto \int dy_1 \int da_1 \, L(x_1, y_1, a_1, \theta_1)$$

$$\propto \int dy_1 \int_0^\infty da_1 \exp\left\{\frac{-M^2}{2\sigma^2}[a_1^2 - 2a_1 d \cos(\theta_1 - \phi)]\right\}$$

$$\propto \int dy_1 \exp\left\{\frac{M^2 d^2}{2\sigma^2}\cos^2(\theta_1 - \phi)\right\}\int_{-\infty}^\infty da_1 \exp\left\{-\frac{M^2}{2\sigma^2}[a_1 - d\cos(\theta_1 - \phi)]^2\right\}$$

$$\propto \int dy_1 \exp\left\{\frac{M^2 d^2}{2\sigma^2}[1 - (\theta_1 - \phi)^2]\right\}.$$

(37)

To complete the integration in (37) we use the Taylor series approximation

$$d^2 \approx d_0^2 s^2(x_1 - x_0) s^2(y_1 - y_0)$$
$$\approx d_0^2\left[1 - \rho(x_1 - x_0)^2 - \rho(y_1 - y_0)^2\right],$$

(38)

where

$$\rho = \frac{1}{2}\frac{d}{dx_1}s^2(x_1 - x_0)\bigg|_{x_1 = x_0} = \frac{\pi^2 M^2 \Delta f^2}{3}.$$

(39)

The result of the integration is to a good approximation

$$p(x_1, \theta_1 | \bar{D}\sigma I) \propto \exp\left\{ -\frac{M^2 d_0^2}{2\sigma^2} \left[\rho (x_1 - x_0)^2 + (\theta_1 - \phi)^2 \right] \right\}. \tag{40}$$

Finally, using (36) and (28)

$$\theta_1 - \phi(x_1, y_1) = \theta_1 - \phi_0 - \frac{4\pi (x_1 - x_0)}{\lambda_0} = \delta\theta \tag{41}$$

and

$$p(x_1, \theta_1 | \bar{D}\sigma I) \propto \exp\left[-\frac{M^2 d_0^2}{2\sigma^2} \left\{ \rho (x_1 - x_0)^2 + \left[\theta_1 - \phi_0 - \frac{4\pi (x_1 - x_0)}{\lambda_0} \right]^2 \right\} \right]$$

$$\propto \exp\left[-\frac{M^2 d_0^2}{2\sigma^2} \left\{ \rho (x_1 - x_0)^2 + (\delta\theta)^2 \right\} \right]. \tag{42}$$

Consider an equiprobability contour in the θ_1, x_1 plane as expressed by

$$\rho (x_1 - x_0)^2 + \left[\theta_1 - \phi_0 - \frac{4\pi (x_1 - x_0)}{\lambda_0} \right]^2 = \text{constant}. \tag{43}$$

Since

$$\rho = \frac{\pi^2 M^2 \Delta f^2}{3} << \frac{16\pi^2}{\lambda_0^2} = 4\pi^2 f_0^2,$$

the contours are highly elongated nearly parallel to the line

$$\theta_1 - \phi_0 = \frac{4\pi (x_1 - x_0)}{\lambda_0}$$

as shown in Figure 3.

Range and phase errors are coupled so as to largely cancel in $\delta\theta$, whose 1-sigma value of $\sigma / M d_0$ is evident in (42). It is therefore not necessary to estimate scatterer range with fractional wavelength precision.

We next determine the conditional expectation of the squared residual error. From (26)

$$r^2 (x, y) = a_1^2 s^2 (x - x_1) s^2 (y - y_1) + d_0^2 s^2 (x - x_0) s^2 (y - y_0)$$

$$- 2a_1 d_0 s (x - x_1) s (y - y_1) s (x - x_0) s (y - y_0) \cos\left[\theta_1 - \phi_0 + \frac{4\pi (x_0 - x_1)}{\lambda_0} \right]$$

$$\approx a_1^2 \left[\alpha_x^2 + \beta_{2x} (x - x_1) + \gamma_{2x} (x - x_1)^2 \right] \left[\alpha_y^2 + \beta_{2y} (y - y_1) + \gamma_{2y} (y - y_1)^2 \right]$$

$$+ d_0^2 \alpha_x^2 \alpha_y^2 - 2a_1 d_0 \alpha_x \alpha_y \left[\alpha_x + \beta_x (x_1 - x_0) + \gamma_x (x_1 - x_0)^2 \right]$$

$$\times \left[\alpha_y + \beta_y (y_1 - y_0) + \gamma_y (y_1 - y_0)^2 \right] \cos\left[\theta_1 - \phi_0 + \frac{4\pi (x_0 - x_1)}{\lambda_0} \right], \tag{44}$$

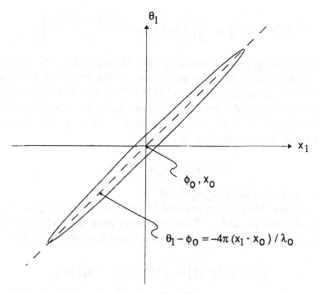

Fig. 3. Contour of equiprobability $p(\theta_1, x_1 | \bar{D}_1 I \sigma)$.

where we have set

$$s(x - x_0) = \alpha_x$$
$$s(y - y_0) = \alpha_y$$

and have expanded $s(x - x_1)$ and $s^2(x - x_1)$ in power series about $x_1 = x_0$ and $s(y - y_1)$ and $s^2(y - y_1)$ in power series about $y_1 = y_0$. The series coefficients are related since

$$\begin{aligned}
\gamma_{2x} &= \left. \frac{1}{2} \frac{d^2}{dx_1^2} s^2(x - x_1) \right|_{x_1 = x_0} = \left. \frac{d}{dx_1} \left\{ s(x - x_1) \frac{d}{dx} s(x - x_1) \right\} \right|_{x_1 = x_0} \\
&= \left. \left\{ s(x - x_1) \frac{d^2}{dx_1^2} s(x - x_1) + \left[\frac{d}{dx_1} s(x - x_1) \right]^2 \right\} \right|_{x_1 = x_0} \\
&= 2\alpha_x \gamma_x + \beta_x^2.
\end{aligned} \tag{45}$$

Similarly

$$\gamma_{2y} = 2\alpha_y \gamma_y + \beta_y^2. \tag{46}$$

The expected squared residual is

$$E\left[r^2(x, y)\right] = \frac{\int\int\int\int r^2(x, y) \, L(x_1, y_1, a_1, \theta_1) \, dx_1 dy_1 da_1 d\theta_1}{\int\int\int\int L(x_1, y_1, a_1, \theta_1) \, dx_1 dy_1 da_1 d\theta_1}. \tag{47}$$

Evaluation of the integrals in (47), retaining only the most significant terms, yields

$$E\left[r^2\left(x,y\right)\right] = \frac{\sigma^2}{M^2}\left[2\alpha_x^2\alpha_y^2 + \frac{\left(\alpha_x^2\beta_y^2 + \alpha_y^2\beta_x^2\right)}{\rho}\right]. \tag{48}$$

This result is readily interpreted in terms of (27) through (29). The first term of (27) is $|\alpha_x\alpha_y|\left(a_1 - d_0\right)$. The variance of $\left(a_1 - d_0\right)$ is simply the expected value of the square of the in-phase component of noise, namely σ^2/M^2. Thus the first term of (27) contributes $\alpha_x^2\alpha_y^2\sigma^2/M^2$ to (48). The second term of (27) represents residual amplitude error associated with the error in position estimates and corresponds to the second of two terms in (48). Finally the phase error dependent term in (29) is

$$a^2\left(\delta\theta\right)^2 \approx d_0^2\alpha_x^2\alpha_y^2\left(\delta\theta\right)^2$$

whose expected value from (42) is $\alpha_x^2\alpha_y^2\sigma^2/M^2$.

The squared residual (48) is symmetrical in $(x - x_0)$ and $(y - y_0)$, i.e., it is insensitive to an interchange of these variables. This symmetry may be exploited to place an upper bound on (48). Any points of stationarity in x lie on a line described by

$$\frac{d}{dx}E\left\{r^2\left(x,y\right)\right\} = g\left(x - x_0, y - y_0\right) = 0. \tag{49}$$

Because of the symmetry, any points of stationarity in y must lie on the line described by

$$g\left(y - y_0, x - x_0\right) = 0$$

and any points of joint stationarity must fall on the intersection of these lines or for $y - y_0 = x - x_0$. As a consequence, if there is a point of joint stationarity which is a maximum over x,

$$E\left[r^2\left(x,y\right)\right] \leq \max_x E\left[r^2\left(x, x - x_0 + y_0\right)\right]. \tag{51}$$

It is straightforward to show that the maximum in (51) occurs for $x = x_0$ and that (x_0, y_0) is a point of joint stationarity. It follows that

$$E\left[r^2\left(x,y\right)\right] \leq E\left\{r^2\left(x_0, y_0\right)\right\} = \frac{2\sigma^2}{M^2}.$$

But this upper bound is equal to the mean squared noise. At points well removed from the scatterer's estimated position, the squared residual will be even lower. Thus system noise rather than residual sidelobe levels should determine the lower limit of system dynamic range. This result suggests that subtractive deconvolution in the form of the proposed CLEAN algorithm may indeed be feasible.

7. The Choice of Scattering Center Models

Subtractive deconvolution as outlined above depends on sequentially fitting the peaks in the IDFT image to scattering center models; it does not require that these models be single scatterer models or even that each peak be matched to the same type of model. One may for example contemplate a model consisting of a group of scatterers, one at the IDFT peak location and others at the vertices of a triangle or square centered at the IDFT peak, all within a single diffraction limited resolution cell. Such additional model complexity may not only improve a CLEAN algorithm's ability to reduce sidelobes; it is in fact necessary to exploit the other great potential advantage of Bayesian analysis, namely superresolution. Resolution is dependent not on aperture extent alone but also on signal-to-noise ratio. As Bretthorst (1988) has shown, when the signal-to-noise ratio is sufficient, it is possible to resolve point sources much closer than the traditional diffraction limit distance.

Consideration of multiple scatterer models leads to an additional problem, that of deciding, based on the data, which of two or more alternative models is to be preferred. This choice too may be made in a Bayesian context (MacKay, 1992) by evaluating, for each alternative hypothesis H_i, the "evidence" $p(\bar{D}|H_i)$ and selecting the a posteriori most probable one using

$$p(H_i|\bar{D}) \propto p(\bar{D}|H_i)p(H_i). \tag{52}$$

8. Simulation and Experimental Confirmation

We have tested the CLEAN algorithm described above on both simulated and real radar data. Figure 4 shows the results of a one-dimensional simulation in which a uniformly spaced array of point scatterers arranged in descending order of amplitude in 5 dB increments was imaged. The scatterer spacing was chosen so that the peaks of each scatterer's first sidelobes (with no windowing of data) coincided with the locations of adjacent scatterers, guaranteeing maximum interaction between scatterers. Figure 4a shows the image produced using the CLEAN algorithm. As the algorithm estimates scatterer positions and strengths it provides an image of δ-functions. This image is convolved with an arbitrarily chosen narrow triangular point spread function to obtain Figure 4a. Each point scatterer may contribute more than a single δ-function because the algorithm iteratively removes only a portion of the strongest scatterer from the IDFT image. (Interference from other scatterers and noise precludes exact removal in one iteration.) The chosen point spread function must be wide enough to encompass the interference produced spread in locations of δ-functions due to a single scatterer. Figure 4a shows all scatterers above noise level to be accurately represented. By contrast an imaging algorithm employing data windowing for sidelobe control produced Figure 5b in which the scatterers are not quite resolved.

Simulations with multiple scatterers in a diffraction limited cell, as expected, produced less reliable results, yet frequently yielded a surprising degree of superresolution. Figure 5, which shows images of two point scatterers separated by 3/4 of a diffraction limited cell width and differing in amplitude by 20 dB, is such an example. In all simulated cases with multiple scatterers in a diffraction limited cell, the δ-functions resulting from a single cell could be "reassembled" by choosing a sidelobeless arbitrary point spread function of cell width, thus preserving diffraction limited resolution and eliminating the resolution loss which would accompany data windowing.

The algorithm was also tested on real radar data from a small ship. Data from 64 aspect angles, with 128 samples per aspect angle, were processed. Figure 6b shows a conventionally processed image exhibiting the 2:1 aspect ratio of resolution cells resulting from the 2:1 rectangular data array. Figure 6a shows the CLEAN image obtained after appending to the data the necessary field of zero samples to create a 1024×1024 square array. In the CLEAN processing, each IDFT peak was removed using the 2:1 aspect ratio point spread function of the aperture and each identified scatterer was replaced with a

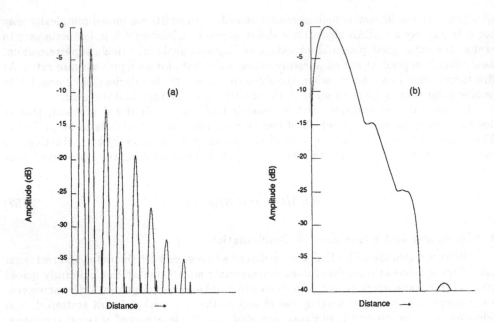

Fig. 4. Image of point scatterer array:(a) Clean image, (b) Windowed data IDFT image.

1:1 aspect ratio point spread function. The improved resolution of the CLEAN image is evident.

9. Conclusions

Conventional coherent radar imaging practice has been examined in the context of Bayesian estimation and point scatterer target modelling. It is shown that good, near maximum likelihood estimates of the locations, amplitudes, and phases of the strongest scatterers are obtainable from the inverse DFT of the unwindowed data. This fact together with the high sidelobe levels of the strong scatterers leads naturally to the consideration of a CLEAN algorithm for sidelobe reduction. It is shown that a CLEAN algorithm is feasible in spite of the 2π radians per half wavelength phase gradient in the aperture point spread function and that it can provide demonstrable improvements in resolution with little or no loss in dynamic range.

ACKNOWLEDGMENTS. The authors wish to acknowledge Dr. Kenneth Knaell for suggesting the applicability of a CLEAN algorithm and Mr. Michael Plumpe for programming the algorithm and obtaining the simulation and test results.

REFERENCES

Bretthorst, G. L.: 1988, 'Bayesian Spectrum Analysis and Parameter Estimation,' in *Maximum Entropy and Bayesian Methods in Science and Engineering*, Vol. 1, 75-145, G. J. Erickson and C. R. Smith (eds.), Kluwer, Dordrecht.

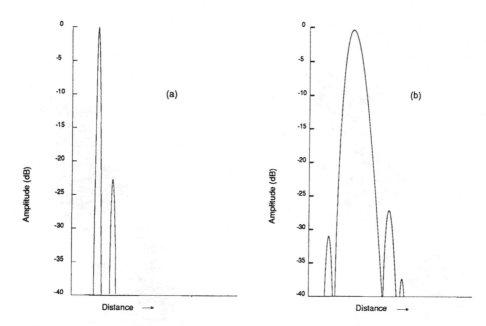

Fig. 5. Image of pair of scatterers with 3/4 cell width separation: (a) Clean image, (b) Windowed data IDFT image.

Högbom, J. A.: 1974, 'Aperture Synthesis with a Non-Regular Distribution of Interferometer Baselines,' *Astron. Astrophys. Suppl.* **15**, 417-426.

Jaynes, E. T.: 1987, 'Bayesian Spectrum and Chirp Analysis,' in *Maximum-Entropy and Bayesian Spectral Analysis and Estimation Problems*, 1-37, C. R. Smith and G. J. Erickson (eds.), Reidel, Dordrecht.

MacKay, D. J. C.: 1992, 'Bayesian Interpolation,' in these proceedings.

Mensa, D. L.: 1981, *High Resolution Radar Imaging*, Artech House, Norwood.

Wehner, D. R.: 1987, *High Resolution Radar Imaging*, Artech House, Norwood.

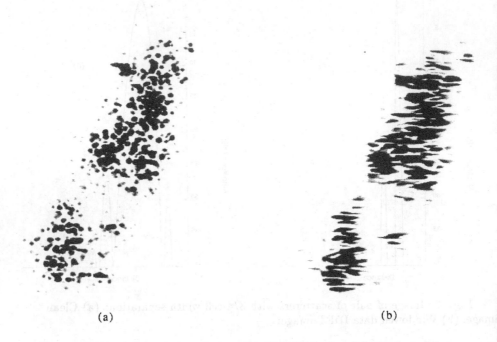

(a)   (b)

Fig. 6. Radar images of a small ship: (a) CLEAN image, (b) Windowed data
IDFT image.

APPLICATION OF MAXIMUM ENTROPY TO
RADIO IMAGING OF GEOLOGICAL FEATURES

Neil Pendock
Department of Computational and Applied Mathematics
University of the Witwatersrand
Johannesburg, South Africa

E. Wedepohl
Chamber of Mines Research Organization
Johannesburg, South Africa

ABSTRACT. Radio waves may be used to image the electrical properties of rock for geological exploration and mining. The tomographic reconstruction of an attenuation image from radio wave survey data may be formulated as an *ill-posed* linear inverse problem. The inversion is ill-posed since the data are incomplete and noisy and an inexact simplified linear model of radio wave propagation through rock is assumed. Such inverse problems will have many 'solutions' which fit the observed data. Since the pixel attenuation values are a positive additive distribution, the reconstructed image with maximum entropy relative to all prior information is a solution with the least amount of structure not indicated by the data.

1. INTRODUCTION

The radio imaging method [RIM] is a relatively new geophysical technique which was first developed in the 1970's. RIM employs the same concepts as the medical CAT-scan technique, but is applied to the mapping of geological features and operates on a much larger spatial scale. Whereas the CAT-scan uses X-rays to produce images of body tissue density, RIM uses radio waves to image the electrical properties of a rockmass between two adjacent boreholes. The reconstructed image represents either radio wave attenuation rates or radio wave velocity. A geological feature which causes changes in either of these two rock properties will be depicted in the image. Most work to date has concentrated on the identification of fractures in potential waste disposal sites (Olsson et al [1987]). Another application for RIM is the mapping of coal seams and base metal ore deposits, as discussed by Nickel *et al* [1988]. Base metal ore invariably has a much higher electrical conductivity than the surrounding rock, resulting in local increases in the radio wave attenuation rate within the ore body. This paper presents results from two RIM surveys in base metal mines where the ore body was successfully mapped in both experiments.

C. R. Smith et al. (eds.), Maximum Entropy and Bayesian Methods, Seattle, 1991, 397–402.
© 1992 Kluwer Academic Publishers.

In cross hole RIM a radio wave antenna is lowered to a fixed point in one of the boreholes and is used to transmit sinusoidal radio waves through the survey area. Another antenna is lowered down the other borehole and is used to collect a profile of electric field strength. The process is repeated for a large number of different transmitter and receiver positions. In this way, a data set is acquired which corresponds to an intersecting network of radio waves.

2. MATHEMATICAL MODEL

Most RIM tomography to date has used the X-ray model to describe the radio wave propagation. The model assumes that radio waves travel along straight lines or raypaths linking the transmitters and receivers. Under this assumption, each data point is the line integral of the radio wave attenuation rate along the straight ray path :

$$E = \frac{E_0 G_{Tx}(\theta) G_{Rx}(\theta)}{r} e^{-\int_l \alpha dr}$$

where

E	= electrical field strength measured at the receiver
E_0	= electrical field strength output by the transmitter
G_{Tx}	= radiation pattern of the transmitting antenna
G_{Rx}	= radiation pattern of the receiving antenna
r	= separation between the receiving and transmitting antennas
$\int_l \alpha dr$	= line integral of attenuation coefficient α

This equation may be re-written in the form

$$y = \int_l \alpha dr$$

where y is the transformed data value $log\left(\frac{E_0 G_{Tx}(\theta) G_{Rx}(\theta)}{Er}\right)$. Discretizing the survey area into pixels yields the standard tomographic matrix equation

$$y = \Lambda x$$

where

$\Lambda = [\lambda_{ij}]$	= length of path i through pixel j
x	= vector of radio wave attenuation coefficients for each pixel
y	= transformed vector of data readings

Iterative methods have been extensively used to solve this equation. The most popular techniques are SIRT [simultaneous iterative reconstruction technique] described by Dines *et al* [1979] and ART [algebraic reconstruction technique] used by Lager *et al* [1977]. Major difficulties were encountered as the system is ill-conditioned and many researchers, including Ivansson [1986] noted that the SIRT and ART image reconstructions were severely corrupted and random and systematic noise in the measured data resulted in image artifacts.

One major problem with RIM is that the angular coverage which can be obtained from two sub-parallel boreholes is severely limited. As a consequence, the method is insensitive to features lying parallel to the boreholes. For practical reasons, the number of data points collected is usually less than the number of pixels to be reconstructed, resulting in an underdetermined linear system of equations with consequently many 'solutions' which fit the data. A further problem is that the straight ray model ignores the diffraction, refraction and reflection behaviour of radio waves. This causes the matrix equations to be inconsistent which may also introduce artifacts into the reconstructed image. Diffracted waves may be generated from the edges of high contrast features, such as ore bodies. Refraction of the radio waves around the [low velocity] ore body may also be significant. Thus the ore body will be typically undersampled.

In order to turn this ill-posed inverse problem into a well-posed one, it is necessary to *choose* a solution which responds in a continuous manner to changes in the measured data. One well-posed solution would be to minimize $(\Lambda x - y)^2 + \lambda(Cx)^2$ where C is some smoothing operator, weighted by a [Lagrange multiplier] λ. Careful choice of λ and C may remove noise related artifacts. Ramirez [1986] used this approach to successfully image fractures in granitic rocks. One criticism of this method is the arbitrary nature of the C and λ chosen and an observation that the resolution of the reconstructed image is degraded.

The image x of attenuation coefficients to be reconstructed is a positive additive distribution and Skilling [1988] has shown that most reasonable choice for the operator C is the relative entropy of x with respect to all prior information about x. By treating the Lagrange multiplier λ as a parameter to be inferred from the data, Skilling and Gull removed the last 'ad-hoc' component from the inversion. Their *quantified Maxent* is thus a general linear image reconstruction methodology which we follow in our inversions.

Much recent work has focused on incorporating diffraction effects into the physical model. The usual approach is to assume a weak-scattering approximation to linearize the problem. This is an invalid assumption for strong scatterers, such as base metal deposits and also ignores the interaction between features, as noted by Slaney *et al* [1984].

3. RESULTS FROM TWO FIELD SURVEYS

Two field surveys were conducted on base metal mines in Namibia, to investigate the ability of RIM to map a real ore body *in situ*. The sites selected had been extensively drilled by means of two borehole fans, so that the ore body outline and grade were well known. Electrical property measurements were also performed to confirm the ore bodies as potential RIM targets. The ore bodies range from a few metres to 20m in thickness and had a lateral extent of about 65m. Grade varies extremely erratically within the ore bodies, with some regions having very high grades, while other sections had no mineralization whatsoever.

The survey was conducted between to outermost two boreholes in both experiments, the aim being to establish whether RIM could provide information similar to the results from the intermediate boreholes. Figure 1 shows the borehole geometries and transmitter and receiver locations for the two surveys. Profiles of electrical field strength were collected at the various locations using five different radio wave frequencies in the HF band [2-30MHz].

To transform the data into the required format suitable for tomographic imaging, values of $E_0, G_{Tx}(\theta)$ and $G_{Rx}(\theta)$ are needed. These were obtained by calibration measurements conducted in homogenous sections of the rockmass. As expected, the data set was significantly influenced by diffracted waves generated from the edges of the ore body. These effects are frequency dependant and could thus be recognized on the data profiles by the presence of frequency dependant diffraction patterns. These diffraction effects were reduced by simple filtering of the data prior to tomographic imaging. Each profile was smoothed

Fig. 1. Survey transmitter and receiver locations.

using a moving window of length three data points. This has the effect of 'averaging-out' the diffraction maxima and minima.

Trial tomographic images were produced for different values of the Lagrange multiplier λ. The range of values was selected by estimating expected noise levels in the data. As λ was increased to an optimum value, an image was obtained which showed very good correlation to the mapped ore body. This optimum value represents the noise level on the data resulting from a combination of random noise and violations of the straight raypath model. Below this value of λ, the ore body was barely discernable. It was thus possible to choose a value for λ so that the effect of diffracted waves was significantly reduced.

Artifacts caused by the diffraction minima are still present on the best reconstructions [figures 2 and 3]. In particular, individual raypaths associated with diffraction peaks stand out. These artifacts may be partially eliminated by removing selected data points from the tomographic inversion.

The reconstructed images correlate with the mapped ore body with an average accuracy of two metres. In addition, the image splits up the ore bodies into discrete high grade blocks - these patterns agree well with the ore grades mapped from boreholes.

4. CONCLUSIONS

RIM can be used to map high contrast features such as ore bodies. The Maximum Entropy method may be used to successfully reconstruct attenuation images, even with the limitations of the physical model discussed above.

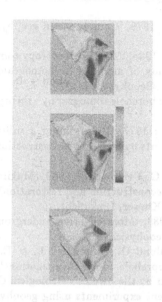

Fig. 2. Survey 1 reconstructed attenuations for three different values of λ.

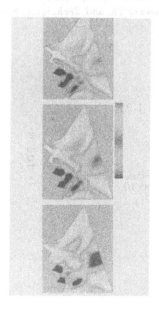

Fig. 3. Survey 2 reconstructed attenuations for three different values of λ.

REFERENCES

Devaney A.J.: 1984, 'Geophysical diffraction tomography', IEEE Trans. Geosci. and Rem. Sens., **GE-22**.

Dines K.A. & R.J. Lytle : 1979, 'Computerized geophysical tomography', IEEE Proceedings, **67**.

Gull S.F. & T.J. Newton : 1986, 'Maximum entropy tomography', Applied Optics, **25**.

Ivansson S. : 1985, 'A study of methods for tomographic velocity estimation in the presence of low velocity zones', Geophysics, **50**.

Ivansson S. : 1986, 'Seismic borehole tomography - theory and computational methods', IEEE Proceedings, **74**.

Lager D.L. & R.J. Lytle : 1977, 'Determining a subsurface electromagnetic profile from high frequency measurements by applying reconstruction technique algorithms', Radio Science, **12**.

Mohammad-Djafari A. & G. Demoment : 1989, 'Maximum entropy and bayesian approach in tomographic image reconstruction and restoration', in J. Skilling (ed.), *Maximum Entropy and Bayesian Methods*, Kluwer, Dordrecht.

Nickel H. & I. Cerny : 1989, 'More effective underground exploration for ores using radio waves', J. Exploration Geophysics [Australia].

Olsson, O., Falk L., Forslund O., Lundmark L. & E. Sandberg : 1987, 'Crosshole investigations - results from borehole radar investigations', RAMAC report - part of a STRIPA experiment conducted by the Swedish Geological Co.

Ramirez A.L.: 1986, 'Recent experiments using geophysical tomography in fractured granite', IEEE Proceedings, **74**.

Skilling J.: 1988, 'The axioms of maximum entropy', in Erickson G.J. & C. Ray Smith (eds.), *Maximum Entropy and Bayesian Methods in Science and Engineering 1 : Foundations'*, Kluwer, Dordrecht.

Slaney M., Kak A.C. & L.E. Larsen : 1984, 'Limitations of imaging with first order diffraction tomography', IEEE Trans. Microwave Th. and Techniques, **MTT-32**.

DETERMINISTIC SIGNALS IN HEIGHT OF SEA LEVEL WORLDWIDE

Robert G. Currie
Institute for Terrestrial and Planetary Atmospheres
State University of New York
Stony Brook, New York 11794, USA

ABSTRACT. Maximum entropy spectrum analysis of 54 worldwide long records of the height of sea level yields evidence for two peaks with periods 19.4 ± 1.6 years in 48 records, and 10.4 ± 0.6 years in 40 instances. Tests by the t–statistic show that the peaks are statistically significant at confidence levels of 99.9 and 99%, respectively. They are identified as the luni–solar 18.6-year M_n term and the solar cycle S_c signal in height of sea level reported previously by Currie (1976, 1981b) only for northern Europe. Amplitude and phase of the luni–solar term are highly nonstationary both with respect to time and with respect to geography. In particular, abrupt $180°$ phase changes in wave polarity are often observed. These phenomena are ubiquitous in climate data and O'Brien and Currie (1991) have developed a model which can explain them.

1. Introduction

Currie (1976) reported evidence for two signals with periods 18.5 ± 1.1 and 10.9 ± 0.6 years in height of sea level along the shores of northern Europe. He identified the first as induced by the 18.6–year luni–solar M_n tidal constituent, and Lisitzin (1974) cites an extensive literature on studies of this term in climate records. The second signal was ascribed as induced by a solar cycle S_c variation in the sun's luminosity of $\approx 0.1\%$ which Currie (1979) estimated was necessary to produce observed amplitudes in air temperature. Currie (1981b) later reexamined heights in sea level data for Europe, and studied wave trains of the S_c term.

The M_n and S_c terms were subsequently found in air temperature and air pressure worldwide (Currie, 1981a, 1987b), in rainfall parameters such as tree–rings, rain gauge records, and floods of rivers (Currie, 1983, 1984a – b, 1987a, 1989, 1991a – c; Currie and O'Brien, 1988, 1989, 1990a – c), and in American agricultural output and macro-economic data (Currie, 1988). Other reports are May and Hitch (1989) and Mitra et al. (1991) on rainfall data from the United Kingdom and India. Tyson (1986, Chap. 3) summarizes his papers and those of his co–workers on an 18–year oscillation in rainfall over southern Africa. There exists a large literature on an 18–19 year variation in fishery data which begins in 1905, and only came to attention recently. A few recent reports are Sindermann (1970), Swinbanks (1982), Winkel (1979), and Wyatt (1984). Currie and Wyatt (unpublished data, 1991) have applied high resolution spectrum analysis to fish yield data from European and North American waters and found both terms.

C. R. Smith et al. (eds.), Maximum Entropy and Bayesian Methods, Seattle, 1991, 403–421.
© 1992 *Kluwer Academic Publishers*.

In this paper spectra of height of sea level records worldwide are presented, and a standard orthodox statistical test shows that the two terms are significant at 99.9% and 99% confidence levels. Examples of M_n waves are given, and they display the same characteristics found previously in air temperature, air pressure, and rain gauge data, as well as in tree–ring chronologies and discharge of rivers. In particular, both amplitude and phase are highly nonstationary. The data base is far more voluminous than in Currie (1976, 1981b), and another novel result is that the signals are now found worldwide. In fact, Currie (1981b) restricted report to northern Europe because, due to an improper procedure discussed herein, he failed to find the terms elsewhere.

2. Spectra and Statistical Test

There are 54 records whose length is 60 years or greater, and 35 of these are along the shores of northern Europe. There are another 73 short records where length varied from 40 to 59 years. The data were obtained from the Permanent Service for Mean Sea Level at Bidston Observatory, Birkenhead, Merseyside L43 7RA in Great Britain.

2.1. MAXIMUM ENTROPY METHOD.

A high pass symmetric filter of length $2N + 1(N = 6)$ was applied to each of the 54 long records sampled yearly followed by MESA or maximum entropy spectrum analysis (Marple, 1987). Since MESA and Yule-Walker (YW) autoregressive AR spectra are compared and contrasted, it will be useful to give the expression used to compute both as a function of frequency f as shown:

$$P(f) = \frac{2P_L}{|\sum_{j=0}^{L} \alpha_j \exp(-2\pi i f j)|^2}. \tag{1}$$

The coefficients α_j are Lagrange multipliers, and are also known as a 'prediction error filter', while P_j is the prediction error power. For $j = 0$, $P(f) = 2P_0$, a constant appropriate for white noise, where P_0 is the zero lag variance of the data. In MESA one avoids computing lagged auto–correlation functions and constructs α_j directly from the data in a bootstrap fashion, using all the data in each iteration from $j = 1$ to $j = L$.

Only recently has Currie (1991a – c) realized that Lagrange multipliers of order $L = 20$ to 23 are near optimum for resolving the M_n and S_c terms whether the record is 100 or 1000 years in length. For each record, spectra were computed for $L = 21$, 22, and 23, and that spectrum is chosen which best exemplified evidence for one or both terms. The 54 spectra were arithmetically averaged, and the mean spectrum is shown in Figure 1a for the first 50 estimates. Peaks with periods 19.4 ± 1.6 and 10.4 ± 0.6 years were found in 48 and 40 out of 54 spectra, respectively, and Figure 1a is not materially changed if one processes all 54 records using a constant order of $L = 21$, or 22, or 23.

The above results are a great improvement over Currie's (1981b) earlier study where L was taken as 30, 40, and 50% record length. Such a procedure applied to the 54 records embodied in Figure 1a yields an enormous range of $14 \leq L \leq 57$ with a mean and standard deviation of $L = 26 \pm 8$; this is quite unrealistic, but it shows that MESA is robust. However, the improper procedure led to Currie's (1981b) failure to find the terms outside of Europe. Woodworth (1985) made the same mistake.

For the 73 short records with lengths 40 to 59 years, spectra were constructed for $L = 19$, 21, and 23, and that spectrum chosen which best exemplified evidence for one or both terms. In order to conserve data the high–pass filter applied was reduced to $2N + 1$ ($N = 3$) points, so some spectra were constructed with only 34 observations. Summarized briefly, 28 records yielded no evidence for either term, 35 yielded a term with period 17.9 ± 1.6 years, and 31 showed a peak with period 10.2 ± 0.6 years.

2.2. STATISTICAL TEST.

Assuming that the two peaks in Figure 1a are random fluctuations, the dashed line from $i = 13$ to 50 is the background spectrum. The rapid rise of the mean power spectrum for $i < 13$ is due to the power response of the high–pass filter originally applied, which at $f = 0$ falls to 10^{-15}. At each f_i ($i = 13$ to 50) of the 54 spectra, the ratio A_i of the spectral estimate to the dashed background spectrum was determined. And at each f_i the mean value $\overline{A}_i$ and its standard deviation were computed. The resulting mean spectral ratios $\overline{A}_i$ are shown in Figure 1b.

In Figure 1b the spectral ratios $\overline{A}_i$ are well above unity near 19 and 10.5 years, so the t–statistic may be used to test whether $\overline{A}_i > 1$ is significant. The mean t–statistic for the peak at 19 years and two adjacent values on each side is $\overline{t}_n = 3.1$, while mean $\overline{t}_s = 2.1$ for the $10 - 11$ year term [the equation used to compute the t–statistic is given by Kendall and Stuart (1967, p 466)]. The degrees of freedom (df) are 106, but to avoid interpolation in the t-statistic table let us take df = 60. For df = 60, and a one-sided test, the peaks with $\overline{t}_n = 3.1$ and $\overline{t}_s = 2.1$ are significant at 99 and 95% confidence levels.

It is now seen that the background noise, shown as a dashed line in Figure 1a, included two band–limited signals and is, therefore, too high. Thus a new background spectrum was obtained by subtracting the power estimates of both peaks and the two adjacent values on each side of each peak. The new background is shown by the lower (solid) horizontal line in Figure 1a, and the new mean ratios $\overline{A}_{ii}$ are plotted in Figure 1c. The mean t–statistics for the peaks in Figure 1c are $\overline{t}_n = 4.7$ and $\overline{t}_s = 3.3$. A value of $t > 3.5$ for df = 60 represents significance at the 99.9% level which $\overline{t}_n$ exceeds, while $\overline{t}_s = 3.3$ is significant at a 99% level. The degrees of freedom could be reduced to 4 and 9, respectively, and the terms would still be significant at a 99%.

2.3. YULE-WALKER METHOD AND HISTORY.

Yule (1927) invented a low order autoregressive time domain AR(2) model to avoid spectrum analysis. He considered Schuster's periodogram to be misleading because the one example then available on economic data (European wheat prices from 1500 to 1869) showed spiky features which Lord Beveridge (1921, 1922) interpreted as deterministic signals. This single time series has been reanalyzed for a half century; and in each instance the unstated implication is that one would be foolish to entertain the idea that economic or meteorological data could possibly contain deterministic signals.

Yule (1927) therefore proposed an autoregressive AR(2) model for sunspot numbers whereby the annual sunspot number observed at year S_t is produced by its values at S_{t-1} and S_{t-2}, with S_t driven by "disturbances" δ_t. Yule's model is written:

$$S_t = \beta_1 S_{t-1} + \beta_2 S_{t-2} + \delta_t, \tag{2}$$

where β_1 and β_2 are constants. The disturbances δ_t came to be known as a 'white noise process' in statistics (see Box and Jenkins, 1970) and 'random shocks' in economics. In Yule's concept the δ_t *somehow* caused or 'drove' the variability seen in sunspot numbers which he claimed was 'pseudo-periodic' with a highly damped autocorrelation function [Currie and O'Brien (1988) have demonstrated that the model is invalid]. And in economics, the 'random shocks' were *supposed somehow* to cause the variability observed in macroeconomic data.

Yule's AR(2) model was propounded by Kendall (1943, 1945, 1946) who claimed it adequately described natural variability seen in British agricultural time series. Currie (unpublished data, 1987) found that MESA spectra of British crop data examined by Kendall (1943), and many other agricultural, economic, and rainfall series for Great Britain and Ireland are dominated by a large term with period near 10 years and a smaller term near 19 years. This is consistent with results of May and Hitch (1989) for British precipitation records where the 10-year term is twice as large as the one near 19 years.

Kendall's (1943, 1945) papers were in response to a book by Davis (1941) on economic time series, who discussed evidence for terms near 10 and 20 years in economic data, signals subsequently found by Currie (1988) using computers and modern methods of signal analysis. Kendall's (1946) book was in response to Lord Beveridge (1944, pp. 275-314), who presented time series of British macro-economic data claiming they contained cyclic components. Currie (unpublished data, 1988) found that the Beveridge data do contain two terms with periods near 10.5 and 19 years.

According to Marple (1987, p 12) it was not until 1948 that Yule's AR(2) model was suggested as a spectral estimator, although this was anticipated by Walker in 1931. In the YW method lagged autocorrelation functions to lag L, ACF(L), are computed and then converted into what will be called Yule coefficients of order Y. The Y coefficients β_j are used in Eq.(1) in place of α_j. In the YW method the term P_j in Eq.(1) is the "driving noise variance," reflecting Yule's belief that δ_t in Eq.(2) 'drove', and thus produced the observed variability in sunspot numbers.

In order to compare results in Figure 1a with the YW method, ACFs to order $Y = L$ were computed for each of the 54 long records using Marple's (1987) subroutines. Each set of ACF(Y) were converted to a set of β_j coefficients $j = 1$ to Y and 54 YW spectra computed. The mean spectrum (not shown) showed a term with period 10.5 ± 0.6 years in 34 instances but no term near 19 years was resolved. This alone invalidates Marple's (1987, p. 12) contention that "The maximum entropy spectrum is very closely related to autoregressive spectral analysis."

Marple (1987, Chap. 8) shows that the YW method has poor resolution for short records relative to MESA and two other modern methods. But the classical FFT could not resolve even the S_c term so YW is a higher resolution method. Its reduced performance follows because, with short records, one loses enormous amounts of information when ACFs are computed. Given 60 observations the lagged ACF(22) is computed with only 16 data points, and so β_{22} will be of very poor quality. But with MESA, the α_{22} Lagrange multiplier is constructed using all 60 observations, as are all the lower order multipliers.

Since Eq.(1) has the same mathematical form for constructing MESA and YW spectra there is widespread belief, including Marple (1987), that MESA is very closely related to AR spectral analysis. This is not so at a practical level as our discussion above shows, nor at a theoretical level. MESA is based on Jaynes' (1983) Principle of Maximum Entropy where one is maximally noncommittal. One does not assume there is a δ_t white noise process which is why Jaynes (1982) calls MESA 'noiseless'. The two methods are thus poles apart conceptually.

3. Time Domain M_n Wave Trains

The Lagrange multipliers employed in constructing the spectra embodied in Figure 1a can be used to generate data outside the range of observation, because they embody the characteristics of the observed data (Ulrych and Clayton, 1976). Therefore, 30 points were generated off each end of each high-pass filtered record. This was followed by convolving a bandpass filter, of length $N + 1$ ($N = 30$) centered at 0.054 cpy, with each extended record and in this manner no observed data are lost. Each resulting wave train was then divided by the amplitude response at 0.054 cpy of the high-pass filter originally applied in order to restore proper amplitude.

3.1. WAVES IN UNITED STATES AND AUSTRALIA.

Figure 2a displays the luni–solar wave at Baltimore, Maryland; vertical bars at maxima are paired with downward pointing arrows and associated dates which mark dates or epochs of maxima in tidal forcing. If a vertical bar is within ±2.5 years of an epoch correlation, or an inphase relation, is deemed to exist based on a criterion discussed at length by Currie and O'Brien (1989), and used extensively in study of 1219 American rainfall records (Currie and O'Brien, 1990a – c). If no vertical bar is shown then the phase relation is deemed to be 'mixed' phase. It is evident that wave maxima for Baltimore were inphase with epochs 1917.5 through 1973.3.

Maxima in height of sea level at epochs for Baltimore implies that the corresponding wave in air pressure recorded minima (assuming that the 19-year term in wind speed and sea surface temperature had no appreciable effect on sea level); minima in air pressure implies that the induced wave in rainfall recorded maxima at epochs. For Baltimore these inferences are valid (Currie and O'Brien, 1989, figures 8-9). The relation should not hold generally because there is an M_n term in sea surface temperature and wind speed which could affect sea level (Loder and Garrett, 1978; Campbell, 1983).

Figure 2b shows the wave in sea level for Halifax in Nova Scotia, Canada, to the northeast of Maine. A wave minimum occurred near 1936.1; this was followed by a transition zone denoted by an asterisk, and at 1954.7 a wave maximum came into phase with the epoch, a correlation which continued at 1973.3. This corresponds to a 180° phase change in wave polarity and was completed within one cycle.

For Honolulu, Hawaii, Figure 3a shows that a wave maximum was inphase with epoch 1917.5. A transition zone then occurred and at 1954.7 and 1973.3 wave minima were correlated with the epochs, a change in phase of 180°.In this instance it took two cycles for the phase switch to become established. Figures 3b – c display waves at Sydney and Newcastle in Australia, which are separated in distance by only about 100 km. After 1930 the waves are remarkably different. At Sydney, wave maxima are correlated with epochs 1936.1 through 1973.3. In contrast, at Newcastle a wave minimum correlated with epoch 1936.1, and by 1973.3 a 180°change in phase had occurred. Thus, at 1936.1 the two harbors were located beneath two stationary air pressure cells which were out of phase with one another. It is clear that the two records are independent time series although the harbors are very near one another.

3.2. Waves in Asia and Europe.

The luni–solar wave at Tonoura, Japan, is shown in Figure 4a where maxima are correlated with epochs 1917.5 through 1973.3; this implies that air pressure should have been a minimum near these epochs, but Currie (1982) found maxima in air pressure over Japan to lead epochs by only about 2 years.

This apparent inconsistency may be due to the strong M_n term in winds over the Himalayas found by Campbell (1983), who reported that at 300 mb heights luni–solar anomalies in wind speed were amplified above equilibrium by a factor of 10,000. Such anomalies would extend to sea level and these winds progress eastwardly over Japan. Results at Bombay in India are shown in Figure 4b, where one observes that maxima in height of sea level are inphase with epochs 1898.9 through 1973.3, the same phase relation found for Japan.

Figures 5a – b display waves from two harbors in the Mediterranean Sea. At Marseille, France, height of sea level reached a maximum at epochs 1898.9 through 1954.7. The wave at Trieste, Italy, to the east of Marseille, displayed the same phasing at 1936.1 and 1954.7, and then switched phase by 180° at epoch 1973.3. In Figure 5c for Cascais, Portugal, a maximum was inphase with epoch 1898.9. The wave then switched phase 180° with minima inphase with epochs 1936.1 through 1973.3. Thus, at epochs 1936.1 and 1954.7 sea level on the Atlantic coast at Cascais was out of phase with both Marseille and Trieste to the east. Again, it is clear that these three time series are independent although the harbors are rather close to one another.

Figures 6a – b are waves at Helsinki and Oulu in Finland. Sea level at Helsinki experienced a 180° phase change at epoch 1917.5. Subsequently, wave minima led epochs by a mean 2.3 years from 1936.1 through 1973.3. However, at epochs 1954.7 and 1973.3 vertical bars would lag epochs by more than 2.5 years and so they are not plotted. At Oulu the distortion in wave form from 1910–30 suggests that it too may have experienced a phase change. Thereafter, wave minima are correlated with epochs.

Figures 7a – b show the luni–solar term at Nedre Sodertalje and Ystad in Sweden. At both harbors wave maxima were inphase with epoch 1898.9, and then both experienced a 180° phase change. At Nedre Sodertalje minima lagged epochs in a mixed phase manner except at 1917.5, whereas at Ystad minima came into phase with epochs 1936.1 and 1954.7. The wave at Aberdeen, Scotland, in Figure 7c is similar to that at Ystad. Mean amplitudes of the M_n wave in the 54 long records varied from 4 to 24 mm with an overall mean and standard deviation of 12 ± 5 mm.

Twenty seven of the 35 long record waves for northern Europe showed unipolar behavior at all epochs. Wave minima led epochs by between 2 and 3 years, so many were of mixed phase. For brevity these are not shown. Wave trains were obtained for the 35 short records where evidence for the luni–solar term was found in terms of spectra. Figure 8a displays the wave for 7 out of 8 records along the Atlantic coast of the United States. At 1936.1, 1954.7 and 1973.3 maxima are inphase with epochs, so these results corroborate the phasing seen in Figure 2a for Baltimore. Figure 8b shows the wave for 3 out of 4 Japanese records where wave minima are inphase with epochs 1936.1 through 1973.3; this is consistent with phasing seen in Figure 4a for Tonoura.

It was seen earlier in Figure 4b that at Bombay, on the western shores of India, wave maxima were inphase with epochs 1898.9 through 1973.3. In Figure 8c the wave at Vishakhapatnam, on the eastern shore of India, is consistent with Bombay with respect to phase. However, in Figure 8d the wave at Calcutta in the Bay of Bengal, northeast of Vishakhapatnam by about 300 km, is remarkably different. At Calcutta wave minima lag the two epochs by a mean 2.5 years, and so the wave is nearly out of phase with the other two harbors. Note too that the amplitude at Calcutta is about 50 mm which is substantially higher than for all the other records.

3.3 DISCUSSION

The 180° changes in phase seen in Figures 2 – 3 and 5 – 7 were found by Currie (1983) in two tree–ring chronologies from South America, and have subsequently been found in nearly 400 chronologies from several continents (Currie, 1991a – c). They are ubiquitous in data for air pressure, air temperature, precipitation, and flood of rivers (Currie, 1987a – b; Currie and Fairbridge, 1985; Currie and O'Brien, 1988, 1989, 1990a – c; O'Brien and Currie, 1991). Therefore, the phase with respect to time at any site is highly nonstationary; amplitudes as a function of time at any site are also highly nonstationary. Moreover, phase and amplitude are also highly nonstationary with respect to geography. Currie and O'Brien (1989, 1990a – c) show that it is common to find American rain gauge records out of phase with one another, even though the sites are separated by only 10's of kilometers. Therefore each climate record is an independent data set.

The M_n constituent tide affects the atmosphere but not significantly, because according to the theoretical model of O'Brien and Currie (1991), which can explain how 180° phase changes occur at air pressure recording stations, the crucial factor is the 18.6-year change in angular velocity of the solid earth itself which occurs in Newton's second law written in a rotating coordinate frame. Pedlosky (1987, p. 18) recognized that this term is "unimportant for *most* oceanographic or atmospheric phenomena except for those time scales that are unusually long". However, the time scales of 8 to 20 years seen in Figure 1 are unusually long because the periods are 10^3 longer than those treated by Pedlosky (1987). It should be noted that the amplitude of the 18.6-year variation in angular velocity is far larger than any other tidal constituent. One thinks of the lunar month as a major tidal constituent, and it is, but the M_n amplitude in change of rotation rate is nearly 200 times larger (Yoder et al., 1982). The O'Brien and Currie (1991) model also implies that a S_c variation in solid earth rotation must exist, and it has been observed (Currie, 1980, 1981c).

4. Broader Iimplications

The M_n and S_c terms in height of sea level seen in Figure 1 are induced by the same signals in air pressure (Currie, 1982; Currie, 1987b; O'Brien and Currie, 1991). These, in turn, induce the terms in rainfall which then induce the terms in crop yield. This has implications in a wide range of scientific disciplines (meteorology, hydrology, biology, agriculture, history), but the most important sciences affected are economics and sociology.

Tyson (1986, Chap. 3) cites about 30 papers on an "18-year oscillation" in rainfall over South Africa, and he and his co–workers in the 1970's forecast serious problems with drought in the early to mid 1980's (mid–epoch 1982.6). Serious drought occurred and Tyson noted that with respect to the droughts of the 1960s (mid-epoch 1964.0): "All too sadly, the great drought of 1982/83 and subsequent dry years have seen a repetition of events on an even greater scale". From his discussion and references there is no question but that these droughts of record had a major impact on the South African economy.

Currie (1988) reported both terms in American crop yield and acres of abandoned crops, as well as in livestock, chicken, and chicken egg production; and these variations are significant in economic terms. At epoch 1936.1 S_c wave minima in corn yield were nearly in phase with M_n wave minima, and the combined shortfall amounted to 11.1×10^8 bushels. This resulted in an economic loss to the nation of 0.7 billion contemporary dollars. To put this loss into perspective the total receipts of the Federal government in 1936 were only 5 billion dollars. The 0.7 billion dollar loss is very much a lower bound because there were corresponding losses in yield for all other crops. And, in addition, there were concomitant maxima in acres of abandoned crops which resulted in further losses.

Currie (1988) therefore proposed that these economic losses (and corresponding gains at times of crop yield maxima) have historically propagated with a multiplier effect into the American economy, and caused what are known in economic science as the "Kuznets long swings" with periods near 19 years (Kuznets, 1930, 1961; Soper, 1975, 1978), and the 10 to 11 year "harvest cycle" hypothesized by the Victorian political economist W. S. Jevons (1964, 1981). Keynes (1936, pp. 329 – 331) stated that the causes for systematic variations in crop yield was not his concern. He was emphatic, however, that if they existed they were the plausible cause of business cycles, and that his *General Theory of Employment, Interest, and Money* would then explain them as due to concomitant systematic variations in investment.

Mitchell (1930, pp. 1 – 60) provides a history of research on business cycles which began soon after the Napoleonic Wars. But the first comprehensive theory was that of Karl Marx who believed he had proven them to be inherent to a capitalist market economy, that he had found, *scientifically*, the "laws of motion" of capitalism. This theory was published in volume III of *Das Kapital* in 1910, while I and II appeared in 1867 and 1883 (Marx died in 1883). Heilbroner (1986, pp. 155 – 170) provides a popular account of Marx's theory which is considered to be his most important achievement. Eventually, Western economists came to agree that capitalist economies were inherently unstable and developed numerous theories cited by Mitchell (1930, pp. 1 – 60). If one accepts Keynes' (1936, pp. 329 – 331) judgement, then Tyson (1986) and Currie (1988) have plausibly found the causes of business cycles. Thus, except for Jevons, Keynes, and two or three others, the ideas of all economists since the Napoleonic Wars (Marxist and non-Marxist alike) on what causes business cycles have plausibly been wrong.

REFERENCES

Beveridge, W. H.: 1921, 'Weather and harvest cycles,' *Econ. J.* **31**, 429-452.

Beveridge, W. H.: 1922, 'Wheat prices and rainfall in western Europe,'*J R Statist Soc* **85**, 412-478.

Beveridge, W. H.: 1944 *Full employment in a free society*, Allen and Unwin, London. Reprinted in 1945, 1953, 1960, and 1967.

Box G.E.P., and Jenkins, G. M.: 1970, *Time series analysis: forecasting and control*, Holden-Day, San Francisco.

Campbell, W. H.: 1983, *Possible tidal modulation of the indian monsoon onset*, Ph.D. thesis, University of Wisconsin, Madison.

Currie, R. G.: 1976 'The spectrum of sea level from 4 to 40 years,' *Geophys J R astr Soc* **46**, 513-520.

Currie, R. G.: 1979, 'Distribution of solar cycle signal in surface air temperature over North America,' *J Geophys Res* **84**, 753-761.

Currie, R. G.: 1980, 'Detection of the 11-yr sunspot cycle in earth rotation,' *Geophys J R astr Soc* **61**, 131-140.

Currie, R. G.: 1981a, 'Evidence for 18.6-year signal in temperature and drought conditions in North America since AD 1800,' *J Geophys Res* **86**, 11055-11064.

Currie, R. G.: 1981b, 'Amplitude and phase of the 11-year term in sea level: Europe,' *Geophys J R astr Soc* **67**, 547-556.

Currie, R. G.:1981c, 'Solar cycle signal in earth rotation: nonstationary behavior,' *Science* **211**, 386-389.

Currie, R. G.: 1982, 'Evidence for 18.6-year term in air pressure in Japan and geophysical implications,' *Geophys J R astr Soc* **69**, 321-327.

Currie, R. G.: 1983, 'Detection of 18.6-year nodal induced drought in the Patagonian Andes,' *Geophys Res Lett* **10**, 1089-1092.

Currie, R. G.: 1984a, 'Evidence for 18.6-year lunar nodal drought in western North America during the past millennium,' *J Geophys Res* **89**, 1295-1308.

Currie, R. G.:1984b, 'Periodic 18.6-year and cyclic 11-year induced drought and flood in western North America,' *J Geophys Res* **8**, 7215-7230.

Currie, R. G.: 1987a, 'On bistable phasing of 18.6-year induced drought and flood in the Nile records since AD 650,' *J Climatol* **7**, 373-389.

Currie, R. G.: 1987b, 'Examples and implications of 18.6- and 11-year terms in world weather records,' in Rampino, M.R,, Newman, W.S., Sanders, J.E., and Konigsson, L.K., eds., *Climate: history, periodicity and predictability*, Van Nostrand Reinhold, New York, pp. 379-403.

Currie, R. G.: 1988, 'Climatically induced cyclic variations in United States crop production: implications in economic and social science,' in Erickson, G.J. and Smith, C.R., eds, *Maximum entropy and Bayesian methods in science and engineering*, vol. 2, Reidel, Dordrecht, pp. 181-241.

Currie, R. G.: 1989, "Comments on 'Power spectra and coherence of drought in the interior plains' by E. 0. Oladipo," *Int J Climatol* **9**, 91-100.

Currie, R. G.:1991a, 'Deterministic signals in tree-rings from Tasmania, New Zealand, and South Africa,' *Ann Geophysicae* **9**, 71-81.

Currie, R. G.: 1991b, 'Deterministic signals in tree-rings from the Corn Belt region,' *Ann Geophysicae*, in press.

Currie, R. G.: 1991c, 'Deterministic signals in tree-rings from North America,' *Int J Climatol*, in press.

Currie, R. G., and Fairbridge, R. W.: 1985, 'Periodic 18.6-year and cyclic 11-year induced drought and flood in northeastern China and some global implications,' *Quat Sci Revs* **4**, 109-134.

Currie, R. G., and O'Brien, D. P.: 1988, 'Periodic 18.6-year and cyclic 10 to 11 signals in northeastern United States precipitation data,' *Int J Climatol* **8**, 255-281.

Currie, R. G., and O'Brien, D. P.: 1989, 'Morphology of bistable 180 phase switches in 18.6-year induced rainfall over the northeastern United States of America,' *Int J Climatol* **9**, 501-525.

Currie, R. G., and O'Brien, D. P.: 1990a, 'Deterministic signals in precipitation records from the American corn belt,' *Int J Climatol* **10**, 179-189.

Currie, R. G., and O'Brien, D. P.: 1990b, 'Deterministic signals in precipitation data in the north-west United States,' *Water Resour Res* **26**, 1649-1656.

Currie, R. G., and O'Brien, D. P.: 1990c, 'Deterministic signals in USA precipitation records: part I,' *Int J Climatol* **10**, 705-818.

Davis, H. T.: 1941, *The analysis of economic time series*, Principia Press, Bloomington.

Heilbroner, R. L.: 1986, *The worldly philosophers: the lives, times and ideas of the great economic thinkers*, Simon and Schuster, New York, 6th edition.

Jaynes, E. T., 'On the rationale of maximum entropy methods,' *Proc IEEE* **70**, 939-952.

Jaynes, E. T.: 1983, *Papers on probability statistics, and statistical physics*, Rosenkrantz, R. D., ed, Reidel, Dordrecht.

Jevons, W. S.: 1884, *Investigations in currency and finance*. Reprint by Kelley, New York, 1964, pp. 206-243.

Jevons, W. S.: 1981, 'The solar influence on commerce and the solar commercial cycle,' in Black, R. D. C., ed., *Papers and correspondence of William Stanley Jevons*, vol 7, Macmillan, London, pp. 90-98 and 108-112.

Kendall, M. G.: 1943, 'Oscillatory movements in English agriculture,' *J R Stat Soc, ser A* **106**, 92-124.

Kendall, M. G.: 1945, 'On the analysis of oscillatory time-series,' *J R Stat Soc, ser A* **108**, 93-141.

Kendall, M. G.: 1946, *Contributions to the study of oscillatory time series*, Cambridge University Press, London.

Kendall, M. G., and Stuart, A.: 1967, *The advanced theory of statistics*, vol 2, 2nd ed, Hafner, New York.

Keynes, J. M.: *The general theory of employment, interest, and money*, Macmillan, London.

Kuznets, S.: 1930, *Secular movements in production and prices*. Reprinted by Kelley, New York, 1967.

Kuznets, S.: 1961, *Capital in the American economy: its formation and financing*, Princeton University Press, Princeton.

Lisitzen, E.: 1974, *Sea-level changes*, Elsevier, Amsterdam.

Loder, J. W., and Garrett, C.: 1978, 'The 18.6 year cycle of sea surface temperature in shallow seas due to variations in tidal mixing,' *J Geophys Res* **83**, 1967-1970.

Marple, S. L., Jr.: 1987, *Digital spectral analysis*, Prentice-Hall, Englewood Cliffs.

May, B. R., and Hitch, T. J.: 1989, 'Periodic variations in extreme hourly rainfalls in the United Kingdom,' *Meteorol Mag* **118**, 45-50.

Mitchell, W. C.: 1930, *Business cycles: the problem and its setting*, National Bureau of Economic Research, New York.

Mitra, K., Mukherji, S., and Dutta, S. N.: 1991, 'Some indications of 18.6 year luni-solar tidal and 10-11 year solar cycles in rainfall in north-west India, plains of Uttar Pradesh and north-central India,' *Int J Climatol*, in press.

O'Brien, D. P., and Currie, R. G.: 1991, 'A theoretical model for *l*80° phase changes in 18.6-year variation of air pressure,' in preparation.

Pedlosky, J.: 1987, *Geophysical fluid dynamics*, Springer-Verlag, New York.

Sindermann, C. J.: 1970, *Principle diseases of marine fish and shell fish*, Academic Press, New York.

Swinbanks, D. D.: 1982, 'Intertidal exposure zones: a way to subdivide the shore,' *J Estuar Mar Biol Ecol* **62**, 69-86.

Soper, J. C.: 1975, 'Myth and reality in economic time series: the long swing revisited,' *Southern Econ J* 41, 570-579.

Soper, J. C.: 1978, *The long swing in historical perspective: an interpretive study*, Arno Press, New York.

Tyson, P. D.: 1986, *Climatic change and variability in southern Africa*, Oxford University Press, New York.

Ulrych, T. J., and Clayton, R. W.: 1976, 'Time series modeling and maximum entropy,' *Phys Earth Planet Interiors* 12, 188-200.

Winkle, W. van, Kirk, B. L., and Rust, N.B.W.: 1979, 'Periodicities in Atlantic coast striped bass *Morone saxatilis* commercial fisheries data,' *J Fish Res Bd Canada* 36, 54-62.

Woodworth, P. L.: 1985, 'A worldwide search for the 11-year solar cycle in mean sea level records,' *Geophys J R astr Soc* 80, 743-755.

Wyatt, T.: 1984, 'Periodic fluctuations in marine fish populations,' *Environ Edu and Infor* 3, 137-162.

Yoder, C. F., Williams, J. G., and Parke, M. E.: 1981, 'Tidal variations of earth rotation,' *J Geophys Res* 86, 881-891.

Yule, G. U.: 1927, 'On a method of investigating periodicities in disturbed series, with special reference to Wolfer's sunspot numbers,' *Phil Trans R Soc, ser A* 226, 267-294. Reprinted in 1971 by Stuart, A., and Kendall, M. G., eds., *Statistical papers of G. Udney Yule*, Griffin, London. Reprinted in 1986 by Kessler, S. B., ed., *Modern spectrum analysis*, vol 2, IEEE Press, New York.

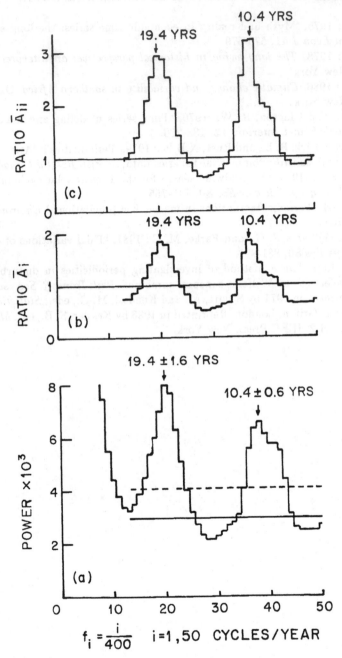

Fig. 1. (a) Maximum entropy power density spectrum of 54 records; (b) and (c) are spectral ratios for statistical tests (see text).

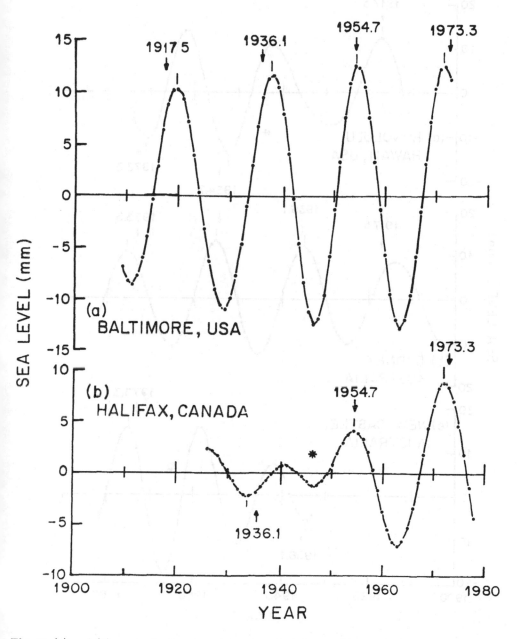

Fig. 2. (a) and (b) are luni–solar wave trains for two harbors along the Atlantic coast of North America.

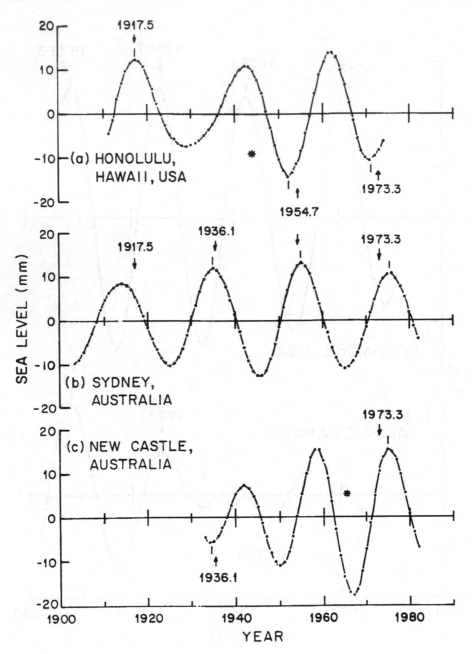

Fig. 3. (a) – (c) show wave trains for three harbors in Hawaii and Australia.

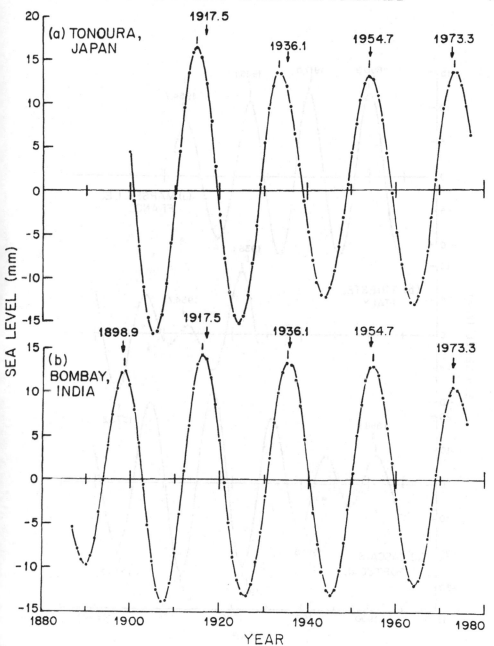

Fig. 4. (a) and (b) depict waves for harbors in Japan and India.

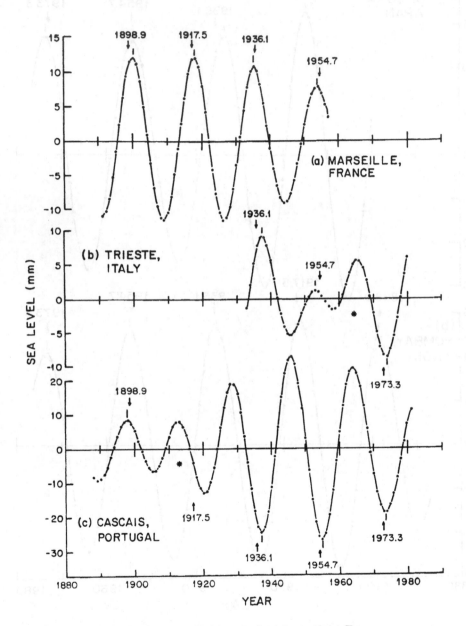

Fig. 5. (a) – (c) are luni–solar waves for three harbors in southern Europe.

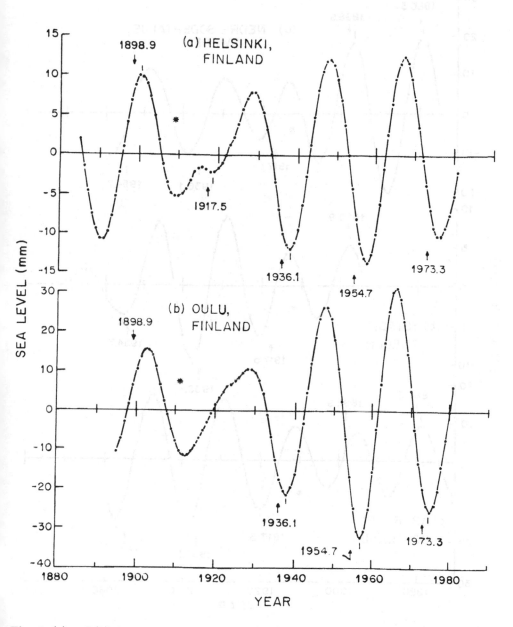

Fig. 6. (a) and (b) show waves at two Finnish harbors.

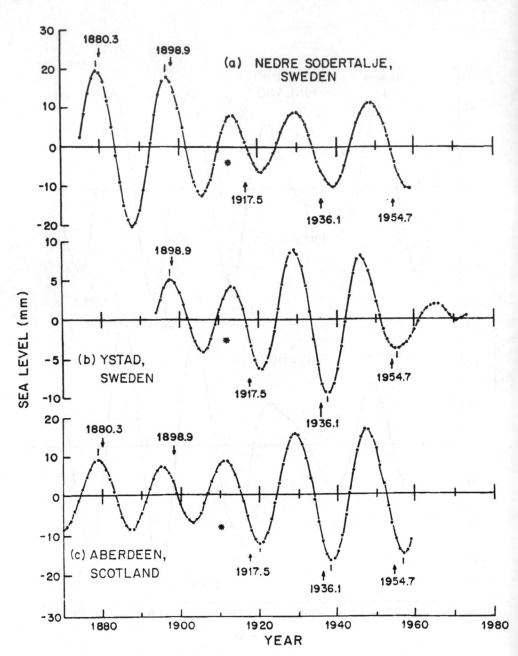

Fig. 7. (a) – (c) depict wave trains at two harbors in Sweden and one in Scotland.

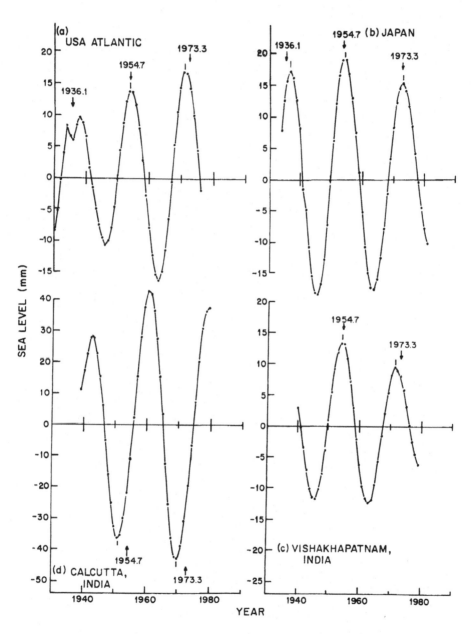

Fig. 8. (a) shows the luni–solar mean wave train for Portsmouth, Washington, D.C., Solomon's Island, Philadelphia, Sandy Hook, Willets Point, and St. John along the Atlantic coast of North America; (b) the mean wave at Mera, Aburatsu, and Wajima in Japan; (c) and (d) are the waves for two harbors in India.

THE GRAND CANONICAL SAMPLER FOR BAYESIAN INTEGRATION

Sibusiso Sibisi
Department of Applied Mathematics and Theoretical Physics
University of Cambridge
England CB3 9EW

ABSTRACT. We discuss a Bayesian analysis of the problem of fitting a number of parametrised components to a dataset. The analysis requires the evaluation of intractable multi-dimensional integrals. Importance sampling is a general Monte Carlo approach to the numerical evaluation of such integrals. We develop an importance sampler, the "Grand Canonical Sampler", for the integrals that arise in our problem and we investigate its performance on numerical tests.

1. Introduction

The evaluation of multi-dimensional integrals is a generic problem in Bayesian analysis, where the structure of the integrand is typically not well-known and cannot be assumed to be smooth. Monte Carlo methods, which sample the integrand at points drawn from some sampling distribution, are extensively used to deal with such problems. The simple Monte Carlo case, which draws samples from a uniform distribution, can be very inefficient for a spiky integrand. Samples should ideally be drawn from a suitably constructed "importance sampler" which concentrates on important regions of the integrand.

In this paper, we develop an importance sampler for a common problem of data analysis—fitting components of a parametric model to a dataset. In spectroscopy, for example, the components may be a set of spectral lines whose number (as well as, typically, some subset of their amplitude, position and shape parameters) is to be determined from the data. We shall, therefore, refer to the problem as the "line-fitting" problem. As has been noted by Bretthorst (1988), a Bayesian approach to the problem involves integrands which can be very spiky with most of the "volume" concentrated in small isolated regions whose locations are not known *a priori*.

We begin with a brief review of importance sampling (for further details, see, for example, Rubinstein, 1981). We proceed to discuss the problem of lines parametrised by amplitude and position parameters and we then describe the sampling scheme for the analytically intractable position marginalisation. We conclude with some numerical investigations.

2. Importance Sampling

We wish to evaluate the d-dimensional integral

$$I = \int_\Omega f(\mathbf{x})d\mathbf{x} \qquad \mathbf{x} \in \mathcal{R}^d, \quad \Omega \subset \mathcal{R}^d$$

423

C. R. Smith et al. (eds.), Maximum Entropy and Bayesian Methods, Seattle, 1991, 423–432.
© 1992 Kluwer Academic Publishers.

An importance sampler is a function $\rho(\mathbf{x})$ satisfying

$$\rho(\mathbf{x}) > 0 \text{ if } f(\mathbf{x}) \neq 0, \mathbf{x} \in \Omega \text{ and } \int_\Omega \rho(\mathbf{x}) d\mathbf{x} = 1$$

Let $\{\mathbf{x}_{(s)}, s = 1 \ldots \mathcal{N}\}$ be a set of uncorrelated samples drawn from $\rho(\mathbf{x})$. Then

$$\tilde{I} = \frac{1}{\mathcal{N}} \sum_{s=1}^{\mathcal{N}} \frac{f(\mathbf{x}_{(s)})}{\rho(\mathbf{x}_{(s)})}$$

is an unbiassed estimate of I (*i.e.* $\langle \tilde{I} \rangle = I$, $\langle \tilde{I} \rangle$ being the expectation of $\tilde{I}$ with respect to ρ) with expected variance $\sigma^2(\tilde{I}) = \left(\int (f^2/\rho) d\mathbf{x} - I^2 \right) / \mathcal{N}$.

Using minimisation of expected variance as the criterion for choosing an optimal importance sampler $\hat{\rho}(\mathbf{x})$, we obtain $\hat{\rho} = |f| / \int |f| d\mathbf{x}$, *i.e.* we should ideally sample from the integrand itself. If $f(\mathbf{x}) \geq 0$ then $\hat{\rho}(\mathbf{x}) = f(\mathbf{x})/I$, for which the expected variance vanishes. This is clearly untenable since, to construct $\hat{\rho}$, we need to know I in advance, thus defeating the object of the exercise. Nonetheless, we should endeavour to construct a sampler that mimics this ideal as closely as the problem allows, so as to obtain a reasonably accurate estimate of I for modest $\mathcal{N}$. While a poor choice of ρ will lead to inefficiency, the estimate $\tilde{I}$ will remain unbiassed; it suffices that $\rho(\mathbf{x}) > 0$ where $f(\mathbf{x}) \neq 0$. Indeed, $\tilde{I}$ will remain unbiassed if successive samples are correlated and drawn from different sampling functions so that, at stage s, the sampler is $\rho_{(s)}(\mathbf{x}|\mathbf{x}_{(1)}, \mathbf{x}_{(2)}, \ldots, \mathbf{x}_{(s-1)})$. The cost of constructing (and sampling from) an efficient sampler should also be modest; it must be possible to generate $\rho_{(s)}(\mathbf{x})$ at every s from a class of "readily computable" functions. However, even in the simplest case of full coordinate independence in $\rho_{(s)}$, we should expect to have to perform $\mathcal{O}(d)$ one-dimensional integrals to carry out appropriate normalisations. Thus, we should typically expect a mimimum cost of $\mathcal{O}(\mathcal{N}d)$ one-dimensional integrals (assuming the cost of actually generating the $\mathcal{N}$ samples $\mathbf{x}_{(s)}$ to be negligible by comparison).

In general, construction of an efficient importance sampler is an elusive task since it may demand detailed knowledge of the structure of the integrand. In adaptive importance sampling, as more samples are taken (thus yielding more information about the integrand), the sampler is modified from some initial function toward $\hat{\rho}$ through, say, an annealing scheme (Evans, 1991). A popular iterative variant of Monte Carlo is the Gibbs sampler which samples from the componentwise full conditional probability distributions of the parameters (Gelfand and Smith, 1990). It is well-suited to the task of generating marginals of the posterior but not to the determination of the normalising factor that we need for model comparison (Gelfand, Dey and Chang, 1991).

3. The Line Fitting Problem

An example of the class of our problems of interest is that of spectrum processing in chemistry. Given noisy samples, $D_k = F(\omega_k) + \text{noise}$, of a spectral trace F, we wish to know the constituent "lines" in F, for which we have a model lineshape. If, for instance, the model lineshape is Lorentzian with the jth line parametrised by its position x_j, amplitude a_j and width λ_j, then, if there are L lines present,

$$F(\omega_k) = \sum_{j=1}^{L} \frac{\lambda_j a_j}{\lambda_j^2 + (\omega_k - x_j)^2} \qquad k = 0 \ldots N - 1$$

In many practical situations, the lines have a common width λ, say, so that the problem amounts to a convolution $F = \phi * f$, ϕ being a Lorentzian blurring function (width λ) and f being a spectrum of L "spikes"; *i.e.*

$$\phi(\omega) = \frac{\lambda}{\lambda^2 + \omega^2} \quad \text{and} \quad f(\omega) = \sum_{j=1}^{L} a_j \delta(\omega - x_j)$$

Although the method we present need, by no means, be restricted to problems of convolution type, we shall use convolution as our exemplar.

Assuming λ to be known, we wish to use the data $\mathbf{D} = (D_0, \ldots, D_{N-1})^T$, to draw inferences about the parameters $\{L, \mathbf{x}, \mathbf{a}\}$ (or some subset thereof) where $\mathbf{a} = (a_1, \ldots, a_L)^T$ and $\mathbf{x} = (x_1, \ldots, x_L)^T$. However, λ may be unknown; indeed, there may be a more appropriate model lineshape (*e.g.* Gaussian) than the Lorentzian form assumed. Therefore, we may also wish to use the data to test our assumptions about the model. Thus we might proceed as follows:

3a) Construct the joint distribution $\Pr(\mathbf{a}, \mathbf{x}, L, \mathbf{D})$. The conditioning on assumptions $\mathcal{A}$ is implicitly understood (*i.e.* $\Pr(\cdot) \equiv \Pr(\cdot | \mathcal{A})$) where $\mathcal{A}$ is taken to include assumptions about the model lineshape as well as the choice of prior for the parameters (*e.g.* we may choose a Gaussian or entropic prior for $\mathbf{a}$).

3b) Marginalise over $\mathbf{a}, \mathbf{x}, L$ to obtain $\Pr(\mathbf{D})$, the evidence for the assumptions $\mathcal{A}$, which we may use to select the best model lineshape amongst other things.

3c) Construct $\Pr(L|\mathbf{D}) = \Pr(L, \mathbf{D}) / \Pr(\mathbf{D})$ which can be used to obtain an estimate $\widehat{L}$ of the true number of lines present.

3d) Construct $\Pr(\mathbf{x}|\mathbf{D}, \widehat{L})$ and use it to obtain an estimate $\widehat{\mathbf{x}}$ of the line positions and the associated uncertainties. Since $\Pr(\mathbf{x}|\mathbf{D}, \widehat{L})$ is an L-dimensional distribution, it is preferable to construct and draw inferences from the one-dimensional marginal $\Pr(x|\mathbf{D}, \widehat{L})$, which can be more readily visualised. (We can legitimately drop the subscript since the coordinate marginals are identical, *i.e.* $\Pr(x_i|\mathbf{D}, \widehat{L}) \equiv \Pr(x_j|\mathbf{D}, \widehat{L})$—the labelling of the coordinates is interchangeable.) We may not be particularly interested in estimating the number of lines present, choosing instead to construct $\Pr(x|\mathbf{D})$, having marginalised over L.

3e) Construct $\Pr(\mathbf{a}|\mathbf{D}, \widehat{\mathbf{x}}, \widehat{L})$ and use it to obtain an estimate $\widehat{\mathbf{a}}$ of the amplitudes and their associated uncertainties.

The amplitude marginalisation will turn out to be straightforward, thus allowing the prompt production of $\Pr(\mathbf{x}, L, \mathbf{D})$. Indeed, in this paper we shall not discuss the estimation of amplitudes. We shall focus on marginalising over line positions $\mathbf{x}$ to obtain $\Pr(L, \mathbf{D})$, for which we shall present an importance sampling scheme. It shall emerge that the process of constructing $\Pr(L^c, \mathbf{D})$ for some currently chosen number of lines L^c, yields all the information needed to deliver $\Pr(L, \mathbf{D})$ and the marginal $\Pr(x|\mathbf{D}, L)$ for all $L \leq L^c$. In particular, if $L^c = L_{\max}$ (the maximum number of lines allowed) $\Pr(\mathbf{D}) = \sum_{L=0}^{L_{\max}} \Pr(L, \mathbf{D})$ follows immediately (thus so do $\Pr(L|\mathbf{D}) = \Pr(L, \mathbf{D}) / \Pr(\mathbf{D})$ and $\Pr(x|\mathbf{D}) = \sum_{L=1}^{L_{\max}} \Pr(x|\mathbf{D}, L) \Pr(L|\mathbf{D}))$. Therefore, the alternative outlook is that the process of constructing $\Pr(\mathbf{D})$, which is needed for model comparison, yields all the information needed to construct marginals of interest at virtually no additional cost.

4. Amplitude Marginalisation

We have $\Pr(\mathbf{a}, \mathbf{x}, L, \mathbf{D}) = \Pr(\mathbf{D}|\mathbf{a}, \mathbf{x}, L)\Pr(\mathbf{a}|L)\Pr(\mathbf{x}|L)\Pr(L)$. Hence $\Pr(\mathbf{x}, L, \mathbf{D}) = \Pr(\mathbf{D}|\mathbf{x}, L)\Pr(\mathbf{x}|L)\Pr(L)$ where

$$\Pr(\mathbf{D}|\mathbf{x}, L) = \int_{-\infty}^{\infty} d\mathbf{a}\,\Pr(\mathbf{D}|\mathbf{a}, \mathbf{x}, L)\Pr(\mathbf{a}|L)$$

To find an explicit expression for the integrand, we first note that

$$F(\omega_k) = \sum_{j=1}^{L} a_j \phi(\omega_k - x_j) \equiv (\mathbf{\Phi a})_k$$

where $\mathbf{\Phi}$ is an $N \times L$ matrix with elements $\Phi_{kj} = \phi(\omega_k - x_j)$. We choose a Gaussian prior for $\mathbf{a}$

$$\Pr(\mathbf{a}|L) = \left(\frac{1}{2\pi\alpha^2}\right)^{L/2} \exp\left(-\mathbf{a}^2/2\alpha^2\right)$$

for some parameter α representing the strength of the lines which we expect to see and, assuming the noise to be $\mathcal{N}(0, \sigma^2)$, we use the Gaussian likelihood

$$\Pr(\mathbf{D}|\mathbf{a}, \mathbf{x}, L) = \left(\frac{1}{2\pi\sigma^2}\right)^{N/2} \exp\left(-\frac{1}{2\sigma^2}(\mathbf{\Phi a} - \mathbf{D})^2\right)$$

The resultant Gaussian integral over $\mathbf{a}$ can be evaluated analytically to obtain

$$\Pr(\mathbf{D}|\mathbf{x}, L) = P_0 (\det \mathbf{B})^{-\frac{1}{2}} \exp\frac{1}{2\sigma^2}\left(\frac{\alpha^2}{\sigma^2}\mathbf{D}^T\mathbf{\Phi B}^{-1}\mathbf{\Phi}^T\mathbf{D}\right)$$

where

$$P_0 = \left(\frac{1}{2\pi\sigma^2}\right)^{N/2} \exp\left(-\mathbf{D}^2/2\sigma^2\right) \quad \text{and} \quad \mathbf{B} = I + \frac{\alpha^2}{\sigma^2}\mathbf{\Phi}^T\mathbf{\Phi}$$

5. Position Marginalisation: The Grand Canonical Sampler

The next step is to marginalise over $\mathbf{x}$ (whose coordinates lie in the finite range $[0, X]$) to get $\Pr(L, \mathbf{D}) = \Pr(\mathbf{D}|L)\Pr(L)$ where

$$\Pr(\mathbf{D}|L) = \int_0^X d\mathbf{x}\,\Pr(\mathbf{x}, \mathbf{D}|L) = \int_0^X d\mathbf{x}\,\Pr(\mathbf{D}|\mathbf{x}, L)\Pr(\mathbf{x}|L) \approx \frac{1}{N}\sum_{s=1}^{N} \frac{\Pr(\mathbf{x}_{(s)}, \mathbf{D}|L)}{\rho(\mathbf{x}_{(s)})}$$

where $\mathbf{x}_{(s)}$ is sampled from $\rho(\mathbf{x})$. We are forced to use an approximation because the form of $\Pr(\mathbf{D}|\mathbf{x}, L)$ given in §4 renders the integral intractable. As pointed out in §2, the ideal of sampling directly (non-iteratively) from the normalised integrand itself—*i.e.* from the posterior $\Pr(\mathbf{x}|\mathbf{D}, L)$—presupposes a knowledge of the integral $\Pr(\mathbf{D}|L)$ in the first place. We nonetheless seek a readily available $\rho(\mathbf{x})$ which mimics $\Pr(\mathbf{x}|\mathbf{D}, L)$.

Let us consider the specific case of $L = 3$. We seek a $\rho(x_1, x_2, x_3)$ which mimics

$$\Pr(x_1, x_2, x_3|\mathbf{D}, 3) = \Pr(x_1|\mathbf{D}, 3)\Pr(x_2|x_1, \mathbf{D}, 3)\Pr(x_3|x_1, x_2, \mathbf{D}, 3)$$

Analogously, $\rho(x_1, x_2, x_3) = \rho_1(x_1)\rho_2(x_2|x_1)\rho_3(x_3|x_1, x_2)$. Accordingly, we would like $\rho_1(x_1)$ to mimic $\Pr(x_1|\mathbf{D}, 3)$, $\rho_2(x_2)$ to mimic $\Pr(x_2|x_1, \mathbf{D}, 3)$ and $\rho_3(x_3)$ to mimic $\Pr(x_3|x_1, x_2, \mathbf{D}, 3)$. We do not, of course, have the latter probabilities at hand because they all derive from $\Pr(x_1, x_2, x_3|\mathbf{D}, 3)$ and its marginals. However, we may expect $\Pr(x_1|\mathbf{D}, 1)$ to mimic $\Pr(x_1|\mathbf{D}, 3)$ in the sense that the two distributions will have peaks at the same positions (although the corresponding peak widths and heights should not be expected to match). Similarly, we may expect $\Pr(x_2|x_1, \mathbf{D}, 2)$ to mimic $\Pr(x_2|x_1, \mathbf{D}, 3)$. Hence we could use $\rho_1(x_1) = \Pr(x_1|\mathbf{D}, 1)$, $\rho_2(x_2|x_1) = \Pr(x_2|x_1, \mathbf{D}, 2)$ and $\rho_3(x_3|x_1, x_2) = \Pr(x_3|x_1, x_2, \mathbf{D}, 3)$.

We can now present the algorithm for computing $\Pr(\mathbf{D}|3)$. In the process, we compute all other quantities that are available at negligible additional cost (*i.e.* no additional evaluations of the joint distribution need be performed). Thus, we shall compute $\Pr(\mathbf{D}|L)$, $L = 0 \ldots 3$, $\Pr(x|\mathbf{D}, L) = \Pr(x, \mathbf{D}|L)/\Pr(\mathbf{D}|L)$, $L = 1 \ldots 3$ as well as $\Pr(\mathbf{D}) = \sum_{L=0}^{3} \Pr(\mathbf{D}|L)\Pr(L)$ and $\Pr(x|\mathbf{D}) = \sum_{L=1}^{3} \Pr(x, \mathbf{D}|L)\Pr(L)/\Pr(\mathbf{D})$.

```
// Initialise:
   Pr(D|2) := 0; Pr(D|3) := 0;
   Pr(x|D,2) := 0; Pr(x|D,3) := 0; Pr(x|D) := 0;
// L = 0:
   Pr(D|0) := P₀;
// L = 1:
   Pr(D|1) := ∫ dx Pr(x, D|1);
   Pr(x|D,1) := Pr(x, D|1)/ Pr(D|1);
   Pr(x|D) := Pr(x|D) + Pr(x, D|1) Pr(1);
   ρ₁(x) := Pr(x|D,1);
   for s := 1 to N
   begin
      Sample x₁ from ρ₁(x);
// L = 2:
      Pr(x₁,D|2) := ∫ dx Pr(x₁, x, D|2);
      Pr(D|2) := Pr(D|2) + Pr(x₁, D|2)/ρ₁(x₁);
      r := Pr(x₁, x, D|2)/ρ₁(x₁);
      Pr(x|D,2) := Pr(x|D,2) + r;
      Pr(x|D) := Pr(x|D) + r Pr(2);
      ρ₂(x|x₁) := Pr(x|x₁, D,2); // [= Pr(x₁, x, D|2)/ Pr(x₁, D|2)]
      Sample x₂ from ρ₂(x|x₁);
// L = 3:
      Pr(x₁,x₂,D|3) := ∫ dx Pr(x₁, x₂, x, D|3);
      Pr(D|3) := Pr(D|3) + Pr(x₁, x₂, D|3)/ρ₁(x₁)ρ₂(x₂|x₁);
      r := Pr(x₁, x₂, x, D|3)/ρ₁(x₁)ρ₂(x₂|x₁);
      Pr(x|D,3) := Pr(x|D,3) + r;
      Pr(x|D) := Pr(x|D) + r Pr(3);
   end
// Normalise:
   Pr(x|D,2) := Pr(x|D,2)/ Pr(D|2); Pr(x|D,3) := Pr(x|D,3)/ Pr(D|3);
   Pr(D|2) := Pr(D|2)/N; Pr(D|3) := Pr(D|3)/N;
   Pr(D) := ∑₃ L=0 Pr(D|L) Pr(L); Pr(x|D) := Pr(x|D)/N Pr(D);
```

In principle $\Pr(x|\mathbf{D})$ could have been, like $\Pr(\mathbf{D})$, constructed only at the end using $\Pr(x|\mathbf{D}, L)$, $L = 1 \ldots 3$. But if interest were to focus on $\Pr(x|\mathbf{D})$ only, we would not want to compute and store a sequence of marginals at each L, opting instead to accumulate $\Pr(x|\mathbf{D})$ directly as presented above. The generalisation from 3 to L lines is clear. The process of computing $\Pr(\mathbf{D}|L)$ involves marching through all values $l = 1, \ldots, L-1$ in order to construct

$$\rho_l(x|x_1, \ldots, x_{l-1}) = \Pr(x|x_1, \ldots, x_{l-1}, \mathbf{D}, l)$$

from which the sample coordinate x_l is drawn. In the process, we may accumulate $\Pr(\mathbf{D}|l)$ (as well as line position marginals of interest). We note that the Gibbs sampler inherited its name from analogy with statistical physics. Pursuing this analogy and noting an analogy between a variable number of lines and a variable number of particles, we call this sampling scheme the Grand Canonical Sampler (GCS).

The computational cost of GCS is given by the number of evaluations of the joint distribution required to perform the one-dimensional normalising integral at each l. In practice, x is finitely digitised to M cells, say. The quadrature scheme for the integral may typically involve using the M values of the integrand to interpolate onto a continuum in x. (Correspondingly, the M values of $\rho_l(x)$ would be the discrete nodes of a sampling distribution defined on the interpolated continuum so that the samples would actually be drawn from the continuum.) However, such interpolation constitutes a minor computational overhead. The accumulation of the marginals at the M-cell resolution does not involve any additional computations of the joint distribution.

Assuming a readily available prior $\Pr(\mathbf{x}|l)$, the cost, $c(l)$, of calculating the joint distribution $\Pr(\mathbf{x}, \mathbf{D}|l)$ is effectively that of calculating $\Pr(\mathbf{D}|\mathbf{x}, l)$. As shown in §4, this involves inverting the $l \times l$ matrix $\mathbf{B}$, which is an $\mathcal{O}(l^3)$ process. But since the integral over x is performed at fixed samples $x_1 \ldots x_{l-1}$, $\mathbf{B}$ changes only in a single row and column for each new x. Correspondingly, the inversion can be reduced to an $\mathcal{O}(l^2)$ process. Thus

$$\text{total cost} = \mathcal{O}\left(Mc(1) + \mathcal{N}M\sum_{l=2}^{L} c(l)\right) = \mathcal{O}\left(M + \mathcal{N}M\sum_{l=2}^{L} l^2\right) = \mathcal{O}(\mathcal{N}L^3 M)$$

This $\mathcal{O}(M)$ behaviour is to be contrasted with the $\mathcal{O}(M^L)$ cost which would result from "exact" integration of $\Pr(\mathbf{x}, \mathbf{D}|L)$ obtained by summing over all permutations of the line position vector $\mathbf{x}$.

6. Priors for α, x and L

6a) We do not wish to detract from the thrust of this paper by dwelling on the treatment of α. Whether we (correctly) introduce a prior $\Pr(\alpha)$ and marginalise over α or simply treat α as an input parameter, the marginalisation over x remains intractable, and that is the problem of primary interest. We shall handle α in the latter manner; a reasonable choice being of the order of the largest peak amplitude. It may be possible to obtain a crude visual estimate of this from the dataset, bearing in mind that the central height of the jth line in the data is a_j/λ (to within the noise). Alternatively, we can choose that α which best fits the data in the sense of maximising $\Pr(\mathbf{D})$. Indeed, we shall be content with choosing α on the basis of maximising $\Pr(\mathbf{D}|L = 1)$.

6b) We take all line positions within the finite range of interest $[0, X]$ to be *a priori* equivalent and we exclude the occupancy of a given position by more than one line. Subject to these minimal assumptions, the prior for **x** is independent of L. In particular, $\Pr(x_1) = 1/X$ and $\Pr(x_1, x_2) = \Pr(x_2|x_1)\Pr(x_1)$ where

$$\Pr(x_2|x_1) = \begin{cases} 1/X & \text{if } x_2 \neq x_1 \\ 0 & \text{if } x_2 = x_1 \end{cases}$$

with similar constructions for a higher number of lines.

6c) There may be physical reasons for an *a priori* preference for certain values of L. For instance, in a particular spectroscopic investigation, it may be reasonable to expect multiplet groups of known multiplicity. We may then wish to discriminate *a priori* against unreasonably high multiplicities. For the present, we choose a uniform $\Pr(L)$ in the finite range $[0, L_{\max}]$

$$\Pr(L) = \begin{cases} 1/L_{\max} & \text{if } L \leq L_{\max} \\ 0 & \text{if } L > L_{\max} \end{cases}$$

7. Numerical Investigations

We work with uniformly spaced data, (*i.e.* $\omega_k = k\delta\omega, k = 0 \ldots N-1$). We also choose a uniform discretisation δx of M cells in $[0, X]$ for each coordinate of **x**. For simplicity, we omit interpolation to the continuum so that the integral over each coordinate simply becomes a sum of the integrand evaluated at the M cells. Samples will correspondingly be restricted to this discretisation. Indeed, in the examples below, we let the data range coincide with the range $[0, X]$ with a common spacing $\delta\omega = \delta x = 1$ so that $M = N$.

Consider a set of 3 lines with (position,amplitude) values (60,5), (120,7) and (160,6) as shown in figure 1 ($M = 250$). The simulated dataset of figure 2 is generated by convolving figure 1 with a Lorentzian of width $\lambda = 1$ and adding Gaussian noise of mean 0 and standard deviation 1.2. We expect a reasonable choice of α to be of the order of 7. Indeed, $\Pr(\mathbf{D}|L = 1)$ is maximised at about $\alpha = 6.85$; this is the value used. Given $L = 3$, exact marginalisation took 35 minutes on a SUN SPARCstation 2 and would get prohibitively expensive for $L > 3$. The exact marginal $\Pr(x|\mathbf{D}, 3)$ is shown in figure 3, the spikes being at the correct line positions (60,120,160). This is the distribution from which the first sample should ideally be drawn in the GCS scheme. As discussed, the sample is, in fact, drawn from $\Pr(x|\mathbf{D}, 1)$, which is shown in figure 4. This distribution does, indeed, mimic the marginal $\Pr(x|\mathbf{D}, 3)$.

We implemented GCS using 20 initialising random seeds. The GCS estimate and associated and error are, respectively, the average and standard deviation of the 20 trials. The following table shows the exact result $E_{\text{exact}} = \log(\Pr(\mathbf{D}|L))$ and the corresponding GCS estimate after N iterates $E_{\text{GCS}}(N)$ with its associated error.

L	E_{exact}	$E_{\text{GCS}}(80)$	$E_{\text{GCS}}(320)$	$E_{\text{GCS}}(1280)$
0	-460.02	-460.02 ± 0.00	-460.02 ± 0.00	-460.02 ± 0.00
1	-445.45	-445.45 ± 0.00	-445.45 ± 0.00	-460.02 ± 0.00
2	-431.98	-432.01 ± 0.09	-431.98 ± 0.04	-431.98 ± 0.01
3	-420.65	-420.70 ± 0.38	-420.67 ± 0.16	-420.65 ± 0.13
4	—	-421.20 ± 0.38	-421.25 ± 0.16	-421.34 ± 0.14
5	—	-421.72 ± 0.37	-421.75 ± 0.18	-421.85 ± 0.14

$\log(\Pr(\mathbf{D}|L))$ rises steeply to $L = 3$ and then gently tails off. This is quite plausible, the data are not well-accounted for by lines fewer than 3; in particular, 0 lines is a very poor model, as borne out by a visual inspection of the data and the smallness of $\log(\Pr(\mathbf{D}|0))$. However, with such noisy data, there can be no strong discrimination against more than 3 lines, thus $\log(\Pr(\mathbf{D}|L))$ should not be expected to fall sharply, if at all, beyond $L = 3$.

The marginals $\Pr(x|\mathbf{D})$ (generated using $L_{max}=3$) for $\mathcal{N} = 80$ and $\mathcal{N} = 1280$ are shown in figure 5 and figure 6 respectively. These are to be compared to the exact marginal which is, in fact, visually identical to $\Pr(\mathbf{x}|\mathbf{D}, 3)$ (plausibly, as $\Pr(\mathbf{x}|\mathbf{D})$ will be dominated by $\Pr(\mathbf{x}|\mathbf{D}, 3)$ because $\Pr(3|\mathbf{D}) >> \Pr(L|\mathbf{D})$, $L < 3$). GCS has picked up all the structure at $\mathcal{N} = 80$ although the relative heights of the peaks still require some adjustment. At $\mathcal{N} = 1280$, the GCS marginal matches the exact marginal almost perfectly. Using $L_{max} = 5$, $\Pr(x|\mathbf{D})$ is shown in figure 7 and figure 8 for $\mathcal{N} = 80$ and $\mathcal{N} = 1280$ respectively. We still obtain high probability at the 3 true positions. But if there is a fourth and a fifth line buried in the noise, there are numerous candidates for their positions, all of which will have relatively low probability.

8. Conclusions

We have presented an efficient importance sampling scheme for marginalisng over line positions in the line fitting problem, where the spiky integrands do not lend themselves to efficient deterministic quadrature. The scheme produces reasonable results within about 100 or less iterates. Thus we can investigate fairly higher numbers of lines without incurring a prohibitive computational penalty. In the process we are able to deliver all marginals of interest at all intermediate line numbers at virtually no additional cost. Future work includes an error analysis on estimated line positions and explicit treatment of amplitude estimation as well as investigation of the use of an entropic prior on the amplitudes.

ACKNOWLEDGMENTS. I wish to thank J Skilling for numerous constructive discussions and comments.

REFERENCES

Bretthorst G.L. (1988). *Lecture Notes in Statistics: Bayesian Spectrum Analysis and Parameter Estimation*. **48**, Springer-Verlag, New York.

Evans M. (1991) 'Chaining via Annealing' *The Annals of Statistics* **19**, 382-393.

Gelfand, A.E., Dey D.K., Chang H. (1991) 'Model Determination Using Predictive Distributions with Implementations via Sampling-Based Methods' *Fourth Valencia International Meeting on Bayesian Statistics*.

Gelfand, A.E., Smith, A.F.M. (1990). 'Sampling Based Approaches to Calculating Marginal Densities', *J. Am. Statist. Assoc.* **85**, 398-409.

Rubinstein R.Y. (1981). *Simulation and the Monte Carlo Method*. Wiley, New York.

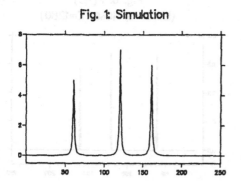

Fig. 1: Simulation

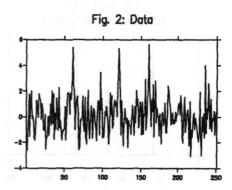

Fig. 2: Data

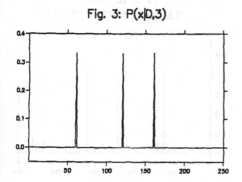

Fig. 3: P(x|D,3)

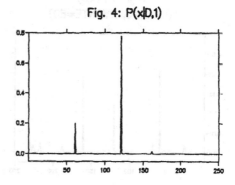

Fig. 4: P(x|D,1)

Fig. 5: P(x|D)
(LMAX=3, NMONTE=80)

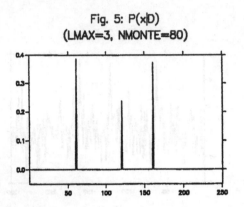

Fig. 6: P(x|D)
(LMAX=3, NOMNTE=1280)

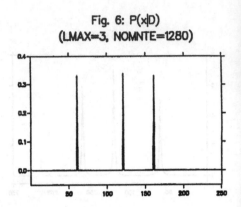

Fig. 7: P(x|D)
(LMAX=5, NMONTE=80)

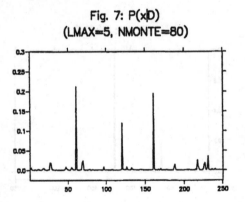

Fig. 8: P(x|D)
(LMAX=5, NMONTE=1280)

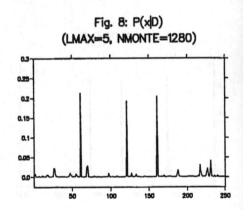

RECENT DEVELOPMENTS IN INFORMATION-THEORETIC STATISTICAL ANALYSIS

Ehsan S. Soofi
School of Business Administration
University of Wisconsin-Milwaukee
PO Box 742, Milwaukee, WI 53201, USA

ABSTRACT. In this paper, first, I provide a sketch of integration of entropy in some areas of statistics during the last four decades. This will be brief and non-exhaustive. Then, I discuss entropy-based quantities that measure information loss or gain due to distributional changes. I provide examples of the information diagnostics that I have developed for measuring information loss in regression analysis and information gain in choice models.

1. Integration of Entropy in Statistics

Entropy first entered the statisticians' world from the fields of communication engineering (Shannon, 1948) and cybernetics (Wiener, 1948). Shannon's derivation of entropy functional as the unique solution to a set of intuitively appealing axioms was echoed by Kullback and Leibler's (1951) generalization of cross- entropy as a measure of information for discrimination between abstract probability spaces. The foundation of information-theoretic approach to statistics was laid by Kullback and Leibler's explication of statistical sufficiency in terms of cross-entropy. Kullback (1954) characterized statistical efficiency in terms of cross-entropy and introduced the minimum discrimination information principle for statistical analysis. Lindley (1956) drew a close analogy between Shannon's model for communication systems and the Bayesian approach to statistical analysis and proposed an entropy-based framework for quantifying the amount of information provided by sample about a parameter. Kullback (1959) provided a unified treatment of many statistical methods in terms of the discrimination information. His book gives a comprehensive account of the research findings on information theory and statistics during the fifties. Two other developments also should be mentioned here. Blyth (1959) discussed minimum variance unbiased estimation of entropy of discrete distribution with finite supports. Stone (1959) discussed comparison of regression experiments along the line of Lindley (1956).

The continual endeavor of Kullback and his associates to apply discrimination information principle in analysis of categorical data drew the attention of a number of statisticians to the topic. By the late seventies information-theoretic approach became an integral part of statistical analysis of categorical data. Gokhale and Kullback (1978) marks the highlight of this effort and provides a comprehensive bibliography of this line of research. (I should add Klotz (1978a and 1978b) to their list.) In section 3 of this paper I present an

C. R. Smith et al. (eds.), Maximum Entropy and Bayesian Methods, Seattle, 1991, 433–444.
© 1992 Kluwer Academic Publishers.

information-theoretic analysis of a probabilistic choice model (Soofi, 1992) which exemplifies the current research along the line of Gokhale and Kullback (1978). During the seventies other streams of information-theoretic approaches to statistical analysis emerged. Zellner (1971) inspired by minimal informative prior principle of Jeffreys (1967) and maximum entropy principle of Jaynes (1968), extended Lindley's work and proposed Maximal Data Information prior procedure for Bayesian econometrics analysis. With further elaborations in Zellner (1982 and 1984), the entropy-based g-prior procedure has now become a part of Bayesian statistics. Zellner (1988) used a number of information- theoretic measures to establish that Bayes' Rule is an optimal information processing procedure. Zellner (1991) gives a thorough review of this line of research.

A number of other authors have provided further insights about foundations of information -theoretic statistical analysis. Akaike (1973) presented the minimum discrimination information method as an extension of the maximum likelihood principle. He showed that the discrimination information function provides suitable criteria for estimating the parameter of a model along with the dimension of the parameter. Akaike Information Criteria (AIC) is now being used as a diagnostic for determining order of time series models in many applications. This line of research is currently being extended by many authors (see, e.g., Hurvich and Tsai (1989), Hurvich, Shumway, and Tsai (1990), and Parzen (1990)). Csiszar (1978) gave geometric interpretation of minimum cross-entropy which was further developed by Loh (1985). Bernardo (1979) characterized cross-entropy as the appropriate utility for determining the risk of using a statistical model as a surrogate for an unknown distribution. This interpretation is becoming very popular among Bayesian statisticians, see, e.g., Polson and Roberts (1991). Larimore (1983) showed that likelihood and sufficiency leads naturally to cross- entropy function as measure of information. Haberman (1984) further elaborated on importance of the minimum discrimination information procedure for continuous distributions. Goel and DeGroot (1979) and Goel (1983) discussed further developments of Lindley's work on comparison of experiments.

Entropy-based procedures have been developed for testing (continuous) distributional hypotheses by many authors including Vasicek (1976), Dudewicz and Van der Meulen (1981), Chandra, De Wet, and Singpurwalla (1982), Gokhale (1983), Arizono and Ohta (1989), and Ebrahimi, Habibullah, and Soofi (1992). Entropy- based procedures for parametric estimation based on non-categorical data have been proposed by many authors including Dey, Ghosh, and Srinivasan (1987), Ghosh and Yang (1988), and Soofi and Gokhale (1991a). Theil in collaboration with a number of associates (see Fiebig and Theil (1984) for references), Vinod (1982), and Brockett (1983) proposed maximum entropy estimation of density, moments, and other distributional characteristics of continuous random variables. This line of research was later continued by Dudewicz and Van der Meulen (1987).

Recent trend in information-theoretic statistical analysis appears to be toward developing entropy-based diagnostics for various statistical problems. Brooks (1982), Hollander, Proschan and Sconing (1987), Turrero (1989), and Ebrahimi and Soofi (1990) have developed information diagnostics useful in reliability analysis. Joe (1989) provides a detailed discussion of information- theoretic measures of associations among random variables. McCulloch (1989) suggests an information diagnostic for sensitivity analysis in Bayesian statistics.

In regression analysis, in addition to the AIC, a number of other information-theoretic diagnostics for model selection have been proposed (see, e.g., Sawa (1978), Young (1987), and Bozdogan (1990)). Information-theoretic diagnostics for the purposes other than model selection have also been developed. Johnson and Geisser (1982, 1983) and Johnson (1987) proposed discrimination information diagnostics for detecting influential observations. Theil and Chung (1988) suggested an information-theoretic measure of fit of the regression model. Soofi (1988 and 1990), Soofi and Soofi (1989), Soofi and Gokhale (1991b) have developed information diagnostics for regression analysis with collinear data. In the next section, I provide a summary of entropy-based diagnostics for analysis collinear data.

2. Information Diagnostics For Regression Analysis

The following basic information quantities are used for defining the information diagnostics. The Expected uncertainty in an outcome of a random variable with distribution f is measured by the entropy

$$H[f] = - \int \log f(u) dF(u); \tag{1}$$

for the continuous distributions $dF(u) = f(u)du$; for the discrete case summation replaces the integral and $dF(u) = f(u)$. Information about the outcome is defined by $-H[f(\cdot)]$.

The change in information due to a distributional change from f_1 to f_2 is measured by the entropy difference

$$\delta H(f_1, f_2) = H[f_1] - H[f_2]. \tag{2}$$

The information loss (gain) due to a distributional change is measured by $\delta(f_1, f_2)$. Let $f_1 = f(x)$ and $f_2 = f(x|y)$ denote the marginal and conditional densities, respectively. Then $\delta H[f(x), f(x|y)]$ measures the amount of information contained in y about an outcome of the random variable X. In general, $\delta H[f(x), f(x|y)]$ may be positive or negative. The expected information in an observation y from the random variable Y is given by

$$\vartheta(x|y) = E_Y\{\delta H[f(x), f(x|y)]\}, \tag{3}$$

where E_Y denotes the expectation with respect to the distribution of Y; Shannon (1948). The expected information $\vartheta(x|y)$ is non-negative. When y represents a parameter of the distribution of X, then $\vartheta(x|y)$ gives Lindley's measure of information.

Another measure of information due to a distributional change is given by the cross entropy between f_1 and f_2,

$$I(f_1 : f_2) = \int \log[f_1(u)/f_2(u)] dF_1(u). \tag{4}$$

Note that the expected amount of information defined in (3) equals to the cross-entropy

$$I(f_{1,2} : f_1 f_2) = \int \int \log[f_{1,2}(u, v)/f_1(u) f_2(v)] dF_{1,2}(u, v), \tag{5}$$

where $f_{1,2}$ is the joint distribution and f_i, $i = 1, 2$ is the marginal distribution. In the information-theoretic tradition, $I(f_{1,2} : f_1 f_2)$ is referred to as the mutual that one random variable provides about another random variable. The mutual information between two sets of random variables is defined similarly.

REGRESSION MODEL

Consider the regression model

$$y = X\beta + \epsilon, \tag{6}$$

where y is an $n \times 1$ vector of observations, X is nonstochastic and nonsingular $n \times p$ regression matrix, β is the $p \times 1$ coefficient vector, and ϵ is the $n \times 1$ noise vector. The least square (LS) estimate of β is $b = (X'X)^{-1}X'y$.

The number of variables in model (6), p, is fixed a priori. The item of interest is the vector of regression coefficients $\beta = (\beta_1, \ldots \beta_p)'$. The focus is in on the effects of a near-collinear structure of X on estimation of coefficients. For this purpose, it is more convenient to rescale X and y such that $X'X$ and $X'y$ are in correlation form, and transform the regression model as

$$y = X\Gamma\Gamma'\beta = Z\alpha + \epsilon, \tag{7}$$

where $\Gamma = [\gamma_1, \ldots, \gamma_p]$ is the orthogonal matrix of the eigenvectors of $X'X$, $Z = X\Gamma$, and $\alpha = \Gamma'\beta$. Model (7) is referred to as the regression in the direction of principal components of $X'X$. Note that $Z'Z = \Lambda = \text{diag}(\lambda_1, \ldots, \lambda_p)$, λ_i, $i = 1, \ldots, p$ are eigenvalues of $X'X$ in descending order. The LS estimate of α is $a = \Lambda^{-1}Z'y$.

Assuming the disturbances $\epsilon_1, \ldots, \epsilon_n$ are uncorrelated, $E(\epsilon_j) = 0$, and $\text{Var}(\epsilon_j) = \alpha^2$, then the least informative (maximum entropy) distribution for the noise vector is the n-variate Guassian $N(0, \sigma^2 I_n)$, where I_n denotes the identity matrix of dimension n. The implied least informative data distribution is $f(y|\beta, \sigma^2) = N(X\beta, \sigma^2 I_n)$. The error variance, σ^2, signifies precision of the assumed model, a priori.

NON-INFORMATIVE PRIOR

Suppose there is no a priori information about the coefficients. Then the unconstrained maximum entropy model for β is the non-informative prior distribution $f(\beta) \propto \text{Constant}$. The posterior distribution of β is

$$f(\beta|b, \sigma^2) = N[b, (X'X)^{-1}\sigma^2]. \tag{8}$$

Under the quadratic loss, the Bayes estimate of β is the posterior mean; i.e., Bayes estimate of β based on the non-informative prior is the LS estimate b. Noting that $a = \Gamma'b$, the posterior distribution of α is

$$f(\alpha|b, \sigma^2) = N[a, \Lambda^{-1}\sigma^2]. \tag{9}$$

Therefore, the elements of α, $\alpha_i = \gamma_i'\beta$, are uncorrelated Guassian with density functions

$$f(\alpha_i|b, \sigma^2) = N[\gamma_i'b, \sigma^2/\lambda_i]. \tag{10}$$

The expected amount of uncertainty in (8) is

$$H[f(\beta|b)] = (p/2)\log(2\pi e) - \frac{1}{2}\log|X'X/\sigma^2| \tag{11}$$

$$= \sum_{i=1}^{p} \frac{1}{2}[\log(\sigma^2/\lambda_i) + \log(2\pi e) \tag{12}$$

$$= \sum_{i=1}^{p} H[f(\alpha_i|b)]. \tag{13}$$

Thus $H[f(\alpha|b)] = H[f(\beta|b)]$. Moreover, the joint posterior entropy decomposes into the posterior entropy of regression coefficients in the direction of principal components, $H[f(\alpha_i|b)]$. When X is ill-conditioned, some eigenvalues are near zero. From (12), observe that $H[f(\beta|b)] \to \infty$, as a $\lambda_i \to 0$. That is, when regression variables are near-collinear, then (8) is quite non-informative about β.

In near-collinear situations small perturbations in the data have large impacts on the posterior distribution (8). Let $X^* = X + dx$, and denote the perturbed posterior is $f(\beta|b^*, \sigma^2) = N[b^*, (X^{*\prime}X^*)^{-1}\sigma^2]$. The cross entropy between the perturbed posterior and the actual posterior is

$$I(f_{X^*} : f_X|b, b^*) = \int f_{X^*}(\beta|b^*) \log[f_{X^*}(\beta|b^*)/f_X(\beta|b)]\, d\beta, \tag{14}$$

$$= \frac{1}{2}\{\text{tr}(X'X)(X^{*\prime}X^*)^{-1} - \log|(X'X)(X^{*\prime}X^*)^{-1}| - p\} + \frac{1}{2}(b - b^*)'X'X(b - b^*)/\sigma^2. \tag{15}$$

This cross-entropy is a diagnostic for examining effects of collinearity on the posterior distribution of β. The first term in (15) measures the perturbation effect on the posterior covariance and the second terms measures the perturbation effect on the posterior mean.

A number of statistical remedies have been proposed in the literature including use of informative prior for β and principal component regression. Information-theoretic diagnostics useful for implementing these procedures are as follows.

INFORMATIVE PRIOR

Suppose the available information is in the form of $E(\beta) = \mu$, and $\text{Var}(\beta) = \tau^2 I_p$. The parameter τ^2 signifies the *prior precision* about the coefficients. Then the least informative (maximum entropy) prior distribution for β is Guassian with density $f(\beta) = N(\mu, \tau^2 I_p)$. This gives the posterior distribution

$$f(\beta|c, \phi) = N[c, D^{-1}\sigma^2], \quad c = D^{-1}(X'y + \phi\mu), \quad D = (X'X + \phi I_p), \tag{16}$$

where $\phi = \sigma^2/\tau^2$ is the prior to model precision ratio. ϕ *indicates the analysts a priori confidence in the prior relative to the model* and constitutes an important quantity in information-theoretic approach to collinearity analysis. A $\phi \approx 0$ indicates a quite high confident in the regression model. The LS procedure uses $\phi = 0$, no prior assumption about the parameters. The Bayes estimate of β based on the family of least informative priors (16) is the posterior mean c.

The amount of information provided by the sample about β is given by

$$\delta_x(\beta|c) = H(\beta) - H(\beta|c) = \frac{1}{2}\log|I_p + X'X/\phi| \tag{17}$$

$$= \sum_{i=1}^{p} \frac{1}{2}\log(1 + \lambda_i/\phi). \tag{18}$$

$$= \sum_{i=1}^{p} \delta_X(\alpha_i|c). \tag{19}$$

Note that since $\delta_X(\beta|c)$ does not depend on y, it is equal to the expected amount of information in the sample about β, namely, $\vartheta_X(\beta|c)$. That is (17) is the mutual information between the random parameter β and the random observation y. In (19), $\delta_X(\alpha_i|c)$ is the information provided by the sample about β in the direction of principal components, $\gamma_i'\beta$.

Based on decomposition (19), a set of information indices for the regression matrix are defined. Note that for a given σ^2, the ordering of the information components are the same as that of the eigenvalues. The information spectrum of the regression matrix is diagnosed by

$$\Delta_i(X;\phi) = \delta_X(\alpha_1|c) - \delta_X(\alpha_i|c) = \frac{1}{2}\log[(\lambda_1 + \phi)/(\lambda_i + \phi)] \tag{20}$$

Particular cases of (20) are :

$$\Delta_i(X;0) = \frac{1}{2}\log(\lambda_1/\lambda_i), \qquad i = 1,\ldots,p, \tag{21}$$

$$\Delta(X;0) = \frac{1}{2}\log(\lambda_1/\lambda_p) = \log\kappa(X); \tag{22}$$

$$\Delta(X;\phi) = \frac{1}{2}\log[(\lambda_1 + \phi)/(\lambda_p + \phi)] = \log\kappa(D) \tag{23}$$

In (22) and (23) κ is the condition number of the respective matrices. Diagnostics (21) display information spectrum of X based on non-informative prior. $\Delta(X;0)$ is the information number of X based on non-informative prior and $\Delta(X;\phi)$ is the information number of X based the Guassian prior for β. When X is orthogonal, $\Delta_i(X;\phi) = \Delta(X;\phi) = \Delta_i(X;0) = \Delta(X;0) = 0$, i.e., all directions are equally informative. The information differences grow as X descends toward singularity. For a singular X, $\Delta(X;0)$ is infinite, whereas $\Delta(X;\phi)$ is finite.

The sample is most informative for the posterior analysis when the regressors are orthogonal. The information difference between the most informative case, when regression matrix is orthogonal X°, and the actual X is given by

$$\delta_{\max}(\beta) - \delta_X(\beta|c) = \sum_{i=1}^{p} \frac{1}{2}\log[(1 + \lambda_i)/(\phi + \lambda_i)] \tag{24}$$

This quantity provides a diagnostic for measuring sample information loss due to non-orthogonality of X. This diagnostic is useful for determining the prior precision required in order to compensate certain amount of information loss due to near-collinearity.

The posterior distribution of β based on an orthogonal regression matrix X° would be

$$f(\beta|c^\circ,\phi) = N[c^\circ, (1+\phi)^{-1}\sigma^2 I_p], \qquad c^\circ = (1 + \phi^{-1}(X^{\circ\prime}y + \phi\mu). \tag{25}$$

The cross-entropy between the actual posterior (16) and the orthogonal reference posterior (25) is

$$I(f_X : f_{X^\circ}|c, c^\circ) = \int f_x(\beta|c)log[f_X(\beta|c)/f_{X^\circ}(\beta|c^\circ)]\,d\beta, \tag{26}$$

$$= \frac{1}{2}\{\text{tr}[(1+\phi)D^{-1} - \log|(1+\phi)D^{-1}| - p\} + (1+\phi)\|c - c^\circ\|^2/2\sigma^2 \tag{27}$$

When X is orthogonal, this cross-entropy is zero and it grows as X descends toward singularity. Thus $I(f_X : f_{X^\circ})$ provides another measure of information loss due to near-collinearity of X. The first term in (27) measures the effect of non-orthogonality of X on the posterior covariance and the second terms measures the effect on the posterior mean.

PRINCIPAL COMPONENT REGRESSION (PCR)

Let Q be a subset of $1, \ldots, p$, and $q \leq p$ be the number of elements in Q. Drop the components which are not in Q from the PCR model (7) and obtain a reduced model

$$y = Z_Q \alpha_Q + \epsilon_Q, \qquad (28)$$

where $Z_Q = X \Gamma_Q, \Gamma_Q$ is the $p \times q$ sub-matrix of Γ formed by the q eigenvectors corresponding to Q, α_Q contains the corresponding elements of α, and ϵ_Q in vector of new disturbances. Obtain the Bayes estimate of α_Q, then transform the Bayes estimate by Γ_Q. Under the Guassian prior, the Bayes estimate of α_Q is c_Q, and the Bayes PCR estimate of β is defined by

$$c_Q^* = \Gamma_Q c_Q. \qquad (29)$$

For the non-informative prior case, the LS estimate a_Q replaces c_Q in (29). The PCR procedure is meaningful only when the prior mean $\mu = 0$, Soofi (1988).

Recall that (from decomposition of information (18)) the amount of sample information about β in the directions of principal components are

$$\delta(\alpha_i | c) = \frac{1}{2} \log(1 + \lambda_i / \phi), \qquad i = 1, \ldots, p. \qquad (30)$$

Since X is non-singular, all eigenvalues are non-zero, and dropping a component implies some loss of information. The loss of information by dropping $p - q$ components that are not in Q is measured by

$$\sum_{i=q}^{p} \delta_X(\alpha_i | c). \qquad (31)$$

This information diagnostic is useful in determining the number of components required for estimation of β without losing a certain amount of sample information.

3. INFORMATION DIAGNOSTICS FOR ANALYSIS OF CHOICE MODELS

Many authors have used mutual information (5) for measuring association between two sets of random variables; see, e.g., Joe (1989). Computation of (5) requires knowledge of the joint distribution $f_{1,2}$ which may not be readily available. A simpler formulation is to quantify the amount of information provided in some fixed (nonstochastic) variables about the probability distribution of a set of random variables. An example of this case from Soofi (1992) follows.

An individual i makes a selection from a set of $J_{(i)}$ competing alternative choices $C_i = \{C_1, \ldots, C_{J(i)}\}$ based on M attributes of the alternatives. All alternatives are evaluated by each decision maker on the M attributes which gives an $M \times J_{(i)}$ matrix $[x_{ijm}]$, where x_{ijm} is the score given by the individual i to the alternative C_j on the mth attribute.

A probabilistic-choice model relates the probability that an individual i will choose an alternative C_{ij} from C_i to the attribute scores in the form of $\pi_{ij} = P(C_{ij} | \beta, x_{ijm}) = F(\beta' x_{ij})$; x_{ij} is the vector of attribute scores for C_j and $\beta = (\beta_1, \ldots, \beta_M)'$ is a vector of unknown coefficients. The model widely used in applied fields is the general multinomial logit

$$\pi_{ij} = P(C_{ij}|\beta, x_{ijm}) = \frac{\exp(\beta' x_{ij})}{\sum_{n=1}^{J(i)} \exp(\beta' x_{ih})} \tag{32}$$

McFadden (1974) derived this model as the optimal solution to a random utility maximization problem and named it "conditional logit."

Various information-theoretic formulation and estimation for special cases of the logit function (32) have been proposed in the literature. The logit models used in analysis of contingency tables (Gokhale and Kullback, 1978) are considered as special cases of (32). These models include, for example, individual-specific attributes such as income or education level of the ith choice maker as determinants of the choice instead of the choice-specific attributes.

The traditional approach for estimating the unknown parameter vector β of (32) is the maximum likelihood (ML) procedure based on a set of N independent choice indicators of the sampled individuals

$$y_{ij} = \begin{cases} 1, & \text{if the individual } i \text{ chooses } C_{ij} \text{ from } C_i \\ 0 & \text{otherwise.} \end{cases} \tag{33}$$

Let $\pi = \{(\pi_{i1}, \ldots, \pi_{iJ(i)}); i = 1 \ldots, N\}$. Then the total entropy of the choice probabilities for N individuals is

$$H(\pi) = -\sum_{i=1}^{N} \sum_{j=1}^{J(i)} \pi_{ij} \log(\pi_{ij}). \tag{34}$$

Maximizing $H(\pi)$ subject to

$$\sum_{j=1}^{J_i} \pi_{ij} = 1 \quad , \quad i = 1, \ldots, N, \tag{35}$$

gives $P(C_{ij}) = 1/J(i)$, $i = 1, \ldots, N$. The total amount of uncertainty is

$$H(U) = \sum_{i=1}^{N} \log[J(i)]; \tag{36}$$

here U denotes the set of N uniform distributions over C_i.

Next consider the following constraints:

$$\sum_{i=1}^{N} \sum_{j=1}^{J(i)} x_{ijm} \pi_{ij} = \sum_{i=1}^{N} \sum_{j=1}^{J(i)} x_{ijm} y_{ij}, \qquad m = 1, \ldots, M. \tag{37}$$

These constraints force the ME probabilities to preserve the observed sum of attribute scores for each of the M attributes. The ME estimate of π_{ij} that satisfies (35) and (37) is

$$\pi_{ij}^* = \frac{\exp(\beta^{*\prime} x_{ij})}{\sum_{h=1}^{J(i)} \exp(\beta^{*\prime} x_{ih})} \tag{38}$$

where, $\beta^* = (\beta_1^*, \ldots, \beta_M^*)'$ is the vector of Lagrange multipliers.

Let $H(\pi^*)$ denote the entropy of the ME logit model with all M attribute constraints present. Then the uncertainty reduction due to inclusion of (37) is

$$\delta H(U, \pi^*) = H(U) - H(\pi^*) > 0 \tag{39}$$

This entropy difference measures the amount of information provided by the M attributes about the choice probabilities.

The above ME formulation is considered as an "Internal Constraint Problem" type, hence π_{ij}^* are equivalent to the ML estimates of the logit model (32); Gokhale and Kullback (1978). $H(\pi^*)$ is equal to minus the value of the log-likelihood function of the model; Soofi (1992). Based on this duality between the ML and the ME procedures, the information gain (39) is readily computed from the standard ML logit outputs.

The information measure (39) is based upon the simple idea of reduction of extreme values of a function by imposing constraints on its domain. It gives diagnostics on the uncertainty reduction power of the given (non-random) attribute scores x_{ijm} about the probability distributions of random choice indicators y_{ij}. Soofi (1992) proposes a number of information indices for logit analysis based on (39).

REFERENCES

Akaike, H.: 1973, 'Information Theory and an Extension of the Maximum Likeli- hood Principle', in B. N. Petrov and F. Csaki, 267-81 (eds.), *2nd International Symposium on Information Theory*, Akademia Kiado, Budapest.

Arizono, I. and H. Ohta: 1989, 'A Test for Normality Based on Kullback-Leibler Information', *The American Statistician*, 34, 20-23.

Bernardo, J. M.: 1979, 'Expected Information as Expected Utility.' *Ann. Statist.*, 7, 686-690.

Blyth, Colin R.: 1959, 'Note on Estimating Information,' *Ann. Math. Statist.*, 30, 71-79.

Bozdogan, H.: 1990, 'On the Information-Based Covariance Complexity and its Application to the Evaluation of Multivariate Linear Models', *Commun. Statist. -Theor. Meth.* 19, 221-278.

Brockett, P. L.: 1983, 'The Unimodal Maximum Entropy Density', *Economics Letters* 12, 261-267.

Brooks, R. J.: 1982, 'On the Loss of Information Through Censoring', *Biometrika* 69, 137-44.

Chandra, M., T. De Wet and N. D. Singpurwalla: 1982, 'On the Sample Redundancy and a Test for Exponentiality,' *Commun. Statist.-Theor. Meth.*, 11, 429-438.

Csiszar, I.: 1975, 'I-divergence Geometry of Probability Distributions and Minimization Problems.' *Ann. Probability*, 3, 146-158.

Dey, D. K., M. Ghosh and C. Srinivasan: 1987, 'Simultaneous Estimation of Parameters Under Entropy Loss', *J. Statist. Plann. Inference*, 15, 347-363.

Dudewicz, E. J. and E. C. Van Der Meulen: 1981, 'Entropy–Based Tests of Uniformity', *J. Am. Statis. Assoc.*, 76, 967-974.

Dudewicz, E. J. and E. C. Van Der Meulen: 1987, 'The Empiric Entropy, a New Approach to Nonparametric Entropy Estimation', in M. L. Puri and J. P. Vilaplana, and W. Wertz (eds.), *New Perspectives in Theoretical and Applied Statistics*, Wiley, New York.

Ebrahimi, N. and E. S. Soofi: 1990, 'Relative Information Loss Under Type II Censored Exponential Data', *Biometrika* **77**, 2, 429-35.

Ebrahimi, N., M. Habibullah and E. S. Soofi: 1992, 'Testing Exponentiality Based on Kullback-Leibler Information', *J. Roy. Statist. Soc.*, **B**. (forthcoming).

Fiebig, D. G., and H. Theil: 1984, *Exploiting Continuity: Maximum Entropy Estimation of Continuous Distributions*, Ballinger, MA.

Ghosh, M. and M. C. Yang: 1988, 'Simultaneous Estimation of Poisson Means Under Entropy Loss', *Ann. Statist.*, **16**, 1, 278-291.

Goel, P. K. and M. H. DeGroot: 1979, 'Comparison of Experiments and Information Measures', *Ann. Statist.* **7**, 1066-1077.

Goel, P. K.: 1983, 'Information Measures and Bayesian Hierarchical Models', *J. Am. Statist. Assoc.*, **78**, 408-410.

Gokhale, D. V.: 1983, 'On Entropy-Based Goodness-of-Fit Tests', *Computational Statist. & Data Analysis*, **1**, 157-165.

Gokhale, D. V., and S. Kullback: 1978, *The Information in Contingency Tables*, Marcel Dekker, New York.

Haberman, S. J.: 1984, 'Adjustment by Minimum Discriminant Information', *Ann. Statist.*, **12**, 971-988.

Hollander, M., F. Proschan and J. Sconing: 1987, 'Measuring Information in Right-Censored Models', *Naval Res. Logist.*, **34**, 669-81.

Hurvich, C. M. and C. L. Tsai: 1989, 'Regression and Time Series Model Selection in Small Samples', *Biometrika*, **76**, 297-307.

Hurvich, C. M., R. Shumway and C. L. Tsai: 1990, 'Improved Estimators of Kullback-Leibler Information for Autoregressive Model Selection in Small Samples', *Biometrika*, **77**, 709-719.

Jaynes, E. T.: 1968, 'Prior Probabilities', *IEEE Transactions on Systems Science and Cybernetics*, **SSC-4**, 227-241.

Jeffreys, H.: 1967, *Theory of Probability* (3rd rev. ed.), Oxford University Press, London.

Joe, H.: 1989, 'Relative Entropy Measures of Multivariate Dependence', *J. Am. Statist. Assoc.*, **84**, 157-164.

Johnson, W. and S. Geisser: 1982, 'Assessing the Predictive Influence of Observations', Kallianpur, Krishnaiah, and Ghosh (eds.) *Essays in Honor of C. R. Rao*, North Holland, Amsterdam.

Johnson, W. and S. Geisser: 1983, 'A Predictive View of the Detection and Characterization of Influential Observations in Regression Analysis', *J. Am. Statist. Assoc.*, **78**, 137-144.

Johnson, W.: 1987, 'The Detection of Influential Observations for Allocation, Separation, and the Determination of Probabilities in a Bayesian Framework', *J. Bus. Econom. Statist.*, **5**, 369-381.

Klotz, J. H.: 1978a, 'Maximum Entropy Constrained Balance Randomization for Clinical Trials', *Biometrics*, **34**, 283-287.

Klotz, J. H.: 1978b, 'Testing a Linear Constraint for Multinomial Cell Frequencies and Disease Screening', *Ann. Statis.*, **6**, 904-909.

Kullback, S.: 1954, 'Certain Inequalities in Information Theory and the Cramer- Rao Inequality', *Ann. Math. Statist.*, **25**, 745-751.

Kullback, S.: 1959, *Information Theory and Statistics*, Wiley, New York, (reprinted in 1968 by Dover, MA).

Kullback, S. and R. A. Leibler: 1951, 'On Information and Sufficiency', *Ann. Math. Statist.*, **22**, 79-86.

Larimore, W. E.: 1983, 'Predictive Inference, Sufficiency, Entropy and an Asymptotic Likelihood Principle', *Biometrika*, **70**, 175-81.

Lindley, D. V.: 1956, 'On a Measure of the Information Provided by an Experiment', *Ann. Math. Statist.*, **27**, 986-1005.

Loh, W.: 1985, 'A Note on the Geometry of Kullback-Leibler Information Numbers', *Commun. Statist. -Theor. Meth.*, **14**, 895-904.

McCulloch, R. E.: 1989, 'Local Model Influence', *J. Am. Statist. Assoc.*, **84**, 473-478.

McFadden, D.: 1974, 'Conditional Logit Analysis of Qualitative Choice Behavior', in P. Zarembka (ed.), *Frontiers in Econometrics*, Wiley, New York.

Parzen, E.: 1990, *Time Series, Statistics, and Information*, IMA Preprint Series #663, Institute for Mathematics and its Applications, University of Minnesota.

Polson, N. G. and G. O. Roberts: 1991, 'A Bayesian Decision Theoretic Characterization of Poisson Processes', *J. Roy. Statist. Soc.*, B, **53**, 675-682.

Sawa, T.: 1978, 'Information Criteria for Discriminating Among Alternative Regression Models', *Econometrica*, **46**, 1273-1292.

Shannon, C. E.: 1948, 'A Mathematical Theory of Communication', *Bell System Tech. J.*, **27**, 379-423; 623-656.

Soofi, E. S.: 1988, 'Principal Component Regression Under Exchangeability', *Commun. Statist. -Theor. Meth.*, **17**, 1717-1733.

Soofi, E. S.: 1990, 'Effects of Collinearity on Information About Regression Coefficients', *J. Econometrics*, **43**, 255-274.

Soofi, E.: 1992, 'A Generalizable Formulation of Conditional Logit With Some Diagnostics', *J. Am. Statist. Assoc.*, (forthcoming).

Soofi, E. S. and D. V. Gokhale: 1991a, 'Minimum Discrimination Information Estimator of the Mean With Known Coefficient of Variation', *Computational Statist. & Data Analysis*, **11**, 165-177.

Soofi, E. S. and D. V. Gokhale: 1991b, 'An Information Criterion for Normal Regression Estimation', *Statist. and Prob. Letters*, **11**, 111- 117.

Soofi, E. S. and A. S. Soofi: 1989, 'A Bayesian Analysis of Collinear Data–The Case of Translog Function', in *Proc. of ASA Bus. and Econ. Stat. Sec.*, 321-326.

Stone, M.: 1959, 'Application of a Measure of Information to the Design and Comparison of Regression Experiments', *Ann. Math. Statist.*, **30**, 55-70.

Theil, H. and C. F. Chung: 1988, 'Information-Theoretic Measures of Fit to Univariate and Multivariate Linear Regressions', *The American Statistician*, **42**, 249-252.

Turrero, A.: 1989, 'On the Relative Efficiency of Grouped and Censored Survival Data', *Biometrika*, **76**, 125-31.

Vasicek, O.: 1976, 'A Test for Normality Based on Sample Entropy', *J. Roy. Statist. Soc.*, Ser. B, **38**, 54-59.

Vinod, H. D.: 1982, 'Maximum Entropy Measurement Error Estimates of Singular Covariance Matrices in Undersized Samples', *J. Econometrics*, **20**, 163-174.

Wiener, N.: 1948, *Cybernetics*, Wiley, New York.

Young, A. S.: 1987, 'On the Information Criterion for Selecting Regressors', *Metrika*, **34**, 185-194.

Zellner, A.: 1971, *An Introduction to Bayesian Inference in Econometrics*, Wiley, New York, (reprinted in 1987 by Krieger Publishing Col., Malabar, Florida).

Zellner, A.: 1982, 'On Assessing Prior Distributions and Bayesian Regression Analysis With g-Prior Distributions', in *Bayesian Decision Techniques: Essays in Honor of Bruno de Finetti*, eds. P. Goel and A. Zellner, Amsterdam: North-Holland, pp. 233-243.

Zellner, A.: 1984, *Basic Issues in Econometrics*, University of Chicago Press, Chicago.

Zellner, A.: 1988, 'Optimal Information Processing and Bayes' Theorem', *American Statistician*, **42**, 278-284.

Zellner, A.: 1991, 'Bayesian Methods and Entropy in Economics and Econometrics', in W. T. Grandy, Jr. and L. H. Schick (eds.), *Maximum Entropy and Bayesian Methods*, pp17-31, 1991 Kulwer Academic Publishers, Netherlands.

MURPHY'S LAW AND NONINFORMATIVE PRIORS

Robert F. Bordley
Social and Economic Sciences Division
National Science Foundation
Washington D.C. 20550

1. Introduction

A client runs into his stockbrocker and has the following conversation:

STOCKBROKER: Friend, we've got to sit down and chat. I've found a fantastic investment for you.
CLIENT: No thanks. I always get burned playing the stockmarket.
STOCKBROKER: But I've go all the probabilities calculated out. You've got an 80% chance of making a fortune!
CLIENT: Somehow whenever you calculate these probabilities, you always make some hidden assumption that turns out to be wrong. What you calculate to be an 80% change of my winning always seems in hindsight to be a 10% chance. Unless you've got an ironclad argument for this investment, I'm not interested.
STOCKBROKER: You're an insufferable pessimist. If I guaranteed you a profit, you'd probably question that.
CLIENT: I sure would.

In this conversation, a stockbroker offers a client some investment opportunity. But the client has very strong priors against the deal paying off. Hence, even though he has not heard anything specific about the investment, the client knows that he will probably discount the investment opportunity and find it unsatisfactory.

Bordley (1991) supposed that the stockbroker offered the client a choice of a lottery whose stated payoffs (in utility units) were normally distributed. But the client does not take these stated payoffs as necessarily true. The client has his own prior over the utility of any lottery offered him yielding various payoffs. The client therefore computes his own estimate of the lottery's payoff probabilities by updating his generic prior over lottery payoffs with the specific information about the lottery's stated payoffs.

To reflect a pessimistic client, this generic prior over utilities was assumed exponential with a lower bound on the minimum level of utility possible. Then the pessimist's assessment of the expected utility of the lottery will equal the lottery's stated expected utility less some fraction of the lottery's stated variance. Hence, behavior that seems incompatible with expected utility if lotteries are taken at face value becomes quite compatible if one allows for 'suspicious' behavior (Bordley and Hazen, 1991).

C. R. Smith et al. (eds.), Maximum Entropy and Bayesian Methods, Seattle, 1991, 445–448.
© 1992 *Kluwer Academic Publishers*.

This idea proved very fruitful in the psychological context. The purpose of this paper is to explore this idea in the context of noninformative priors.

2. Noninformative Priors

Suppose we are developing a model estimating the probability of some decision opportunity having various possible payoffs. To complete the model, we need to specify a probability distribution for the values of some key parameter. Unfortunately, we have no intuitions about that parameter and what its possible values might be. In this case, it is often customary to assign that parameter a noninformative probability distribution (Berger, 1985), i.e., a distribution which presumably conveys 'no information' about the parameter. Maximum entropy techniques provide one possible way to assign such a noninformative distribution; other approaches are based on notions of group invariance.

Such an approach is reasonable if we really have no information at all about the parameter. But there is one bit of information which we do have about the parameter, namely that it is relevant in computing the utility of some decision opportunity. As the introduction noted, we often do have intuitions about the utility of decision opportunities likely to confront us. These intuitions about what the utility of the decision opportunity ought to be imply a probability density over this unknown parameter. Hence, instead of assigning a noninformative prior to our parameter, we might choose a prior over that parameter which causes the resulting posterior for the decision's utility to be very close to our prior opinions about the utility of decision opportunities offered us.

As an example, suppose that our stockbroker does get his client to sit down with him. Consider the following conversation:

STOCKBROKER: As you can see, you can buy this stock and resell it in 60 days. If its price increases in 60 days, you make a profit. If it goes down in price, the firm agrees to buy back the stock any time in these 60 days at the stock's original price. You can't lose!

CLIENT: Will the firm buy back the stock even if it's gone bankrupt in the meantime?

STOCKBROKER: Well no, but the odds of a company going bankrupt within 60 days of a new stock issuance are small.

CLIENT: I am convinced that this deal cannot possibly be as attractive as you say. I've been talking with you to try to figure out what the catch is. I think I see what the catch is. The only aspect of the deal which you haven't carefully researched is the odds of a company going bankrupt within 60 days. Hence if there is a catch, it must be with that variable. In other words, if the deal isn't as attractive as it seems, then the odds of the firm going bankrupt must be a lot higher than either one of us realized. So I conclude that the odds of the firm going bankrupt are very high. Hence, I can stick with my original feeling that this deal isn't any good.

STOCKBROKER: YOU PIG-HEADED FOOL! I've been giving you an honest description of an excellent deal and all you've been doing is looking for an excuse to reject it. Do you really think I have the time to provide definitive estimates of every uncertain aspect of this deal?

CLIENT: If you want me to invest, then I'm afraid you'll have to. As I told you, my gut instincts incline me against any kind of deal. Hence, I am convinced the deal isn't as good as you say. So, unless you can provide a proof for what you're claiming, I'm probably not going to invest.

STOCKBROKER: I give up! Socrates was wrong. There is a limit to the powers of rational persuasion.

CLIENT: Yes, my friend, commonsense is a powerful shield against even the most eloquent displays of verbal sophistry. (Of course, hemlock works too.)

More formally, there is one parameter in the model, namely the firm's being bankrupt or not being bankrupt, about which both client and stockbroker know next to nothing. Because the client expects the deal to be a bad one, he assigns this parameter that probability distribution which makes the deal look as sour as possible.

Before we sympathize too much with the frustrated stockbroker, suppose we replace the stockbroker by an alleged telepath. This gives us Jaynes' example (1991) in which a telepath is providing arguments and demonstrations trying to convince the client of his powers. The client's prejudices are against believing the telepath's arguments so he looks for ways in which the telepath's evidence could have been created by subterfuge. If there is some possibility of an effect arising from subterfuge, then the client will assume that subterfuge is at work and dismiss the telepath's claims.

3. The Murphy Law Prior and Noninformative Priors

To formalize this, let $p(u)$ be the client's prior probability over the utility of decision opportunities offered him. Let $p(u|M)$ be the posterior utility on the basis of the decision model, M. If we let x be the critical parameter about which the decision maker has no intuition, then

$$p(u|M) = \sum_x p(u|x, M) p(x|M).$$

What our client wants to do is choose $p(x|M)$ so that $p(u|M)$ is as close as possible to $p(u)$, i.e., he wants to reinforce his prior expectations. Hence, he wants to choose $p(x)$ to minimize

$$\sum_u p(u) \log \left[\sum_x p(u|x, M) p(x|M) / p(u) \right],$$

subject to

$$\sum_x p(x) = 1.$$

The first-order conditions become

$$\sum_u \frac{p(u) p(u|x, M)}{\sum_x p(u|x, M) p(x)} - k = 0.$$

4. Conclusions: The Garbage-In/Garbage-Out Principle

We say that a model parameter is critical if it is possible to choose $p(x)$ so that $p(u|M) = p(u)$. Hence, our results indicate that when we have no intuition about a critical parameter, our model is totally uninformative — in the sense that it does not change our opinions. This, of course, is just a restatement of the GARBAGE-IN/GARBAGE-OUT principle. Hence, a Bayesian using our Murphy's Law utility prior might agree with the classifical statistician in the uselessness of models when we lack any information on critical model parameters.

Lindley and Good discussed 'Bayes/non-Bayes compromises', i.e., areas of statistics on which both Bayesians and non-Bayesians will agree. This paper shows that taking prior expectations into account similarly leads to a Bayes/non-Bayes compromise over the pivotal issue of parameters about which the decision maker has no intuitions.

REFERENCES

(1) Berger, James, *Statistical Decision Theory and Bayesian Analysis*, Springer-Verlag, New York, 1985.
(2) Bordley, Robert F., 'An Intransitive Expectations-Based Bayesian Variant of Prospect Theory,' *J. Risk and Uncertainty* (in press).
(3) Bordley, Robert, and Gordon Hazen, 'Weighted Linear Utility and SSB as Expected Utility with Suspicion,' *Management Science* 4, 396–408, 1991.
(4) Jaynes, Edwin T., *Probability Theory: The Logic of Science*, Washington University, 1992.

A SCIENTIFIC CONCEPT OF PROBABILITY

Jin Yoshimura*
Department of Mathematics
University of British Columbia
Vancouver BC Canada V6T 1Z2

Daniel Waterman
Department of Mathematics
Syracuse University
Syracuse, New York 13244-1150 USA

ABSTRACT. A notion of probability has been developed as a scientific tool to describe uncertain phenomena in science. However, the basic concept of probability is still controversial. Here we propose a scientific definition of probability space and probability. Probability space is defined as a piece of (incomplete) information about a specific matter that has a domain in time and space. Probability is then defined as a countably additive measure of such incomplete information. This probability measure will satisfy the Kolmogorov axioms without certain of the defects found in systems derived from these axioms.

1. Introduction

A notion of probability has been developed as a scientific tool to describe uncertain phenomena in science (Krüger et al, 1987). It has been widely used in almost all branches of science. However, the foundation of probability is still controversial (Raiffa, 1968; Savage, 1974). We have no consensus about the meaning of probability; three major meanings are logical symmetry, frequency and degree of belief. Among them, the first two interpretations are criticized in terms of their limited applicability. Contrarily, the last one, the degree of belief used in Bayesian statistics, is often criticized or ignored because of its introduction of subjectivity into the mathematical sciences, despite its wide usage and applicability.

Apart from the meaning of probability, current theories in probability and statistics are, in the main, based on the Kolmogorov axioms (Kolmogorov, 1950). The attractiveness of this axiomatic foundation is due primarily to the ability it provides to exploit measure theory as a useful tool. However, the Kolmogorov axioms, without further restrictions, exhibit flaws and lead to difficulties (Savage, 1974). There is no meaning to probability

---

*Now at: Department of Biological Sciences, State University of New York at Binghamton, Binghamton, New York 13902-6000 USA

C. R. Smith et al. (eds.), Maximum Entropy and Bayesian Methods, Seattle, 1991, 449–452.

inherent in these axioms. Also, the unrestricted use of these axioms leads to many paradoxes and problems in probability measure.

In this paper we do not discuss details of these problems. Rather we propose a definition of probability space and of probability as an attempt to establish a scientific concept of probability. To address the meaning of probability, we define probability space as information (Yoshimura, 1987). We shortly discuss the meaning of Bayes theorem and an application to quantum mechanics. We then define probability as a countably additive measure of such incomplete information. This probability measure will satisfy the Kolmogorov axioms without certain of the defects found in many systems derived from these axioms.

In this paper, the terms event, outcome, and result are used synonymously in probability and statistics to denote a set of specific states of matter.

2. A Definition of Probability Space

DEFINITION OF PROBABILITY SPACE. *Let m be a specific matter, t, the time of inference, and s, the space from which the inference is made. Then a probability space Φ is defined as a piece of information that is a function of m, t, and s, i.e.,*

$$\Phi = \Phi(m, t, s).$$

Note that the space s and time t are dimensions of the domain of information. Here the space s could be a physical space with an arbitrary finite measure or human knowledge. In the latter case, our definition agrees with Jaynes' interpretation of probability, a state of knowledge (Jaynes, 1979; 1990; 1991).

If a probability space has a unique state (outcome) with probability one, it becomes complete information; it is no longer a probabilistic inference. Thus probability space is an expression of incomplete information as in fuzzy theory (see Jaynes, 1991).

This definition of probability space is also related to the Shannon-Wiener definition of information, a measure of the amount of information in a probability space (Gal-Or, 1981). We simply define a probability space as information itself (content).

The inherent meaning of probability in this definition is thus quite similar to that given by Jaynes (1991); however our approach is quite different from his methods. He considers plausible (probabilistic) logic by implementing Bayesian methods (e.g. Bayesian updates). We instead focus on the domain of probability space (i.e. information or knowledge); we consider probability space as a function on a space-time structure. For example, Bayes theorem could arise in at least two completely different contexts: with/without a temporal shift (t). A Bayesian update is always accompanied with a temporal shift in the domain of the probability space (see Jaynes, 1991). In contrast, we could have an 'if' assumption of conditional probability without a change in the temporal domain (with/without a change in spatial domain).

With this definition, probability space is now considered as an inference; therefore it will end or be completed when the state (result or outcome) becomes available (known) to the domain. Consider betting on a die in an inverted cup. When a bettor overturns the cup, he knows the result and the probabilistic inference ends. This termination of inference should be the same in any probabilistic inference. It then applies to a quantum probability space; a wave packet is a mere probabilistic inference (Yoshimura, 1987; Jaynes, 1990). Then, the reduction of wave packet is simply the end of this inference, i.e. the observation of the states (results or outcomes). We will not discuss this further, but it has been pointed out that, with this interpretation, Schrodinger's cat and the EPR paradox never occur (Yoshimura, 1987; Jaynes, 1989; 1990).

3. A Definition of Probability

A probability measure based on the Kolmogorov axioms requires that the "size" of our measure spaces be restricted. It is not possible to define a countably additive probability measure on the σ-algebra of *all* subsets of the unit interval which agrees with Lebesgue measure and assigns the same measure to congruent sets (Savage, 1974). Further, the existence of any nonatomic countably additive probability measure on that σ-algebra is inconsistent with the continuum hypothesis (Ulam, 1930). These and other problems have led many to replace the requirement of countable additivity with finite additivity (Dubins and Savage, 1965).

Here we shall give a definition of probability which is simple, appears to avoid many of these problems, and provides enough structure for most purposes. The probability measure thus defined will satisfy the Kolmogorov axioms.

DEFINITION OF PROBABILITY. *Let Ω be the set of states. A decomposition of Ω, $D(\Omega) = \{E_i\}$ will be a finite or countably infinite collection of mutually exclusive events from Ω such that*

$$\Omega = \cup E_i.$$

Let P be a nonnegative additive function on $D(\Omega)$ such that

$$\sum P(E_i) = 1.$$

Let $w(D(\Omega))$ be the σ-algebra generated by $D(\Omega)$. Then the probability of $A \in w(D(\Omega))$, $P(A)$, is the value at A of the unique extension of P to $w(D(\Omega))$ as a measure.

We note that the elements of $w(D(\Omega))$ are exactly of the form $\cup E_{i_j}$, where $\{i_j\} \subset \{i\}$, and that $P(\cup E_{i_j}) = \sum P(E_{i_j})$.

Clearly, if $\{A_n\}$ is a finite or countably infinite collection of disjoint elements of $w(D(\Omega))$, then for each n, $A_n = \cup E_{i(n)}$ uniquely, and when $\{i(n)\}$ is the sequence $\{i_j\}$ corresponding to A_n,

$$P(\cup A_n) = P(\underset{n}{\cup} \underset{i(n)}{\cup} E_{i(n)}) = \sum_n \sum_{i(n)} P(E_{i(n)}) = \sum_n P(A_n).$$

As an example, let Ω be the interval [0,1]. Suppose $D(\Omega) = \{E_i\}$, $i = 1, 2, \ldots$, is a sequence of pairwise disjoint Lebesgue measurable sets such that $\overset{\infty}{\underset{1}{\cup}} = [0,1]$. Let $P(E_i) = |E_i|$, the Lebesgue measure of E_i. Then $w(D(\Omega))$ is the collection of unions of E_i, and P on $w(D(\Omega))$ is just the restriction of Lebesgue measure to that σ-algebra.

Note that with this definition, an event of positive probability cannot be the union of uncountably many events of zero probability.

Our definition is not so restrictive as to disable ordinary methods of analysis. Consider, for example, the collection of σ-algebras which are generated by countably many Lebesgue measurable subsets of the line. These form a lattice L with the join of algebras A and B defined as the smallest σ-algebra containing A and B and the meet of A and B as the largest σ-algebra contained in both A and B. A function f is Lebesgue measurable if and only if for every real c, there is an $A \in L$ such that $\{f > c\} \in A$. Given any countable collection of mutually exclusive events (disjoint Lebesgue measurable sets), there will be an algebra in L generated by the collection. Thus a distribution on the Lebesgue measurable sets (in the usual sense) can be replaced by its restriction to a suitable subalgebra in L.

Even though the set of all possible states is unrestricted (e.g. continuous), this probability measure is discrete; this is natural (necessary), since it is a measure of (incomplete) information (i.e. probability space) which has to be discrete by definition. A discrete probability measure is known to avoid the defects accompanied with the Kolmogorov axioms. This probability measure is such a discrete measure. We believe that it is natural for a probability measure to be based on the decomposition of possible states.

REFERENCES

Dubins, L.E. and Savage, L.J.: 1965, *How to Gamble If You Must: Inequalities for Stochastic Processes*, McGraw-Hill, New York.

Gal-Or, B.: 1981, *Cosmology, Physics, and Philosophy*, Springer-Verlag, New York.

Kolmogorov, A.: 1950, *Foundations of the Theory of Probability*, Chelsea, New York.

Krüger, L, Daston, L.J. and Heidelberger, M. (eds.): 1987, *The Probabilistic Revolution, Volume 1 Ideas in History*, MIT Press, Cambridge, Massachusetts.

Jaynes, E.T.: 1979, 'Where Do We Stand on Maximum Entropy?', in R.D. Levine and M. Tribus (eds.), *The Maximum Entropy Formalism*, MIT Press, Cambridge, Massachusetts, pp. 15-118.

Jaynes, E.T.: 1989, 'Clearing Up Mysteries-The Original Goal', in J. Skilling (ed.), *Maximum Entropy and Bayesian Methods*, Kluwer, Dordrecht, pp. 1-27.

Jaynes, E.T.: 1990, 'Probability in Quantum Theory', in W.H. Zurek (ed.), *Complexity, Entropy, and the Physics of Information*, Addison-Wesley, Redwood City, CA, pp. 380-403.

Jaynes, E.T.: 1991, *Probability Theory-The Logic of Science*, Kluwer, Dordrecht (in press).

Raiffa, H.: 1968, *Decision Analysis: Introductory Lectures on Choices under Uncertainty*, Addison-Wesley, Reading, Massachusetts.

Savage, L. J.: 1974, *The Foundations of Statistics*, Dover, New York.

Ulam, S.: 1930, 'Zur Masstheorie in der allgemeinen Mengenlehre', *Fund. Math.* 16, p. 140.

Yoshimura, J.: 1987, 'The Announcement: the Solution for the Measurement Problem in Quantum Mechanics', in R. Gilmore (ed.), *XV international colloquium on group theoretical methods in physics*, World Scientific, Teaneck, New Jersey, pp. 704-709.

BAYESIAN LOGIC AND STATISTICAL MECHANICS - ILLUSTRATED BY A QUANTUM SPIN 1/2 ENSEMBLE

Everett G. Larson
Department of Physics and Astronomy
Brigham Young University
Provo, Utah 84602 USA

1. Introduction

The greatest challenge for the appropriate use of Bayesian logic lies in the need for assigning a "prior"[1-13]. In physical applications, this prior is often strongly molded by the physical "boundary conditions"[1,12,13], and thereby begins to take on some aspects of "objective" meaning. Many claim that the prior usually plays only a minor role in the interpretation of the data. We shall show situations in which a prior, objectively justified by the physical constraints, can lead to a very strong molding of our interpretation of the measured data. We shall also show how the concept of a "probability distribution function over probability distribution functions"[8,9] has practical utility in clarifying and taking advantage of the distinction between the "randomness" intrinsic in the preparation of an ensemble of physical systems, and the uncertainty in our limited knowledge of the values of the parameters that characterize this preparation procedure. In this paper we consider three physical situations in which Bayesian concepts lead to a useful, non-conventional interpretation of the information contained in a small number of physical measurements. The first two of these relate to an ensemble of quantum mechanical spin-1/2 systems, each prepared by the same procedure[13]. The third relates to the energies of molecules sampled from a fluid that obeys the classical statistical mechanics of Gibbs[1,8,9]. In the first example, we show how a prior chosen to match the physical constraints on the system leads to a very conservative "rule of succession"[7,10,3] that discounts any significant deviation from "random" frequencies (for measured values) as being nonrepresentational of the distribution being sampled. In the second example, we show how an appropriate, physically motivated prior leads to a rule of succession that is less conservative than that of Laplace[5], and even, in most circumstances, is less conservative than the maximum likelihood inference[1-3] from the observed frequency. This example also shows how a prior that is uniform in one parameter space can yield a rather peaked prior in the space of another related (and physically important) parameter, (in this case, the temperature). In the third example, we show how Bayesian logic together with the knowledge of the results of only a small number of physical measurement events augments equilibrium statistical mechanics with non-Maxwellian distribution functions and probability distribution functions over temperature. These distribution functions are a more faithful representation of our state of knowledge of the system than is a Maxwellian distribution with a temperature inferred by the principle of maximum likelihood [1-3,14,9].

C. R. Smith et al. (eds.), Maximum Entropy and Bayesian Methods, Seattle, 1991, 453–466.
© 1992 Kluwer Academic Publishers.

2. Mathematical Development

We now wish to obtain a setting suitable for the description of the density operator of an ensemble of spin 1/2 systems, our images of this density operator (or density matrix)[15-23], and the relationship of the assumed or inferred orientation of the axis of quantization of the system state to the axis of resolution of the measurement. (A more complete exposition is given in our earlier articles[12,13].) We start by considering a density operator aligned such that its "up" spin direction is oriented in the direction given by the spherical angles (θ, ϕ) with respect to an original coordinate system.

PARAMETRIZATION OF THE DENSITY OPERATORS

The original Cartesian coordinate system may be transformed into one whose z-axis is pointing in the direction given by the spherical angles (θ, ϕ) of the original one in the following manner: Rotate the system counterclockwise about the original z-axis by an angle ϕ; then rotate it counterclockwise about the new y-axis by θ. Using the transformation properties of the spin states we may relate the spin states $\hat{v}(\theta, \phi) = \{\hat{v}_\uparrow(\theta, \phi), \hat{v}_\downarrow(\theta, \phi)\}$, aligned with respect to the z-axis of the new coordinates, to the spin states $\hat{v}(0) = \{\hat{v}_\uparrow(0), \hat{v}_\downarrow(0)\}$ aligned with respect to the z-axis of the original coordinates, via the relations:

$$\{\hat{v}_\uparrow(\theta, \phi), \hat{v}_\downarrow(\theta, \phi)\} = \{\hat{v}_\uparrow(0), \hat{v}_\downarrow(0)\} \, \hat{U}(\theta, \phi) \tag{1a}$$

$$\hat{U}(\theta, \phi) = \begin{pmatrix} \exp(-i(\phi/2))\cos(\theta/2) & -\exp(-i(\phi/2))\sin(\theta/2) \\ \exp(i(\phi/2))\sin(\theta/2) & \exp(i(\phi/2)\cos(\theta/2) \end{pmatrix}. \tag{1b}$$

The density operator diagonal with respect to the original coordinates and with eigenvalues $\{\lambda_\uparrow = 1/2(\mu_0 + 1), \lambda_\downarrow = 1/2(-\mu_0 + 1)\}$ (where $\mu_0 = \langle \sigma_z \rangle$, the expectation value of the z-component of the spin (normalized to unity) for this density operator) may be written:

$$\hat{\rho}(\mu_0, 0) = |\, \hat{v}(0) \rangle \, \lambda \, \langle \hat{v}(0) \,|. \tag{2a}$$

where λ is the diagonal matrix formed from the eigenvalues $\lambda_\uparrow$ and $\lambda_\downarrow$.

This density operator, when rotated to become aligned with a new z-axis pointing in the (θ, ϕ) direction becomes:

$$\hat{\rho}(\mu_0, \theta, \phi) = |\, \hat{v}(\theta, \phi) \rangle \, \lambda \, \langle \hat{v}(\theta, \phi) \,| = |\, \hat{v}(0) \rangle \, U(\theta, \phi) \, \lambda \, U^\dagger(\theta, \phi) \, \langle \hat{v}(0) \,|$$
$$= |\, \hat{v}(0) \rangle \, \rho(\mu_0, \theta, \phi) \langle \, \hat{v}(0) \,| \tag{2b}$$

where the density matrix $\rho(\mu_0, \theta, \phi)$ is expressible as

$$\rho(\mu_0, \theta, \phi) = U(\theta, \phi) \, \lambda \, U^\dagger(\theta, \phi)$$
$$= \begin{pmatrix} \rho_{\uparrow\uparrow}(\theta, \phi) & \rho_{\uparrow\downarrow}(\theta, \phi) \\ \rho_{\downarrow\uparrow}(\theta, \phi) & \rho_{\downarrow\downarrow}(\theta, \phi) \end{pmatrix} = \begin{pmatrix} 1/2(\mu_0 \cos\theta + 1) & (\mu_0/2)\sin\theta \exp(-i\phi) \\ (\mu_0/2)\sin\theta \exp(i\phi) & 1/2(-\mu_0 \cos\theta + 1) \end{pmatrix} \tag{2c}$$

The expectation value for the z-component of the spin for this rotated density operator is:

$$\langle \sigma_z \rangle = \rho_{\uparrow\uparrow}(\theta, \phi) - \rho_{\downarrow\downarrow}(\theta, \phi) = \mu_0 \cos\theta \tag{3a}$$

Without loss of generality we may consider the system to be originally aligned such as to make $\langle \sigma_z \rangle \geq 0$, thereby making $\lambda_\uparrow - \lambda_\downarrow \equiv \mu_0 \geq 0$. Positivity of $\hat{\rho}$ then requires $0 \leq \mu_0 \leq 1$, where $\mu_0 = 1$ corresponds to the set of pure quantum mechanical states and $\mu_0 = 0$ corresponds to the density operator of unconstrained maximum entropy[12].

GEOMETRICAL REPRESENTATION OF THE DENSITY OPERATORS

Let each density operator $\hat{\rho}(\mu_0, \theta, \phi)$ be mapped onto its corresponding point in ordinary position space, which has the spherical coordinates (r, θ, ϕ), with $r = \mu_0$, the expectation value of the spin component in the direction of (θ, ϕ). (This is the direction of alignment of the density operator $\hat{\rho}(\mu_0, \theta, \phi)$ and its eigenspinor basis $\hat{v}(\theta, \phi)$). Several features of this mapping are discussed in our previous articles[12,13]. Among the most important of these features is that the metric distance s between two different density operators, one specified by (r_1, θ_1, ϕ_1) the other specified by (r_2, θ_2, ϕ_2), is obtained from geodesic integration of the infinitisimal metric distance ds whose square is given by:

$$(ds)^2 = Tr[(d\hat{\rho})^\dagger d\hat{\rho}] = Tr[(d\rho)^\dagger d\rho] = 2[(dr)^2 + (rd\theta)^2 + (r \sin\theta \, d\phi)^2] \tag{4}$$

Thus, the distance between the representative points, in the unit sphere, corresponding to two different density operators, is (to within a multiplicative constant) the metric distance between the density operators as prescribed by the theory of operators and matrices [24,25]. Therefore, this mapping is both topologically and metrically faithful, thus expediting the choice of physically meaningful prior probability densities.

THERMODYNAMIC ENSEMBLES OF SPIN-1/2 SYSTEMS

In Examples I and II below, we shall be dealing with a thermodynamic ensemble of free and independent spin-1/2 systems polarized by a uniform magnetic B field for which the density operator is aligned along the direction (θ, ϕ) opposite to B with the probability of measuring the spin aligned in this direction given by $p_\uparrow^0 = \lambda_\uparrow = 1/(1 + exp(-\beta\varepsilon))$, $\beta = 1/(k_B T)$, $\varepsilon = \frac{e}{2m} g \frac{\hbar}{2} |B \cdot \sigma|$ where, for electrons, $g = 2$ and $|B \cdot \sigma| = |B|$. This density operator, when expressed in terms of spinor states aligned with respect to our original coordinate system gives $p_\uparrow = \rho_{\uparrow\uparrow}(\theta, \phi) = \frac{1}{2}(\mu_0 \cos\theta + 1)$, $p_\downarrow = \rho_{\downarrow\downarrow}(\theta, \phi) = \frac{1}{2}(-\mu_0 \cos\theta + 1)$, whereas the eigenvalues are $\lambda_\uparrow \equiv p_\uparrow^0 = \frac{1}{2}(\mu_0 + 1)$, $\lambda_\downarrow \equiv p_\downarrow^0 = \frac{1}{2}(-\mu_0 + 1)$.

3. Example I – A Physical System Supporting A Strong Prior

Consider a system of identical spin 1/2 particles polarized by a uniform magnetic field of known strength but unknown orientation and in thermal equilibrium at a known temperature. The density operator for this system therefore has known eigenvalues, $\lambda_\uparrow \equiv p_\uparrow^0$ and $\lambda_\downarrow \equiv p_\downarrow^0$ determined, as shown above, by the ratio of the magnetic field strength to the absolute temperature; but the (θ, ϕ) alignment of its eigenspinors is unknown. The probability of measuring, in any specified order, n "up" spin values and m "down" spin values, along the z-axis of our original coordinate system, from a density operator aligned along (θ, ϕ) is therefore $(p_\uparrow)^n (p_\downarrow)^m$, where $p_\uparrow$ and $p_\downarrow$ are given above. We seek a probability distribution over the parameters of this density operator, conditioned by such measurements. We shall reduce this probability distribution to one over $p_\uparrow$. A prior, appropriate to this density operator may be chosen, in the geometrical parameter space, as being confined to the surface of a sphere of radius μ_0, and uniform over this surface. Such a prior corresponds to a prior on $p_\uparrow$ that is uniform over the interval $p_{\uparrow_{min}} \leq p_\uparrow \leq p_{\uparrow_{max}}$, where $p_{\uparrow_{min}} = p_\downarrow^0 = \frac{1}{2}(-\mu_0 + 1)$ and $p_{\uparrow_{max}} = p_\uparrow^0 = \frac{1}{2}(\mu_0 + 1)$. For typical field strengths and temperatures, $\mu_0 << 1$. By Bayes' theorem, the probability over $p_\uparrow$ conditioned by the measurement, upon the axis at $\theta = 0$, of n "up" spin values and m "down" spin values, is given by

$$\Pi(p_\uparrow)d(p_\uparrow) = \frac{(p_\uparrow)^n(p_\downarrow)^m dp_\uparrow}{\int_{p_{\uparrow min}}^{p_{\uparrow max}} (p_\uparrow)^n(p_\downarrow)^m dp_\uparrow} = \frac{(p_\uparrow)^n(p_\downarrow)^m dp_\uparrow}{B_{p_{\uparrow max}}(n+1, m+1) - B_{p_{\uparrow min}}(n+1, m+1)} \quad (5)$$

The incomplete beta function, $B_p(j, k)$, has the representation:

$$B_p(j, k) = B_{1-p}(k, j) = \sum_{l=0}^{k-1}(-1)^l \binom{k-1}{l} \frac{p^l}{j+l} \quad (6)$$

The mean of $p_\uparrow$, over this distribution, $\Pi(p_\uparrow)$, is

$$\langle p_\uparrow \rangle = \int_{p_{\uparrow min}}^{p_{\uparrow max}} p_\uparrow \Pi(p_\uparrow) d(p_\uparrow) = \frac{B_{p_{\uparrow max}}(n+2, m+1) - B_{p_{\uparrow min}}(n+2, m+1)}{B_{p_{\uparrow max}}(n+1, m+1) - B_{p_{\uparrow min}}(n+1, m+1)} \quad (7)$$

which for $m = 0$ becomes

$$\langle p_\uparrow \rangle = \left(\frac{n+1}{n+2}\right) \left(\frac{p_{\uparrow max}^{n+2} - p_{\uparrow min}^{n+2}}{p_{\uparrow max}^{n+1} - p_{\uparrow min}^{n+1}}\right) \quad (8)$$

Since $\langle p_\uparrow \rangle \leq p_{\uparrow max}$, this leads to the following bounding inequalities (for $m = 0$)

$$\left(\frac{n+1}{n+2}\right) p_{\uparrow max} \leq \langle p_\uparrow \rangle \leq p_{\uparrow max} \quad (9)$$

Since the set of points corresponding to allowed density operators forms the surface of a sphere, and this sphere is a convex set of points, the mean, over this surface, with any probability distribution, leads to a density operator whose representative point lies within the volume of this sphere. Thus, no matter how large or how small the fraction of "up" spin measurements is of the total number of spin measurement events, the prior will not allow the density operator to yield an expectation value $\langle p_\uparrow \rangle$ for the probability of measuring spin "up" that is outside the range $\frac{1}{2}(-\mu_0 + 1) \leq \langle p_\uparrow \rangle \leq \frac{1}{2}(\mu_0 + 1)$.

This is like the following scenario. A teacher gives a true/false test to a person she considers to be a very poor student. She is firmly convinced that, given enough questions of the same level of difficulty, there is no way that this student will get more than 55% of them correct. In her opinion, his ignorance will make him also incapable of getting more than 55% of them incorrect. She therefore would analyze his response of getting 1000 correct answers to 1000 questions, using a Bayesian prior that is uniform over the range of 45% to 55% correct, and zero for all other values. Thus she would conclude that, with an unlimited number of similar questions, the student would likely get almost 55% of them correct. This extreme interpretation would be significantly softened if the teacher allowed, in her prior, even a small non-zero probability for the true value to be outside of the range defined by her prejudice, (such as would be accomplished by replacing her step-like prior with a Gaussian prior). However, in our thermodynamic analog, such a modification is inappropriate.

A graph of $\Pi(p_\uparrow)$ for $(\mu_0 = 0.1, n = 8, m = 0)$ is shown in Figure 1, and one for $(\mu_0 = 1, n = 8, m = 0)$ is shown in Figure 2.

4. Example II – A Strong Prior Induced By A Uniform Prior

Consider a system of identical spin 1/2 particles polarized by a uniform magnetic field, **B**, of known strength and known orientation (to within a very small angle γ of uncertainty), but in thermal equilibrium at an unknown temperature $T = 1/(k_B \beta)$. In the geometrical parameter space of the density operators, the representative points of the density operators that satisfy these physical constraints lie within that part of the parameter sphere which also lies within the volume enclosed by the very narrow, needle-like cone with apex at the origin, axis extending along the $-\mathbf{B}$ direction, and apex angle γ. Each of the density operators whose representative points lie within this allowed region has its density matrix virtually diagonal (with eigenvalues $p_\uparrow^0 = \frac{1}{2}(\mu_0 + 1)$ and $p_\downarrow^0 = \frac{1}{2}(-\mu_0 + 1)$) in the spinor basis $\{\hat{v}_\uparrow(0), \hat{v}_\downarrow(0)\}$ aligned along the $-\mathbf{B}$ direction (here taken also as the direction of spin-component measurement and the z-axis of our coordinate system). The probability of measuring, in any specified order, n "up" spin values and m "down" spin values, along the z-axis of our original coordinate system, from a density operator aligned along (θ, ϕ) is therefore $(p_\uparrow^0)^n (p_\downarrow^0)^m$, where $p_\uparrow^0$ and $p_\downarrow^0$ are given above. We seek a probability distribution over the parameters of this density operator, conditioned by such measurements. We shall reduce this probability distribution to one over $p_\uparrow^0$ (which we shall now call $p_\uparrow$). Thus, by Bayes' theorem, the probability over $p_\uparrow$ conditioned by the measurement, upon the z-axis, of n "up" spin values and m "down" spin values, is given by

$$\Pi(p_\uparrow)d(p_\uparrow) = \frac{(p_\uparrow)^n (p_\downarrow)^m (p_\uparrow - p_\downarrow)^2 dp_\uparrow}{\int_{\frac{1}{2}}^{1} (p_\uparrow)^n (p_\downarrow)^m (p_\uparrow - p_\downarrow)^2 dp_\uparrow} = \frac{(p_\downarrow)^m (p_\uparrow)^n (p_\downarrow - p_\uparrow)^2 (-dp_\downarrow)}{\int_0^{\frac{1}{2}} (p_\downarrow)^m (p_\uparrow)^n (p_\downarrow - p_\uparrow)^2 dp_\downarrow}$$

$$= \frac{(p_\uparrow)^n (p_\downarrow)^m (p_\uparrow - p_\downarrow)^2 dp_\uparrow}{B_{\frac{1}{2}}(m+1, n+3) - 2B_{\frac{1}{2}}(m+2, n+2) + B_{\frac{1}{2}}(m+3, n+1)} \tag{10}$$

The volume element associated with $[dp_\uparrow = d\{\frac{1}{2}(\mu_0 + 1)\} = d\{\frac{1}{2}(r+1)\}]$ is $\frac{\pi}{6} \sin^2 \gamma [d(r^3) = d\{(p_\uparrow - p_\downarrow)^3\} = 3(p_\uparrow - p_\downarrow)^2 d(p_\uparrow - p_\downarrow) = 6(p_\uparrow - p_\downarrow)^2 dp_\uparrow]$. The incomplete beta function, $B_p(j, k)$, is shown in Eq.(6). The mean of $p_\uparrow$, over this distribution, $\Pi(p_\uparrow)$, is

$$\langle p_\uparrow \rangle = \int_{\frac{1}{2}}^{1} p_\uparrow \Pi(p_\uparrow) d(p_\uparrow) = \frac{B_{\frac{1}{2}}(m+1, n+4) - 2B_{\frac{1}{2}}(m+2, n+3) + B_{\frac{1}{2}}(m+3, n+2)}{B_{\frac{1}{2}}(m+1, n+3) - 2B_{\frac{1}{2}}(m+2, n+2) + B_{\frac{1}{2}}(m+3, n+1)}$$

$$\tag{11}$$

Because of our prior, the "rule of succession" for the mean value, $\langle p_\uparrow \rangle$, belonging to this probability distribution function is significantly less conservative than that of Laplace. The prior for $\Pi(p_\uparrow)$ is shown in Figure 3. The distribution function $\Pi(p_\uparrow)$ conditioned by $(n = 8, m = 0)$ is shown in Figure 4, and that for $(n = 71, m = 7)$ is shown in Figure 5. The value of $\langle p_\uparrow \rangle$ for $(n = 8, m = 0)$ is ≈ 0.93246 (and the value from maximum likelihood is 1). The value of $\langle p_\uparrow \rangle$ for $(n = 71, m = 7)$ is ≈ 0.90538 (and the value from maximum likelihood is $71/78 \approx 0.91026$). For each of these, the value of $\langle p_\uparrow \rangle$ from Laplace's rule of succession is $9/10$. Also, the value of $\langle p_\uparrow \rangle$ for $(n = 5, m = 3)$ is ≈ 0.766205, which is larger than the Laplace value $3/5$, and even is significantly larger than the maximum likelihood value $5/8$. The value of $\langle p_\uparrow \rangle$ for the prior is $7/8$, (a bit larger than the "random value", $1/2$).

Probability Distributions Over β and T

We may transform our distribution function $\Pi(p_\uparrow)$ (where $p_\uparrow = 1/(1 + exp(-\beta\varepsilon))$) to a distribution on $\beta^* \equiv \beta\varepsilon$, and also to one on the reduced temperature, $T^* \equiv 1/\beta^*$, via the relations

$$\Pi(p_\uparrow) |dp_\uparrow| = P_{\beta^*}(\beta^*) |d\beta^*| = P_{T^*}(T^*) |dT^*| \tag{12}$$

Our prior probability densities on β^* and T^* therefore become:

$$P_{\beta^*}(\beta^*) = 6(p_\uparrow - p_\downarrow)^2 \left| \frac{dp_\uparrow}{d\beta^*} \right| = 6e^{-\beta^*} \frac{\left(1 - e^{-\beta^*}\right)^2}{\left(1 + e^{-\beta^*}\right)^4} \tag{13a}$$

$$P_{T^*}(T^*) = P_{\beta^*} \left| \frac{d\beta^*}{dT^*} \right| = 6e^{-\frac{1}{T^*}} \frac{\left(1 - e^{-\frac{1}{T^*}}\right)^2}{\left(1 + e^{-\frac{1}{T^*}}\right)^4 (T^*)^2} \tag{13b}$$

These prior probability densities, P_{β^*} and P_{T^*} are each a unique function of its argument and are shown in Figure 6 and Figure 7, respectively. Our uniform prior over the allowed volume of the geometrical parameter space has thus induced a monotonic but non-uniform prior on $p_\uparrow$, and priors with a definite "peaked" maximum on each of β^* and T^*.

The maximum of $P_{\beta^*}(\beta^*)$ and that of $P_{T^*}(T^*)$ do not correspond to the same reduced temperature. They correspond, instead, to the maximum in a weighted sensitivity of the expected value of the spin measurement to a fixed small change in β^* (for $P_{\beta^*}(\beta^*)$) or in T^* (for $P_{T^*}(T^*)$). The regions of large amplitude of these priors are therefore the regions of temperature for which such a spin system could serve as a suitable thermometer. The fact that the prior is a unique function of β^* or T^* and thus, when expressed in terms of β or T, contains the scale factor, ε, shows that systems of different ε are suitable for measuring different ranges of temperature, (a well-known fact that appears here in a Bayesian context).

5. Example III – Bayesian Logic and Classical Statistical Mechanics

The quantum statistical mechanical examples presented above can be appropriately related to and contrasted with a problem of Bayesian inference of temperature in classical statistical mechanics, originally presented by Tikochinsky and Levine[8] and criticized by Cyranski[9]. We choose to explicate this problem as our third example, with particular attention to the induced non-Maxwellian energy distribution function, and show how the results of the Bayesian inference make logical and physical sense, thus limiting the domain of relevance of Cyranski's criticism.

Bayesian Logic and the Classical Statistical Mechanics of Gibbs

Consider the following problem. We are sampling molecules from a fluid in thermal equilibrium, and measuring their translational kinetic energies. The temperature is considered to be high enough that the classical statistical mechanics of Gibbs applies to the translational kinetic energies, so the translational kinetic energy of each molecule follows a Maxwell probability distribution given by

$$f(\varepsilon)d\varepsilon = g(\varepsilon)\beta^{3/2} exp(-\beta\varepsilon)d\varepsilon = 2\pi^{-1/2}\varepsilon^{1/2}\beta^{3/2} exp(-\beta\varepsilon)d\varepsilon \tag{14}$$

where $\beta = 1/(k_B T)$, T is the absolute temperature, and k_B is Boltzmann's constant. This distribution gives a mean translational kinetic energy $\langle \varepsilon \rangle = 3/(2\beta)$. Our problem is that we do not know the value of β that applies to this thermodynamic system (i.e., we do not know its temperature), and we want, nevertheless to characterize this system by an appropriate probability distribution function on the energy. Suppose that we make measurements of the translational kinetic energies of N molecules sampled from this system, and that the values of these energies are $\varepsilon_1, \varepsilon_2 ... \varepsilon_N$, with the arithmetic mean of this set being $\langle \varepsilon \rangle_N$. The probability of making these measurements to within the intervals $d\varepsilon_1, d\varepsilon_2 ... d\varepsilon_N$ is given by

$$\prod_{k=1}^{N} f(\varepsilon_k)d\varepsilon_k = \left(\prod_{k=1}^{N} g(\varepsilon_k)d\varepsilon_k \right) \beta^{3N/2} exp(-\beta N \langle \varepsilon \rangle_N) \tag{15}$$

This may also be interpreted as the likelihood function on β. By Bayes' theorem, when multiplied by our prior probability that β will be found to be within $d\beta$ of the chosen value, this likelihood function becomes a non-normalized conditional probability that β will be found to within $d\beta$ of the chosen value conditioned by the measured energies $\varepsilon_1, \varepsilon_2 ... \varepsilon_N$. Thus, we may obtain a normalized probability distribution function on β, conditioned by our measured energies as:

$$P(\beta \mid \varepsilon_1, \varepsilon_2 ... \varepsilon_N)d\beta = \frac{\beta^{3N/2} exp(-\beta N \langle \varepsilon \rangle_N)w(\beta)d\beta}{\int_0^{\infty} \beta^{3N/2} exp(-\beta N \langle \varepsilon \rangle_N)w(\beta)d\beta} \tag{16}$$

where $w(\beta)$ is our prior probability density on β. We want our rule for choosing $w(\beta)$ to be invariant to changing our parametrization from β to $T = 1/(k_B\beta)$, or to any power of T or β. Therefore, in agreement with Jeffreys[1,11], we choose $w(\beta) = 1/\beta$, which supports $w(T) = 1/T$ through the relation $\mid w(\beta)d\beta \mid = \mid w(T)dT \mid$, as desired. This choice is also invariant to any scaling of β or T, giving equal prior probability values for each β (or T) interval in the successive "doubling" of β or T. We find that $\langle \beta \rangle_N$, the mean value of β given by this probability distribution, satisfies $\langle \varepsilon \rangle_N = 3/(2\langle \beta \rangle_N)$, in correspondence with the matching $\langle \varepsilon \rangle = 3/(2\beta)$ given by the Maxwell distribution function $f(\varepsilon)$. This further confirms the suitability of our choice of $w(\beta)$[8]. (As Jaynes[1,2] has shown, there are even more convincing reasons for this choice of $w(\beta)$ that arise from group invariance and marginalization theory.) Also the variance of β is given by $(\sigma_\beta)^2 = \langle (\beta - \langle \beta \rangle)^2 \rangle = \langle \beta^2 \rangle - \langle \beta \rangle^2 = (2/(3N))\langle \beta \rangle^2$ so that the expected fractional deviation has the measure $\sigma_\beta/\langle \beta \rangle = \sqrt{2/(3N)}$. Using this probability distribution on β, we may take as our normalized probability distribution on ε, the mean of $f(\varepsilon)$ weighted by $P(\beta \mid \varepsilon_1, \varepsilon_2 ... \varepsilon_N)$. Thus

$$P(\varepsilon \mid \varepsilon_1, \varepsilon_2 ... \varepsilon_N)d\varepsilon = \left[\int_0^{\infty} P(\beta \mid \varepsilon_1, \varepsilon_2 ... \varepsilon_N)f(\varepsilon)d\beta \right] d\varepsilon$$

$$= g(\varepsilon)\frac{(N\langle \varepsilon \rangle_N)^{\frac{3N}{2}}}{(N\langle \varepsilon \rangle_N + \varepsilon)^{\frac{3(N+1)}{2}}} \frac{\Gamma(\frac{3}{2}(N+1))}{\Gamma(\frac{3}{2}N)}d\varepsilon \tag{17}$$

where the Γ function satisfies $\Gamma(x+1) = x\Gamma(x)$, $\Gamma(1) = 1$, and $\Gamma(\frac{1}{2}) = \sqrt{\pi}$. For a single measured value, ε_1, this becomes

$$P(\varepsilon \mid \varepsilon_1)d\varepsilon = g(\varepsilon)\frac{(\varepsilon_1)^{\frac{3}{2}}}{(\varepsilon_1 + \varepsilon)^3}\frac{\Gamma(3)}{\Gamma(\frac{3}{2})}d\varepsilon \tag{18}$$

The distribution function $f(\varepsilon)$, for $\langle \varepsilon \rangle = 5$, is shown in Figure 8. The distribution function $P(\varepsilon \mid \varepsilon_1)$ for $\varepsilon_1 = 5$ is shown in Figure 9. The distribution function $P(\beta \mid \varepsilon_1, \varepsilon_2 ... \varepsilon_N)$ for $\langle \varepsilon \rangle_N = 5$ and $N = 5$ is shown in Figure 10. The distribution function $P(\varepsilon \mid \varepsilon_1)$ gives as $\langle \langle \varepsilon \rangle \rangle$, (its mean value for ε), the value $3\varepsilon_1$, and as its variance, $(\sigma_\varepsilon)^2$, the value $2\langle \langle \varepsilon \rangle \rangle^2/3$, so that the expected fractional deviation for ε has the measure $\sigma_\varepsilon/\langle \langle \varepsilon \rangle \rangle = \sqrt{2/3}$, in complete symmetry with the corresponding fractional deviation in β for a distribution function induced by a single measurement on ε. The value of ε for which this distribution function has a maximim is $\varepsilon = \varepsilon_1/5 = \langle \langle \varepsilon \rangle \rangle/15$, showing that this distribution function is very asymmetric about its maximum, having a very long and "thick" "tail" extending well into the larger values of ε, in contrast with the exponentially "decaying" tail of the Maxwell distribution function $f(\varepsilon)$ (The Maxwell distribution function induced by the principle of maximum likelihood has its maximum at $\varepsilon_1/3$ and its mean at ε_1). The long tail of $P(\varepsilon \mid \varepsilon_1)$, arises because the Bayesian formalism gives appropriate recognition to the possibility that the temperature of the system may actually be considerably larger than the temperature matched by maximum likelihood. The "closer-in" maximum of $P(\varepsilon \mid \varepsilon_1)$ arises to recognize that the temperature also may be considerably smaller. By using Stirling's asymptotic formula for the Γ functions, and the appropriate representation for the exponential function, one can easily show that, as the number of measured energy values increases without limit, the distribution function $P(\beta \mid \varepsilon_1, \varepsilon_2 ... \varepsilon_N)$ approaches the Maxwell distribution function $f(\varepsilon)$, with β replaced by $3/(2\langle \varepsilon \rangle_N)$. Thus, for a very large number of measurements, the procedure yields, as the distribution function conditioned by the measured energy values, the Maxwell distribution function whose mean energy is equal to the arithmetic mean of the measured energy values. But for a much smaller number of measurements it yields the non-Maxwellian distribution function $P(\beta \mid \varepsilon_1, \varepsilon_2 ... \varepsilon_N)$, with a power-law tail (not an exponential tail), reflecting the fact that the smaller number of measurements leaves the temperature of the system incompletely determined. This distribution function, $P(\beta \mid \varepsilon_1, \varepsilon_2 ... \varepsilon_N)$, which should be "updated" with each successive energy measurement, is the Bayesian probability distribution on the translational kinetic energy of the molecules, induced by the measured energy values and by our prior. It therefore should be more appropriate[9] for predicting the distribution of molecular energies in this thermal system than would be a Maxwell distribution function whose temperature would be inferred, for example, by the maximum likelihood inference, (which is also the temperature inferred from the mean value of β from our Bayesian probability distribution function on β).

6. Conclusions

Bayesian concepts can be used to give new and useful insight into the relationship of experimental measurements to the true physical properties of the system being studied. In *Example I* we saw how a Bayesian prior, appropriate to the physical constraints on the system, can very strongly mold our interpretation of the experimentally measured values of the physical quantities of the system, (in this case by discounting any significant deviation from "random" frequencies for these measured values as being nonrepresentational of the distribution being sampled). In *Example II* we saw how a uniform prior over the physically acceptable geometrical parameters, in a space with a faithful and geometrically uniform metric, can induce very non-uniform, non-random, and non-monotonic priors in the spaces of the parameters of primary physical interest, (in this case, the energy, the temperature, and the reciprocal temperature). We saw how the regions of large amplitude for these priors correspond to regions of high sensitivity of the system, for measuring the corresponding physical quantity. In *Example III* we saw how Bayseian concepts help us to discriminate between uncertainties intrinsic to the preparation procedure for the system being analyzed, and uncertainties due to our limited knowlege of the preparation procedure. This leads to new probability distrbution functions that faithfully represent the totality of our knowledge (and of our ignorance) about the state of the system being studied. It is hoped that the perceptions reported here might be relevant in the broader context of Bayesian methods for practical problems in science and engineering[26].

ACKNOWLEDGMENTS.

The author is grateful to Phillip R. Dukes for his help and support and for his friendly appraisal and helpful criticism of this work. He wishes to thank Mingsheng Li for his help with the calculations. He also wishes to thank Prof. E. T. Jaynes for his inspiration and his leadership (and for making available some chapters from his new book[2]). Finally, he wishes to thank Professors Gary Erickson, Paul Neudorfer, and C. Ray Smith for their untiring scientific and logistic support of this conference.

REFERENCES

[1] E. T. Jaynes, *Papers on Probability, Statistics, and Statistical Physics*, (D. Reidel, 1983), esp. pp. (xv,81,221-226,351).

[2] E. T. Jaynes, *Probability Theory –The Logic of Science*, (in press) esp. Chaps. 4&13.

[3] R. A. Fisher, *Statistical Methods and Scientific Inference*, Third Edition, (Hafner Press, New York, 1973), esp. pp. 11, 12, 24, 56, 115, and 133.

[4] R. D. Rosenkranz, *Inference, Method and Decision*, (D. Reidel, 1977), esp. pp. 48,52&62.

[5] P. S. de Laplace, *Theorie analytique des probabilitié*, 3rd Edition, (Paris, 1820) p. 178.

[6] G. Pólya, *Patterns of Plausible Inference*, (Princeton Univ. Press, 1954), esp. pp. 132-139.

[7] R. Carnap, *The Continuum of Inductive Methods*, (Univ. of Chicago Press, 1952).

[8] Y. Tikochinsky and R. D. Levine, *J. Math. Phys.* **25**, 2160 (1984).

[9] J. F. Cyranski, 'The Probability of a Probability' in *Maximum Entropy and Bayesian Methods in Applied Statistics*, Proceedings of the Fourth Maximum Entropy Workshop, Univ. of Calgary, 1984, J. H. Justice, ed. (Cambridge Univ. Press, 1986), 101-116.

[10] N. C. Dalkey, 'Prior Probabilities Revisited' in *Maximum Entropy and Bayesian Methods in Applied Statistics*, Proceedings of the Fourth Maximum Entropy Workshop, Univ. of Calgary, 1984, J. H. Justice, ed. (Cambridge Univ. Press, 1986), 117-130.

[11] H. Jeffreys, *Theory of Probability* Third Edition, (Oxford University Press, 1961).

[12] P. R. Dukes and E. G. Larson, 'Assignment of Prior Expectation Values and a More Efficient Means of Maximizing $-Tr\hat{\rho}ln\hat{\rho}$ Constrained to Measured Expectation Values for Quantum Systems with Total Angular Momentum J', in *Maximum Entropy and Bayesian Methods*, Proceedings of the Tenth Maximum Entropy Workshop, Univ. of Wyoming, 1990, W. T. Grandy and L. H. Schick, (eds.) (Kluwer Academic, Dordrecht, 1990), 161-168.

[13] E. G. Larson and P. R. Dukes, 'The Evolution of Our Probability Image for the Spin Orientation of a Spin - 1/2 - Ensemble', in *Maximum Entropy and Bayesian Methods*, Proceedings of the Tenth Maximum Entropy Workshop, Univ. of Wyoming, 1990, W. T. Grandy and L. H. Schick, (eds.) (Kluwer Academic, Dordrecht, 1990), 181-189.

[14] L. Tisza and P. M. Quay, *Annals of Physics* **25**, 48 (1963), p. 74.

[15] J. v. Neumann, *Math. Grundlagen der Quantenmechanik* (Dover, New York, 1943), p.41.

[16] A. Messiah, *Quantum Mechanics* (North-Holland, Amsterdam 1962), esp. Chaps. 5&7.

[17] P.-O. Löwdin, *Linear Algebra and the Fundamentals of Quantum Theory,* Technical Note 125, Uppsala Quantum Chemistry Group, 1964 (unpublished).

[18] P.-O. Löwdin, 'Some Comments about the Historical Development of Reduced Density Matrices' (Contribution to *Density Matrices and Density Functionals (Proceedings of the A. John Coleman Symposium)*), R. E. Erdahl and V. H. Smith, Eds., (D. Reidel, 1987).

[19] U. Fano, 'Description of States in Quantum Mechanics by Density Matrix and Operator Techniques', *Rev. Mod. Phys.* **29**, 74(1957).

[20] A. J. Coleman, 'Reduced Density Matrices from 1930 to 1989' (Contribution to *Density Matrices and Density Functionals (Proceedings of the A. John Coleman Symposium)*), R. E. Erdahl and V. H. Smith, Eds., (D. Reidel, Dordrecht, 1987), 5-20.

[21] R. McWeeny, 'Some Recent Advances in Density Matrix Theory' *Rev. Mod. Phys.* **32**, 335(1960).

[22] E. G. Larson, 'Reduced Density Operators, Their Related von Neumann Density Operators, Close Cousins of These, and Their Physical Interpretation' (Contribution to *Density Matrices and Density Functionals (Proceedings of the A. John Coleman Symposium)*), R. E. Erdahl and V. H. Smith, Eds., (D. Reidel, Dordrecht, 1987), 249-274.

[23] E. G. Larson, *Int. J. Quantum Chem.* **7**, 853 (1973).

[24] L. K. Hua, *Harmonic Analysis of Functions of Several Complex Varibles in the Classical Domains,* (Amer. Math. Soc., Providence, 1963).

[25] F. H. Fröhner, *Applications of the Maximum Entropy Principle in Nuclear Physics,* (Kernforschungszentrum, Karlsruh GmbH, 1989), esp. pp. 16&17.

[26] G. J. Erickson and C. R. Smith, eds., *Maximum Entropy and Bayesian Methods in Science and Engineering,* Volume 2: Applications (Kluwer, Dordrecht, Netherlands, 1988).

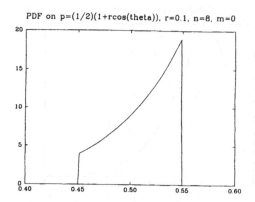

Figure 1. Probability Distribution Function on $p_\uparrow$ for ($\mu_o = 0.1$, $n = 8$, $m = 0$ (for Example I.)

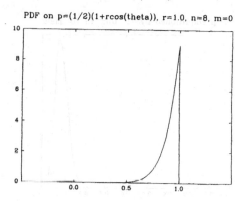

Figure 2. Probability Distribution Function on $p_\uparrow$ for ($\mu_o = 1$, $n = 8$, $m = 0$). (for Example I)

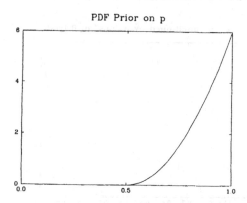

Figure 3. Prior Probability Distribution Function on $p_\uparrow$. (for Example II)

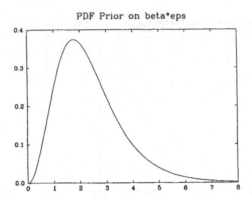

Figure 4. Probability Distribution Function on $p_\uparrow$ conditioned by 8 "up" values in 8 trials. (for Example II)

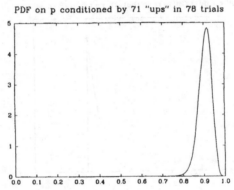

Figure 5. Probability Distribution Function on $p_\uparrow$ conditioned by 71 "up" values in 78 trials. (for Example II)

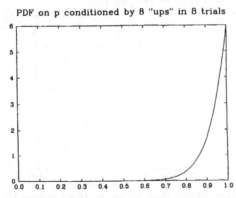

Figure 6. Prior Probability Distribution Function on β^*. (for Example II)

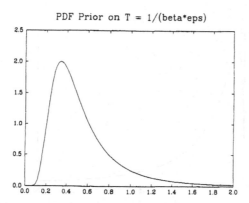

Figure 7. Prior Probability Distribution Function on T^*. (for Example II)

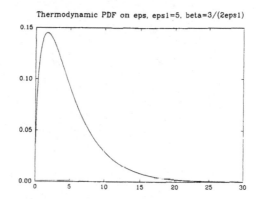

Figure 8. Thermodynamic Probability Distribution Function $f(\epsilon)$ for $\epsilon_1 = 5$, $\beta = 3/2\epsilon_1$. (for Example III)

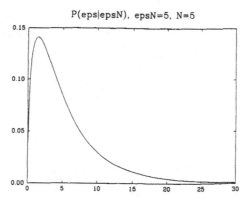

Figure 9. Probability Distribution Function $P(\epsilon|\epsilon_1)$ for $\epsilon_1 = 5$. (for Example III

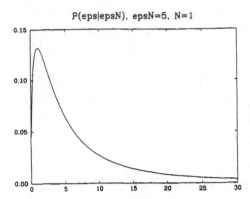

P(eps|epsN), epsN=5, N=1

Figure 10. Probability Distribution Function $P(\beta|\epsilon_1, \epsilon_2, \ldots \epsilon_n)$ for $< \epsilon >_N = 5$ and $N = 5$. (for Example III)

SUBJECT INDEX

Fundamental Theories of Physics

Series Editor: Alwyn van der Merwe, *University of Denver, USA*

Fundamental Theories of Physics

22. A.O. Barut and A. van der Merwe (eds.): *Selected Scientific Papers of Alfred Landé.* [*1888-1975*]. 1988   ISBN 90-277-2594-2
23. W.T. Grandy, Jr.: *Foundations of Statistical Mechanics.*
    Vol. II: *Nonequilibrium Phenomena.* 1988   ISBN 90-277-2649-3
24. E.I. Bitsakis and C.A. Nicolaides (eds.): *The Concept of Probability.* Proceedings of the Delphi Conference (Delphi, Greece, 1987). 1989   ISBN 90-277-2679-5
25. A. van der Merwe, F. Selleri and G. Tarozzi (eds.): *Microphysical Reality and Quantum Formalism, Vol. 1.* Proceedings of the International Conference (Urbino, Italy, 1985). 1988   ISBN 90-277-2683-3
26. A. van der Merwe, F. Selleri and G. Tarozzi (eds.): *Microphysical Reality and Quantum Formalism, Vol. 2.* Proceedings of the International Conference (Urbino, Italy, 1985). 1988   ISBN 90-277-2684-1
27. I.D. Novikov and V.P. Frolov: *Physics of Black Holes.* 1989   ISBN 90-277-2685-X
28. G. Tarozzi and A. van der Merwe (eds.): *The Nature of Quantum Paradoxes.* Italian Studies in the Foundations and Philosophy of Modern Physics. 1988
                ISBN 90-277-2703-1
29. B.R. Iyer, N. Mukunda and C.V. Vishveshwara (eds.): *Gravitation, Gauge Theories and the Early Universe.* 1989   ISBN 90-277-2710-4
30. H. Mark and L. Wood (eds.): *Energy in Physics, War and Peace.* A Festschrift celebrating Edward Teller's 80th Birthday. 1988   ISBN 90-277-2775-9
31. G.J. Erickson and C.R. Smith (eds.): *Maximum-Entropy and Bayesian Methods in Science and Engineering.*
    Vol. I: *Foundations.* 1988   ISBN 90-277-2793-7
32. G.J. Erickson and C.R. Smith (eds.): *Maximum-Entropy and Bayesian Methods in Science and Engineering.*
    Vol. II: *Applications.* 1988   ISBN 90-277-2794-5
33. M.E. Noz and Y.S. Kim (eds.): *Special Relativity and Quantum Theory.* A Collection of Papers on the Poincaré Group. 1988   ISBN 90-277-2799-6
34. I.Yu. Kobzarev and Yu.I. Manin: *Elementary Particles. Mathematics, Physics and Philosophy.* 1989   ISBN 0-7923-0098-X
35. F. Selleri: *Quantum Paradoxes and Physical Reality.* 1990   ISBN 0-7923-0253-2
36. J. Skilling (ed.): *Maximum-Entropy and Bayesian Methods.* Proceedings of the 8th International Workshop (Cambridge, UK, 1988). 1989   ISBN 0-7923-0224-9
37. M. Kafatos (ed.): *Bell's Theorem, Quantum Theory and Conceptions of the Universe.* 1989   ISBN 0-7923-0496-9
38. Yu.A. Izyumov and V.N. Syromyatnikov: *Phase Transitions and Crystal Symmetry.* 1990   ISBN 0-7923-0542-6
39. P.F. Fougère (ed.): *Maximum-Entropy and Bayesian Methods.* Proceedings of the 9th International Workshop (Dartmouth, Massachusetts, USA, 1989). 1990
                ISBN 0-7923-0928-6
40. L. de Broglie: *Heisenberg's Uncertainties and the Probabilistic Interpretation of Wave Mechanics.* With Critical Notes of the Author. 1990   ISBN 0-7923-0929-4
41. W.T. Grandy, Jr.: *Relativistic Quantum Mechanics of Leptons and Fields.* 1991
                ISBN 0-7923-1049-7
42. Yu.L. Klimontovich: *Turbulent Motion and the Structure of Chaos.* A New Approach to the Statistical Theory of Open Systems. 1991   ISBN 0-7923-1114-0

Fundamental Theories of Physics

KLUWER ACADEMIC PUBLISHERS – DORDRECHT / BOSTON / LONDON